Encyclopedia of Drugs and Alcohol

Editorial Board

Encyclopedia of Drugs and Alcohol

VOLUME 3

Jerome H. Jaffe, M.D.

Editor in Chief
University of Maryland, Baltimore

MACMILLAN LIBRARY REFERENCE USA
SIMON & SCHUSTER MACMILLAN
NEW YORK

SIMON & SCHUSTER AND PRENTICE HALL INTERNATIONAL
LONDON   MEXICO CITY   NEW DELHI   SINGAPORE   SYDNEY   TORONTO

Macmillan Library Reference
Simon & Schuster Macmillan
1633 Broadway
New York, NY 10019-6785

Library of Congress Catalog Card Number: 94-21458

Printed in the United States of America

printing number
  2 3 4 5 6 7 8 9 10

Library of Congress Cataloging-in-Publication Data
Encyclopedia of drugs and alcohol / Jerome H. Jaffe, editor-in-chief.
        p.   cm.
    Includes bibliographical references and index.
    ISBN 0-02-897185-X (set)
    1. Drug abuse—Encyclopedias.   2. Substance abuse—Encyclopedias.
  3. Alcoholism—Encyclopedias.   4. Drinking of alcoholic beverages—
  Encyclopedias.   I. Jaffe, Jerome.
  Ref   HV5804.E53   1995   v.3
        362.29′03—dc20   95-2321
                                                          CIP

This paper meets the requirements of ANSI-NISO Z39.48-1992 (Permanence of Paper). ♾™

Encyclopedia of
Drugs and Alcohol

R

RATIONAL AUTHORITY Drug addicts are reported to have a low tolerance for ANXIETY. As a result, few are able to voluntarily sustain an extended period of drug treatment, which is necessary for meaningful intervention. Instead, they tend to disengage themselves from treatment programs once the anxiety has been brought to the surface (Brill & Lieberman, 1969). "Rational authority," a late 1960s euphemism for mandatory (but not necessarily punitive) treatment, became a basis for holding addicts in a long-term treatment program.

The philosophy behind rational authority justifies the development of coercive mechanisms or strategies that permit assigning to treatment those addicts who ordinarily would not voluntarily seek assistance. Rehabilitation programs based upon this philosophy derive their legitimate coercive powers through the *authority* of the courts. The authority is considered *rational* because it is utilized in a humane and constructive manner, and it does this by relating the means of authority to the ends of rehabilitation.

This conceptualization represents an evolutionary change from the emphasis on the use of authority as a punitive end in itself. Rational authority also suggests combining the authority of the probation or parole officer with the techniques of social casework. As such, authority becomes a means for the officer or associated rehabilitation worker to implement desired behavioral changes. In addition to being required to obey the usual conditions of probation, addicts can be involuntarily held in a therapeutic setting until they have acquired a tolerance for abstinence and the conditioning processes thought to maintain addiction have been reversed. Evaluations of programs in New York, California, and Pennsylvania that are based upon rational authority indicate that when addicts are thus supervised, they are often less likely to relapse into addictive behavior (Brill & Lieberman, 1969).

(SEE ALSO: *California Civil Commitment Program; Civil Commitment; Coerced Treatment for Substance Offenders; Contingency Contracts; New York State Civil Commitment Program; Treatment Alternatives to Street Crime Treatment/Treatment Types*)

BIBLIOGRAPHY

BRILL, L., & LIEBERMAN, L. (1969). *Authority and addiction.* Boston: Little, Brown.

LEUKEFELD, C. G., & TIMS, F. M. (EDS.). (1988). *Compulsory treatment of drug abuse: Research and clinical practice* (NIDA Research Monograph 86). Rockville, MD: U.S. Department of Health and Human Services.

HARRY K. WEXLER

RATIONAL RECOVERY (RR) Rational Recovery (RR) is one of a number of self-help movements that have emerged as alternatives to ALCOHOLICS ANONYMOUS (AA) for those with drug and alcohol problems. Rational Recovery began with the publication of *Rational Recovery from Alcoholism: The Small Book* by Jack Trimpey in 1988. The program is based on Rational Emotive Therapy, a mental-health treatment with a cognitive orientation developed by the psychologist Albert Ellis. It is premised on the assumption that psychological difficulties are caused by irrational beliefs that can be understood and overcome, not by existential or spiritual deficits. The emphasis is on rational self-examination rather than on religiosity.

An RR "coordinator" leads a group of five to ten members, who meet once or twice weekly for ninety minutes. Each coordinator maintains contact with an adviser, a mental-health professional familiar with the RR program. RR emphasizes cognitive devices for securing abstinence, such as discussion of "the Beast," a term used to personify the compulsive thoughts that drive an individual to drink. Members use a "Sobriety Spreadsheet" on which they write out irrational beliefs that activate their desire to drink. They also read Trimpey's *The Small Book* to develop the proper attitude toward abstinence. These devices are used in RR meetings as well as outside to examine vulnerability to drinking and to overcome it. At meetings these issues are also addressed in a less formal way in "cross-talk," an open, face-to-face exchange among participants.

RR differs from AA in that it does not encourage supportive exchanges and phone calls between meetings, nor does the enrollee solicit a sponsor among established members. Also in contrast to AA, there is no equivalent of "working" the TWELVE-STEPS, and a spiritual or religious orientation to treatment is explicitly eschewed. Like SECULAR OR-GANIZATIONS FOR SOBRIETY (SOS), RR encourages study of its methods and outcome. One such study by Galanter and coworkers sent follow-up questionnaires to seventy RR groups in nineteen states and received sixty-three responses. Ninety-seven percent of participants in the responding groups filled out questionnaires. They were mostly men about forty-five years old, each with about a twenty-five-year history of alcohol problems. The majority were employed, had attended college, and had heard about the program through the media or by word of mouth. A majority had used marijuana, a substantial minority had also used cocaine, and a small minority had used heroin.

At the time of the study (the early 1990s), RR was a much younger organization than AA. Most of the coordinators had been members for only nine months, most groups had been meeting for about a year, and the implementation of the movement's specific techniques (use of the Sobriety Spreadsheet and discussion of "the Beast") was not consistent. Nevertheless, the members' commitment to the central tenet of the movement, sobriety, was considerable. Although 75 percent had previously attended AA meetings, the majority (82%) rated RR principles higher than AA principles in helping them achieve sobriety. However, it seems quite likely that RR benefits considerably from the experience these former AA members bring with them. A sizable percentage of RR participants who returned questionnaires were involved with mental-health care as well as with RR. Thirty-six percent had seen a psychotherapist the week before the survey, and 21 percent were currently taking medication prescribed for psychiatric problems. Many group coordinators had formal mental-health training, and 24 percent had graduate degrees or certificates in mental health. It is likely that, just as AA derives some legitimacy from its spiritual roots, RR derives some of its influence from the credibility of the professional psychology with which it is associated. Without carefully controlled studies that adjust for differences in patient backgrounds, it is hazardous to compare outcome studies from RR to studies of AA and other self-help groups. The data that do exist, however, tentatively suggest that RR may do at least as well.

An RR group can be formed at no cost by a recovering substance abuser in consultation with the executive office of the Rational Recovery movement (Box 800, Lotus, California 95651).

(SEE ALSO: *Sobriety; Treatment Types: Self-Help and Anonymous Groups*)

BIBLIOGRAPHY

GALANTER, M., EGELKO, S., & EDWARDS, H. (1993). Rational Recovery: Alternative to AA for addiction? *American Journal of Drug and Alcohol Abuse 19*, 499–510.

GELMAN, D., LEONARD, E. A., & FISHER, B. (1991). Clean and Sober and Agnostic. *Newsweek*, July 8, pp. 62–63.

TRIMPEY, J. (1988). *Rational recovery from alcoholism: The small book.* Lotus, CA: Lotus Press.

MARC GALANTER
JEROME H. JAFFE

RECEPTOR: DRUG A receptor is any cellular molecule or molecular grouping on any cell in the body where a drug is recognized and binds to initiate physiological effects—it is the part of the organism where the chemical agent interacts to produce its effects. Receptors are normally formed from the proteins in a cell. These proteins may function as receptors for hormones or NEUROTRANSMITTERS found naturally within the organism (i.e., endogenous ligands) to perform regulatory roles. Receptors can also be found on enzymes that are important in various metabolic or regulatory pathways in the body, on proteins involved in the transport of materials across cellular membranes, or on proteins that provide structure for the cell. Receptors can be found anywhere a chemical (ligand) can bind to alter physiological processes.

An AGONIST is a drug that binds to the receptor (i.e., it has affinity for the receptor-binding site); it mimics the properties of an endogenous neurotransmitter or hormone (i.e., it possesses intrinsic activity). An ANTAGONIST is a drug that binds to the receptor (i.e., it has affinity for the receptor binding site); but it competes with an endogenous (internal) or exogenous (external) ligand for binding to the receptor to interfere with the normal effects of the agonist. In theory, therefore, antagonists have no intrinsic activity by themselves, but produce their effects by inhibiting the action of an agonist. A drug that acts on a receptor is usually potentially capable of only altering the rate of a normal physiological function. In other words, drugs do not create effects or produce new functions in the cell—they simply modulate ongoing biological functioning. Under certain circumstances, however, this may not be the case, since not all drugs necessarily act at receptors. Some drugs (i.e., caustic agents and other similar toxic substances) act directly on cellular structures, resulting in cell damage or cell death.

The binding of drugs to receptors can involve all the known types of chemical bonds, including ionic, hydrogen, hydrophobic, van der Waals, and covalent bonds. In most cases of drug-receptor binding, many types of bonds are important. Usually, if the binding involves covalent bonds, the duration of binding is increased. A drug exhibiting noncovalent, but high affinity, binding can also function essentially as an irreversible ligand for the receptor. The chemical structure of the drug can obviously affect the configuration of the drug-receptor complex and can, there-

fore, affect the formation of chemical bonds. This structure-activity relationship between a drug and its receptor can, in turn, affect the affinity of a drug for the receptor as well as its intrinsic activity.

Slight modifications in the chemical structure of a drug molecule can render the drug less efficacious or even inactive in some instances. Changes in the chemical structure of endogenous or exogenous ligands for a receptor can often result in the discovery of more potent drugs or drugs with an increased incidence of therapeutic effects compared to toxic side effects. Chemical modifications can also affect the tissue selectivity of a drug, so that the drug binds to receptors located in one part of the organism (e.g., the brain) while having little or no binding affinity for different receptors localized in other parts of the body. If a drug binds to a receptor that serves functions common to most cells, then the effects of the drug will be extremely widespread, and it may be less useful to dangerous therapeutically. If the drug binds to a receptor that is localized on only a few types of cells, then the drug's therapeutic effects will be much more specific, while its side effects will be reduced. Even drugs with relatively specific binding profiles can produce serious toxic effects. Nevertheless, many important drugs have been developed from the modification of the chemical structure of endogenous neurotransmitters or hormones for a given receptor-binding site.

FUNCTIONS

Many protein components of the cell serve as receptors for endogenous regulatory ligands such as neurotransmitters or hormones. These physiological receptors are often found in the plasma membrane of the cell and function to bind the regulatory hormone or neurotransmitter and then to amplify the signal carried by this endogenous ligand in the target cell. This amplification of transmission of the receptor-mediated signal can occur either through direct intracellular effects or through the synthesis and release of another intracellular regulatory molecule called a *second messenger*. Therefore, the receptor serves two functions. The first function is the recognition and binding of the ligand, and the second function is the transmission or propagation of the signal or message in the target cell. This, in turn, gives rise to the concept that receptors have two distinct functional domains: (1) a ligand-binding domain, involved in the recognition and binding of the ligand, and (2) an effector domain, involved with the propagation of the message. A receptor-effector system can consist of an individual receptor protein that interacts with other closely associated cellular proteins to generate the physiological response. One well-known example of a receptor-effector system is the hormone-sensitive adenyl cyclase system (see Figure 1).

In this case, receptors regulate the activity of adenyl cyclase, an intracellular enzyme and effector that synthesizes the second messenger, adenosine 3',5'-monophosphate (cyclic AMP). Two separate guanine nucleotide-binding regulatory proteins (known as *G proteins*) serve as intermediates between the receptors and the enzyme. One type of G protein acts to transduce stimulatory signals, while the other type of G protein is involved with inhibitory processes. Another type of receptor is one that opens an ion channel in the cell membrane following the binding of an agonist. A well-known example of this type of receptor is the GAMMA-AMINOBUTYRIC ACID (GABA) receptor (Figure 2). When GABA binds to the $GABA_A$ receptor, a chloride ion channel is opened. The resulting influx of chloride anions (negatively charged ions) inhibits the cell by producing a hyperpolarization. Drugs called BENZODIAZEPINES increase the frequency of the opening of the chloride ion channel. Receptors can also integrate extracellular information as they coordinate signals from multiple ligands, both with each other and with other metabolic activities within the cell. This is particularly evident when one considers that the differ-

ent receptors for a wide variety of chemically unrelated ligands use relatively few biochemical mechanisms to produce their effects—and that even these few biochemical events often share many common elements.

While receptors often initiate and mediate the regulation of many physiological and biochemical processes within the cell, receptors are themselves also subject to regulatory and homeostatic control. The continued sustained binding of an agonist to a receptor generally results in receptor desensitization or refractoriness (down-regulation). In this case, the effect that follows continued or additional exposure to the same concentration of the drug is often decreased when compared to the original effect. When only the signal from the stimulated receptor is affected, *homologous* receptor desensitization occurs. With *heterologous* receptor desensitization, receptors from different neurotransmitters or hormones that act through a single signal become less effective. This type of desensitization can result from the modification of each receptor through a common feedback mechanism, or it can result from effects exerted at some point in the effector pathway that is downstream (i.e., inside the cell) from the receptors themselves. If, however, receptor stimulation is reduced from normal, a hyperactivity or supersensitivity (up-regulation) to receptor agonists is often observed. In some cases, receptor supersensitivity can result from feedback regulation, resulting in the increased synthesis of receptors.

Drugs of abuse can obviously also produce their effects through interactions with specific receptor-

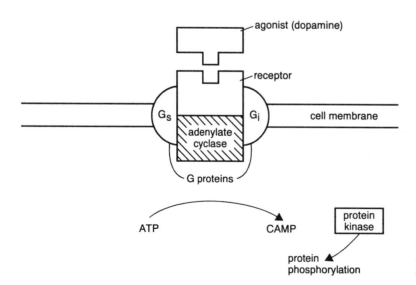

Figure 1
Receptor Regulates Adenyl Cyclase,
an Intracellular Enzyme

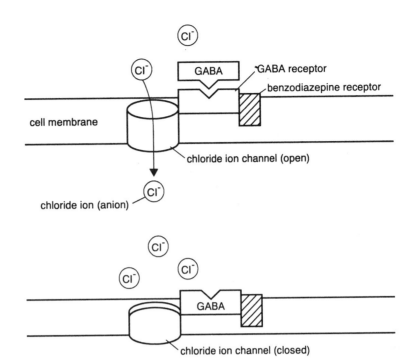

Figure 2
*GABA (gamma-aminobutyric acid)
Receptor Inhibits Chloride Anions*

binding sites. The following is a brief description of the receptor-mediated effects of selected drugs that are often used for nonmedical purposes (and for which the receptors involved are relatively well-characterized).

OPIOIDS

MORPHINE, HEROIN, and other OPIOIDS (both endogenous opioid peptides and some exogenous drugs) produce their various pharmacological effects through interactions with opioid receptors. Receptor-binding studies have suggested the existence of a number of distinct subtypes of opioid receptors that can interact with opioid drugs or with the endogenous opioid peptides. These receptors can be classified into three distinct categories in the central nervous system, which are designated μ (mu), κ (kappa), and δ (delta). A given opioid drug may interact to a variable degree with all three types of receptors to act as an agonist, as a partial agonist, or as an antagonist. Opioids are usually divided into three main groups: (1) morphine-like opioids, which function as agonists primarily at mu and perhaps at kappa and delta receptors; (2) opioid antagonists, which bind to opioid receptors but do not have any intrinsic activity on their own; and (3) opioids with mixed agonist-antagonist actions, which bind to

opioid receptors and function as agonists at some receptors and as antagonists or weak partial agonists at others. The opioid-receptor antagonist *NALOXONE* binds with high but variable affinity to all three categories of opioid receptors.

In general, analgesia is thought to result from the opioid agonist-induced activation of mu receptors at supraspinal sites and at kappa receptors within the spinal cord. Morphine-induced analgesia is likely produced through interactions with the mu receptor. At the same time, these mu-opioid receptors also likely mediate the euphoria, feelings of well-being, or pleasure associated with morphine use and abuse. Other effects that appear to be mediated through interactions of opioid drugs with mu receptors include respiratory depression, miosis (pinpoint pupils), and reduced motility in the gastrointestinal tract (i.e., constipation).

Receptor-binding studies have also identified two distinct types of mu receptors, based on their relative affinities for opioid agonists. The mu_1 subtype (i.e., the higher affinity mu receptor) appears to mediate supraspinal analgesia, whereas mu_2 subtype (i.e., the lower affinity mu receptor) may be more involved with respiratory depression and gastrointestinal activity. Agonists for the kappa receptor produce spinal analgesia but less intense miosis and respiratory depression than the mu agonists. Furthermore, in contrast to the euphoria that can result from the administration of mu agonists (e.g., morphine), the stimulation of kappa receptors can produce dysphoria (negative feelings) and other related psychomimetic effects, such as feelings of disorientation and depersonalization. Some of these psychomimetic effects are not effectively blocked by the opioid-receptor antagonist naloxone, suggesting that other binding sites may be involved in the mediation of these effects. In fact, pentazocine and related compounds also bind to two other distinct binding sites in the central nervous system. The first is known as the PHENCYCLIDINE (PCP)-binding site, since it displays an affinity for PCP. The PCP-binding site appears to be involved in the inhibitory regulation of cation channels (e.g., Ca^{2+} [calcium ion]) associated with the amino-acid neurotransmitter GLUTAMATE and N-methyl-D-aspartate (NMDA). The second site is known as the sigma site. The sigma site also binds drugs that interact with D_2 dopaminergic receptors, suggesting that the sigma site may be involved in some of the naloxone-insensitive psychomimetic effects of opioid drugs. Since there are few selective agonists for delta receptors that effectively cross the blood-brain barrier, the effects of delta receptor stimulation in humans are not clear. All three categories of opioid receptors (i.e., the mu, kappa, and delta) appear to produce an inhibitory effect on synaptic neurotransmission within the central nervous system. These binding sites are generally found on presynaptic nerve terminals, where stimulation leads to the decreased release of excitatory neurotransmitters. All three types of opioid receptors are also coupled to G proteins to regulate second messenger systems. The preceding description of the actions of opioids illustrates two important concepts: (1) the existence of multiple-receptor subtypes and (2) the fact that not all drugs called opioids interact exclusively with opioid receptors.

BENZODIAZEPINES

The BENZODIAZEPINES (e.g., diazepam [Valium], alprazolam [Xanex]) also interact with specific receptors. Normally these receptors respond to the inhibitory neurotransmitter gamma-aminobutyric acid (GABA), by opening a chloride ion channel in the nerve membrane. The entrance of chloride ions into the cell decreases cellular activity (i.e., the actions of GABA are inhibitory). Benzodiazepines reduce the spontaneous or evoked electrical activity of major neurons (nerve cells) in all regions of the brain and spinal cord. The activity of these large neurons is normally regulated by small inhibitory interneurons that primarily release GABA. Benzodiazepines prolong the inhibitory period that follows the activation of these GABA-containing interneurons, and this effect is blocked with the GABA-receptor antagonist bicuculline. The behavioral and electrophysiological effects of benzodiazepines are also blocked or decreased by pretreatment with drugs that inhibit the effects of GABA or block the synthesis of this neurotransmitter.

The benzodiazepine receptor appears to be a component of a larger macromolecular complex containing distinct binding sites for the benzodiazepines, BARBITURATES, and GABA—all three of which are associated with a chloride ion channel. In vitro studies have demonstrated that benzodiazepines increase the frequency of the opening of chloride ion channels induced by GABA, but they have no effect on these channels in the absence of GABA. The affinity of benzodiazepines for their binding sites is enhanced by either GABA or chloride; additionally, benzodiazepines enhance the binding of GABA to its receptor. The sedative-barbiturates also increase the binding of both GABA and benzodiazepines in a chloride-dependent fashion, and these effects can be blocked with convulsant agents such as picrotoxin.

A relatively good correlation exists for the ability of benzodiazepine agonists to alleviate anxiety or to block seizure activity and the affinity of these drugs for benzodiazepine-binding sites. Benzodiazepine-receptor antagonists, such as flumazenil, compete with agonists for these binding sites to block or inhibit their biological effects. Some antagonists produce a limited degree of biological activity, similar to that seen with benzodiazepine agonists, and are thus termed *partial agonists*. Other compounds (e.g., certain beta-carbolines) compete with benzodiazepine-receptor agonists for binding, but they actually produce opposite biological and neurochemical effects. These beta-carbolines inhibit the effects of GABA on chloride ion channels and can induce seizure activity and anxiety. These compounds have therefore been named benzodiazepine-receptor *inverse agonists*.

Benzodiazepine-binding sites are found throughout the central nervous system, although the density of these binding sites varies in the different brain regions. The highest densities are present in the cerebral cortex, the limbic system, and the cerebellar cortex; the lowest densities occur in the thalamus and the lower brain stem. The existence of several subtypes of benzodiazepine-receptor-binding sites has been demonstrated pharmacologically. Two subtypes of benzodiazepine receptors, identified as BZD_1 and BZD_2 (or omega$_1$ and omega$_2$), are present in the central nervous system and appear to mediate the neuropharmacological effects of the benzodiazepines. Zolpidem, a novel imidazopyridine hypnotic with rapid onset and short duration of action, has a high and preferential affinity for BZD_1-binding sites in the rodent brain. The distribution in the brain of BZD_1 and BZD_2 subtypes of receptors varies—with BZD_1 sites localized primarily in the cerebellum and sensorimotor cortex and BZD_2 sites in the hippocampus and the striatum. A second class of benzodiazepine-binding sites is localized both within the central nervous system and in peripheral tissues and is both pharmacologically and physiologically distinct from the two central-type benzodiazepine binding sites. Binding sites in this second class have been referred to as "peripheral-type," since they were initially identified in peripheral tissues (e.g., lung, kidney).

OTHER ABUSED DRUGS

Receptors for a number of other drugs commonly used for nonmedical purposes have also been identified, although much less is known about the physiological properties of these binding sites. As mentioned above, PHENCYCLIDINE (PCP) an arylcyclohexylamine, binds to the sigma receptor (which also binds pentazocine and related opioids), although the function for this binding site remains unknown. PCP also binds to the N-methyl-D-aspartate (NMDA) RECEPTOR to inhibit the flux of cations (e.g., Ca^{2+}) that is initiated by glutamate and/or aspartate to affect neuronal activity (e.g., long-term potentiation), especially in the hippocampus.

A binding site for cocaine has also recently been identified and cloned, and this site appears to be closely related to the dopamine transporter. When dopamine is released by nerve cells, the major process to terminate the cellular effects of this neurotransmitter involves its reabsorption into the neurons that released it. This process requires the active transport of dopamine back into the presynaptic neurons. COCAINE interacts with its binding site to inhibit this reabsorption of DOPAMINE so that more of the neurotransmitter remains in the synaptic cleft for a longer period of time, and this effectively potentiates the activity of dopamine. The ability of cocaine and related drugs to support intravenous self-administration (a measure of reward or reinforcing properties of these drugs) is highly correlated with the affinity of these drugs for the dopamine transporter, but not to similar sites associated with other biogenic amine neurotransmitters (e.g., norepinephrine, serotonin). Since a variety of other studies have implicated dopamine in cocaine reward, the binding site associated with the dopamine transporter (especially in cortical and limbic brain regions) may be the site responsible for the reinforcing properties of the drug (i.e., for euphoria).

LYSERGIC ACID DIETHYLAMIDE (LSD) and related psychedelic drugs have actions at multiple sites in the central nervous system. There is considerable evidence that suggests that the effects of LSD involve agonistic activity at SEROTONIN (5-HT) receptors in the midbrain. There are at least five different subtypes of 5-HT receptors, but current data suggest that LSD is relatively selective for the 5-HT_2 receptor where it functions as a partial agonist (the maximum response to LSD is only 25% of that to serotonin).

A CANNABINOID receptor that binds Δ^9-THC (l-Δ^9-TETRAHYDROCANNABINOL, the chemical responsible for most of the characteristic psychological effects of marijuana) has also recently been identified. The highest levels of binding are in brain regions such as the hippocampus, suggesting that some of the effects of marijuana on memory may result from the interactions of the drug with cannabinoid receptors in this brain region. As can be seen above from the relationship between morphine-like drugs and opioid receptors, more efficient and effective treatments for acute intoxication, as well as withdrawal from various drugs used nonmedically for their subjective purposes, will likely arise from the knowledge gathered from the continued investigation of receptor mechanisms potentially mediating these effects.

(SEE ALSO: *Brain Structures and Drugs; Pharmacodynamics; Reward Pathways and Drugs*)

BIBLIOGRAPHY

AKIL, H., ET AL. (1984). Endogenous opioids: Biology and function. *Annual Review of Neuroscience, 6*, 223–255.

ANILINE, O., & PITTS, F. N., JR. (1982). Phencyclidine (PCP): A review and perspectives. *CRC Critical Reviews in Toxicology, 10*, 145–177.

COLQUHOUN, D. (1979). The link between drug binding and response: Theories and observations. In R. D. O'Brien (Ed.), *The receptors: A comprehensive treatise.* New York: Plenum.

DEWEY, W. L. (1986). Cannabinoid pharmacology. *Pharmacology Review, 38*, 151–178.

GARDNER, C. R. (1988). Functional in vivo correlates of the benzodiazepine agonist-inverse agonist continuum. *Progress in Neurobiology, 31*, 425–476.

GILMAN, A. G. (1987). G Proteins: Transducers of receptor-generated signals. *Annual Review of Biochemistry, 56*, 615–649.

HAYNES, L. (1988). Opioid receptors and signal transduction. *Trends in Pharmacological Science, 9*, 309–311.

JAFFE, J. H. (1990). Drug addiction and drug abuse. In A. G. Gilman et al. (Eds.), *Goodman and Gilman's the pharmacological basis of therapeutics*, 8th ed. New York: Pergamon.

JAFFE, J. H., & MARTIN, W. R. (1990). Opioid analgesics and antagonists. In A. G. Gilman et al. (Eds.), *Goodman and Gilman's the pharmacological basis of therapeutics*, 8th ed. New York: Pergamon.

MARTIN, W. R. (1986). Cellular effects of cannabinoids. *Pharmacology Reviews, 38*, 45–74.

MARTIN, W. R. (1983). Pharmacology of opioids. *Pharmacology Reviews, 35*, 283–323.

McKENNA, D. J., ET AL. (1989). Common receptors for hallucinogens in rat brain: A comparative autoradiographic study using [125I]LSD and [125I]DOI, a new psychomimetic radioligand. *Brain Research, 476*, 45–56.

MOLINOFF, P. B., WOLFE, B. B., & WEILAND, G. A. (1981). Quantitative analysis of drug-receptor interactions, II. Determination of the properties of receptor subtypes. *Life Sciences, 29*, 427–443.

OLSEN, R. W., ET AL. (1986). Barbiturate and benzodiazepine modulation of GABA receptor binding and function. *Life Sciences, 39*, 1969–1976.

RALL, T. W. (1990). Hypnotics and sedatives: ethanol. In A. G. Gilman et al. (Eds.), *Goodman and Gilman's the pharmacological basis of therapeutics*, 8th ed. New York: Pergamon.

RAZDAN, R. K. (1986). Structure-activity relationships in cannabinoids. *Pharmacology Reviews, 38*, 75–150.

RITZ, M. C., ET AL. (1987). Cocaine receptors on dopamine transporters are related to self-administration of cocaine. *Science, 237*, 1219–1223.

ROSS, E. M. (1990). Pharmacodynamics: Mechanisms of drug action and the relationship between drug concentration and effect. In A. G. Gilman et al. (Eds.), *Goodman and Gilman's the pharmacological basis of therapeutics*, 8th ed. New York: Pergamon.

SIMONDS, W. F. (1988). The molecular basis of opioid receptor function. *Endocrinology Review, 9*, 200–212.

WROBLEWSKI, J. T., & DANYSZ, W. (1989). Modulation of glutamate receptors: Molecular mechanisms and functional implications. *Annual Review of Pharmacology and Toxicology, 29*, 441–474.

NICK E. GOEDERS

RECEPTOR: NMDA (N-Methyl D-Aspartic Acid)

The NMDA receptor is a protein on the surface of neurons (nerve cells). When the major excitatory NEUROTRANSMITTER, GLUTAMATE, binds to this protein, the central pore of the NMDA receptor channel opens—then cations (the ions of sodium, potassium, and calcium) are able to cross the cell membrane. The movement of cations through the pore results in neuronal excitation.

The NMDA receptor is one of several cell receptor surface proteins activated by glutamate. The HALLUCINOGEN PHENCYCLIDINE (PCP) blocks the open channel of the NMDA receptor preventing cation flow. It is believed that overactivation of the NMDA receptor could be responsible for the neuronal cell death observed following some forms of stroke; it may even be involved in the cell death associated with neurodegenerative diseases.

(SEE ALSO: *Neurotransmission; Receptor: Drug*)

BIBLIOGRAPHY

CHOI, D. (1988). Glutamate Neurotoxicity and Diseases of the Nervous System. *Neuron, 1*, 623–634.

COLLINGRIDGE, G., & LESTER, R. (1989). Excitatory amino acid receptors in the vertebrate central nervous system. *Pharmacology Reviews, 40*(2), 145–210.

MAYER, M. L., & WESTBROOK, G. L. (1987). The physiology of excitatory amino acids in the vertebrate central nervous system. *Progress in Neurobiology, 28*, 197–276.

GEORGE R. UHL
VALINA DAWSON

RECIDIVISM   Recidivism is defined as the tendency to relapse to a previous behavior that may be harmful or delinquent in nature. In the substance-abuse field, it is generally accepted that ceasing to consume an abused substance is only one aspect of the TREATMENT process; maintaining this state is more difficult.

There have been numerous outcome studies on recidivism rates of various treatment programs. One review of alcohol-treatment outcome studies showed a 75 percent rate of recidivism at one year. Return to OPIATE use has been found to range from 39 to 97 percent and NICOTINE recidivism to range from 75 to 80 percent at one-year follow-up. In the early 1990s much of the work in substance-abuse treatment focused on relapse prevention in efforts to decrease these high recidivism rates. Cognitive-behavioral approaches are frequently used to enable patients to identify their high-risk relapse situations and to develop strategies for dealing with them more appropriately.

(SEE ALSO: *Coerced Treatment for Substance Offenders; Crime and Drugs; Opioid Dependence: Course of the Disorder over Time; Prisons and Jails: Drug Treatment in; Relapse; Relapse Prevention; Shock Incarceration and Boot Camp Prisons*)

BIBLIOGRAPHY

ANNIS, H. (1986). A relapse prevention model for treatment of alcoholics. In W. Miller & N. Heather (Eds.), *Treating addictive behaviors: Process of change*. New York: Plenum.

CATALANO, R., ET AL. (1988). Relapse in the addictions: Rates, determinants, and promising strategies. In *1988 Surgeon General's Report on Health Consequences of Smoking*. Washington, DC: Office of Smoking and Health.

MARLATT, G. A., & GORDON, J. (EDS.). (1985). *Relapse prevention: A self-control strategy for the maintenance of behaviour change*. New York: Guilford.

MILLER, W., & HESTER, R. (1980). Treating the problem drinker: Modern approaches. In *The addictive behaviours: Treatment of alcoholism, drug abuse, smoking and obesity*. New York: Pergamon.

MYROSLAVA ROMACH
KAREN PARKER

RECREATIONAL DRUG USE   *See* Addiction: Concepts and Definitions; Policy Alternatives: Safer Use of Drugs.

REINFORCEMENT   Although the term *reinforcement* has many common usages and associated misunderstandings, its meaning is precise with regard to the principles of behavior analysis and applications to drug-abuse treatment and methodology. Imprecision in defining terms can result in a flawed application of techniques; when failure results, the researcher may inappropriately reject a concept that was never initially clear. Success, however, can perpetuate a distortion and lead to continued attempts to validate a concept that was not adequately defined. Carefully defined terms, therefore, represent a crucial initial step in understanding basic principles and also dictate the way fundamental concepts are evaluated and validated in scientific investigations.

In brief, *reinforcement* is a process by which an *increase* in a particular response occurs following the presentation or termination of some event. Often reinforcement is contrasted with *punishment*, which refers to the *decrease* in response following the presentation or termination of an event. There is often confusion between (positive) reinforcement and negative reinforcement. Both refer to *increases* in a particular response but differ in whether the stimulus is presented *following* the response (positive reinforcement) or whether the response results in the *termination* of a stimulus (negative reinforcement). Negative reinforcement is also referred to as escape; or if the response can completely avoid the stimulus, the process is called avoidance.

It is important to note that reinforcement and punishment are descriptive, empirical processes that refer specifically to behavior and its consequences. *Reinforcement* is not an explanatory term; nothing is implied about *why* an event is reinforcing, whether there are specific neuroanatomical regions associated with reinforcing activities, or, indeed, whether reinforcement and pleasure coexist, although this assumption is made frequently and incorrectly. The defining characteristics of reinforcement (and punishment) depend on *how* a behavior is changed and not on the types of consequences that serve as reinforcing (or punishing) events (see Morse & Kelleher, 1977 for a more complete description of these terms). Among those factors that help determine whether an event is reinforcing or punishing are the individual's previous experience, the type of behavior, and the features of the environment that exist at the time the event occurs. The apparent idiosyncrasy of these conditions may make it difficult to establish common reinforcers (or punishers) that will work uniformly and effectively for all individuals.

The view that drugs of abuse could function as reinforcers in experimental animals, increasing and sustaining the behavior that led to their presentation, provided a strong and continuing impetus for the examination of drugs as reinforcing stimuli. This development brought the study of drug-seeking behavior and drug abuse into a framework for carefully controlled behavioral analyses embedded within well-established and objective principles (Schuster & Johanson, 1981).

(SEE ALSO: *Addiction: Concepts and Definitions; Causes of Substance Abuse: Learning; Research, Animal Model; Reward Pathways and Drugs; Wikler's Pharmacologic Theory of Drug Abuse*)

BIBLIOGRAPHY

MORSE, W. H., & KELLEHER, R. T. (1977). Determinants of reinforcement and punishment. In W. K. Honig & J. E. R. Staddon (Eds.), *Handbook of operant behavior.* Englewood Cliffs, NJ: Prentice Hall.

SCHUSTER, S. R., & JOHANSON, C. E. (1981). An analysis of drug-seeking behavior in animals. *Neuroscience and Biobehavioral Reviews, 5,* 315–323.

JAMES E. BARRETT

RELAPSE An individual who has recovered from an illness or has entered a period of stability in a chronic illness and who subsequently suffers a recurrence of symptoms is said to have experienced a relapse. At times this may be the result of insufficient treatment, either of intensity or duration (e.g., inadequate antidepressant dosage in depression). In other cases, it may be part of the natural history of the illness. Relapse has been studied most extensively in the addictive disorders, primarily from the perspective of social learning and behavioral theories. The basic hypothesis of relapse proposes that certain situations pose a high risk for the individual reengaging in the problem behavior (i.e., drug taking); generally, these are situations in which the behavior has occurred at a high frequency in the past. If the individual manages to successfully resist repeating the behavior, usually because of having learned new coping skills with which to master the situation, then there will be an increase in confidence or "self-efficacy." This increase in self-efficacy will lead to a lower probability of relapse each time the high risk situations recur. However, if the person has not developed alternate coping skills or has not made a commitment to changing the behavior, then the expectation of a positive experience resulting from use of the drug will return in the high-risk situations and a "lapse" to substance use may occur. Many researchers and some programs make a distinction between a single use or episode of use (a "slip") and a relapse, but if the level of consumption or pattern of use is comparable to prior use, it is termed "relapse."

Relapse represents a dynamic process. Many factors are felt to be important determinants of relapse, including pharmacological factors, conditioned responses to environmental cues, and negative emotional states such as depression, anxiety, and anger. Similarly, there are a number of attributional, cognitive, and affective consequences of relapse. Many substance-use treatment programs incorporate relapse-prevention techniques. Circumstances in which drug use has occurred—events, places, moods, cognitions—are carefully analyzed. Individuals are then taught to problem-solve, so as to develop new behaviors (e.g., build appropriate social supports) and cognitions in circumstances that are incompatible with drug use.

(SEE ALSO: *Causes of Substance Abuse; Relapse Prevention; Tobacco: Dependence; Treatment/Treatment Types*)

BIBLIOGRAPHY

BROWNELL, K. D., ET AL. (1986). Understanding and preventing relapse. *American Psychologist, 41,* 765–782.

MACKAY, P. W., DONOVAN, D. M., & MARLATT, G. A. (1991). Cognitive and behavioural approaches to alcohol abuse. In R. J. Frances & S. I. Miller (Eds.), *Clinical textbook of addictive disorders.* New York: Guilford.

SAUNDERS, B., & ALLSOP, S. (1987). Relapse: A psychological perspective. *British Journal of Addiction, 82,* 417–429.

MYROSLAVA ROMACH
KAREN PARKER

RELAPSE PREVENTION Relapse prevention (RP) is a cognitive–behavioral treatment program designed to teach individuals self-management skills to cope effectively with relapse risks. The three elements of RP are (1) behavioral skills training for clients to employ alternative behaviors in high-risk situations; (2) cognitive restructuring of expectancies and attributions about alcohol; and (3) lifestyle balance (Marlatt & Gordon, 1985).

Derived from social learning and behavioral principles, this cognitive–behavioral approach grew out of the desire by many psychologists to find a way of viewing addictive behavior different from the prevailing disease model of addiction. Frustrated with the lack of scientific support for the disease model, social learning theorists examined psychosocial and environmental factors rather than biological processes as important precursors to addictive behavior. Social learning theory views addictive behavior as a maladaptive coping response and focuses on the similarity across different addictive behaviors, including behaviors not typically viewed as medical— compulsive GAMBLING, compulsive sexual activity, excessive exercise, or OVEREATING. This theory also proposes that both mild and severe rates of an addictive behavior stem from the same learning processes.

Social learning theory proposes that people learn to associate events in the environment not only with physical sensations, but also with thoughts and emotions. This process is thought to occur through observing other people as well as through learning from their own experiences. For example, a child may grow up in a family and learn that Dad comes home from work and has a drink at the end of each day. The child, through watching the father, learns that ALCOHOL is associated with relaxation. Thus, when the child begins using alcohol as an adolescent or an adult, EXPECTANCIES exist for feeling relaxed, in addition to the experience of the pharmacological effects of alcohol as a relaxant. Therefore, the components of the learning process include environmental cues, beliefs and expectations about a substance or behavior, family history with the substance, and the consequences of the behavior or use. While the behavior is being learned, the positive consequences of the substance typically outweigh the negative. The immediate gratification of the substance use and the social support a person receives when engaging in the behavior are viewed as more "positive" or rewarding than any negative or punishing consequences. Social factors are therefore important in both the development of and maintenance of the habit. Once the negative consequences of the substance use outweigh the positive, however, the individual seeks to change the habit. An addictive behavior is therefore viewed as an overlearned, maladaptive habit that leads to delayed negative consequences, such as poor health or the loss of one's friends.

In the effort to understand the environmental contributions to relapse, Marlatt (1978) investigated the circumstances under which individuals initially used a substance. This study resulted in a categorization of the most common high-risk relapse situations, including accompanying thoughts and feelings that may trigger relapse. Situations were divided into two main categories:

1. Experiences that were primarily within the individual; those not associated with a specific event involving another person at the point of the slip. These experiences included coping with negative or unpleasant feelings, such as frustration or anger; coping with unpleasant bodily sensations, such as having a physical illness or going through withdrawal symptoms; wanting to have a positive mood, such as to "feel high"; testing one's willpower as to whether the substance can be used in a "controlled" manner; or giving in to a temptation in response to seeing items in the environment that remind the person of their prior use.

2. Experiences that were primarily associated with the influence of other people on the individual. These included coping with conflict with other people, leading to feelings of anger, frustration, worry, anxiety, or fear; dealing with social pressure, such as a friend who offers a drink or cigarette, or just being in the presence of someone using a drug, although no direct offer is made; and the desire to experience more positive feelings in a social situation, like wanting to "party."

These same categories were replicated with different populations, including various types of alcoholics, cigarette smokers, heroin addicts, participants in weight-loss programs, and compulsive gamblers (Marlatt & Gordon, 1980). Relapse prevention was therefore designed as a self-management strategy to teach people to assess their own high-risk situations and to learn alternative coping skills in the attempt to reduce the likelihood of a relapse. Should a relapse occur, RP also teaches skills to minimize the slip, enabling the individual to return quickly to the goal of abstinence rather than experiencing the ABSTINENCE VIOLATION EFFECT (AVE). Chaney, O'Leary, and Marlatt (1978) first applied these techniques with male chronic alcoholics who achieved over the course of a year fewer days of excessive drinking, fewer days of continuous drinking, and fewer total drinks consumed than alcoholics who did not receive the same training.

First, RP teaches people how to anticipate or assess their own high-risk situations by monitoring their own behavior and thoughts, by gaining descriptions of their past relapse experiences, by teaching them to pay close attention to any current fantasies about relapsing, and by determining what kinds of situations tend to hold the most current risk to abstinence.

Second, the therapist tries to use behavioral skill training to teach the client more effective coping strategies. Skills include muscle relaxation, visual imagery, anger management, ASSERTIVENESS TRAINING, or, on rare occasions, actually planning a relapse in the presence of the therapist so that the individual can learn how to rapidly regain control of the slip.

Third, if the person has positive outcome expectancies with the substance, cognitive restructuring is acquired as the therapist educates the client to differentiate immediate versus long-term effects of substance use, encouraging the person to think through the entire chain of events of use, rather than just focusing on the initial effects. The therapist also makes efforts to build a sense of self-efficacy, or personal control, so that clients believe they can successfully master a high-risk situation.

Fourth, the therapist and client may create an informal contract, or agreed-upon steps, should a relapse occur. The purpose of this contract is to limit the extent of use after an initial slip, and may include a reminder card the person can carry with specific steps to take in the event of a relapse. For example, the card can have a list of friends to con-

tact, the therapist's phone number, places to go that are supportive of abstinence, or phrases to think about that will aid in the cessation of use.

Fifth, the therapist teaches COPING skills in the event of the abstinence violation effect, where an individual slips to uncontrolled use of the substance. By renaming the relapse a "slip" or a "mistake," from which to learn, rather than an unrecoverable error reflecting a person's inability to abstain, the individual is taught to focus on the changing of an addictive behavior as a learning process, full of ups and downs, with easy and difficult times, which can end in the obtained goal of abstinence.

Finally, RP is geared toward overall lifestyle moderation. The occurrence of addictive behavior is viewed as the result of an imbalance between the "shoulds" and the "wants" in one's life. Successful abstinence is therefore more likely if a person can derive satisfaction from areas in life that do not include the substance. "Positive" addictions, such as exercise, can help the client alleviate STRESS, and thus become another coping strategy, while overall moderation can be viewed as reducing high-risk situations that result in an urge to use the substance.

RP can be applied either as a maintenance program for a specific activity, or for general lifestyle change. The techniques are applied after any initial intervention method (inpatient treatment or voluntary cessation) and are aimed toward maintenance of the behavior change within the context of a moderate lifestyle. RP is presented to the individual as an overview of the process of relapse, with alternative skills available at each step of the journey toward abstinence.

Research on the effectiveness of RP has shown that it is at least as effective as traditional disease-model, abstinence-oriented approaches, and that with some client populations (people who show elevated levels of depression or evidence of antisocial behavior) it has improved the abstinence rates compared with these other approaches. RP has also been shown to be effective with a variety of addictive behaviors, including smoking, opiate use, and overeating. Research studies have been designed to delineate more clearly (1) which types of individuals do best with RP, (2) whether the original taxonomy of situations is still valid, and (3) whether the taxonomy holds with a more diverse sample of people.

Other RP models place greater emphasis on the role of expectancies or self-efficacy. Each model contains its own techniques aimed at reducing the reoccurrence of the targeted addictive behavior (Daley, 1988).

(SEE ALSO: *Addiction: Concepts and Definitions; Causes of Substance Abuse: Psychological (Psychoanalytic) Perspective; Contingency Contracts; Disease Concept of Alcoholism and Drug Addiction; Overeating and Other Excessive Behaviors; Rational Recovery; Tolerance and Physical Dependence; Treatment: Alcohol, Psychological Approaches; Treatment Types; Wikler's Pharmacologic Theory of Drug Addiction*)

BIBLIOGRAPHY

BROWNELL, K. D., ET AL. (1986). Understanding and preventing relapse. *American Psychologist, 41*, 762–765.

CHANEY, E. F., O'LEARY, M. R., & MARLATT, G. A. (1978). Skill training with alcoholics. *Journal of Consulting and Clinical Psychology, 46*, 1092–1104.

DALEY, D. C. (1988). *Relapse: Conceptual, research and clinical perspectives.* Binghamton, NY: Haworth Press.

MARLATT, G. A. (1978). Craving for alcohol, loss of control, and relapse: A cognitive behavioral analysis. In P. E. Nathan, G. A. Marlatt, & T. Loberg (Eds.), *Alcoholism: New directions in behavioral research and treatment.* New York: Plenum.

MARLATT, G. A., & GEORGE, W. H. (1984). Relapse prevention: Introduction and overview of the model. *British Journal of Addiction, 79*, 261–273.

MARLATT, G. A., & GORDON, J. R. (1985). *Relapse prevention: Maintenance strategies in the treatment of addictive behaviors.* New York: Guilford Press.

MARLATT, G. A., & GORDON, J. R. (1980). Determinants of relapse: Implications for the maintenance of behavior change. In P. O. Davidson & S. M. Davidson (Eds.), *Behavioral medicine: Changing health lifestyles.* New York: Brunner/Mazel.

MOLLY CARNEY
ALAN MARLATT

RELIGION AND DRUG USE Drug use and religion have been intertwined throughout history, but the nature of this relationship has varied over time and from place to place. Alcohol and other drugs have played important roles in the religious rituals of numerous groups. For example, among a number of South American Indian groups, TOBACCO was considered sacred and was used in religious ritual, including the consultation of spirits and the initiation of religious leaders. Similarly, wine, representing the blood of Christ, has been central in the Holy Communion observances of both Roman Catholic and some Protestant churches. Considered divine by the Aztecs of ancient Mexico, the PEYOTE cactus (which contains a number of psychoactive substances, including the psychedelic drug mescaline) is used today in the religious services of the contemporary Native American Church (Goode, 1984).

Although tobacco, ALCOHOL, peyote, and other drugs have been important in the religious observances and practices of numerous groups, many religious teachings have opposed either casual use or the abuse of psychoactive drugs—and some religious groups forbid any use of such drugs, for religious purposes or otherwise. Early in America's history, the Protestant religious groups were especially prominent in the TEMPERANCE MOVEMENT, many of the ministers preached against the evils of drunkenness, and well-known Protestant leaders, such as John Wesley, called for the prohibition of all alcoholic beverages (Cahalan, 1987). The Latter-Day Saints' (Mormons) leader Joseph Smith prohibited the use of all common drugs, including alcohol, tobacco, and caffeine (no coffee or tea)—as did other utopian groups founded during the Great Awakening of the early 1800s. Religious groups and individuals were also active in America's early (1860s to 1880s) anti-smoking movement (U.S. Department of Health and Human Services, 1992). In contemporary American society, certain religious commitments continue to be a strong predictor of either use or abstinence from drugs, whether licit or illicit (Gorsuch, 1988; Cochran et al., 1988; Payne et al., 1991). For example, Islam forbids alcohol and opium use but coffee, tea, tobacco, khat, and various forms of marijuana were not prohibited, because they came into the Islamic world after the prohibitions were laid down. Indulgence in any debilitating substance is, however, not considered proper or productive. Christianity, Judaism, and Buddhism may not prohibit specific drugs but they and most other widespread, mainstream religious traditions also caution against indulgence in most things. In our society, many who have indulged have sought the help of ALCOHOLICS ANONYMOUS (AA) or NARCOTICS ANONYMOUS (NA)—both self-help groups founded on strong religious underpinnings.

In this discussion, our attention is limited to recent conditions in the United States, focusing on potentially dangerous, abusive, and/or illicit patterns of drug use. Since such drug use is widely disapproved by most religious teachings and leaders, it is not surprising to find that those with strong religious commitments are less likely to be drug users or abusers. Moreover, research findings clearly show that religious involvement has been a protective factor, helping some adolescents resist the drug epidemics of the 1970s and 1980s.

Because religion has been found to be a protective factor against drug use and dependence, and because our society is concerned with drug use among young people, much of the research linking religion with drug use focuses on adolescents and young adults. This age range is particularly important for several reasons. First it is the period during which most addiction to NICOTINE begins; the majority of people who make it through their teens as nonsmokers do not take up the habit during their twenties or later. Second, ADOLESCENCE and young adulthood is the period during which abusive alcohol consumption is most widespread. Third, recent EPIDEMICS in the use of illicit drugs have been most pronounced among teenagers and young adults. Fourth, during this portion of the life span, many changes, opportunities, and risks occur; thus the strictures and guidelines provided by religious commitment may be especially important in helping young people resist the temptation to use and abuse drugs. Finally, evidence that religious conversion is most likely to occur during adolescence (Spilka et al., 1991) makes this period particularly appropriate for research on the link between religion and drug use.

THE RELATIONSHIP BETWEEN RELIGIOUS COMMITMENT AND DRUG USE

Research investigating the relationship between religious commitment and drug use consistently indicates that those young people who are seriously involved in religion are more likely to abstain from drug use than those who are not; moreover, among users, religious youth are less likely than nonreligious youth to use heavily (Gorsuch, 1988; Lorch & Hughes, 1985; Payne et al., 1991).

Examples from 1982 and 1992. Figure 1 shows how drug use was related to religious commitment among high school seniors in 1982 and in 1992. Individuals with the highest religious commitment were defined as those who usually attend services once a week or more often and who describe religion as being very important in their lives; individuals with low commitment are those who never attend services and rate religion as not important. Figure 1 clearly indicates that those with low religious involvement were more likely than average to be frequent cigarette smokers, occasional heavy drinkers, and users of MARIJUANA and COCAINE; conversely, those highest in religious commitment were much less likely to engage in any of these behaviors. Other analyses have shown that similar relationships exist for other illicit drugs (Bachman et al., 1986) and for other age groups (Cochran et al., 1988; Gorsuch, 1988).

Recent Changes in Drug Use and Religious Commitment. Figure 1 presents data from the two points in time, 1982 and 1992. It is obvious in the figure that the proportion of seniors using the illicit drugs marijuana and cocaine declined markedly during that decade; the proportion reporting instances of heavy drinking declined appreciably; and the proportion of frequent smokers declined somewhat. It is clear also that religion was linked to drug use in both years, although the relationships appear a bit more dramatic during the period of heavier use.

Because high religious commitment is associated with low likelihood of drug use, it is reasonable to ask whether any of the decline in illicit drug use during the 1980s could be attributed to a heightened religious commitment among young people during that period. The answer is clearly negative. The same annual surveys that showed declines in drug use also showed that religious commitment, rather than rising during that period, was actually declining to a slight degree among high school seniors. It thus appears that other factors accounted for the declines in illicit drug use, factors such as the increasing levels of risk and the heightened disapproval associated with such behaviors (Bachman et al., 1988; Bachman et al., 1990; Johnston, 1985; Johnston et al., 1992).

Religion as a Protective Factor. The most plausible interpretation of the relationship between religion and drug use during recent years, in our view, is that religion (or the lack thereof) was not primarily responsible for either the increases or the subsequent decreases in illicit drug use. Rather, it

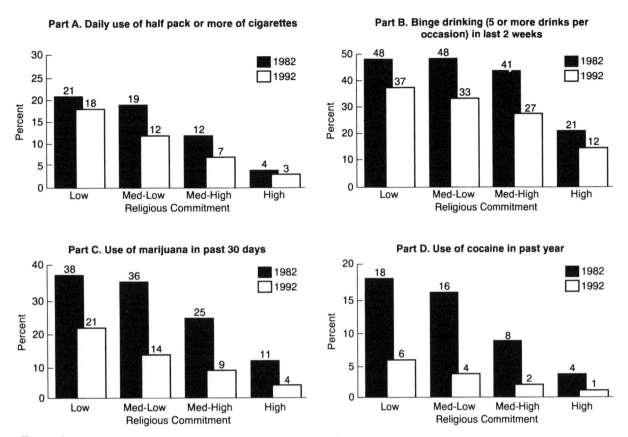

Part A. Daily use of half pack or more of cigarettes

Part B. Binge drinking (5 or more drinks per occasion) in last 2 weeks

Part C. Use of marijuana in past 30 days

Part D. Use of cocaine in past year

Figure 1

Drug Use among High School Seniors Shown Separately for Four Levels of Religious Commitment, 1982 and 1992

appears that those with the strongest religious commitment were least susceptible to the various epidemics in drug use. Figure 2 provides one example in support of that interpretation. The figure shows trends in cocaine use from 1976 through 1988, distinguishing among the four different degrees of religious commitment. Cocaine use roughly doubled between 1976 and 1979 among high school seniors and began to decline sharply after 1986 (and, although the figure shows data only through 1988, has continued to decline). But the most important pattern in the figure, for present purposes, is that these historical trends in cocaine use were much more pronounced among those with little or no religious commitment. Put another way, it seems that strong religious commitment operated as a kind of protective factor sheltering many youths from the waves of drug use sweeping the nation.

Denominational Differences. There are important differences among religious groups in the stress laid upon drug use (Lorch & Hughes, 1988). In particular, the more fundamentalist Protestant denominations, as well as Latter-Day Saints (Mormons) and Black Muslims, rule out the use of alcohol and tobacco and disdain illicit drug use. Research examining differences in drug use among young people finds that those who belong to fundamentalist denominations are more likely to abstain from drug use than are youth who belong to more liberal denominations (Lorch & Hughes, 1985). Analyses of the data on high school seniors (unpublished) corroborate the findings of earlier research; individuals strongly committed to fundamentalist denominations (e.g., Baptists) are much lower than average in their use of drugs and lower than those strongly committed to other religious traditions.

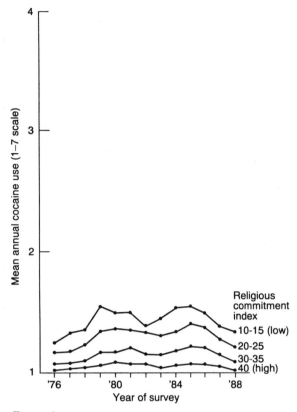

Figure 2

Trends in Annual Cocaine Use Shown Separately for Four Levels of Religious Commitment, High School Seniors, 1976–1988

POSSIBLE CAUSAL PROCESSES

Since religious commitment is negatively related to drug use, it becomes important to understand the possible causal processes underlying that relationship. A number of possibilities can be briefly considered.

Overlaps with Other Causes. Religious commitment among young people is correlated with a number of other factors known to relate to drug use. In particular, students who get good grades, who plan to go to college, and who are not truant are also less likely to use drugs, as well as more likely to display high levels of religious commitment. These various factors are closely interrelated in a common syndrome (Dryfoos, 1990; Jessor & Jessor, 1977), and thus it is difficult to disentangle causal processes. Indeed, it could be argued that religious commitment

is probably one of the root causes, contributing both to educational success and to the avoidance of drug use. Analyses of possible multiple causes of drug use (or abstention) have shown that religious commitment overlaps with other predictors, but only partially. In other words, although religious commitment may be part of a larger syndrome, it also appears to have some unique (i.e., nonoverlapping) impact on drug use. It is, therefore, important to explore the special features of religious involvement that tend to suppress drug and alcohol use.

Content of Religious Teaching. One obvious explanation is that most religious traditions teach followers to avoid the abuse of drugs. Restrictions vary, of course, from one tradition to another, and the greater emphasis on prohibition in fundamentalist denominations seems the most likely explanation for the lower levels of use among adherents. But even in traditions that do not explicitly or completely ban drug use, there is still much teaching ranging from respect for one's own body to family responsibilities to broader social responsibilities, all arguing against the abuse of drugs. Because all drugs, including cigarettes and alcohol, are illicit for minors, young people who are strongly committed to religion may abstain from drug use simply in obedience to the laws of the nation; but even more important, they are likely to act in obedience to what they perceive to be God's laws.

Parental Examples and Precepts. In addition to the direct teachings associated with attendance at religious services, young people raised in religious traditions are likely to be exposed to parents and other relatives who follow such teachings. Thus a part of the explanation for lower drug use among religiously involved young people may be that their families reinforce the religious strictures against use and abuse. A further factor may simply be availability; religious parents who do not drink, smoke, or use drugs will not have these substances in their homes, thus reducing the opportunity for young people to experiment with them.

Peer Group Factors. The dynamics operating within the family probably have their parallel in broader social contacts. That is, those who are strongly committed to religion probably associate with others holding similar views. Thus the strongly religious are less likely to belong to peer groups that encourage experimentation with cigarettes, alcohol, and other drugs and more likely to participate in peer networks and activities that do not involve drugs. Given the strong relationship between drug use by peers and an adolescent's own drug use, the norms of the peer group are especially important as predictors of whether a particular teenager will initiate drug use (Jessor & Jessor, 1977).

CONCLUSION

The relationship between religion and drug use among young people is not completely straightforward. On the one hand, a considerable amount of research indicates that young people who are strongly committed to religion are less likely than their uncommitted counterparts to use drugs. On the other hand, data presented here and elsewhere suggest that religion has had relatively little impact on recent national declines in drug use among young people. Further examination of this relationship reveals that America's drug epidemic occurred primarily among those not affected by religion; highly religious youth were relatively immune to the plague that infected a significant portion of the nation's youth. Accordingly, we conclude that religious commitment has been, and continues to be, an effective deterrent to the use and abuse of licit and illicit drugs.

This work was supported by Research Grant no. 3 R01 DA 01411 from the National Institute on Drug Abuse and a National Science Foundation Minority Postdoctoral Fellowship to the second author. We thank Dawn Bare for her contribution in data analysis and figure preparation.

(SEE ALSO: *Ethnic Issues and Cultural Relevance in Treatment; Jews, Drug and Alcohol Use among; Prevention Movement; Vulnerability: An Overview*)

BIBLIOGRAPHY

BACHMAN, J. G., ET AL. (1990). Explaining the recent decline in cocaine use among young adults: Further evidence that perceived risks and disapproval lead to reduced drug use. *Journal of Health and Social Behavior, 31,* 173–184.

BACHMAN, J. G., ET AL. (1988). Explaining the recent decline in marijuana use: Differentiating the effects of perceived risks and disapproval, and general lifestyle factors. *Journal of Health and Social Behavior, 29,* 92–112.

BACHMAN, J. G., ET AL. (1986). *Change and consistency in the correlates of drug use among high school seniors: 1975–1986.* Monitoring the Future Occasional Paper no. 21. Ann Arbor: University of Michigan, Institute for Social Research.

BOCK, E. W., ET AL. (1987). Moral messages: The relative influence of denomination on the religiosity-alcohol relationship. *Sociological Quarterly, 28*(1), 89–103.

CALAHAN, D. (1987). *Understanding America's drinking problem.* San Francisco: Jossey-Bass.

COCHRAN, J. K., ET AL. (1988). Religiosity and alcohol behavior: An exploration of reference group theory. *Sociological Forum, 3*(2), 256–276.

DRYFOOS, J. G. (1990). *Adolescents at risk: Prevalence and prevention.* New York: Oxford University Press.

GOODE, E. (1984). *Drugs in American society,* 2nd ed. New York: Knopf.

GORSUCH, R. L. (1988). Psychology of religion. *Annual Review of Psychology, 39,* 201–221.

JESSOR, R., & JESSOR, S. L. (1977). *Problem behavior and psychosocial development: A longitudinal study of youth.* New York: Academic Press.

JOHNSTON, L. D. (1985). The etiology and prevention of substance use: What can we learn from recent historical changes? In C. L. Jones & R. J. Battjes (Eds.), *Etiology of drug abuse: Implications for prevention.* Washington, DC: U.S. Government Printing Office.

JOHNSTON, L. D., ET AL. (1992). *Smoking, drinking, and illicit drug use among American secondary school students, college students, and young adults, 1975–1991,* vols. 1 and 2. Washington, DC: U.S. Government Printing Office.

LORCH, B. R., & HUGHES, R. H. (1988). Church, youth, alcohol, and drug education programs and youth substance abuse. *Journal of Alcohol and Drug Education, 33*(2), 14–26.

LORCH, B. R., & HUGHES, R. H. (1985). Religion and youth substance use. *Journal of Religion and Health, 24*(3), 197–208.

PAYNE, I. R., ET AL. (1991). Review of religion and mental health: Prevention and the enhancement of psychosocial functioning. *Prevention in Human Services, 9*(2), 11–40.

SPILKA, B. (1991). Religion and adolescence. In R. M. Lerner et al. (Eds.), *Encyclopedia of adolescence.* New York: Garland.

U.S. DEPARTMENT OF HEALTH AND HUMAN SERVICES. (1992). *Smoking and health in the Americas.* Atlanta:

Public Health Service, Centers for Disease Control, National Center for Chronic Disease Prevention and Health Promotion, Office of Smoking and Health.

JERALD G. BACHMAN
JOHN M. WALLACE, JR.

REMOVE INTOXICATED DRIVERS (RID-USA, Inc.)

Founded in 1978, this volunteer grass-roots organization (P.O. Box 520, Schenectady, NY 12301; 518-372-0034) is devoted to efforts to deter impaired driving, help victims seek justice and restitution, close loopholes in DWI (driving while impaired) laws, and educate the public on the scope of impaired-driving tragedies. RID activists have played a key role in the passage of reforms of the impaired-driving laws in many states, enabled passage of more than 500 anti-DWI laws, and monitored more than 15,000 court cases.

RID's victim-support activities, which are free, include providing long-term emotional support to victims of drunk-driving crashes and to their families; counseling victims and accompanying them throughout all phases of criminal prosecution of the offender; assisting victims in obtaining compensation; and referring victims and their families to appropriate supportive agencies. Court monitoring and research activities include monitoring the efforts of police, prosecutors, magistrates, and judges in drunk-driving cases through research and analysis of local court records, and reporting these findings to the public. RID's public awareness and education activities are extensive. Members organize public meetings; present educational talks to community and religious organizations; participate in forums, exhibits, and media events; supplement high school driver-education classes; and support SADD (STUDENTS AGAINST DRIVING DRUNK) and other similar student groups. They study and report on alcohol-related vehicle and traffic laws; support concepts such as designated-driver and alcohol-server education; and promote SNAP (a Sane National Alcohol Policy), which advocates raising taxes on alcohol, curbing campus beer promotions, and airing public-service advertising to counter all broadcast alcohol commercials.

RID is organized into autonomous chapters, with more than 150 chapters in at least 41 states in the United States and a national group in France. Financial support comes from member dues, government and corporate grants, charitable contributions, and memorial gifts. Information on how to organize a RID chapter is available from the national office in Schenectady, New York.

(SEE ALSO: *Accidents and Injuries from Alcohol, Dramshop Liability Laws; Drunk Driving; Mothers Against Drunk Driving*)

FAITH K. JAFFE

REPEAL OF PROHIBITION

See Prohibition of Alcohol.

RESEARCH

This section is devoted primarily to detailed explanations of the ways in which behavioral psychologists and psychopharmacologists explore the interactions between drug actions and behavior in laboratory settings. The section begins with an overview article, *Aims, Description, and Goals.* The article *Developing Medications to Treat Substance Abuse and Dependence* ties basic research directly to clinical applications. The articles on *Drugs as Discriminitive Stimuli; Measuring Effects of Drugs on Behavior; Measuring Effects of Drugs on Mood;* and *Motivation* describe these general research techniques and concepts and their applicability to understanding drug abuses.

Research in the field of drug dependence, however, is much broader and more diverse than the topics included in this section. In fact, research is conducted on most of the topics contained in this encyclopedia—from epidemiological studies to new methods for detecting drug smuggling; from herbicides that can target specific plant sources of illicit drugs to how to target prevention messages to subgroups within the population; from how certain drugs produce their toxic effects to developing new drugs to reduce drug craving or prevent relapse; from how the interactions of environment and genetics make certain individuals more vulnerable to drug use to the relative effectiveness of different treatment programs. Many of these research issues are touched upon in such diverse articles as those on controlling illicit drug supply; on TREATMENT; or PREVENTION; and on VULNERABILITY AS A CAUSE OF SUBSTANCE ABUSE.

Clinical, behavioral, epidemiological, and basic research is carried out primarily by researchers at universities, government research centers, and research institutes. It is funded both publicly and privately. The work of a representative few of these centers is described elsewhere in the encyclopedia (see *Addiction Research Foundation (Canada); Addiction Research Unit (U.K.); Center on Addiction and Substance Abuse (CASA); Rutgers Center of Alcohol Studies; U.S. Government/U.S. Government Agencies (SAMHSA, NIAAA, NIDA, CSAP, CSAT).* In 1992, worldwide, there were more than 80 research centers devoted to problems of drugs and alcohol. Fifty-eight of the centers were in the United States; thirteen were in Europe and the U.K.; the others were in Central and South America, Asia, Australia, and New Zealand.

For more information on research, see also *Imaging Techniques: Visualizing the Living Brain; Pain: Behavioral Methods for Measuring the Analgesic Effects of Drugs; Research, Animal Models.*

Aims, Description, and Goals In a Chinese book on pharmacy, which dates to 2732 B.C., references are found to the properties of MARIJUANA (a type of Old World HEMP, *Cannabis sativa* of the mulberry family). In an Egyptian papyrus from about 1550 B.C., there is a description of the effects of OPIUM (a product of the opium poppy, *Papaver somniferum*). In every culture, the uses of ALCOHOL are documented in both oral and written tradition, often going back into antiquity—the Bible, for example, mentions both the use and abuse of wine. Although people have made observations on PSYCHOACTIVE substances for thousands of years, much remains to be learned about both alcohol and drugs of abuse; much research remains to be done before these substances and their effects can be fully understood.

WHAT WE NEED TO KNOW

Most substance-abuse research carried out today is a consequence of public health and social concerns. With millions of people using and abusing many different substances, and because of the close association between AIDS and drug abuse, it is imperative to know just how dangerous—or not dangerous—any given drug is to public health and safety. For economic as well as medical reasons, it is essential to find the most effective ways to use our health resources for preventing and treating substance abuse. So many questions still exist that no one scientific discipline can answer them all. The answers must be found through studies in basic chemistry, molecular biology, genetics, pharmacology, neuroscience, biomedicine, physiology, behavior, epidemiology, psychology, economics, social policy, and even international relations.

From a social standpoint, the first question for research must be: How extensive is the problem? Surveys and other indicators of drug and alcohol usage are the tools used by epidemiologists to determine the extent and nature of the problem, or to find out how many people are abusing exactly which drugs, how often, and where. As the dimensions of the problem are defined, basic scientists begin their work, trying to discover the causes and effects of substance abuse at every level, from the movement of molecules to the behavior of entire human cultures. Chemists determine the physical structure of abused substances, and then molecular biologists try to determine exactly how they interact with the subcellular structures of the human body. Geneticists try to determine what components, if any, of substance abuse are genetically linked. Pharmacologists determine how the body breaks down abused substances and sends them to different sites for storage or elimination. Neuroscientists examine the effects of drugs and alcohol on the cells and larger anatomical structures of the brain and other parts of the nervous system. Since these structures control our thoughts, emotions, learning, and perception, psychologists and behavioral pharmacologists study the drugs' effects on their functions. Cardiologists and liver and pulmonary specialists study the responses of heart, liver, and lungs to drugs and alcohol. Immunologists examine the consequences of substance abuse for the immune system, a study made critical by the AIDS epidemic. The conclusions reached through these basic scientific inquiries guide clinicians in developing effective treatment programs.

In considering drug abuse, people have long wondered why so many plants contain substances that have such profound effects on the human brain and mind. Surely people were not equipped by nature with special places on their nerve cells (called RECEPTORS) for substances of abuse—on the off chance that they would eventually smoke marijuana or take COCAINE or HEROIN. The discovery in the late 1960s that animals would work to obtain injections or drinks of the same drugs that people abuse was an important scientific observation; it contributed to the hypothesis that there must be a biological basis for substance abuse. These observations and this reasoning led scientists to look for substances produced by people's own bodies (endogenous substances) that behave chemically and physiologically like those people put into themselves from the outside (exogenous substances)—like alcohol, NICOTINE, marijuana, cocaine, and other drugs of abuse. When receptors for endogenous substances were discovered—first for the OPIATES in the 1970s and only recently for PCP, cocaine, marijuana, and LSD—their existence helped establish the biological basis for drug abuse. So did the discovery of a genetic component for certain types of ALCOHOLISM. These discoveries by no means negate the extensive behavioral and social components of substance abuse, but they do suggest a new weapon in dealing with the problem—that is, the possibility of using medication, or a biological therapy, as an adjunct to psychosocial therapies. Asserting a biological basis for substance abuse also removes some of the social stigma attached to drug and alcohol addiction. Since drug dependence is a disorder with strong biological components, we begin to understand that it is not merely the result of weak moral fiber.

Armed with information that was derived from basic research, clinical researchers in hospitals and clinics test and compare treatment modalities, looking for the best balance of pharmacological and psychosocial methods for reclaiming shattered lives. Finding the right approach for each type of patient is an important goal of treatment research, since patients frequently have a number of physical and mental problems besides substance abuse. The development of new medications to assist in the treatment process is an exciting and complex new frontier in substance-abuse research.

The best way to prevent the health and social problems that are associated with substance abuse has always been a significant research question. Insights gained from psychological and social research enable us to design effective prevention programs targeted toward specific populations that are particularly vulnerable to substance abuse for both biomedical and social reasons. Knowing the consequences of substance abuse often helps researchers to formulate prevention messages. For example, the identification of the FETAL ALCOHOL SYNDROME (FAS), a pattern of birth defects among children of mothers who drank heavily during pregnancy, was a major research contribution to the prevention of alcohol abuse. Drug-abuse-prevention research has

assumed a new urgency with the realization, brought about by epidemiologists and others, that the AIDS virus is blood-borne—spread by sexual contact and by drug abusers who share contaminated syringes and needles. HIV-positive drug users then spread the disease through unprotected sexual intercourse. Public education about drug abuse and AIDS must use the most powerful and carefully targeted means of reaching the populations at greatest risk for either disease, and these means can be determined only by the most careful social research and evaluation methodologies.

Substance-abuse research is no different from any other sort of scientific endeavor: the process is not always orderly. Critical observations by clinicians frequently provide basic researchers with important insights, which guide the research into new channels. Observations in one science often lead to breakthroughs in other areas.

METHODS

The range of methods employed by scientists studying substance abuse is as wide as the range of methods in all the biological and social sciences. One important method is the use of animal models of behavior to answer many of the questions raised by drug and alcohol use. Animal models are used in biomedical research in virtually every field, but the discovery that animals will, for the most part, self-administer alcohol and the same drugs of abuse that humans do, meant that there was a great potential for behavioral research uncontaminated by many of the difficult-to-control social components of human research. The results of animal studies have been verified repeatedly in human research and in clinical observation, thus validating this animal model of human drug-seeking behavior.

Research Personnel. Drug- and alcohol-abuse research is conducted by many different types of qualified professionals, but mostly by medical researchers (MDs) and people with advanced degrees (PhDs) in the previously mentioned sciences. They work with animals and with patients in university and federally funded laboratories, as well as in privately funded research facilities, in offices, and in clinical treatment centers. Other sites include hospitals, clinics, and sometimes schools, the streets, and even advertising agencies when prevention research is under way.

FUNDING

Who pays for substance-abuse research has always been an important issue. In the late 1980s and early 1990s, most of the drug- and alcohol-abuse research in the world was supported by the U.S. government. One of the federally funded National Institutes of Health—the NATIONAL INSTITUTE ON DRUG ABUSE (NIDA)—funds over 88 percent of drug-abuse research conducted in the United States and abroad. In 1992, this amounted to over 362 million dollars, which supported NIDA's own intramural research at the Addiction Research Center and the research done in universities under grants awarded by the institute. NIDA's sister institute, the NATIONAL INSTITUTE ON ALCOHOL ABUSE AND ALCOHOLISM (NIAAA), plays a parallel role in funding alcohol-abuse research. In 1992, it funded 175 million dollars in alcohol-research grants. Many other U.S. government agencies also have important roles in sponsoring and conducting substance-abuse research. For the most part, state and local governments do not sponsor substance-abuse research, although they do much of the distribution of funds for treatment and prevention programs.

Other countries, most notably Canada, sponsor basic clinical and epidemiological substance-abuse research within their own universities and laboratories, but none does so on a scale that is comparable to that of the United States. Private foundations and research institutions like the Salk Institute for Biological Studies, Rockefeller University, and the Scripps Clinic and Research Foundation use their own funds, as well as federal grant support, to pay for their research endeavors. Pharmaceutical companies also support some substance-abuse research—mostly clinical work related to medications that might be used as part of treatment programs for drug and alcohol abuse. Again, much of this work is sponsored, in part, by the U.S. government.

(SEE ALSO: *National Household Survey; Substance Abuse and HIV/AIDS; Research, Animal Model; U.S. Government/U.S. Government Agencies*)

BIBLIOGRAPHY

Alcohol and Health. (1990). Seventh Special Report to the U.S. Congress. DHHS Publication no. (ADM) 90-1656. Washington, DC: U.S. Government Printing Office.

BARINAGA, M. (1992). Pot, heroin unlock new areas for neuroscience. *Science, 258,* 1882–1884.

COOPER, J. R., BLOOM, F. E., & ROTH, R. H. (1986). *The biochemical basis of pharmacology.* New York: Oxford University Press.

Drug Abuse and Drug Abuse Research III. (1991). Third Triennial Report to Congress. DHHS Publication no. (ADM) 91-1704. Washington, DC: U.S. Government Printing Office.

GERSHON, E. S., & RIEDER, R. D. (1992). Major disorders of mind and brain. *Scientific American, 267*(3), 126–133.

JAFFE, J. H. Drug addiction and drug abuse. (1990). In A. G. Gilman et al. (Eds.), *Goodman and Gilman's the pharmacological basis of therapeutics,* 8th ed. New York: Pergamon.

CHRISTINE R. HARTEL

Developing Medications to Treat Substance Abuse and Dependence
Dependence on drugs, ALCOHOL, or TOBACCO is difficult to treat, and practitioners have tried many approaches in their attempts to arrive at successful treatments. One approach is to develop medications, or pharmacological treatments. This approach is most effective when the medication is given along with behavioral treatments. These behavioral treatments help the individual cope with the underlying etiology of his or her drug use and the problems associated with drug use; they may also help ensure compliance in taking the medication that is prescribed.

PERPETUATION OF DRUG ABUSE: EUPHORIA AND WITHDRAWAL

Many people who are drug- or alcohol-dependent want to stop their habit, but often they have a difficult time doing so. There are at least two reasons for this difficulty. First, the drugs produce pleasant or euphoric feelings that the user wants to experience again and again. Second, unpleasant effects can occur when the drug use is stopped. The latter effect, commonly known as WITHDRAWAL, has been shown after prolonged use of many drugs, including alcohol, OPIATES (such as HEROIN), SEDATIVE HYPNOTICS, and anxiety-reducing drugs. Other drugs, such as COCAINE and even CAFFEINE (COFFEE and COLA drinks) and NICOTINE (cigarettes), are also believed to be associated with withdrawal effects after prolonged use. These unpleasant withdrawal effects are alleviated by further drug use. Thus drugs are used and abused because they produce immediate pleasant effects and because the drug reduces the discomfort of withdrawal.

The symptoms of withdrawal are fairly specific for each drug and include physiological effects and psychological effects. For example, alcohol withdrawal can be associated with shaking or headaches, and opiate withdrawal with anxiety, sweating, and increases in blood pressure, among other effects. Withdrawal from cocaine may cause depression or sadness, withdrawal from caffeine is associated with headaches, and withdrawal from nicotine often produces irritability. All drug withdrawals are also associated with a strong craving to use more drugs. Much work has been done to document the withdrawal effects from alcohol, opiates, BENZODIAZEPINES, and tobacco; however, documentation of withdrawal from cocaine or other stimulant drugs has only recently begun to be examined.

NEURAL CHANGES WITH CHRONIC DRUG USE

Both withdrawal and the pleasant or euphoric effects from drug use occur, in part, as a result of the drug's action on the brain. The immediate or acute effects of most drugs of abuse affect areas of the brain that have been associated with "reward" or pleasure. These drugs stimulate areas normally aroused by natural pleasures such as eating or sexual activity. Long-term, or chronic, drug use alters these and other brain areas. Some brain areas will develop TOLERANCE to the drug effects, so that greater and greater amounts are needed to achieve the original effects of the drug. Some examples of drug effects that develop tolerance are the ANALGESIC or pain-killing effect of opiates and the euphoria- or pleasure-producing effect of most drugs of abuse, which are probably related to its abuse potential.

Because some brain areas may also become sensitized, an original drug effect will either require a lesser amount of the drug to elicit the effect when the drug is used chronically or the effect becomes greater with chronic use. This phenomenon has been studied most extensively in cocaine use. Cocaine is associated with behavioral sensitization of motor activity in animals and paranoia (extreme delusional fear) in humans. There are physiological effects that develop tolerance or sensitization as well. For example, the chronic use of cocaine will sensitize some brain areas so that seizures are more easily

induced. Other health risks of drug use will be addressed below.

In addition to these more direct acute and chronic drug effects, another phenomenon occurs with long-term drug use. This phenomenon is the *conditioned drug effect*, in which the environmental or internal (mood states) cues commonly presented with drug use become conditioned or psychologically associated with drug use. For example, when angry, a drug addict may buy or use drugs in a certain place with certain people. After frequently taking drugs under similar conditions, the individual can experience a strong craving or even withdrawal when in the environment in which he or she has taken drugs or feels angry. When the individual tries to stop using drugs, exposure to these conditioned cues can often lead to relapse because the craving and withdrawal effects are so powerful. Very little research has been done on the neural bases of these conditioned effects; thus it is not known whether these effects are mediated by similar or different neural mechanisms.

RESEARCH ON DRUG EFFECTS

Many of these acute and chronic effects of drugs on the brain have been investigated in animal research, which allows greater control over the research, including manipulations of drug exposure. A number of animal models are used to assess drug preferences, and, since most drugs that humans abuse are also preferred by animals, these models are useful for understanding human drug abuse. Moreover, animal research allows scientists to study directly the various areas of the brain that are involved in drug use. In addition, recent technological advancements on noninvasive IMAGING have allowed scientists to look at pictures of the brains of humans while they are being administered drugs or while they are withdrawing from drugs. This human work has also enhanced our knowledge of the drug effects on the brain as well as validated the information gained from animal research.

Another useful line of research in assessing the effects of drugs involves human laboratory studies. In one type of study, research volunteers who have had experience with the abused drugs are given a specific drug (e.g., morphine), and various psychological and physiological measurements are obtained. The psychological measurements can include reports from the subject on the effects of the drug as well as more sophisticated behavioral measures that tell the experimenter how much the drug is preferred. Another type of human laboratory study is to study the effects of drug withdrawal. For opiates, withdrawal can be precipitated by an opiate ANTAGONIST drug (NALTREXONE), and withdrawal signs and symptoms are measured. For other drugs (such as cocaine), withdrawal is more difficult to measure because little is known about their withdrawal syndromes.

Some of the information that scientists have learned from such studies includes delineating specific brain areas as well as the NEUROTRANSMITTERS (the chemicals released by the brain cells) involved in drug use and withdrawal. Thus, when specific neurotransmitters become identified as playing an important role in drug use or withdrawal, scientists can administer experimental drugs that act on these neurotransmitters to see if the animals will alter their drug preference or show less severe withdrawal signs. Researchers can also give these experimental drugs to the human research volunteers to see if the medication alters the subject's perception of or behavior toward the abused drug or if it alleviates withdrawal symptoms. If the results from these animal and human laboratory studies are promising, then these agents can be tested on treatment-seeking, drug-dependent individuals in clinical trials. This latter type of research is more time-consuming and expensive than the laboratory studies, but it helps provide an answer to the ultimate question: Does this medication help the individual stop abusing drugs?

APPROACHES TO DEVELOPING MEDICATIONS FOR DRUG ABUSE

Researchers can use the knowledge gained from animal and human studies of the effects of drugs on the brain as they develop medications for alcohol and drug dependence. Most likely, one medication will be needed to help detoxify the drug-dependent individual and a second medication to help sustain abstinence from drug use. This two-phase medication regimen is used for opiate and alcohol treatment, and it may ultimately be the approach used for countering dependency on other drugs, such as cocaine, sedatives, and nicotine. In theory, a pharmacological treatment agent or medication would block or reduce either the acute, rewarding effect of the drug or the discomfort of withdrawal. In practice, few treatment drugs have been found to be very effective in sustaining abstinence from drugs or alcohol.

Any pharmacological agent should be able to be given orally, as this is much easier than other routes of administration, such as injections. The agent itself must be medically safe and not enhance any of the health risks associated with illicit drug use, since the individual may illicitly use drugs while being maintained on the treatment agent. Finally, the pharmacological treatment agent must be acceptable to the patient. That is, if the agent causes undesirable side effects, individuals will likely not take it.

Current research with alcohol and drug effects on the brain and with treatment outcome hold great promise for effective pharmacological agents. This search process will necessarily include the animal and human laboratory studies mentioned as well as medicinal chemistry research. Medicinal chemistry research is used to develop new compounds that have similar but slightly altered chemical structures to the abused drugs or to the neurotransmitters that mediate the drug or alcohol effects. These new compounds are then tested in animals to see if they produce therapeutic effects. These effects include having a low potential for being another drug of abuse and attenuating the effects of the abused drug under study, preferably in a way that would lead to decreased drug abuse.

EXAMPLES OF MEDICATIONS USED TO TREAT DRUG ABUSE

Several types of medications have been developed for countering various kinds of dependencies.

Opioid Dependence. Some of the best examples of pharmacotherapies for drug abuse were developed for opiate addicts. One of the first pharmacological agents used to treat opiate addicts is METHADONE. Methadone itself is an opiate drug and effectively reduces or blocks the withdrawal discomfort brought on by discontinued use of heroin or other illegal opiate. Although methadone is itself addicting, it is delivered to the opiate-dependent patients in a facility with psychological and other medical and support treatments and services. The methadone is safer than opiates obtained illegally—in part because it is given orally. Because illegal opiates are often injected by addicts, they can lead to many diseases—including AIDS and hepatitis, if the needles are shared with an infected person. Illegal drug use is expensive, and many addicts steal to support their habit. Moreover, since drugs obtained illegally vary in their quality and purity, there is a greater chance of getting an overdose that produces severe medical problems and, perhaps, death. Thus methadone decreases the need to use illegal opiates, as a result of its ability to relieve withdrawal as well as to block the effects of other opiates by cross-tolerance. Moreover, it reduces the health risks and social problems associated with illegal opiate use.

Another treatment drug that was developed for opiate dependence and abuse is naltrexone. This agent blocks the ability of the opiate drug to act on the brain. Thus, if a heroin addict maintained on naltrexone injects heroin, he or she will not feel the pleasant or other effects of the heroin. The principle behind this approach is based on research suggesting that drug use is continued, despite the dire consequences, because of euphoria associated with its use. Once maintained on naltrexone, the addict may forget this association, because the drug can no longer produce these effects. Unfortunately, although naltrexone works well for some, others will simply discontinue using the naltrexone in order to get high from drugs again.

Before opiate abusers can be maintained on the medication naltrexone, they must be detoxified from the opiate drugs in their systems. Although abstaining ("cold turkey") from heroin use for several days will accomplish detoxification, the withdrawal process is difficult because of the physical distress it causes. Thus, another DETOXIFICATION method was developed in which the withdrawal is precipitated, or triggered, with naltrexone while the symptoms are treated with another medication, CLONIDINE. When withdrawal is precipitated, the symptoms are worse than that seen with natural withdrawal, but the symptom course is much briefer. Moreover, clonidine helps alleviate the symptoms, to make this shorter-term withdrawal process less severe.

Alcohol Dependence. An example of another type of medication is one used to treat alcoholism: DISULFIRAM. The basis for this agent's therapeutic effect is different from that of methadone or naltrexone. When someone is maintained on disulfiram, future alcohol ingestion will cause stomach distress and, possibly, vomiting, because the disulfiram prevents the breakdown of a noxious alcohol metabolite by the liver. Patients maintained on disulfiram

should come to forget the pleasant effects of alcohol use, which is similar to the psychological basis of naltrexone maintenance. Moreover, they should begin to develop an aversion to alcohol use. Another similarity to the use of naltrexone is that disulfiram treatment of alcoholism has not been very successful, because the patient who wants to use alcohol again can simply stop using the disulfiram.

Some pharmacological agents have been tested to reduce craving for alcohol and thus help the alcoholic abstain from drinking. These drugs include naltrexone, which was developed for opiate addicts, and fluoxetine. The former medication is a potential treatment drug because most drugs of abuse are believed to be mediated, in part, through the brain's natural opiate system (ENDORPHINS, etc.). Based on research that implicates a specific neurotransmitter system (SEROTONIN) in alcohol craving, the latter medication and others of this type may be useful. However, as in the treatment of opiate abuse, alcoholics must be detoxified before any of these medications are used as maintenance agents.

Tobacco Dependence. One commonly used pharmacological treatment for tobacco dependence is NICOTINE GUM (Nicorette). The main reason to quit smoking is that it is linked to lung cancer, emphysema, and other serious illnesses. Yet the active ingredient in cigarettes, NICOTINE, is associated with pleasant effects and with withdrawal discomfort, thereby making it an extremely addicting drug. Providing smokers with nicotine replacement in the form of a gum will help them avoid the health risks associated with cigarettes. One problem with nicotine gum is that it is difficult to chew correctly; people need to be shown how to chew it in order to get the therapeutic effect. Recently, a patch has been made available that is placed on the arm and automatically releases nicotine. This method shows good treatment potential. Detoxification from nicotine may also be facilitated with the medication clonidine, the same agent used to help alleviate opiate withdrawal symptoms.

Stimulant Dependence. Developing pharmacological treatment agents for stimulant (e.g., cocaine) dependence is a difficult task but has been the focus of a great deal of research. One of the difficulties for treating cocaine abuse is that cocaine affects many different neurotransmitter systems in various ways. Thus one approach may be to develop a treatment drug or regimen of drugs that affects a variety of neurotransmitter systems. However, the exact nature of the neural effects of cocaine are still not entirely understood.

Another difficulty is that it is not clear what approach to take in developing a treatment drug. One obvious technique in developing a medication for cocaine abuse is to use an agent that blocks the rewarding aspects of cocaine use. This type of drug would, presumably, decrease cocaine use because the rewarding effects are no longer experienced. However, this approach is similar to having opiate addicts use naltrexone, which has not been well accepted by heroin addicts. Clinical work with some treatment agents that were suggested to block the rewarding effects of cocaine did not prove to be useful in the treatment of abuse and dependence. Whether this lack of treatment effect resulted from a flaw in the method or from the limitations in our knowledge of cocaine's effects on neurotransmitter systems is not clear. One problem is that the potential blocking agents for cocaine may produce dysphoria, or an unpleasant feeling.

Another approach to treating cocaine abuse and dependence is based on a premise similar to that of methadone for opiate abuse. That is, a pharmacological agent similar in its effects to cocaine, but one that is not addicting, may be a useful anticraving agent. Just as methadone helps alleviate drug withdrawal, an agent of this type for cocaine abuse may alleviate the distress and craving associated with abstinence from cocaine. Several medications of this type have been tried, including bromocriptine and AMANTADINE. Thus far, these and other agents have shown some limited treatment promise.

Most of the approaches to developing pharmacological treatments for cocaine abuse have been based on research suggesting that one specific neurotransmitter (DOPAMINE) is important for cocaine's rewarding effects. Yet other neurotransmitters are activated during cocaine use and may be better targets for developing new treatment drugs. That is, although dopamine is critical for the rewarding aspects of cocaine use, other neurotransmitter systems may be more important in withdrawal distress. Although withdrawal distress from cocaine has been difficult to document, depression is thought to be one aspect of abstaining from chronic cocaine use. Antidepressant medications, such as desipramine and imipramine, have shown some, albeit limited, treatment potential.

Sedative Dependence. Current treatments for sedative dependence include detoxification agents, not anticraving agents. Detoxification is accomplished by tapering the dosage of BENZODIAZEPINES over 2 to 3 weeks. More recently, carbamazepine, an antiseizure analgesic medication, has been shown to relieve alcohol and sedative withdrawal symptoms, including seizures and delirium tremens. Future work with agents that block the actions of benzodiazepines may hold promise as a maintenance or anticraving agent to help the sedative abuser abstain from drug abuse.

CONCLUSION

One of the greatest lessons learned from the practice of giving medications to drug-abusing individuals is that these medications must be accompanied by psychological and social treatments and support. Medications do not work on their own. Moreover, medications that are developed based on theoretical principles of altering or blocking the drug's effects in the brain may not be useful in the practice of treating drug abuse and dependence, because the premises of how to develop a pharmacological treatment agent may not be accurate. Yet the largest research challenge is to understand the etiology and mechanisms of drug abuse. Thus more research in many fields is needed to identify potential medications in order to develop more effective treatments for the difficult problem of drug abuse and dependence.

(SEE ALSO: *Addiction: Concepts and Definitions; Imaging Techniques: Visualizing the Living Brain; Treatment/Treatment Types*)

BIBLIOGRAPHY

JAFFE, J. H. (1985). Drug addiction and drug abuse. In A. G. Gilman, et al. (Eds.), *Goodman and Gilman's the pharmacological basis of therapeutics*, 7th ed. New York: Macmillan.

KOSTEN, T. R., & KLEBER, H. D. (EDS.). (1992). *Clinician's guide to cocaine addiction*. New York: Guilford Press.

LIEBMAN, J. L., & COOPER, S. J. (EDS.). (1989). *The neuropharmacological basis of reward*. Oxford: Clarendon Press.

LOWINSON, J. H., RUIZ, P., & MILLMAN, R. B. (EDS.). (1992). *Substance abuse: A comprehensive textbook*. Baltimore: Williams & Wilkins.

MILLER, N. S. (ED.). (1991). *Comprehensive handbook of drug and alcohol addiction.* New York: Marcel Dekker.

THERESE A. KOSTEN

Drugs as Discriminative Stimuli Human behavior is influenced by numerous stimuli in the environment. Those stimuli acquire behavioral control when certain behavioral consequences occur in their presence. As a result, a particular behavioral response becomes more or less likely to occur when those stimuli are present. For example, several laboratory experiments have demonstrated that it is possible to increase a particular response during a stimulus (such as a distinctively colored light) by arranging for reinforcement (such as a preferred food or drink) to be given following that response when the stimulus is present; when that stimulus is absent, however, responses do not produce the reinforcer. Over a period of time, responding will then occur when the stimulus is present but not when it is absent. Stimuli that govern behavior in this manner are termed *discriminative stimuli* and have been widely used in behavioral and pharmacological research to better understand how behavior is controlled by various stimuli, and how those stimuli, in turn, might affect the activity of various drugs.

It is important to recognize that there are differences between discriminative stimuli that merely set the occasion for a response to be reinforced and other types of stimuli that directly *produce* or *elicit* responses. Discriminative stimuli do not coerce a response from the individual in the same way that a stimulus such as a sharp pierce evokes a reflexive withdrawal response. Instead, discriminative stimuli may be seen as providing guidance to behavior because of the unique history of reinforcement that has occurred in their presence.

DRUGS AS DISCRIMINATIVE STIMULI

Although the stimuli that typically govern behavior are external (i.e., located in the environment outside the skin), it is also possible for internal or subjective stimuli to influence behavior. One of the more popular methods to emerge in the field of behavioral pharmacology has been the use of drugs as discriminative stimuli. The procedure consists of establishing a drug as the stimulus, in the presence of which a particular response is reinforced. Typically,

to establish a drug as a discriminative stimulus, a single dose of a drug is selected and, following its administration, one of two responses are reinforced; with rodents or nonhuman primates, this usually entails pressing one of two simultaneously available levers, with reinforcement being scheduled intermittently after a fixed number of correct responses. Alternatively, when saline or a placebo is administered, responses on the other device are reinforced. Over a number of experimental sessions, a discrimination develops between the administration of the drug and saline, with the interoceptive (subjective) stimuli produced by the drug seen as guiding or controlling behavior in much the same manner as any external stimulus such as a visual or auditory stimulus. Once the discrimination has been established, as indicated by the selection of the appropriate response following either the training drug or the saline administration, it is possible to investigate aspects of the drug stimulus in the same way as one might investigate other physical stimuli. It is thus possible to determine gradients of intensity or dose-effect functions with the training drug as well as generalization functions aimed at determining how similar the training drug dose is to a different dose or to another drug substituted for the training stimulus.

BASIC EXPERIMENTAL RESULTS

One of the more striking aspects of the drug discrimination technique is the strong relationship that has been found between the stimulus-generalization profile and the receptor-binding characteristics of the training drug. For example, animals trained to discriminate between a BENZODIAZEPINE anxiolytic, such as CHLORDIAZEPOXIDE, and saline solution typically respond similarly to other drugs that also interact with the receptor sites for benzodiazepine ligands. Anxiolytic drugs that produce their effects through other brain mechanisms or receptors do not engender responses similar to those occasioned by benzodiazepines. This suggests that it is activity at a specific RECEPTOR that is established when this technique is used and not the action of the drug on a hypothetical psychological construct such as anxiety (Barrett & Gleeson, 1991).

Several studies have examined the effects of drugs of abuse by using the drug discrimination procedure, and they have established COCAINE and numerous other drugs—such as an OPIATE, PHENCYCLIDINE

(PCP), or MARIJUANA—as a discriminative stimulus in an effort to help delineate the neuropharmacological or brain mechanisms that contribute to the subjective and abuse-liability effects of these drugs. As an example, Figure 1 shows the results obtained

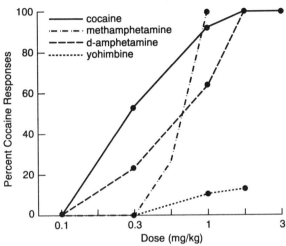

Figure 1

Discriminative Stimuli. Effects of establishing a dose of 1.7 mg/kg cocaine, administered intramuscularly, as a discriminative stimulus in pigeons. Following the administration of the training dose of cocaine, 30 consecutive pecks on one illuminated response key resulted in food reinforcement, whereas following the administration of saline, 30 consecutive pecks on a different key produced food. Once the discrimination was established, various doses of other drugs were substituted for cocaine. The discriminative stimulus effects of cocaine were dose-dependent, with doses from 0.1 to 1.7 producing increases in responding on the key correlated with the training dose of cocaine. Similarly, d-amphetamine and methamphetamine also resulted in responding on the cocaine key, thereby showing that these drugs have some of the same subjective stimulus properties and presumably neuropharmacological effects as cocaine. A drug that does not produce generalization, yohimbine, an α₂-adrenoreceptor antagonist, resulted only in modest levels of responding on the cocaine-associated response key, which suggests that this is not a mechanism by which cocaine produces its subjective behavioral and pharmacological effects.

SOURCE: Adapted from Johanson & Barrett, 1993.

in pigeons trained to discriminate a 1.7 milligram per kilogram (mg/kg) dose of cocaine from saline. The dose-response function demonstrates that doses below the training dose of cocaine yielded a diminished percentage of responses on the key correlated with cocaine administration, which suggests that the lower doses of cocaine were less discernible than the training dose. In addition, other psychomotor stimulants such as AMPHETAMINE and METHAMPHETAMINE also produced cocaine-like responding, and this suggests that these drugs share some of the neurochemical properties of cocaine. In contrast, other drugs, such as the α₂-adrenoreceptor antagonist yohimbine, along with several other drugs such as morphine, PCP, or marijuana (that are not illustrated) do not produce responding on the key correlated with cocaine administration—thereby suggesting that the mechanisms of action underlying those drugs, as well as their subjective effects, are not similar to those of cocaine and the other psychomotor stimulants in this figure.

IMPLICATIONS

The use of drugs as discriminative stimuli has provided a wealth of information on the way drugs are similar to more conventional environmental stimuli in their ability to control and modify behavior. The procedure has also increased our understanding of the neuropharmacological mechanisms that operate to produce the constellation of effects associated with those drugs. The technique has wide generality and has been studied in several species, including humans—in whom the effects are quite similar to those of nonhumans.

Because it is believed that the subjective effects of a drug are critical to its abuse potential, the study of drugs of abuse as discriminative stimuli takes on added significance. A better understanding of the effects of drugs of abuse as pharmacologically subjective stimuli provides a means by which to evaluate possible pharmacological as well as behavioral approaches to the treatment of drug abuse. For example, a drug that prevents or antagonizes the discriminative-stimulus effects (and presumably the neuropharmacological actions) of an abused drug might be an effective medication to permit individuals to diminish their intake of abused drugs, because the stimuli usually associated with its effects will no longer occur. Similarly, although little work has been performed on the manipulation of environmental stimuli correlated with the drug stimulus, it might be possible to design innovative treatment strategies in which other stimuli compete with the subjective discriminative-stimulus effects of the abused drug. Thus, a basic experimental procedure such as drug discrimination has provided a useful experimental tool for understanding the behavioral and neuropharmacological effects of abused drugs.

Further work may help design and implement novel treatment approaches to modifying the behavioral and environmental conditions surrounding the effects of abused drugs and thus result in diminished behavioral control by substances of abuse.

(SEE ALSO: *Abuse Liability of Drugs; Drug Types; Research, Animal Model*)

BIBLIOGRAPHY

BARRETT, J. E., & GLEESON, S. (1991). Anxiolytic effects of 5-HT$_{1A}$ agonists, 5-HT$_3$ antagonists and benzodiazepines. In R. J. Rodgers & S. J. Cooper (Eds.), *5-HT$_{1A}$ agonists, 5-HT$_3$ antagonists and benzodiazepines: Their comparative behavioral pharmacology.* New York: Wiley.

JOHANSON, C. E., & BARRETT, J. E. (1993). The discriminative stimulus effects of cocaine in pigeons. *Journal of Pharmacology and Experimental Therapeutics, 267,* 1–8.

JAMES E. BARRETT
JUNE STAPLETON

Measuring Effects of Drugs on Behavior

People throughout the world take drugs such as HEROIN, COCAINE, and ALCOHOL because these drugs alter behavior. For example, cocaine alters general activity levels; it increases wakefulness and decreases the amount of food an individual eats. Heroin produces drowsiness, relief from pain, and a general feeling of pleasure. Alcohol's effects include relaxation, increased social interactions, marked sedation, and impaired motor function. For the most part, the scientific investigations of the ways drugs alter behavior began in the 1950s, when chlorpromazine was introduced as a treatment for SCHIZOPHRENIA. As a result of this discovery, scientists became interested in the development of new medications to treat behavioral disorders as well as in the development of procedures for studying behavior in the laboratory.

HOW IS BEHAVIOR STUDIED?

The simplest way to study the effects of drugs on behavior is to pick a behavior, give a drug, and observe what happens. Although this approach sounds very easy, the study of a drug's effect on behavior is not so simple. Like any other scientific inquiry, research in this area requires careful description of the behaviors being examined. If the behavior is not carefully described, it is difficult to determine whether a change in behavior following drug administration is actually due to the drug.

Behavior is best defined by describing how it is measured. By specifying how to measure a behavior, an *operational definition* of that behavior is developed. For example, to study the way in which a drug alters food intake, the following procedure might be used: First, select several people and present each with a box of cereal, a bowl, a spoon, and some milk after they wake up in the morning. Then measure how much cereal and milk they each consume within the next thirty minutes. To make sure the measurements are correct, repeat the observations several times under the same conditions (i.e., at the same time of day, with the same foods available). From these observations, determine the average amount of milk and cereal consumed by each person. This is the baseline level. Once the baseline level is known, give a small amount of drug and measure changes in the amount of milk and cereal consumed. Repeat the experiment, using increasing amounts of the drug. This concept of baseline level and change from baseline level is common to many scientific investigations.

In addition to defining behavior by describing how it is measured, a good behavioral procedure is also (1) sensitive to the ways in which drugs alter behavior and (2) is reliable. Sensitivity refers to whether a particular behavior is easily changed as the result of drug administration. For example, food consumption may be altered by using cocaine, but other behaviors may not be. Reliability refers to whether a drug produces the same effect each time it is taken. In order to say that cocaine reliably alters the amount of food consumed, it should decrease food consumption each time it is given, provided that the experimental conditions surrounding its administration are the same.

WHAT FACTORS INFLUENCE A DRUG'S EFFECTS ON BEHAVIOR?

Although good behavioral procedures are necessary for understanding a drug's effects on behavior, pharmacological factors are also important determinants of a drug's effect. Pharmacological factors include the amount of drug given (the *dose*), how quickly the drug produces its effects (its *onset*), the time it takes for its effects to disappear (its *duration*), and whether the drug's effects are reduced (*tolerance*) or increased (*sensitivity*) if it is taken several times. Although this point may seem obvious, it is often overlooked. It is impossible to describe the behavioral effects of a drug on the basis of just one dose of the drug, since drugs can have very different effects, depending on how much of the drug is taken. Moreover, the probability that a drug will produce an effect also depends on the amount taken. As an example, consider Figure 1, which shows the risk of being involved in a traffic accident as a function of the amount of alcohol in a person's blood.

The way in which a drug is taken is also important. Cocaine can be taken by injection into the veins, by smoking, or by sniffing through the nose. Each of these routes of administration can produce different effects. Environmental factors also influence a drug's effect. Cocaine might change the amount of cereal and milk consumed in the morning but it might not change the amount consumed at a different time of day or if other types of food are available. Finally, individual factors also influence the drug effect. These include such factors as how many times an individual has taken a particular drug; what happened the last time it was taken; or what one may have heard from friends about a drug's effects.

HOW IS BEHAVIOR STUDIED IN THE LABORATORY?

Human behavior is very complex, and it is often difficult to examine. Although scientists do conduct studies on people, many investigations of drug ef-

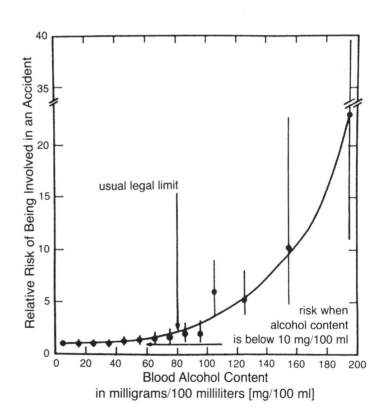

Figure 1
Risk of Being Involved in a Traffic Accident as a Function of the Amount of Alcohol in the Blood

fects on behavior are carried out using animals. With animals, investigators have better control over the conditions in which the behavior occurs as well as better information about the organism's past experience with a particular drug. Although animal experiments provide a precise, controlled environment in which to investigate drug effects, they also have their limitations. Clearly, they cannot research all the factors that influence human behavior. Nevertheless, many of the effects that drugs produce on behavior in animals also occur in humans. Moreover, behavioral studies sometimes require a large number of subjects with the same genetic makeup or with no previous drug experience. It is easier to meet these requirements in animal studies than in studies with people.

Since animals are often used in research studies, it is important to remember that behavioral scientists are very concerned about the general welfare of their animals. The U.S. Animal Welfare Act set standards for handling, housing, transporting, feeding, and veterinary care of a wide variety of animals. In addition, all animal research in the United States is now reviewed by a committee that includes a veterinarian experienced in laboratory-animal care. This committee inspects animal-research areas and reviews the design of experiments to ensure that the animals are treated well.

WHAT APPROACHES ARE USED TO EXAMINE DRUG EFFECTS?

In general, there are two ways to examine drug effects on behavior in the laboratory. One approach relies on observation of behavior in an animal's home cage or in an open area in which the animal (or person) can move about freely. When observational approaches are used, special precautions are necessary. First of all, the observer's presence should not disrupt the experiment. Televison-monitoring systems and videotaping make it possible for the observer to be completely removed from the experimental situation. Second, the observer should not be biased. The best way to insure that the observer is not biased is to make the observer "blind" to the experimental conditions; that is, the observer does not know what drug is given or which subject received the drug. If the study is done in human subjects, then they also should be blind to the experimental conditions. An additional way to make sure observations are reliable is to use more than one observer

and compare observations. If these precautions are taken, observational approaches can produce interesting and reliable data. Indeed, much of what is known about drug effects on motor behavior, food or water intake, and some social behaviors comes from careful observational studies.

Another approach uses the procedures of classical and operant conditioning. This involves training animals to make specific responses under special conditions. For example, in a typical experiment of this sort, a rat is placed in an experimental chamber and trained to press a lever to receive food. The number and pattern of lever presses are measured with an automatic device, and changes in responding are examined following drug administration. These procedures have several advantages. First, they produce a very consistent measure of behavior. Second, they can be used with human subjects as well as with several different animal species. Third, the technology for recording behavior eliminates the need for a trained observer.

WHAT BEHAVIORS DO DRUGS ALTER?

Some of the behaviors that drugs alter are motor behavior, sensory behavior, food and water intake, social behavior, and behavior established with classical and operant conditioning procedures. By combining investigations of these behaviors, scientists classify drugs according to their prominent behavioral effects. For example, drugs such as AMPHETAMINE and cocaine are classified as PSYCHOMOTOR STIMULANTS because they increase alertness and general activity in a variety of different behavioral procedures. Drugs such as morphine are classified as analgesics because they alter the perception of pain, without altering other sensations such as vision or audition (hearing).

Motor Behavior. Most behaviorally active drugs alter motor behavior in some way. MORPHINE usually decreases motor activity, whereas with cocaine certain behaviors occur over and over again (that is, repetitively). Other drugs, such as alcohol, may alter the motor skills used in DRIVING a car or operating various types of machinery. Finally, some drugs alter exploratory behavior, as measured by a decrease in motor activity in an unfamiliar environment. Examination of the many ways in which drugs alter motor behavior requires different types of procedures. Some of these procedures examine fine motor control or repetitive behavior; others simply measure spontaneous motor activity.

Although changes in motor behavior can be observed directly, most studies of motor behavior use some sort of automatic device that does not depend on human observers. One of these devices is the running wheel. The type of running wheel used in scientific investigations is similar to the running wheel in pet cages. This includes a cylinder of some sort that moves around an axle when an animal walks or runs in it. The only difference between a running wheel in a pet cage and a running wheel in the laboratory is its size and the addition of a counter that records the number of times the wheel turns. Another device for measuring motor behavior uses an apparatus that is surrounded by photocells. If the animal moves past one of the photocells, a beam of light is broken and a count is produced. Yet another way to measure motor behavior is with video tracking systems. An animal is placed in an open area and a tracking system determines when movement stops and starts as well as its speed and location. This system provides a way to look at unique movement patterns such as repetitive behaviors. For example, small amounts of amphetamine increase forward locomotion, whereas larger amounts produce repetitive behaviors such as head bobbing, licking, and rearing. Until recently, this type of repetitive behavior was measured by direct observation and description.

Although technology for measuring motor behavior is very advanced, it is important to remember that how much drug is given, where it is given, and the type of subject to whom it is given will also influence a drug's effect on motor behavior. Whether a drug's effects are measured at night or in the day is an important factor. The age, sex, species, and strain of the animal is also important. Whether food and water are available is another consideration as well as the animal's previous experience with the drug or test situation. As an example, see Table 1, which shows how the effects of alcohol on motor behavior differ depending on the amount of alcohol in a person's blood.

Sensory Behavior. The integration and execution of every behavior an organism engages in involves one or more of the primary senses, including hearing, vision, taste, smell, and touch. Obviously, a drug can affect sensory behavior and thereby alter a number of different behaviors. For example, drugs such as LYSERGIC ACID DIETHYLAMIDE (LSD) pro-

TABLE 1
Blood Alcohol Level and Behavioral Effect

Percent Blood Alcohol	Behavioral Effect
0.05	alertness reduced
0.10	reaction time prolonged
0.20	motor function impaired
0.30	severe motor impairment
0.40	consciousness lost

duce visual abnormalities and HALLUCINATIONS. PHENCYCLIDINE (PCP) produces a numbness in the hands and feet. Morphine alters sensitivity to painful stimuli.

It is difficult to investigate drug effects on sensory behavior, since changes in sensory behavior cannot be observed directly. In order to determine whether someone hears a sound, one must report having heard it. In animal studies, rats or monkeys are trained to press a lever when they hear or see a given stimulus. Then a drug is given and alterations in responding are observed. If the drug alters responding, it is possible that the drug did so by altering sensory behavior; however, care must be taken in coming to this conclusion since a drug might simply alter the motor response used to measure sensory behavior without changing sensory behavior at all.

One area of sensory behavior that has received considerable attention is pain perception. In most procedures for measuring pain perception, a potentially painful stimulus is presented to an organism and the time it takes the organism to respond to that stimulus is observed. Once baseline levels of responding are determined and considered reliable, a drug is given. If the time it takes the organism to respond to the stimulus is longer following drug administration and if this change is not because the animal is too sedated to make a response, then the drug probably has altered pain perception.

Among the most common procedures used to measure pain perception is the tail-flick procedure in which the time it takes an animal to remove its tail from a heat source is measured prior to and after administration of a drug such as morphine. Another commonly used procedure measures the time it takes an animal to lick its paws when placed on a warm plate or to remove its tail from a container of warm water. Thus, an alteration in pain perception is operationally defined as a change in responding in the presence of a painful stimulus. It is also important to note that the animal, not the experimenter, determines when to respond or remove its tail. Also, these procedures do not produce long-term damage or discomfort that extends beyond the brief experimental session.

Food and Water Intake. The simplest way to measure food and water intake is to determine how much an organism eats or drinks within a given period of time. A more thorough analysis might also include counting the number of times an organism eats or drinks in a single day, or measuring the time between periods of eating and periods of drinking. Several factors are important in accurately measuring food and water intake. For example, how much food or water is available to the organism and when is it available? Is it a food the organism likes? When did the last meal occur?

In animals, food intake is often measured by placing several pieces of pelleted food of a known weight in their cages. The food that remains after a period of time is weighed and subtracted from the original amount to get an estimate of how much was actually eaten. Water intake is usually measured with calibrated drinking tubes clipped to the front of the animal's cage or with a device called a drinkometer, which counts the number of times an animal licks a drinking tube. An accurate measure of fluid intake also requires a careful description of the surrounding conditions. For example, was fluid intake measured during the day or during the night? Was food also available? What kind of fluid was available? Was there more than one kind of fluid available? These procedures are also used to examine drug intake. If rats are presented with two different drinking tubes, one with alcohol, another with water, they will generally drink more alcohol than water; however, the amount they drink is generally not sufficient to produce intoxication or physical dependence. Rats will drink a large amount of alcohol as well as other drugs of abuse such as morphine and cocaine when these drugs are the only liquid available. Indeed, most animals will consume sufficient quantities to become physically dependent on alcohol or morphine.

Social Behavior. Behaviors such as aggression, social interaction, and sexual behavior are usually measured by direct experimenter observation. Aggressive behavior can be measured by observing the number of times an animal engages in attack behavior when another animal is placed into its cage. In some cases, isolation is used to produce aggressive

behavior. Sexual behavior is also measured by direct observation. In the male rat or cat, the frequency of behaviors such as mounting, intromission, and ejaculation are observed. Another interesting procedure for measuring social behavior is the social interaction test. In this procedure, two rats are placed together and the time they spend in active social interaction (sniffing, following, grooming each other) is measured under different conditions. In one condition, the rats are placed in a familiar environment; in another condition, the environment is unfamiliar. Rats interact more when they are in a familiar environment than when they are in an unfamiliar environment. Moreover, antianxiety drugs increase social interaction in the unfamiliar area. These observational techniques can produce interesting data, provided that they are carried out under well-controlled conditions, the behavior is well-defined, and care is taken to make sure the observer neither disrupts the ongoing behavior nor is biased.

Classical Conditioning. Classical conditioning was made famous by the work of the Russian scientist Ivan Pavlov in the 1920s. In those experiments, Pavlov used the following procedure. First, dogs were prepared with a tube to measure saliva, as shown in Figure 2. Then Pavlov measured the amount of saliva that was produced when food was given. The amount of saliva not only increased when food was presented but also when the caretaker arrived with the food. From these careful observations, Pavlov concluded that salivation in response to the food represented an inborn, innate response that did

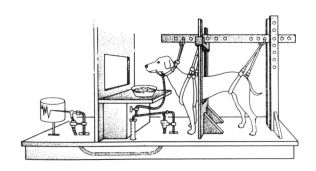

Figure 2

Diagram of Pavlov's Classical Conditioning Experiment. A tube is attached to the dog's salivary duct, and saliva drops into a device that records the number of drops

not require any learning. Because no learning was required, he called this an unlearned (unconditioned) response and the food itself an unlearned (unconditioned) stimulus. The dogs did not automatically salivate, however, when the caretaker entered the room; but after the caretaker and the food occurred together several times, the presence of the caretaker was paired with (or conditioned to) the food. Pavlov called the caretaker the *conditioned stimulus* and he called the salivation that occurred in the presence of the caretaker a *conditioned response*.

Events in the environment that are paired with or conditioned to drug delivery can also produce effects similar to the drug itself, much in the same way that Pavlov's caretaker was conditioned to food delivery. For example, when heroin-dependent individuals stop taking heroin, they experience a number of unpleasant effects, such as restlessness, irritability, tremors, nausea, and vomiting. These are called withdrawal or abstinence symptoms. If an individual experiences withdrawal several times in the same environment, then events or stimuli in that location became paired with (or conditioned to) the withdrawal syndrome. With time, the environmental events themselves can produce withdrawal-like responses, just as the caretaker produced salivation in Pavlov's dogs.

Operant Conditioning. About a decade after Pavlov's discovery of classical conditioning, another psychologist, B. F. Skinner, was developing his own theory of learning. Skinner observed that certain behaviors occur again and again. He also observed that behaviors with a high probability of occurrence were behaviors that produced effects on the environment. According to Skinner, behavior "operates" on the environment to produce an effect. Skinner called this process *operant conditioning*. For example, people work at their jobs because working produces a paycheck. In this situation, working is the response and a paycheck is the effect. In other situations, a person does something to avoid a certain effect. For example, by driving a car within the appropriate speed limit, traffic tickets are avoided and the probability of having a traffic accident is reduced. In this case, the response is driving at a given speed and the effect is avoiding a ticket or an accident.

If the effect that follows a given behavior increases the likelihood that the behavior will occur again, then that event is called a *reinforcer*. Food, water, and heat are common reinforcers. Drug administration is also a reinforcer. It is well known that animals will respond on a lever to receive intravenous injections of morphine, cocaine, and amphetamine, as well as a number of other drugs. Not all drugs are self-administered, however. For example, animals will respond to avoid the presentation of certain nonabused drugs such as the ANTI-PSYCHOTICS (medications used in the treatment of schizophrenia). Because there is a good correlation between drugs that are self-administered by animals and those that are abused by people, the self-administration procedure is often used to examine drug-taking behavior.

In most operant conditioning experiments, animals perform a simple response such as a lever press or a key peck to receive food. Usually the organism has to make a fixed number of responses or to space responses according to some temporal pattern. The various ways of delivering a reinforcer are called *schedules of reinforcement*. Schedules of reinforcement produce very consistent and reliable patterns of responding. Moreover, they maintain behavior for long periods of time, are easily adapted for a number of different animals, and provide a very accurate measure of behavior. Thus, they provide a well-defined, *operational measure* of behavior, which is used to examine the behavioral effects of drugs.

Motivation, Learning, Memory, and Emotion. One of the biggest challenges for behavioral scientists is to develop procedures for measuring drug effects on processes such as motivation, emotion, learning, or memory since these behaviors are very difficult to observe directly. Drugs certainly alter processes such as these. For example, many drugs relieve anxiety. Other drugs produce feelings of pleasure and well-being; still others interfere with memory processes. Given the complexity of devised procedures, they are not described in detail here; however, it is important to emphasize that the approach for examining the effects of drugs on these complex behaviors is the same as it is for any behavior: First, carefully define the behavior and describe the conditions under which it occurs. Second, give a drug and observe changes in the behavior. Third, take special care to consider pharmacological factors, such as how much drug is given, when the drug is given, or the number of times the drug is given. Fourth, consider behavioral factors, such as the nature of the behavior examined, the conditions under which the behavior is examined, as well as the individual's past experience with the behavior.

SUMMARY

To find out how drugs alter behavior, several factors are considered. These include the PHARMACOLOGY of the drug itself as well as an understanding of the behavior being examined. Indeed, the behavioral state of an organism, as well as the organism's past behavior and experience with a drug contribute as much to the final drug effect as do factors such as the dose of the drug and how long it lasts. Thus, the examination of drug effects on behavior requires a careful description of behavior with special attention to the way in which the behavior is measured. Behavioral studies also require a number of experimental controls, which assure that changes in behavior following drug administration are actually due to the drug itself and not the result of behavioral variability.

(SEE ALSO: *Addiction: Concepts and Definitions; Aggression and Drugs; Causes of Drug Abuse; Pharmacodynamics; Psychomotor Effects of Alcohol and Drugs; Reinforcement; Research, Animal Model; Sensation and Perception and Effects of Drugs; Tolerance and Physical Dependence*)

BIBLIOGRAPHY

CARLTON, P. L. (1983). *A primer of behavioral pharmacology.* New York: W. H. Freeman.

DOMJAN, M., & BURKHARD, B. (1982). *The principles of learning and behavior.* Pacific Grove, CA: Brooks/Cole Publishing Co.

GREENSHAW, A. J., & DOURISH, C. T. (EDS.). (1987). *Experimental psychopharmacology.* Clifton, NJ: Humana Press.

GRILLY, D. M. (1989). *Drugs and human behavior.* Boston: Allyn & Bacon.

JULIEN, R. M. (1988). *A primer of drug action.* New York: W. H. Freeman.

McKIM, W. A. (1986). *Drugs and behavior.* Englewood Cliffs, NJ: Prentice-Hall.

MYERS, D. G. (1989). *Psychology.* New York: Worth.

RAY, O., & KSIR, C. (1987). *Drugs, society, & human behavior.* St. Louis: Times Mirror/Mosby.

SEIDEN, L. S., & DYKSTRA, L. A. (1977). *Psychopharmacology: A biochemical and behavioral approach.* New York: Van Nostrand Reinhold.

LINDA A. DYKSTRA

Measuring Effects of Drugs on Mood

Subjective effects are feelings, perceptions, and moods that are the personal experiences of an individual. They are not accessible to other observers for public validation and, thus, can only be obtained through reports from the individual. Subjective-effect measures are used to determine whether the drug is perceived and to determine the quantitative and qualitative characterization of what is experienced. Although subjective effects can be collected in the form of narrative descriptions, standardized questionnaires have greater experimental utility. For example, they may be used to collect the reports of individuals in a fashion that is meaningful to outside observers, can be combined across subjects, and can provide data that are reliable and replicable. The measurement of subjective effects through the use of questionnaires is scientifically useful for determining the pharmacologic properties of drugs—including time course, potency, abuse liability, side effects, and therapeutic utility. Many of the current methods used to measure subjective effects resulted from research aimed at reducing drug abuse.

HISTORY

Drug abuse and drug addiction are problems that are not new to contemporary society; they have a long-recorded history, dating back to ancient times. For centuries, various drugs including ALCOHOL, TOBACCO, MARIJUANA, HALLUCINOGENS, OPIUM, and COCAINE, have been available and used widely across many cultures. Throughout these times, humans have been interested in describing and communicating the subjective experiences that arise from drug administration. Although scientists have been interested in the study of PHARMACOLOGY for many centuries, reliable procedures were not developed to measure the subjective effects of drugs until recently.

Throughout the twentieth century, the U.S. GOVERNMENT has become increasingly concerned with the growing problem of drug abuse. To decrease the availability of drugs with significant ABUSE LIABILITY, the government has passed increasingly restrictive laws concerning the possession and sale of existing drugs and the development and marketing of new drugs. The pressing need to regulate drugs that have potential for misuse prompted the government to sponsor research for the development of scientific methodologies that would be useful in assessing the abuse liability of drugs.

Two laboratories that made major contributions to the development of subjective-effect measures were Henry Beecher and his colleagues at Harvard University and the government-operated Addiction Research Center (ARC) in Lexington, Kentucky. Beecher and his colleagues at Harvard conducted a lengthy series of well-designed studies that compared the subjective effects of various drugs—opiates, sedatives, and stimulants—in a variety of subject populations that included patients, substance abusers, and normal volunteers and highlighted the importance of studying the appropriate patient population. Additionally, this group laid the foundation for conducting studies with solid experimental designs, which include double-blind and placebo controls, randomized dosing, and characterization of dose-response relationships. Investigators at the ARC conducted fundamental studies of both the acute (immediate) and chronic (long-term) effects of drugs, as well as physical dependence and withdrawal symptoms (e.g., Himmelsbach's opiate withdrawal scale). A number of questionnaires and procedures now in use to study the subjective effects of drugs were developed, including the Addiction Research Center Inventory and the Single Dose Questionnaire. Although many of the tools and methods developed at the ARC are still in use, other laboratories have since modified and expanded subjective-effect measures and their applications.

MEASURES

Question Format. Subjective-effects measures are usually presented in the form of groups of questions (questionnaires). These questions can be presented in a number of formats, the most frequently used of which are ordinal scales and visual analog scales. The ordinal scale is a scale of ranked values in which the ranks are assigned based upon the amount of the measured effect that is experienced by each individual. Subjects are usually asked to rate their response to a question on a 4- or 5-point scale (e.g., to rate the strength of the drug effect from 0 to 4, with 0 = not at all; 1 = a little; 2 = moderately; 3 = quite a bit; and 4 = extremely). A visual-analog scale is a continuous scale presented as a line without tick marks or sometimes with tick marks to give some indication of gradations. A subject indicates

the response by placing a mark on that line, according to a particular reference point; for example, lines are usually anchored at the ends with labels such as "not at all" and "extremely." Visual-analog scales can be unipolar (example: "tired," rated from no effect to extremely), or they may be bipolar (example: "tired/alert," with "extremely tired" at one end, "extremely alert" at the other, and "no effect" in the center). Another frequently used format is the binomial scale, usually in the form of yes/no or true/false responses, such as the Addiction Research Center Inventory. A fourth format utilizes a nominal scale, in which the response choices are categorical in nature and mutually exclusive of each other (e.g., drug class questionnaire).

Questionnaires. Frequently used subjective-effect measures include investigator-generated scales, such as adjective-rating scales, and standardized questionnaires, such as the Profile of Mood States and the Addiction Research Center Inventory. A description of a number of questionnaires follows; however, this list is illustrative only and is not meant to be exhaustive.

Adjective Rating Scales. These are questionnaires on which subjects rate a list of symptoms, describing how they feel or effects associated with drug ingestion. The questionnaires can be presented to subjects with either visual-analog or ordinal scales. Items can be used singly or grouped into scales. Some adjective-type scales are designed to measure global effects, such as the strength of drug effects or the subject's liking of a drug, while other adjective rating scales are designed to measure specific drug-induced symptoms. In the latter use, the adjectives used may depend on the class of drugs being studied and their expected effects. For example, studies of amphetamine include items such as "stimulated" and "anxious," while studies of opioids include symptoms such as "itching" and "talkative." To study physical dependence, symptoms associated with drug withdrawal are used; for example, in studies of opioid withdrawal, subjects might rate "watery eyes," "chills," and "gooseflesh." Most adjective-rating scales have not been formally validated; investigators rely on external validity.

Profile of Mood States (POMS). This questionnaire was developed to measure mood effects in psychiatric populations and for use in testing treatments for psychiatric conditions such as depression and anxiety. It is a form of an adjective-rating scale. This scale was developed by Douglas McNair, Ph.D., and has been modified several times. It exists in two forms—one consisting of 65 and another of 72 adjectives describing mood states that are rated on a 5-point scale from "not at all" (0) to "extremely" (4). The item scores are weighted and grouped by factor analysis into a number of subscales, including tension–anxiety, depression–dejection, anger–hostility, vigor, fatigue, confusion–bewilderment, friendliness, and elation. This questionnaire has been used to measure acute drug effects, usually by comparing measures collected before and after drug administration. Its use in drug studies has not been formally validated; however, it has been validated by replication studies in normal and psychiatric populations and in treatment studies.

Single Dose Questionnaire. This was developed in the 1960s at the ARC to quantify the subjective effects of opioids. It has been used extensively and has been modified over time. This questionnaire consists of four parts; (1) a question in which subjects are asked whether they feel a drug effect (a binomial yes/no scale); (2) a question in which subjects are asked to indicate which among a list of drugs or drug classes is most similar to the test drug (a nominal scale); (3) a list of symptoms (checked yes or no); and (4) a question asking subjects to rate how much they like the drug (presented as an ordinal scale). The list of drugs used in the questionnaire includes placebo, opiate, stimulant, marijuana, sedative, and other. Examples of symptoms listed are turning of stomach, skin itchy, relaxed, sleepy, and drunken. While this questionnaire has not been formally validated, it has been used widely to study opioids, and the results have been remarkably consistent over three decades.

Addiction Research Center Inventory (ARCI). This is a true/false questionnaire containing more than 550 items. The ARCI was developed by researchers at the ARC to measure a broad range of physical, emotive, and subjective drug effects from diverse pharmacological classes. Subscales within the ARCI were developed to be sensitive to the acute effects of specific drugs or pharmacological classes (e.g., morphine, amphetamine, barbiturates, marijuana); feeling states (e.g., tired, excitement, drunk); the effects of chronic drug administration (Chronic Opiate Scale); and drug withdrawal (e.g., the Weak Opiate Withdrawal and Alcohol Withdrawal Scale). The ARCI subscales most frequently used in acute drug-effect studies are the Morphine–Benzedrine Group (MBG) to measure euphoria; the Pentobarbital–Chlorpromazine–Alcohol Group (PCAG) to measure apathetic sedation; and the Lysergic Acid Diethylamide Group (LSDG) to measure dysphoria or somatic discomfort. The use of the MBG, PCAG, and LSDG scales has remained standard in most studies of abuse liability. Subscales on this questionnaire were developed empirically, followed by extensive validation studies.

Observer-rated Measures. These may frequently accompany the collection of subjective effects and are often based on the subjective-effect questionnaires. Ratings are made by an observer who is present with the subject during the study, and items are limited to those drug effects that are observable. Observer-rated measures may include drug-induced behaviors (e.g., talking, scratching, activity levels, and impairment of motor function), as well as other drug signs such as redness of the eyes, flushing, and sweating. Observer-rated measures can be designed using any of the formats used in subject-rated measures. Examples of observer-rated questionnaires that have been used extensively are the Single Dose Questionnaire, which exists in an observer-rated version, and the Opiate Withdrawal Scale developed by Himmelsbach and his colleagues at the ARC.

USES OF SUBJECTIVE-EFFECT MEASURES

The methodology for assessing the subjective effects of drugs was developed, in large part, to characterize the abuse liability, the pharmacological properties, and the therapeutic utility of drugs. *Abuse liability* is the term for the likelihood that a drug will be used illicitly for nonmedical purposes. The assessment of the abuse-liability profile of a new drug has historically been studied by comparing it with a known drug, whose effects have been previously characterized. Drugs that produce euphoria are considered more likely to be abused than drugs that do not produce euphoria.

Subjective-effects measures may also be used to characterize the time course of a drug's action (such as time to drug onset, time to maximal or peak effect, and the duration of the drug effect). These procedures can provide information about the pharmacological properties of a particular drug, such as its drug class, whether it has AGONIST or ANTAGONIST effects, and its similarity to prototypic drugs within a given drug class. Subjective-response reports are also

TABLE 1
Typical Response Profiles for Sedatives, Stimulants, and Opiates on Selected Subjective-Effect Measures

	Global Effects	ARCI	POMS	Adjectives
Sedatives	Drug Effect Liking High	PCAG	Fatigue (increase) Vigor (decrease)	Tired Sleepy Relaxed Drunk
Stimulants	Drug Effect Liking High	MBG	Vigor (increase) Fatigue (decrease)	Stimulated Nervous Thirsty Jittery
Opiates	Drug Effect Liking High	MBG PCAG		Nauseous Itchy Nodding Energetic

useful in assessing the efficacy (the ability of a drug to produce its desired effects), potency (amount or dose of a drug needed to produce that effect), and therapeutic utility of a new drug. Subjective reports provide information regarding the potency and efficacy of a new drug in comparison to available treatment agents. Subjective-effect measures may be useful in determining whether a drug produces side effects that are dangerous or intolerable to the patient. Drugs that produce unpleasant or dysphoric mood-altering effects may have limited therapeutic usefulness.

DESCRIPTION OF MAJOR FINDINGS OBTAINED WITH DIFFERENT DRUG CLASSES

Drugs of different pharmacological classes generally produce profiles of subjective effects that are unique to that class of drugs and that are recognizable to individuals. The subjective effects of major pharmacological classes have been characterized using the questionnaires described above. Table 1 lists some major pharmacological classes and their typical effects on various instruments. While global measures provide quantitative information regarding drug effects, they tend not to differentiate among different types of drugs. Nevertheless, the more specific subjective-effect measures, such as the ARCI and the Adjective Rating Scales, yield qualitative information that can differentiate among drug classes.

CONCLUSION

Measures of the subjective effects of drugs have been extremely useful in the study of pharmacology. Questionnaires have been developed that are sensitive to both the global effects and the specific effects of drugs; however, research is still underway to develop even more sensitive subjective-effect measures and new applications for their use.

(SEE ALSO: *Abuse Liability of Drugs; Addiction: Concepts and Definitions; Causes of Substance Abuse; Drug Types*)

BIBLIOGRAPHY

BEECHER, H. K. (1959). *Measurement of subjective responses: Quantitative effects of drugs.* New York: Oxford University Press.

DEWIT, H., & GRIFFITHS, R. R. (1991). Testing the abuse liability of anxiolytic and hypnotic drugs in humans. *Drug and Alcohol Dependence, 28*(1), 83–111.

FOLTIN, R. W., & FISCHMAN, M. W. (1991). Assessment of abuse liability of stimulant drugs in humans: A methodological survey. *Drug and Alcohol Dependence, 28*(1), 3–48.

MARTIN, W. R. (1977). *Drug addiction I.* Berlin: Springer-Verlag.

MARTIN, W. R. (1977). *Drug addiction II.* Berlin: Springer-Verlag.

PRESTON, K. L., & JASINSKI, D. R. (1991). Abuse liability studies of opioid agonist-antagonists in humans. *Drug and Alcohol Dependence, 28*(1), 49–82.

KENZIE L. PRESTON
SHARON L. WALSH

Motivation   Motivation is a theoretical construct that refers to the neurobiological processes responsible for the initiation and selection of such goal-directed patterns of behavior as are appropriate to the physiological needs or psychological desires of the individual. *Effort* or *vigor* are terms used to describe the intensity of a specific pattern of motivated behavior. Physiological "drive" states, caused by imbalances in the body's homeostatic regulatory systems, are postulated to be major determinants of different motivational states. Deprivation produced by withholding food or water is used routinely in studies with experimental animals to establish prerequisite conditions in which nutrients or fluids can serve as positive reinforcers in both operant and classical conditioning procedures. In more natural conditions, the processes by which animals seek, find, and ingest food or fluids are divided into appetitive and consummatory phases. Appetitive behavior refers to the various patterns of behavior that are used to locate and bring the individual into direct contact with a biologically relevant stimulus such as water. Consummatory behavior describes the termination of approach behavior leading subsequently to ingestion of food, drinking of fluid, or copulation with a mate.

Incentive motivation is the term applied to the most influential psychological theory that explains how the stimulus properties of biologically relevant stimuli, and the environmental stimuli associated with them, control specific patterns of appetitive behavior (Bolles, 1972). According to this theory, the initiation and selection of specific behaviors are triggered by external (incentive) stimuli that also guide the individual toward a primary natural incentive, such as food, fluid, or a mate. Drugs of abuse and electrical brain-stimulation reward can serve as artificial incentives. In a further refinement of this theory, Berridge and Valenstein (1991) defined incentive motivation as the final stage in a three-part process. The first phase involves the activation of neural substrates for pleasure, which in the second

phase are associated with the object giving rise to these positive sensations and the environmental stimuli identified with the object. The critical third stage involves processes by which salience is attributed to subsequent perceptions of the natural incentive stimulus and the associated environmental cues. It is postulated that this attribution of "incentive salience" depends upon activation of the mesotelencephalic dopamine systems. The sensation of pleasure and the classical associative learning processes that mediate stages 1 and 2 respectively are subserved by different neural substrates.

In the context of drive states as the physiological substrates of motivation, the level of motivation is manipulated by deprivation schedules in which the subject is denied access mainly to food or water for fixed periods of time (e.g., twenty-two hours of food deprivation). An animal's increased motivation can be inferred from measures such as its running speed in a runway to obtain food reward. Under these conditions, speed is correlated with level of deprivation. Another measure of the motivational state of an animal is the amount of work expended for a given unit of food, water, or drug. Work here is defined as the number of lever presses per reinforcer. If one systematically obtains an increase in the number of presses, one can identify a specific ratio of responses per reward beyond which the animal is unwilling to work. This final ratio is called the break point. In the context of drug reinforcement, the break point in responding for COCAINE can be increased or decreased in a dose-dependent manner by dopamine agonists and antagonists respectively.

Appetitive behavior also can be measured directly in animal behavior studies either by an animal's latency (the time it takes) in approaching a source of food or water during presentation of a conditioned stimulus predictive of food, or simply by measuring the animal's latency approaching a food dispenser when given access to it. The fact that these appetitive behaviors are disrupted by dopamine antagonists has been interpreted as evidence of the role of mesotelencephalic dopamine pathways in incentive motivation.

In extending these ideas to the neural bases of drug addiction, Robinson and Berridge (1993) emphasized the role of sensitization, or enhanced behavioral responses to fixed doses of addictive drugs, that occurs after repeated intermittent drug treatment. Neurobiological evidence indicates that sensitization is directly related to neuroadaptations in the mesotelencephalic dopamine systems. As a result of these neural changes, a given dose of amphetamine, for example, causes enhanced levels of extracellular dopamine and an increase in the behavioral effects of the drug. Given the role proposed for the mesotelencephalic DOPAMINE systems in incentive salience, it is further conjectured that craving, or exaggerated desire for a specific object or its mental representation, is a direct result of drug-induced sensitization. In this manner, repeated self-administration of drugs of abuse, such as AMPHETAMINE, produce neural effects that set the stage for subsequent craving for repeated access to the drug.

(SEE ALSO: *Brain Structures and Drugs; Causes of Substance Abuse; Research, Animal Model*)

BIBLIOGRAPHY

BERRIDGE, K. C., & VALENSTEIN, E. S. (1991). What psychological process mediates feeding evoked by electrical stimulation of the lateral hypothalamus? *Behavioral Neuroscience, 105,* 3–14.

BOLLES, R. C. (1992). Reinforcement, expectancy and learning. *Psychol. Rev., 79,* 394–409.

ROBINSON, T. E., & BERRIDGE, K. C. (1993). The neural basis of drug craving: An incentive-sensitization theory of addiction. *Brain Res. Reviews, 18,* 247–291.

ANTHONY G. PHILLIPS

RESEARCH, ANIMAL MODEL The articles in this section describe studies of the effects of drugs on animals in the laboratory. These studies are important because many of our current beliefs about the nature of drug dependence involve concepts of learning and reinforcement, and many recently developed treatments are founded on these beliefs. The section contains *An Overview of Drug Abuse* research using animal models and detailed articles on various research concepts beings explored in this way: *Conditioned Place Preference; Conditioned Withdrawal; Drug Discrimination Studies; Drug Self-Administration; Environmental Influences on Drug Effects; Learning, Conditioning, and Drug Effects— An Overview; Learning Modifies Drug Effects; Learning Modifier Drug Effects; Operant Learning Is Affected by Drugs.*

See also *Aggression and Drugs: Research Issues; Motivation and Incentives;* and the articles in the section entitled *Research.*

An Overview of Drug Abuse A great deal of biomedical research is based on the belief that only through careful scientific analysis will we achieve a sound understanding of the problem of drug abuse and how to control it. Animal models of a human condition are an integral part of that analysis. Animal models were developed to help us understand the factors that control drug abuse. Under laboratory conditions it is possible to control many factors, such as the environment, genetics, drug history, and behavioral history, that cannot easily be controlled outside the laboratory. When these factors can be controlled, their influence on drug abuse can be precisely determined. As always, the use of animals involves the assumption that the behavior of animals is a valid model of a human disorder. The drug abuse research that has been conducted to this point makes it clear that this is a valid assumption.

There are three major animal models of aspects of drug abuse to consider: PHYSICAL DEPENDENCE, drug self-administration, and drug discrimination. Each of these has provided basic information about the fundamental processes that control drug abuse. In addition, each has provided practical information about the abuse potential of new drugs. Information on both of these topics represents an important contribution of animal research to solving the problems of drug abuse.

PHYSICAL DEPENDENCE

Often when a drug is administered repeatedly, TOLERANCE develops to its effects. That is, the dose of drug that is taken must be increased to achieve the same effect. With prolonged exposure to high doses, physical (or physiological) dependence may develop. That is, the person is dependent on the drug for normal physiological functioning. The existence of physical dependence is revealed when drug administration is stopped. When the drug is no longer administered, various physical changes begin to appear. Depending on the specific drug, these could include autonomic signs (e.g., diarrhea and vomiting), somatomotor signs (e.g., exaggerated reflexes, convulsions), and behavioral signs (e.g., decreases in food and water intake). These effects have been called withdrawal but in the literature are also known as abstinence syndrome.

Historically, it was believed that physical dependence was the cause of drug addiction. That is, it was felt that one had to become physically dependent on a drug before abuse would occur and that the drug dependence or addiction was motivated by the need to relieve the abstinence syndrome. One of the major contributions of modern drug abuse research has been to make it clear that this is not true. In fact, much drug abuse occurs in people who are not physically dependent. Nevertheless, since the need to avoid the abstinence syndrome can increase the likelihood that a person will continue to abuse a drug, it is important that we understand physical dependence. Also, it would be desirable for new drugs that are developed not to produce physical dependence.

The development of physical dependence is most common with the OPIOIDS (morphine and morphine-like drugs) and central nervous system (CNS) depressants (e.g., BARBITURATES and ALCOHOL). Since opioids are very valuable painkillers but produce physical dependence when used repeatedly, there has been great interest in the development of drugs that can kill pain but do not produce physical dependence. Standard approaches to testing new opioids in animals for their potential for inducing physical dependence have been developed. In the early stages of testing, a new drug that has been found to be an effective ANALGESIC is given to an animal that is physically dependent on morphine (mice, rats, dogs, and monkeys have been used). After giving the drug, a trained observer scores the occurrence, intensity, and duration of abstinence signs such as shivering, restlessness, irritability, abdominal cramps, vomiting, diarrhea, and decreased eating and drinking. The drug may not affect the abstinence syndrome; it may relieve it or it may make the syndrome worse. A drug that relieves morphine abstinence probably will produce morphine-like physical dependence and may not be considered for further development on this basis. On the other hand, a drug that has no effect on abstinence, or even makes it worse, probably will not produce morphine-like physical dependence and may be worth pursuing. Often such a drug will be evaluated for its ability to produce physical dependence when it is administered repeatedly. A drug that produces no physical dependence of its own is clearly a candidate for further development.

Literally hundreds of new opioid drugs have been evaluated in animals for their capacity to produce physical dependence, and, in the process, we have learned much about physical dependence. It is clear that the higher the drug dose and the more frequent the exposure, the more intensive the physical dependence that develops. But recent research with human subjects has strongly suggested that even a single dose of an opioid may produce some level of physical dependence. Research has also shown that drugs that suppress the signs of morphine abstinence in a dependent animal generally have morphine-like effects themselves. That is, they suppress respiration and cough, kill pain, and have the potential to be abused and produce physical dependence. These drugs are known as opioid agonists. Other drugs, known as opioid ANTAGONISTS, may cause abstinence signs and symptoms to appear. Opioid antagonists do not have morphine-like effects themselves but are capable of blocking or reversing the effects of morphine and morphine-like drugs. Still other drugs, called mixed AGONIST-ANTAGONISTS, can have either type of effect, depending on dose and whether the animal is physically dependent. This group of drugs has proven particularly interesting in terms of its contribution to our understanding of how opioids work. In addition, many of them are effective analgesics with apparently low potential to produce physical dependence.

Other classes of drugs besides opioids produce physical dependence in animals as well. Many of the basic findings about physical dependence on CNS depressants (e.g., dose and frequency) are similar to what has been found with opioids. However, the abstinence syndrome can be even more severe than that seen with opioids. HALLUCINATIONS and even life-threatening convulsions can develop when long-term abuse of a barbiturate or alcohol is stopped. Abstinence syndromes have also been found after long-term exposure to TETRAHYDROCANNABINOL (THC), the active ingredient in MARIJUANA, and PHENCYCLIDINE (PCP). On the other hand, the abstinence syndrome that follows long-term exposure to such CNS stimulants as AMPHETAMINE or COCAINE is, by comparison, mild.

DRUG SELF-ADMINISTRATION

The distinguishing characteristic of drug abuse is the behavior of drug self-administration. When that behavior becomes excessive and has adverse consequences for the individual or society, the individual is considered to be a drug abuser. Therefore, the development of animal models for studying drug self-administration was the essential first step toward identifying factors that control the behavior. Humans consume drugs by several different routes of administration, including oral (e.g., alcohol), intravenous (e.g., cocaine and heroin), and inhalation (e.g., nicotine and crack cocaine). Although some of the factors that control drug abuse may be independent of the route of administration, others may not. Therefore, it has been important to develop models in which animals self-administer drugs by each of these routes.

Early attempts to study drug self-administration in animals involved oral self-administration. Oral self-administration of drugs has proven difficult to establish in laboratory animals, probably because most drug solutions have a bitter taste. Also, when consumed orally, the onset of the drug effect is relatively slow, making it difficult for the animal to make the association between drinking and drug effect. For these reasons, when first given a choice between water and a drug solution, most animals choose the water. However, conditions can be arranged so that the animal drinks large amounts of the drug solution in relatively short periods, by making the drug solution available when food is available, either as a meal or delivered repeatedly as small pellets of food. After a period of drug consumption in association with food, food can be removed from the experiment and the animal will continue to consume the drug orally. When given a choice between the drug solution and water, the animal will prefer the drug solution. This approach has been particularly important for research with alcohol, since humans abuse this drug orally.

To study intravenous self-administration, an animal is surgically implanted with a chronic intravenous catheter through which a drug can be administered. The animal wears a backpack and tether that protect the catheter and attach to a wall of the cage. The cage usually has levers that the animal can press to receive a drug injection and lights that can be turned on to signal that a drug injection is available. At that time, a lever press turns on an electric pump and injects a drug solution through the catheter into the vein. In this way, the animal model mimics intravenous drug injection by humans using a syringe. Since taste is not a factor and onset of drug action is rapid, conditioning animals to inject drugs by the intravenous route has proven relatively straightforward.

Reliable methods for administering drugs to animals by inhalation are important for studying the abuse of drugs that are inhaled, such as TOBACCO,

SOLVENTS, or CRACK. Methods for studying solvent inhalation have been available for several years. Usually an animal is given the opportunity to press a lever to deliver a brief bolus of solvent vapor to the area around its nose. Methods for studying crack cocaine smoking in monkeys have only recently been developed. Monkeys are first trained to suck on a drinking tube; then the apparatus is arranged so that sucking on the tube delivers crack smoke to the monkey. Although the technique is new, it shows promise for the study of smoking in laboratory animals.

Research using these animal models has shown that, with few exceptions, animals self-administer the same drugs that humans abuse and show similar patterns of intake. For example, when given unlimited access to stimulants like amphetamine, both humans and animals alternate periods of high drug intake with periods of no drug intake. In the case of heroin, both animals and humans gradually increase drug intake to levels that are then stable for months and even years. In addition, animals do not self-administer drugs that humans do not abuse (e.g., aspirin) and even avoid those that humans report to be unpleasant (e.g., ANTIPSYCHOTIC DRUGS). These basic findings validate this as an excellent animal model of drug abuse by humans. The exceptions are the hallucinogens and marijuana, which animals do not readily self-administer.

Research using the self-administration model has increased our understanding of drug abuse in several different areas. It has become clear that drug self-administration is controlled by events that are initiated inside (e.g., a drug-induced change in brain chemistry) or outside (e.g., stress) the organism. With regard to events initiated inside the organism, we have begun to learn about the NEUROTRANSMITTER systems in the brain that are activated when drugs are self-administered. These changes are probably responsible for producing the drug effect that people find desirable and that maintains their self-administration (the reinforcing effect). A substantial amount of recent research has focused on the neurotransmitter changes that are involved in the reinforcing effect of cocaine. It has been known for some time that cocaine increases the concentration of certain neurotransmitters in synapses. Research indicates that it is this effect on certain synapses in the CNS that use the neurotransmitter DOPAMINE in the brain that almost certainly plays the primary role in cocaine's reinforcing effect. Similar research suggests that the neurotransmitter SEROTONIN may play a primary role in the effects of alcohol.

Even though neurotransmitter changes occur when an individual self-administers a drug, they are not always sufficient to maintain drug self-administration or to make it excessive. Events initiated outside the organism—that is, environmental events—can increase or decrease drug self-administration. In the case of alcohol, for example, consumption can be increased in animals simply by presenting other things of value (e.g., food pellets) every few minutes. Although it is not known exactly why this occurs, analogous conditions may increase the consumption of alcohol and other drugs by some humans. Drug self-administration can also be decreased by environmental conditions. For example, increasing the cost of a drug or the effort required to obtain it decreases consumption. Drug self-administration can also be decreased by imposing punishment or by making valuable alternatives to drug self-administration available.

Animal research has also made it clear that certain individuals may, because of their genetic makeup, be more susceptible to the effects of alcohol or other drugs. For example, genetically different strains of rats can differ in their sensitivity to the effects of codeine, morphine, or alcohol. Also, animals can be selectively bred to be more or less sensitive to the effects of a drug. These findings clearly demonstrate a genetic component to drug sensitivity. Research suggests that these animals differ in the amounts of these drugs that they will self-administer. How broadly this conclusion cuts across drugs of abuse is unknown but is an active area of research.

In short, drug abuse research with animals has made it clear that whether drug self-administration occurs depends on an interaction between a drug, an organism, and an environment. A susceptible individual in an environment in which a drug is available and in which conditions encourage drug self-administration is more likely to be a drug abuser than one in which environmental conditions discourage drug abuse.

DRUG DISCRIMINATION

When a person takes a drug of abuse, it has effects that the person feels and can describe. These effects are called *subjective effects* (versus *objective effects* that can be seen by an observer), and they play an important role in drug abuse. A person is more likely to abuse a drug that has effects that the person describes as pleasant than one that the person describes as unpleasant.

The subjective effects of drugs of abuse have been studied in humans for many years and in several different ways. Early research involved administering drugs, usually morphine-like drugs, to former heroin addicts who then answered questionnaires that were designed to detect and classify the subjective effects of the drug. The single-dose opiate questionnaire asks the subject whether he or she can feel the drug, to identify the drug, to describe the symptoms, and to rate how much he or she likes it. The Addiction Research Center Inventory consists of a series of true-false statements that describe internal states that might be produced by drugs. The Profile of Mood States is a list of adjectives that can be used to describe mood. Responses to these questionnaires depend on variables such as type of drug and drug dose. Recent research has examined the subjective effects of a wider variety of drugs (including stimulants and depressants) not only in experienced but also inexperienced subjects. The purpose of this research is to understand the factors that can influence a person's subjective response to drugs of abuse.

Since subjective effects require a verbal description of an internal state, they can be directly studied only in humans. Over the past 20 to 30 years, however, it has become clear that animals can be trained to respond in a way that suggests they can detect the internal state produced by a drug. The behavioral paradigm is called DRUG DISCRIMINATION, and the drug effect is called *a discriminative stimulus effect*. Although a number of drug-discrimination paradigms have been developed, the most common is a two-lever paradigm in which the animal is trained to press one lever after it has received a drug injection and the second lever after an injection of the drug vehicle or, in some cases, another drug. Responding on the lever that is appropriate to the injection is reinforced, usually by presenting a food pellet, while responding on the incorrect lever is not. If this is done repeatedly over a period of several weeks, the animal learns to respond almost exclusively on the lever associated with the injection. Although it is impossible to know what an animal feels, it seems as if the animal is reporting whether it feels the drug by the lever it presses. The animal can then be asked to "tell" us whether a new drug "feels" like the training drug. It will respond on the drug lever if the new drug is similar to the training drug and on the vehicle lever if it is not. It can also be "asked" whether a drug blocks the effects of the training drug. If the test drug blocks the effect of the training drug, it will respond on the vehicle lever when given both drugs.

There is a strong correspondence between the classification of drugs by humans based on their subjective effects and those by animals based on their discriminative stimulus effects. Research using the drug-discrimination model has increased our understanding of control of behavior by drugs in several different ways. First, this research has made it clear that behavior that is learned under the influence of a drug is more likely to occur again when the drug or a similar drug is taken again. This is a fundamental mechanism by which drugs control behavior. As with drug self-administration, a substantial amount of recent research has focused on the neurotransmitter changes that are involved in the discriminative stimulus effects of cocaine and alcohol. Again, dopamine seems to play a prominent role in this effect of cocaine, while serotonin may mediate the effects of alcohol. Environmental events, by contrast, do not seem to alter the discriminative stimulus effects of drugs substantially. However, little research has been done in this area.

ABUSE LIABILITY TESTING AND TREATMENT RESEARCH

One important application of animal models of drug abuse is the prediction of the likelihood that a new drug will be abused if it is made available to people. Clearly, the prevalence of abuse of a drug can be reduced by restricting its availability, and drugs with high potential for abuse should be the least available. All the models discussed here are used for predicting some aspect of the abuse liability of new drugs. However, the task is not simply a matter of detecting abuse liability and making the drug unavailable. ABUSE LIABILITY must be considered in the context of any potential therapeutic use of the drug, and a cost-benefit analysis that weighs liability for abuse against therapeutic benefits should be made.

Opioids are an excellent example of these considerations. Morphine is often the only appropriate analgesic for intense PAIN. However, it produces physical dependence and has a high potential for abuse. A drug that produces analgesia equivalent to or greater than that of morphine but does not pro-

duce physical dependence would be a highly desirable compound. Techniques for establishing this have been described in related articles. A new drug can be tested for its ability to suppress abstinence syndrome in monkeys that are dependent on morphine and for its ability to produce physical dependence of its own type in naive animals. A similar approach is taken with the drug in drug self-administration experiments. We may ask whether the drug maintains self-administration in experienced monkeys or whether naive monkeys will initiate self-administration. In addition, we can evaluate whether the drug is likely to be preferred to morphine by allowing an animal to choose between morphine and the new drug or determining how hard the animal will work to receive an injection of the drug relative to how hard it will work for morphine. Finally, we can ask whether the drug has discriminative stimulus effects that are similar to those of morphine or of any other drug of abuse. A drug that supports physical dependence, is self-administered, and has morphine-like discriminative stimulus effects is likely to have high potential for abuse in humans and unlikely to be a viable substitute for morphine. On the other hand, a drug that lacks one or more (preferably all) of these effects may be worth pursuing.

Animal models of drug abuse have been used for the development of drugs that may be useful in the treatment of drug abuse. In some ways it seems unusual to suggest treating a drug abuse problem with another drug. However, in the case of opioids, METHADONE, a morphine-like agonist, has proven to be quite useful in the treatment of opioid dependence. Although the drug is still self-administered and physical dependence is maintained, treatment with methadone allows the person to lead a relatively normal life that does not require the high-cost behaviors (e.g., crime, intravenous injection) associated with abuse of illicit opioids.

The animal models described here, particularly drug self-administration and drug discrimination, are now being applied to the development of drugs that may be useful in treatment. These approaches are based on the reasonable but as yet unvalidated assumption that blocking or mimicking the reinforcing and subjective effects of drugs will decrease drug abuse. In the case of cocaine, dopamine antagonists and, surprisingly, opioids have shown some promise in animal models as potential treatment compounds. It is not yet clear whether these compounds will be effective in humans. Nevertheless, this is an area of active research that shows promise for helping with treatment of drug abuse for development as treatment compounds.

(SEE ALSO: *Abuse Liability of Drugs; Reinforcement; Research*)

BIBLIOGRAPHY

BRADY, J. V., & LUKAS, S. E. (1984). Testing drugs for physical dependence potential and abuse liability. *NIDA research monograph 52.* Washington, DC: U.S. Government Printing Office.

COLPAERT, F. C., & BALSTER, R. L. (1988). *Transduction mechanisms of drug stimuli.* Berlin: Springer-Verlag.

WOOLVERTON, W. L., & NADER, M. N. (1990). Experimental evaluation of the reinforcing effects of drugs. In *Testing and evaluation of drugs of abuse.* New York: Alan R. Liss.

WILLIAM WOOLVERTON

Conditioned Place Preference A procedure called conditioned place preference has been used to study the "rewarding" effects of drugs. The procedure is designed to ask the question "When given a choice, will an animal prefer an environment in which it has experienced a drug to one in which it has not?" To answer this question, an animal is placed in an experimental chamber that is divided into two compartments that are different in some way. For example, they may have different floors and/or distinctive odors. Initially, the animal is placed in the chamber for several preconditioning trials and the time spent in each compartment is measured. Usually, a rat exhibits some preference for one or the other side in these trials. At this point, the experimenter can do one of two things—(1) modify the compartments in some way, perhaps by changing the lighting, so that equal time is spent in the two chambers before proceeding (balanced procedure), (2) go ahead with the experiment with unequal preferences (unbalanced procedure). With either procedure, conditioning trials are conducted next.

To run conditioning trials, a barrier is placed in the middle of the chamber that does not allow the animal to switch sides. The drug of interest is then administered to the animal and it is confined to one compartment for usually fifteen to thirty minutes. If

the unbalanced procedure is used, the animal is usually placed in the compartment that was initially avoided. A second group may be given a placebo (a substance that has no effect) under these same conditions or a placebo may be given to these same animals before placing them in the second compartment in alternating sessions. In this way, the effect of the drug is associated with a particular environment. After several—three to ten—conditioning sessions, the animal is placed in the chamber without being given the drug, and the door is removed so that the animal can spend time in either compartment. The length of time spent in each chamber is recorded and used as a measure of preference for that chamber.

The hypothesis underlying this sort of experiment is that the length of time spent in an environment should increase if that environment is associated with the effects of a drug of abuse. In fact, many studies have shown that this does happen with drugs such as HEROIN, COCAINE, and AMPHETAMINES. In the balanced procedure, animals spend more time in the drug-associated side than in the other side. In the unbalanced procedure, the animals spend more time in the drug-associated side than they did previously, but only rarely demonstrate an actual preference for it. As would be expected, preference is greater with higher doses of the drug and does not occur with placebo injections. In addition, it does not occur with drugs that are not typically abused, such as antipsychotic drugs, antidepressant drugs, and opioid antagonists. Thus, it seems likely that the technique measures a drug effect that is related to drug abuse.

Like other models for studying drug abuse, conditioned place preference has strengths and weaknesses. Among its strengths is the fact that animals are tested in a drug-free state. Therefore, the measure of preference is not influenced by the direct effects of drugs. The procedure can be done with drug injections given by routes other than intravenous, therefore surgical preparation is not involved. Moreover, the procedure is rapid, with maximum effect usually evident within three conditioning sessions.

The major weakness relates to the drug effects that it is measuring. Since drug administration is not due to the behavior of the animal (i.e., self-administration), it is by definition not a reinforcing effect. Although many of the same drugs that are self-administered induce place preferences, it is not clear whether the drug effect studied in conditioned place preference is the same as that studied in procedures that directly measure reinforcing effects. Another weakness is that is it not known whether it is meaningful to compare drugs in terms of their ability to engender place preferences. That is, if drug X induces a greater place preference than drug Y, does it have more abuse potential? Finally, because the procedure involves the simple behavioral response of moving from one chamber to another, it is not known whether it can be used to study some of the complex behavioral variables that are known to be determinants of drug self-administration. Despite these ambiguities, however, the simplicity of the procedure makes it likely that it will continue to be useful for studying drug abuse.

(SEE ALSO: *Abuse Liability of Drugs; Reinforcement*)

BIBLIOGRAPHY

BOZARTH, M. A. (1987). Conditioned place preference: A parametric analysis using systemic heroin injections. In *Methods of assessing the reinforcing properties of abused drugs*, pp. 241–273. New York. Springer-Verlag.

HOFFMAN, D. C. (1989). The use of place conditioning in studying the neuropharmacology of drug reinforcement. *Brain Research Bulletin, 23*, 373–387.

WILLIAM WOOLVERTON

Conditioned Withdrawal Upon cessation from drug taking, many individuals experience unpleasant effects (i.e., WITHDRAWAL), which can include both physiological and psychological symptoms. For example, for OPIATE drugs such as MORPHINE and HEROIN, withdrawal symptoms can include restlessness, anorexia, gooseflesh, irritability, nausea, and vomiting. Withdrawal symptoms are most pronounced following a long history of exposure to ALCOHOL and opiates, but a variety of withdrawal symptoms can occur after exposure to most psychoactive drugs.

As with most other drug effects, researchers have shown that these withdrawal symptoms can be conditioned or linked by learning to environmental cues. This research on *conditioned withdrawal* has included both human case reports and laboratory

animal research. For example, Vaillant (1969) reported that individuals who had been abstinent from opiates for months would experience "acute craving and withdrawal symptoms" upon reexposure to situations previously associated with opiate use. Further, Goldberg and Schuster (1967) showed that withdrawal symptoms also can be conditioned in laboratory animals. In their experiment, rhesus monkeys were addicted to morphine by giving them the drug repeatedly. The monkeys were then given an occasional injection of nalorphine, an opiate antagonist, which immediately led to the monkeys exhibiting signs characteristic of withdrawal. The injection of nalorphine was always given in the presence of a specific environmental stimulus, in this case a tone. Following several exposures to the tone paired with nalorphine, Goldberg and Schuster found that presentation of the tone itself was sufficient to produce the withdrawal signs.

The behavioral mechanism most likely to account for the phenomenon of conditioned withdrawal is *classical conditioning* (also known as *Pavlovian*). In Pavlov's classic experiments on this type of conditioning, a neutral stimulus such as a bell, is repeatedly paired with a nonneutral stimulus such as food. Eventually the bell itself elicited salivation, which was initially observed only to the food. In conditioned withdrawal, a neutral stimulus (e.g., a bell, a needle, a room, a friend, a street corner, or certain smells) is paired with the nonneutral stimulus of withdrawal until eventually those neutral stimuli will also elicit withdrawal symptoms.

The phenomenon of conditioned withdrawal can have important implications for drug-abuse treatment. The experience of drug withdrawal is often an important factor in the long-term maintenance of drug abuse. That is, as individuals experience withdrawal, they are likely to seek out a new drug supply to relieve withdrawal symptoms. An important aspect of drug-abuse treatment is relieving the symptoms of withdrawal during the period immediately following the cessation of drug use. Conditioned effects, however, are often long-lasting and do not depend on the continued presentation of the initial nonneutral stimulus (in this case withdrawal). Even after a patient has been withdrawn from a drug, stimuli that have been conditioned to elicit withdrawal symptoms may still be effective. Therefore, upon reexposure to those stimuli a patient may be much more likely to relapse to drug abuse. Thus, to be successful, any treatment regimen for drug abuse must deal with conditioned withdrawal.

(SEE ALSO: *Causes of Substance Abuse; Wekler's Pharmacologic Theory of Drug Addiction*)

BIBLIOGRAPHY

GOLDBERG, S. R., & SCHUSTER, C. R. (1967). Conditioned suppression by a stimulus associated with nalorphine in morphine-dependent monkeys. *Journal of the Experimental Analysis of Behavior, 10,* 235–242.

VAILLANT, G. E. (1969). The natural history of urban narcotic drug addiction—Some determinants. In H. Steinburg (Ed.), *Scientific basis of drug dependence.* New York: Grune & Stratton.

CHARLES SCHINDLER
STEVEN GOLDBERG

Drug Discrimination Studies   When a person takes a drug of abuse, it has effects that a person feels and can describe. These are termed *subjective effects* and they play an important role in drug abuse. People are more likely to abuse a drug that has effects they describe as pleasant than one they describe as unpleasant.

The subjective effects of drugs of abuse have been studied in humans for many years and in several different ways. Early research involved administering drugs, usually morphine-like drugs, to former HEROIN addicts—who then answered questionnaires that were designed to detect and classify the subjective effects of the drug. The single-dose OPIATE questionnaire asks subjects whether they can feel the drug, to identify the drug, to describe the symptoms, and to rate how much they like it. The Addiction Research Center Inventory consists of a series of true/false statements that describe internal states that might be produced by drugs. The Profile of Mood States is a list of adjectives that can be used to describe mood. Responses to these questionnaires depend on variables such as type of drug and drug dose. Recent research has examined the subjective effects of a wider variety of drugs (including STIMULANTS and DEPRESSANTS) in both experienced and inexperienced subjects. The purpose of this research is to understand the factors that can influence a person's subjective response to drugs of abuse.

Since subjective effects require a verbal description of an internal state, they can only be studied directly in humans. Since the 1960s, however, it has become clear that animals can be trained to respond in a way that suggests they can detect the internal state produced by a drug. The behavioral paradigm is called DRUG DISCRIMINATION, and the drug effect is called a *discriminative stimulus effect*. Although a number of drug-discrimination paradigms have been developed, the most common is a two-lever paradigm. Here the animal is trained to press one lever after it has received a drug injection and the second lever after an injection of the drug vehicle or, in some cases, another drug. Responding on the lever that is appropriate to the injection is reinforced, usually, by a food pellet; responding on the incorrect lever is not reinforced. If this is done repeatedly over a period of several weeks, the animal learns to respond almost exclusively on the lever associated with the injection.

Although it is difficult to know what an animal feels, it seems as if the animal is telling us whether it feels the drug or not by the lever it presses. The animal can then be asked to "tell" us whether a new drug "feels" like the training drug. It will respond on the drug lever if it does and on the vehicle lever if it does not. It can also be "asked" whether a drug blocks the effects of the training drug. If the test drug does block the effect of the training drug, the animal will respond on the vehicle lever when given both drugs.

CONCLUSIONS

What makes this area of research so exciting are the striking similarities between the classification of drugs by humans, based on their subjective effects, to those by animals, based on their discriminative stimulus effects. Therefore, this animal model can be used to investigate the influence of factors such as genetics, drug history, and behavioral history—factors that cannot be easily controlled in human subjects—on the subjective effects of drugs. It also allows us to predict whether a new drug is likely to have subjective effects, like a known drug of abuse, or is likely to block the subjective effects of the drug of abuse, without giving the drug to humans. If an animal is trained to discriminate a drug of abuse and presses the drug lever when given the new drug, then it is highly likely that the new drug will have subjective effects in humans similar to those of the drug of abuse. Its availability might then be restricted. If the animal responds on the vehicle lever when given the combination of the new drug and the drug of abuse, the new drug may block the subjective effects of the drug of abuse. Such a drug might then be useful for treating drug abuse.

(SEE ALSO: *Abuse Liability of Drugs; Drug Types; Sensation and Perception*)

BIBLIOGRAPHY

COLPAERT, F. C. (1986). Drug discrimination: Behavioral, pharmacological, and molecular mechanisms of discriminative drug effects. In *Behavioral analysis of drug dependence*. Orlando, FL: Academic.

COLPAERT, F. C., & BALSTER, R. L. (1988). *Transduction mechanisms of drug stimuli*. Berlin: Springer-Verlag.

WILLIAM WOOLVERTON

Drug Self-Administration   One factor that distinguishes a drug of abuse from a drug that is not abused is that taking the drug of abuse increases the likelihood that it will be taken again. In such a case, we say that this drug has reinforced the drug self-administration response and that it has reinforcing effects. Factors that influence reinforcing effects, therefore, profoundly influence drug self-administration and drug abuse. Knowing the reinforcing effects of drugs is essential to understanding drug abuse.

Techniques developed on laboratory animals allow us to study the reinforcing effects of drugs, using the intravenous and oral routes as well as smoking. To study intravenous self-administration, the researcher surgically implants a chronic intravenous line (a catheter) through which a drug can be administered. Laboratory animals (rats, mice, monkeys, and so on) live in cages in which they can operate some device, usually a lever press, that turns on an electric pump to send some drug solution through the catheter. Oral self-administration is harder to establish, since drugs are usually bitter; however, by arranging conditions so that large amounts of drug solution are ingested in relatively short periods—usually by adding the drug to water when food is

available—researchers can condition animals to self-administer drugs orally. Research on the smoking of TOBACCO or CRACK-COCAINE is important and this too needs conditioning for reliable study.

Animals used in research studies have shown that, with few exceptions, they abuse the same drugs that humans abuse and show similar patterns of intake. (Exceptions include MARIJUANA and HALLUCINOGENS, such as LSD, which animals do not seem to find reinforcing.) Drug self-administration studies have been used to predict whether a new drug is likely to be abused by humans if it becomes easily available. More important, such research has allowed us to understand some factors that can increase or decrease the reinforcing effects of drugs that contribute to human drug abuse. Some of these factors relate to the drug itself; others to the environment. For example, drugs that increase the concentration of the NEUROTRANSMITTER DOPAMINE in the synapses of the brain (e.g., cocaine) are more likely to have abuse potential than those that do not.

Research has made it clear that even the most preferred drug—cocaine—will be self-administered differently depending on environmental conditions. If more lever presses are required to obtain it (it "costs" more), less is consumed. Drug self-administration can also be decreased by punishment or by making valuable alternatives available. In short, drug self-administration research has shown that whether a drug will be abused is determined by a complex interaction between the drug, the environment, and the organism. Current research is aimed at understanding the dynamics of that interaction in a quantitative way.

(SEE ALSO: *Abuse Liability of Drugs; Adjunctive Drug Testing*)

WILLIAM WOOLVERTON

Environmental Influences on Drug Effects   More than any other discipline, the field of behavioral PHARMACOLOGY has attempted to understand the influence of nonpharmacological, or environmental, factors on the effects of abused drugs. Since the classic demonstration by Dews (1955, 1958) showing that the effects of pentobarbital and METHAMPHETAMINE depend on the manner in which behavior is controlled by the schedule of REINFORCE-

MENT, researchers have been interested in various environmental influences on the effects of drugs. Some of these effects are described elsewhere in this encyclopedia (and see Barrett, 1987, for a more detailed review). This article reviews additional influences to illustrate the overwhelming conclusion that the effects of a drug depend on complex environmental variables that may override the typical pharmacological effects of a compound. Indeed, the evidence for environmental influences on drug action is so compelling that when the effects of abused drugs are characterized, "susceptible to environmental modulation" should be a salient distinguishing description along with physiological features.

BEHAVIORAL CONSEQUENCES

The specific manner in which behavior is controlled by its consequences may often represent a strong influence on drug action. In research situations, this is apparent in the effects of AMPHETAMINE or COCAINE on punished and nonpunished responses maintained by the presentation of food. Low rates of nonpunished responses are typically increased by these drugs (PSYCHOMOTOR STIMULANTS), whereas comparable low rates on punished responses are not affected by these drugs or are only decreased further. In the Dews studies (1955, 1958), the effects of the drugs differed depending on whether behavior was maintained at relatively high response rates under a fixed-ratio schedule that provided food following every nth response or whether responses occurred at lower rates under a fixed-interval schedule that provided food for the first response after t minutes. Explanations of the differential effects of the drugs could not be based on different levels of motivation, since these schedule conditions alternated sequentially within the same experimental session. Although these and similar results were obtained under carefully controlled experimental conditions, such findings document the essential point that environmental conditions surrounding and/or supporting behavior play a very important role in determining the effects of drugs.

BEHAVIORAL CONTEXT

The environmental modulation of drug effects has been shown repeatedly, by using schedule-controlled responses and various types of events. These findings represent two areas of research demonstrating how drug effects are modified directly by existing environmental conditions:

(1) More remote influences can also influence drug action. In behavioral history, for example, consequences that have occurred in the distant past can significantly alter the effects of abused drugs even though no traces of that experience are apparent in current behavior.

(2) In other studies in which environmental influences helped determine the effects of an abused drug, behavioral consequences occurring under one experimental condition alter the action of drugs occurring under different conditions. In this case, the conditions that interact are relatively close in time. For example, in an experiment with monkeys, exposure to a procedure in which responses avoided the delivery of a mild electric shock completely reversed the effects of amphetamine on punished responses that had occurred in a different and adjacent context (i.e., under different stimulus conditions from the avoidance schedule and separated by only a few minutes).

Comparable results, although with different species, different schedule conditions, and different environmental events, have also been arrived at with ALCOHOL, cocaine, and CHLORDIAZEPOXIDE (Barrett, 1987). The findings show the generality of this phenomenon—that the environment is an important variable contributing to the effects of drugs on behavior. The actions of a drug at its receptor site and the transduction mechanisms that ensue can be affected by events occurring in the environment.

SUMMARY

The studies described here indicate the powerful influences that exist in the environment that can alter the course of the effects of abused drugs. Such findings illustrate the need to examine those influences and the manner in which they occur, although it is often tempting to attribute all changes in behavior to the abused drug. Consequences that are immediate, as in the existing environment, or remote, such as in the individual's past experience, help determine the acute effects of drugs and may also contribute to long-term abuse and persistent drug use.

(SEE ALSO: *Adjunctive Drug Taking; Causes of Substance Abuse; Reward Pathways and Drugs; Tolerance and Physical Dependence*)

BIBLIOGRAPHY

BARRETT, J. E. (1987). Nonpharmacological factors determining the behavioral effects of drugs. In H. Y. Meltzer (Ed.), *Psychopharmacology: The third generation of progress.* New York: Raven Press.

DEWS, P. B. (1958). Studies on behavior: IV. Stimulant actions of methamphetamine. *Journal of Pharmacology and Experimental Therapeutics, 122,* 137–147.

DEWS, P. B. (1955). Studies on behavior: I. Differential sensitivity to pentobarbital of pecking performance in pigeons depending on the schedule of reward. *Journal of Pharmacology and Experimental Therapeutics, 113,* 393–401.

JAMES E. BARRETT

Learning, Conditioning, and Drug Effects—An Overview

The effects of abused drugs can be examined at many levels ranging from the molecular to the cellular to the behavioral. Each of these research areas contributes significant information to understanding the mechanisms by which drugs of abuse and alcohol produce their diverse effects. The most tangible sign of both immediate and long-terms actions of abused drugs is their effects on behavior. Often it is assumed, and wrongly, that behavior is a passive reflection of more significant events occurring at a different and (usually) more molecular level. Understanding those cellular events is occasionally viewed as the key to understanding drug abuse and to intervention strategies. In fact, however, behavior itself and the variables that control it play a prominent and often profound role in directly determining drug action and, most likely, those cellular and molecular events that participate in behavior and in the effects of drugs. The variables that guide and influence behavior also affect molecular substructures—therefore, behavioral and neurobiological processes are interdependent.

EXPERIMENTAL ANALYSIS OF BEHAVIOR AND DRUGS OF ABUSE

The progression of behavioral approaches in the study of the effects of abused drugs is characteristic of the cumulative and evolutionary nature of scientific progress. A number of techniques are now available that permit the development and maintenance of a variety of behaviors that are remarkably stable over time, sensitive to a number of interventions, and reproducible within and across species. These procedures have evolved over the past several years and reflect the combined efforts of individuals in different disciplines ranging from psychology, pharmacology, physiology, and ethology. For the most part, research studying the effects of abused drugs on behavior has been conducted by two basic procedures. One procedure uses *unconditioned behavior,* such as locomotor activity that is more spontaneous in its occurrence (but still influenced by environmental conditions) and requires no specific training before it can be studied. Many PSYCHOMOTOR STIMULANTS such as COCAINE and AMPHETAMINE, for example, produce large and consistent increases in locomotor activity in laboratory animals. Frequently, however, unconditioned behavior is produced or *elicited* by the presentation of specific stimuli, and it is then brought under experimental control by arranging for the production of a response to a stimulus other than that originally responsible for its occurrence. The Russian physiologist Ivan Pavlov, for example, performed extensive studies in 1927 in which he used the unconditioned salivary response to food and to conditioned stimuli paired with food to study processes of classical or *respondent* conditioning. Although this approach has been used somewhat less often than other techniques, respondent conditioning procedures still serves as a very useful method for studying drug action (Barrett & Vanover, 1993).

The second procedure, which is designated as *operant conditioning,* uses the methods and techniques developed by the pioneering American psychologist B. F. Skinner (1938) to investigate behavior controlled by its consequences. The body of experimental research using operant conditioning techniques to study the effects of abused drugs is extensive (see Iversen & Lattal, 1991, for general reviews of the techniques and applications).

Unconditioned and Conditioned Respondent Behavior. Respondent behavior is elicited by specific stimuli and usually involves no specific training or conditioning, since the responses studied are typically part of the behavioral repertoire of the species and are expressed under suitable environmental conditions. Although the factors responsible for the occurrence of these behaviors presumably lie in the organism's distant evolutionary past, certain unconditioned responses can be brought under more direct and immediate experimental control through the use of procedures first discovered and systematically explored by Pavlov. These procedures consist of expanding the range of stimuli capable of producing an elicited response. In respondent conditioning, previously noneffective stimuli acquire the ability to produce or elicit a response by virtue of their temporal association with an unconditional stimulus, such as food, which is capable of eliciting a response without prior conditioning. Thus, when a distinctive noise, such as a tone, is repeatedly presented at the same time that or shortly before food is given, the tone acquires the ability to elicit many of the same responses originally limited to food.

Respondent behaviors depend primarily on *antecedent* events that elicit specific responses. Typically, these responses do not undergo progressive differentiation in that the responses to either a conditioned or an unconditioned stimulus are generally quite similar. These procedures also do not establish new responses but simply expand the range of stimuli to which that response occurs.

Operant Behavior. In contrast to respondent behavior, operant behavior is controlled by *consequent* events in that it is established, maintained, and further modified by its consequences. Operant behavior occurs for reasons that are not always known. The responses may have some initially low probability of occurrence or they may never have occurred previously. Novel or new responses are typically established by the technique of "shaping," in which a behavior resembling or approximating some final desired form or characteristic is selected, increased in frequency, and then further differentiated by the provision of a suitable consequence, such as food presentation to a food-deprived organism. This technique embodies the principle of reinforcement and has been widely used to develop operant responses such as lever pressing by rodents, humans, and nonhuman primates. Behavior that has evolved under such contingencies may bear little or no resemblance to its original form and can perhaps only be understood by careful examination of the organism's history. Although some behaviors often appear unique or novel, it is likely that the final product emerged as a continuous process directly and sequentially related to earlier conditions. The manner in which operant responses have been developed and maintained, as well as further modified, has been the subject of extensive study in the behavioral pharmacology of abused drugs and has had a tremendous impact on this field. Many of the potent variables that influence behavior, such as reinforcement, punishment, and precise schedules under which these events occur, also are of critical import in determining how a drug will affect behavior.

Respondent versus Operant Behavior. Although it is possible to tell operant behavior from respondent behavior in a number of ways, these processes occur concurrently and blend almost indistinguishably. For example, the administration of a drug may *elicit* certain behavioral and physiological responses such as increased heart rate and changes in perception that are respondent in nature; stimuli associated with the administration of that drug may also acquire some of the same ability to elicit those responses. If the administration of the drug followed a response and if the subsequent frequency of that response increased, then the drug also could be designated a reinforcer of the operant response. Thus, these important behavioral processes frequently occur simultaneously and must be considered carefully in experimental research, and also in attempting to understand the control of behavior by abused drugs. The primary distinctions between operant and respondent behavior now appear to be the way these behaviors are produced and the possible differential susceptibility to modification by consequent events. Respondent behavior is produced by the presentation of eliciting stimuli; characteristic features of these behaviors are rather easily changed by altering the features of the eliciting stimulus such as its intensity, duration, or frequency of presentation. Under all of these conditions, however, the response remains essentially the same.

In contrast, operant behavior depends to a large extent on its consequences, and with this process, complex behavior can develop from quite simple relationships. One has only to view current behavior as an instance of the organism's previous history acting together with more immediate environmental consequences to gain some appreciation for the continuity and modification of behavior in time. Current behavior is often exceedingly difficult to understand because of the many prior influences or consequences that no longer operate but which may leave residual effects. The effects of a particular consequence or intervention can be quite different depending on the behavior that exists at the time the event occurs. An individual's prior history, then, is important not only because it has shaped present behavior but also because it will undoubtedly determine the specific ways in which the individual responds to the current environment. Accordingly, prior behavioral experience can have a marked effect in determining how a drug will change behavior.

BEHAVIORAL METHODOLOGY AND THE EVALUATION OF ABUSED DRUGS

Experiments with drugs and behavior were initiated in Pavlov's laboratory in Russia during the time that Pavlov was studying the development of conditioned respondent procedures (see Laties, 1979, for a review of this early work). Early experiments with the effects of drugs on operant behavior were initiated shortly after Skinner began his pioneering work (Skinner & Heron, 1937). More intensive studies using drugs and operant-conditioning techniques were not conducted, however, until effective drugs for the treatment of various psychiatric disorders such as SCHIZOPHRENIA were introduced in the 1950s. These discoveries prompted the development and extension of behavioral techniques to study these drugs, and many of the procedures were subsequently used in the study of abused drugs. From these combined efforts, several key principles evolved that have served as the foundation for understanding and evaluating the effects of abused drugs.

Environmental Events. As already discussed, behavior can be controlled by a wide range of environmental events. One question that arose early in the study of the behavioral effects of drugs was whether the type of environmental event that controlled behavior contributed to the effects of a drug—that is, whether a behavior controlled by a positive event, such as food presentation, would be affected in the same manner as a behavior controlled by a more negative event, such as escape from an unpleasant noise or bright light. Although seemingly straightforward, the issue is not easily addressed because other known factors contribute to the actions of drugs, such as the rate at which a behavior controlled by the event occurs. If rates are not similar, any comparison between drug effects on behavior controlled by those different events might be spurious. Indeed, when such comparisons have been conducted in nonhuman primates under carefully controlled conditions, it has been shown that the type of environmental event controlling behavior can play an important role in determining the qualitative effects of a drug on behavior (Barrett & Wit-

kin, 1986; Nader, Tatham, & Barrett, 1992). For example, when the effects of certain drugs such as ALCOHOL or MORPHINE were studied by using behavioral responses of monkeys who were similarly maintained by a food stimulus or a mild electric-shock stimulus, the drugs produced different effects depending on the maintaining event (Barrett & Katz, 1981). These findings suggest that the manner in which behavior is controlled by its environmental consequences—that is, the characteristics of the environment—can be of considerable importance in determining how an individual will be affected by a particular drug. This was one of the experiments that supported the view that a drug is not a static molecule with uniform effects, but rather that the way the substance interacts with its receptor and initiates the cascade of biochemical processes depends very much on the dynamic interaction of behavior within its environment. When the issue is viewed in this light, it is clear that environmental events and the way they impinge on behavior contribute substantially to the specific effects of a drug and its impact on the individual organism.

Examples of similar types of environmental control over pharmacological effects of drugs also come from studies that employed respondent conditioning procedures to demonstrate that stimuli paired with morphine or heroin injections can influence the development and manifestation of fundamental pharmacological processes such as tolerance and lethality (Siegel, 1983). These studies add to the rather convincing body of evidence that environmental conditions accompanying the administration and effects of the drug can be of considerable importance in determining the effects of that drug when it is administered, as well as when it is subsequently administered.

Behavioral and Pharmacological History. In addition to pointing to the contribution of the immediate environment in determining the effects of abused drugs, a number of studies demonstrated that the consequences of *past* behavior could also contribute significantly to the effects of drugs, often by resulting in an action that is completely opposite to that shown in organisms without that history. These findings convey the complexity involved in understanding the effects of drugs of abuse, and the difficulties in attempting to understand their actions in humans with more complex life histories than those of experimental animals. In addition, related studies showed that prior experience with one drug could also directly affect the manner in which behavior is influenced by other drugs.

Early studies using different training conditions to develop a visual discrimination in pigeons demonstrated that an antipsychotic drug, Thorazine (chlorpromazine), and an antidepressant drug, imipramine, had different effects on that discriminative behavior, depending on how the training occurred (Terrace, 1963). Similarly, studies that used exploratory behavior of rats in mazes demonstrated that the effects of a mixture of amphetamine (STIMULANT) and a BARBITURATE drug (DEPRESSANT) depended on whether the rats had previously exposed to the maze (Steinberg, Rushton, & Tinson, 1961). More recently, studies with squirrel monkeys showed that prior behavioral experience can influence the effects of a wide range of drugs, including morphine, cocaine, and amphetamine, as well as alcohol, under a variety of experimental conditions (summarized by Barrett, Glowa, & Nader, 1989; Nader et al., 1992). In one study, for example, the effects of amphetamine were studied on behavior reinforced by food that was also suppressed by punishment. Under these conditions, amphetamine produced a further decrease in punished responding. If those same monkeys, however, were then exposed to a procedure in which responding postponed or avoided punishing shock and were then returned to the punishment condition, amphetamine no longer decreased responding; instead, it produced large *increases* in suppressed responding. Thus, the effects of amphetamine in this study depended on the prior behavioral experience of the animal.

These findings raise a number of issues surrounding the etiology of drug abuse as well as issues pertaining to an individual's risk for or vulnerability to abusing particular drugs. If, as seems likely, certain drugs are abused because of their effects on behavior, and those behavioral effects are related to past history, then the historical variables become exceptionally important in eventually understanding and treating, as well as preventing, drug abuse. Perhaps previous behavioral experience generates conditions under which a drug may have quite powerful actions on behavior and on the subjective effects that drug produces; by virtue of their previous history, the susceptible individuals may be predisposed to drug abuse. If these arguments are valid, it should be possible, after achieving a better understanding of the factors, to develop behavioral strategies for "inoculating" or "immunizing" individuals against particular drug effects. Although such possibilities may seem remote at this time, it is very clear that behavioral variables can direct the effects of abused drugs in striking and significant ways.

SUMMARY

Although drugs of abuse have a reliable and predictable spectrum of effects under a broad range of conditions, the implications from studies are that many of the more characteristic effects of abused drugs can be altered by the organism's history and by the environmental conditions under which the drug is and has been administered. As Falk (1983) said so eloquently, "Pharmacological structure does not imply motivational destiny"; the reasons for the effects of an abused drug depend on more than the static molecular properties of that drug. Both past and present environmental factors can play an overwhelming role in determining the behavioral effects of abused drugs, and they may indeed be a major source of the momentum behind the continued use and abuse of those substances.

(SEE ALSO: *Abuse Liability of Drugs; Addiction: Concepts and Definitions; Adjunctive Drug Taking; Causes of Substance Abuse; Reinforcement; Vulnerability As Cause of Substance Abuse*)

BIBLIOGRAPHY

BARRETT, J. E., GLOWA, J. R., & NADER, M. A. (1989). Behavioral and pharmacological history as determinants of tolerance- and sensitization-like phenomena in drug action. In M. S. Emmett-Oglesby & A. J. Goudie (Eds.), *Tolerance and sensitization to psychoactive agents: An interdisciplinary approach.* Clifton, NJ: Humana Press.

BARRETT, J. E., & KATZ, J. L. (1981). Drug effects on behaviors maintained by different events. In T. Thompson, P. B. Dews, & W. A. McKim (Eds.), *Advances in behavioral pharmacology* (Vol. 3). New York: Academic Press.

BARRETT, J. E., & VANOVER, K. E. (1993). 5-HT receptors as targets for the development of novel anxiolytic drugs: Models, mechanisms and future directions. *Psychopharmacology, 112,* 1–12.

BARRETT, J. E., & WITKIN, J. M. (1986). The role of behavioral and pharmacological history in determining the effects of abused drugs. In S. R. Goldberg & I. P. Stolerman (Eds.), *Behavioral analysis of drug dependence.* New York: Academic Press.

FALK, J. L. (1983). Drug dependence: Myth or motive? *Pharmacology Biochemistry and Behavior, 19,* 385–391.

IVERSEN, I. H., & LATTAL, K. A. (1991). *Experimental analysis of behavior* (Parts 1 and 2). New York: Elsevier.

LATIES, V. G. (1979). I. V. Zavodskii and the beginnings of behavioral pharmacology: An historical note and translation. *Journal of the Experimental Analysis of Behavior, 32,* 463–472.

NADER, M. A., TATHAM, T. A., & BARRETT, J. E. (1992). Behavioral and pharmacological determinants of drug abuse. *Annals of the New York Academy of Sciences, 654,* 368–385.

PAVLOV, I. (1927). *Conditioned reflexes: An investigation of the physiological activity of the cerebral cortex.* London: Oxford University Press.

SIEGEL, S. (1983). Classical conditioning, drug tolerance, and drug dependence. In Y. Israel et al. (Eds.), *Research advances in alcohol and drug problems* (Vol. 7). New York: Plenum.

SKINNER, B. F. (1938). *Behavior of organisms.* New York: Appleton-Century-Crofts.

SKINNER, B. F., & HERON, W. T. (1937). Effects of caffeine and benzedrine upon conditioning and extinction. *Psychological Record, 1,* 340–346.

STEINBERG, H., RUSHTON, R., & TINSON, C. (1961). Modification of the effects of an amphetamine-barbiturate mixture by the past experience of rats. *Nature, 192,* 533–535.

TERRACE, H. S. (1963). Errorless discrimination learning in the pigeon: Effects of chlorpromazine and imipramine. *Science, 140,* 318–319.

JAMES E. BARRETT

Learning Modifies Drug Effects A general framework within which to understand the basic processes and principles of respondent conditioning (as first discovered in the 1920s by Russian physiologist Ivan Pavlov [1849–1936] and subsequently elaborated in many laboratories over the next six decades) is described elsewhere. Here, specific examples of the role of conditioned drug effects are provided in an effort to more fully develop the point that conditioned or learned responses come about as a reaction to stimuli that have been associated with drug injections. These stimuli can play a powerful role in governing subsequent behavior in the absence of the drug.

CONDITIONED EFFECTS OF DRUGS

In addition to studies described previously showing that tolerance to the effects of a drug, as well as lethality, can depend on respondent-conditioning phenomena, a number of additional studies have demonstrated the conditioning of WITHDRAWAL and other responses that are typically associated only with the presentation or removal of the drug. For example, by pairing a tone stimulus with the administration of nalorphine, an OPIOID ANTAGONIST that precipitates withdrawal signs or the abstinence syndrome (agitation, excessive salivation, and emesis) in morphine-dependent subjects, it was possible to show in rhesus monkeys that the tone acquired the ability to elicit withdrawal responses when presented in the absence of nalorphine (Goldberg & Schuster, 1967; 1970). Striking illustrations of similar conditioned withdrawal responses in HEROIN addicts, as well as CRAVING, in which environmental stimuli trigger the disposition to self-administer the drug, also have been described. These behavioral responses to stimuli that have been previously associated with drug withdrawal or administration often occur after a prolonged period of time spent without drugs (O'Brien, 1976).

In some cases, drugs also acquire stimulus control over behavior in a procedure known as *state-dependent learning.* This procedure is different in some ways from that used to study drugs as discriminative stimuli. State-dependent learning refers to the finding that subjects exposed to a particular procedure when injected with a drug often are impaired upon reexposure to that condition if the drug is not present. Thus, the drug can be viewed as part of the original context in which a response was learned. One concern that stems from the finding that behavior learned during a drug-related condition is impaired in the absence of the drug is that of the potentially enduring problems related to frequent abuse of drugs during adolescence — a period often associated with major developmental and cognitive growth.

REINFORCING EFFECTS OF DRUG-PAIRED STIMULI

Thus far, the focus has been on the effects of environmental stimuli paired with the administration of a drug rather than on stimuli paired with a drug as a reinforcer. As has been frequently demonstrated, and as is true of many stimuli, drugs can have multiple functions. These include *discriminative* effects, which set the occasion for certain responses to occur, and they also include *reinforcing* effects, whereby a response is increased in probability when a reinforcing drug follows the occurrence of that response. Drug self-administration techniques have been very informative and useful in the study of the effects of abused drugs.

One additional experimental procedure that has been used in this field of research is that of repeatedly pairing a rather brief visual or auditory stimulus (e.g., a light or a tone, respectively) with the reinforcing administration of the drug and then using that stimulus also as a reinforcer to maintain behavior without drug administration. Perhaps the most compelling work in this area stems from a procedure in which a stimulus was presented according to a schedule to follow a particular response. On certain occasions, that stimulus also was associated with the administration of a drug—that is, the stimulus occurred at various times without the drug and then also just preceding the drug. Known technically as a "second-order schedule," this technique exerts powerful control over the occurrence and patterning of behavior, and it results in sustained responding for extended time periods in the absence of anything but the stimuli that have been paired with the administration of the drug itself (Katz & Goldberg, 1991). In other words, conditioned stimuli that have been paired with a drug can exert considerable control over behavior.

SUMMARY

To summarize, conditioned drug effects play an important role in the behavior stemming from drug abuse. Stimuli correlated with the administration of a drug, as well as behavior in the presence of that drug, frequently result in those stimuli gaining considerable control over the discriminative effects or reinforcing effects of that drug (or both). Perhaps this is one of the main reasons that drug effects are so compelling and problematic: Not only does the drug itself have powerful effects, but stimuli correlated with the drug also acquire the ability to produce similar effects.

(SEE ALSO: *Addiction: Concepts and Definitions; Causes of Substance Abuse; Memory and Drugs: State Dependent Learning; Research*)

BIBLIOGRAPHY

GOLDBERG, S. R., & SCHUSTER, C. R. (1970). Conditioned nalorphine-induced abstinence changes: Persistence in post morphine-dependent monkeys. *Journal of the Experimental Analysis of Behavior, 14*, 33–46.

GOLDBERG, S. R., & SCHUSTER, C. R. (1967). Conditioned suppression by a stimulus associated with nalorphine in morphine-dependent monkeys. *Journal of the Experimental Analysis of Behavior, 10*, 235–242.

KATZ, J. L., & GOLDBERG, S. R. (1991). Second-order schedules of drug injection: Implications for understanding reinforcing effects of abused drugs. In N. K. Mello (Ed.), *Advances in substance abuse, behavior and biological research* (Vol. 4). London: Jessica Kingsley.

O'BRIEN, C. P. (1976). Experimental analysis of conditioning factors in human narcotic addiction. *Pharmacological Review, 27*, 533–543.

JAMES E. BARRETT

Operant Learning Is Affected by Drugs

According to psychologist B. F. Skinner, behavior that is rewarded or reinforced is more likely to occur again. The family dog soon learns that hanging around the kitchen table brings food. In this example, the food is a *reinforcer* because it increases the likelihood that the dog will spend time near the kitchen table. Thus, the dog's behavior "operates" on the environment to produce an effect. This process is called *operant conditioning*. The techniques of operant conditioning are used widely to establish new behaviors both in humans as well as in animals. Because behavior that is operantly conditioned is very sensitive and reliable, it is often used to examine drug effects.

A TYPICAL OPERANT CONDITIONING EXPERIMENT

In most operant conditioning experiments, an animal is placed in a special chamber which is called a Skinner box after the man that developed operant

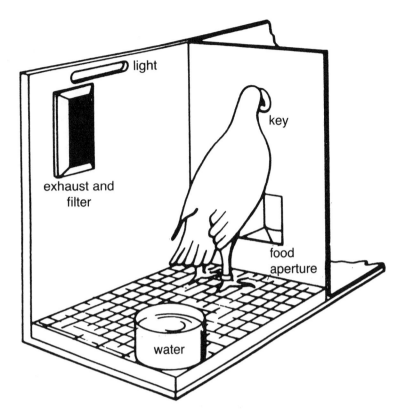

Figure 1
Diagram of an Operant Conditioning Chamber. When the pigeon presses the key, food is delivered. A separate device counts the number of times the pigeon pecks the key.
SOURCE: L. S. Seiden, & L. A. Dykstra (1977).

conditioning. A typical operant chamber, which is shown in Figure 1, has a response key or lever and a place for delivering food. The animal's responses are counted by a computer and also recorded on a roll of paper that shows the distribution of responses over time. Although the experimental chamber in Figure 1 is designed for animals, operant conditioning procedures are also used to examine drug effects in humans. In these studies, the person may sit in a chair and respond by moving a joystick or perhaps sit at a keyboard and respond to stimuli on a computer screen.

SCHEDULES OF FOOD DELIVERY

In most operant conditioning experiments in animals, responses on a lever or key produce food according to some schedule. Behavior maintained by a schedule of reinforcement is called *schedule-controlled behavior*. For example, the pigeon or rat may have to make a specific number of responses in order to receive food. When this occurs, the organism is responding under a *fixed ratio schedule*. A similar schedule is the *variable ratio schedule* in which reinforcement occurs after an unpredictable number of responses. With both the fixed ratio and the variable ratio schedules, animals respond very quickly, in fact, under a fixed ratio schedule that requires thirty responses for food delivery, pigeons may respond as fast as five times a second. Another operant schedule is the *fixed interval schedule* in which the first response that occurs after a specified period of time produces food. With this schedule, rates of responding increase as the time for food delivery approaches. For example, in a fixed interval five-minute schedule, responding is very low during the first two minutes of the interval; responding picks up speed during the third and fourth minutes of the interval and becomes very rapid during the last minute, just before the food is delivered.

By comparing drug effects on different schedules of REINFORCEMENT, scientists have shown that the way in which a drug alters responding depends on the rate of responding produced by a given schedule of reinforcement as well as the amount (or dose) of drug given. Thus, a drug's effects are *rate-dependent* as well as *dose-dependent*. The rate-dependency theory of drug action was first proposed by Peter Dews in the early 1960s and is best exemplified by the effects of amphetamine. Dews noted that amphetamine alters responding differently under a schedule of reinforcement that produces low rates of responding than under a schedule of reinforcement that produces high rates of responding. Specifically, a small amount of AMPHETAMINE increases very low rates of responding, whereas the same amount of amphetamine either decreases or does not change high rates of responding. Other drugs in the amphetamine class such as COCAINE and METHYLPHENIDATE (Ritalin) also alter responding in a rate-dependent manner.

One of the most interesting aspects of the rate-dependency theory of drug action is that it emphasizes the importance of behavioral as well as pharmacological factors in determining a drug's effect on behavior. Thus, the rate at which an animal responds is an important determinant of the way in which a drug alters behavior. It also helps to explain why drugs such as amphetamine and methylphenidate, which generally increase motor activity, might be useful in treating hyperactivity. Since hyperactive children respond at a very high rate, amphetamine would be expected to decrease these high rates of responding.

In contrast to the rate-dependent effects observed for amphetamine-like drugs, the most notable effect of drugs such as MORPHINE is that they decrease rates of responding under several different schedules of reinforcement. The extent to which morphine decreases rate of responding depends on how much morphine is given. Thus, its effects are dose-dependent. Moreover, like all schedule-controlled behavior, these decreases in rate of responding are very consistent and reliable. If a rat is trained to respond under a fixed ratio schedule of food presentation and then given morphine, morphine will decrease rates of responding by about the same amount each time it is given; however, if morphine is given daily for a week or more, its rate-decreasing effects diminish. In other words, TOLERANCE develops. Interestingly, the development of tolerance depends on the behavior examined as well as how much drug is given.

Morphine's effects on responding under schedules of reinforcement is also used as a baseline to investigate the biochemical and physiological events that occur when morphine is given. Opioid antagonists, which block the binding of morphine to opioid receptors, are able to reverse morphine's effects on schedule-controlled behavior. Since morphine's effects on responding are blocked when opioid receptors are blocked, these data suggest that morphine produces its behavioral effects by interacting with opioid receptors. Responding under schedules of reinforcement is also used to examine the biochemical and physiological effects of other drugs. For example, amphetamine's effects on schedules of reinforcement are altered by drugs that interfere with the neurotransmitter dopamine, suggesting that the dopamine system is involved in amphetamine's behavioral effects.

SCHEDULES OF PUNISHMENT

Although schedule-controlled behavior generally is maintained by the delivery of food, in some situations, responding is punished by the presentation of an unpleasant event. In a typical punishment procedure, responding is first maintained by a schedule of food delivery. Brief periods are then added during which responding is both reinforced by food and also punished by an unpleasant event. As a result, responding occurs at a lower rate during periods in which responding is punished than during unpunished periods. Figure 2 shows the design of a typical punishment procedure. First, note in the first panel that responding maintained by food alone occurs at a high rate. In the second panel, responding is punished by the addition of an unpleasant event and, as a result, rate of responding is decreased during the punishment period. The third panel shows that a drug such as alcohol selectively affects responding during the punishment period by restoring rates of responding to their baseline levels. Because these increases in punished responding occur following alcohol as well as a number of other antianxiety agents, but not following drugs such as morphine or amphetamine, increases in punished responding may reflect the antianxiety properties of these drugs. Indeed, the punishment procedure is used by a number of pharmaceutical companies to predict whether a drug might be useful in treating anxiety.

SCHEDULES OF REINFORCEMENT AS A WAY TO MEASURE LEARNING

Schedules of reinforcement are also used to examine the rate at which new behaviors are learned. Clearly, it takes some time to train an animal to respond under a schedule of reinforcement. This period of training is called the acquisition period and provides a measure of learning. One way to design a

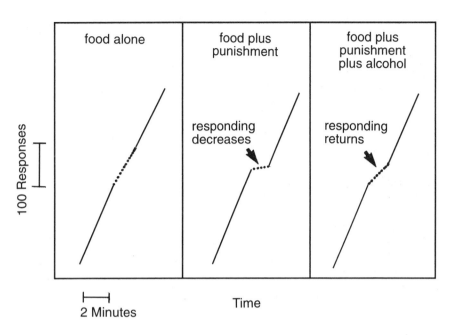

Figure 2
Diagram of a Typical Punishment Procedure. In the first panel, responding is maintained by food alone. The second panel shows responding maintained by food as well as responding during a period in which responding is punished. During this period, responding is decreased. The third panel shows the effects of ethanol on punished responding.

learning experiment is to measure how long it takes a group of rats to learn to respond under a schedule of reinforcement when a drug is given and compare that to how long it takes another group of rats to learn the same task without a drug. In experiments such as these, animals are usually trained to respond under very complicated schedules of reinforcement. Sometimes the animal has to complete the requirements of several different schedules in order to obtain food; in other procedures, the animal responds differently in the presence of different kinds of stimuli. In another procedure, the time it takes an animal to learn a pattern of responses is determined when a drug is given and compared to the time it takes the same animal to learn a different pattern of responses without a drug. ETHANOL, the BARBITURATES, and several antianxiety drugs all increase the number of errors animals make in learning new response sequences. Studies using a similar procedure in humans show that ethanol and certain antianxiety drugs also increase the number of errors people make when they learn new response sequences.

SUMMARY

Schedules of reinforcement offer several advantages for studying the behavioral effects of drugs. First, schedule-controlled responding is very consis-

tent and remains unchanged for long periods of time. This consistency makes it easy to examine changes in behavior after a drug is given. Second, schedule-controlled behavior can be used with human subjects as well as with several different animal species, including mice, rats, pigeons, and monkeys. Finally, schedule-controlled behavior is recorded with automatic devices so that the experimenter is completely removed from the experiment and the nature of the behavior is easy to measure. From these studies, several important concepts have emerged. Scientists have shown that the behavioral effects of drugs depend not only on the amount of drug given, but they also depend on the nature of the behavior being examined. Both the rate of occurrence of a behavior as well as the presence of punishing stimuli are very important determinants of how drugs alter behavior.

The PSYCHOMOTOR STIMULANTS increase responding under schedules of reinforcement when responding occurs at a low rate; when responding occurs at higher rates, the psychomotor stimulants decrease rates of responding. The most notable effect of morphine is that it decreases overall rates of responding. Alcohol and the antianxiety agents are unique in that they increase responding that is suppressed by the presentation of a punishing stimulus. Finally, several drugs interfere with the learning of complex patterns of responding.

(SEE ALSO: *Adjunctive Drug Taking; Behavioral Tolerance; Memory and Drugs: State Dependent Learning; Memory, Effects of Drugs on; Reinforcement; Tolerance and Physical Dependence*)

BIBLIOGRAPHY

CARLTON, P. L. (1983). *A primer of behavioral pharmacology.* New York: W. H. Freeman.

McKIM, W. A. (1986). *Drugs and behavior.* Englewood Cliffs, NJ: Prentice-Hall.

SEIDEN, L. S., & DYKSTRA, L. A. (1977). *Psychopharmacology: A biochemical and behavioral approach.* New York: Van Nostrand Reinhold.

LINDA A. DYKSTRA

RESEARCH AND THE U.S. GOVERNMENT *See* U.S. Government: Agencies Supporting Substance Abuse Research.

REWARD PATHWAYS AND DRUGS

The observation that animals would work in order to receive electrical stimulation to discrete brain areas was first described by Olds and Milner (1954). In this paper, they stated, "It is clear that electrical stimulation in certain parts of the brain, particularly the septal area, produces acquisition and extinction curves which compare favorably with those produced by conventional primary reward." This phenomenon is usually referred to as *brain-stimulation reward* (BSR), *intracranial self-stimulation* (ICSS), or *intracranial stimulation* (ICS).

Most abused substances increase the rate of response (lever pressing) for rewarding ICS, and this has been interpreted as an increase in the reward value of the ICS. Because changes in rate of response could also be a function of the effects of the drug on motor performance, a number of methods have been developed that control for the confounding nonspecific effects of the drugs under study, at least in part. The three most commonly used procedures are phase shifts (Wise et al., 1992), two-level titration (Gardner et al., 1988), and the psychophysical discrete-trial procedure (Kornetsky & Porrino, 1992). Using these threshold methods for determining the sensitivity of an animal to BSR, there is general agreement that most of the commonly abused

substances do in fact increase the sensitivity of animals to the rewarding action of the electrical stimulation and this action is independent of any motor effects of the substances.

PHASE SHIFTS

In this method, rates of response are determined at various intensities of stimulation. Data are usually presented as rate-intensity (rate-frequency) functions. If a drug shifts the rate-intensity function to the left, it is interpreted as an increase in sensitivity of the animal to the rewarding stimulation. A shift to the right is interpreted as a decrease in sensitivity. Threshold (sometimes called *locus of rise*) is defined as the intensity that yields half the maximum rate of response for the animal. If the maximum rate becomes asymptotic at approximately the same stimulus intensity as observed after saline, it is assumed that any phase shift is a direct effect of the drug on the reward value of the stimulation, not the result of a nonspecific motor effect of the drug.

TWO-LEVER TITRATION

In this procedure, rats are placed in a chamber with two levers; pressing one of the levers results in rewarding stimulation, but at the same time the response attenuates the intensity of stimulation by a fixed amount. A response on the second lever resets the intensity to the original level. The threshold is defined as a mean intensity at which the reset response is made.

PSYCHOPHYSICAL DISCRETE TRIAL METHOD

A wheel manipulandum is usually used, although the method has been employed using a response lever. In this method, discrete trials are used, each demanding only a single response by the rat in order to receive the rewarding stimulation. A trial consists of an experimenter-delivered (noncontingent) stimulation. If the animal responds by turning the manipulandum within 7.5 seconds, it receives a second stimulation at the identical stimulation intensity as the first stimulus. Current intensities are varied in a stepwise fashion or descending and ascending order. This yields a response-intensity function, with the threshold defined as the intensity at which the animal responds to 50 percent of the trials. Of the methods currently used, this is the only one that does not make use of the response rate as an integral part of the procedure for the determination of the reward threshold—thus it is independent of the rate of response and the possible confounding motor effects of the drug.

DOPAMINE AND ICS

Although most abused drugs lower the threshold for ICS for some drugs, the findings have not always been consistent, particularly with HALLUCINOGENS and the SEDATIVE-HYPNOTICS, including ALCOHOL (ethanol). For the most part, the threshold-lowering effects caused by the abused substances are compatible with the hypothesis that facilitation of DOPAMINE is involved in their rewarding effects. Drugs that increase dopamine availability at the synapse facilitate ICS, and those that block dopamine transmission decrease ICS (i.e., they raise the threshold—or the amount of current—needed to produce rewarding effects).

SUMMARY

Because abused substances clearly enhance the rewarding value of the intracranial stimulation and not simply cause a general increase in motor behavior, the brain-stimulation-reward model directly allows for the study of the neuronal mechanisms involved in the rewarding effects of abused substances. Although this is not as homologous a model of drug-taking behavior as is the self-administration model, it predicts as well as the self-administration model the ABUSE LIABILITY of compounds, and it readily lends itself to analysis of the mechanisms involved in the rewarding effects of abused substances.

(SEE ALSO: *Research, Animal Model*)

BIBLIOGRAPHY

GARDNER, E. L., ET AL. (1988). Facilitation of brain stimulation reward by Δ^9-tetrahydrocannabinol. *Psychopharmacology, 96,* 142–144.

KORNETSKY, C., & PORRINO, L. J. (1992). Brain mechanisms of drug-induced reinforcement. In C. P. O'Brien & J. H. Jaffe (Eds.), *Addictive States.* New York: Raven Press.

OLDS, J., & MILNER, P. (1954). Positive reinforcement produced by electrical stimulation of septal area and other regions of rat brain. *Journal of Comparative Physiology and Psychology, 47,* 419–427.

WISE, R. A., ET AL. (1992). Self-stimulation and drug reward mechanisms. In P. W. Kalivas & H. H. Samson (Eds.), *The neurobiology of drug and alcohol addiction.* New York: Annals of the New York Academy of Sciences, Vol. 654.

CONAN KORNETSKY

RITALIN   *See* Methylphenidate.

ROCKEFELLER DRUG LAW   The Rockefeller drug law—the best-known MANDATORY SENTENCING statute for drug crimes in the United States, was proposed by New York's Governor Nelson A. Rockefeller in reaction to a HEROIN epidemic in his state. The law, which took effect on September 1, 1973, required that judges impose lengthy prison sentences on drug traffickers, and it forbade prosecutors from dismissing charges covered by the law. The goal was to publicize the situation during the legislative process in an effort to deter people from both drug use and trafficking by imposing tough and certain punishments. The law did not have the desired effect and was repealed as of July 1, 1976.

A major evaluation concluded that neither drug use nor drug trafficking was reduced after the law was passed. The likelihood that a defendant, once arrested, would be incarcerated did not increase—although the likelihood that a defendant, once convicted, would be imprisoned did increase (Joint Committee, 1977).

Because many lawyers and judges believed that the penalties were too severe in many cases, case dismissals at every stage before conviction increased steadily during the years the law was in effect. For example, in 1972—before the law took effect—39 percent of arrests resulted in indictments (formal felony charges). In 1976, only 25 percent of arrests resulted in indictments. Of cases indicted in 1972, convictions resulted from 87 percent but in 1979, only 79 percent did.

The processing of cases became much more expensive for New York State. For every crime affected by the law, the percentage of defendants pleading guilty fell, and the proportion of trials increased. The evaluators concluded that it "took between ten and 15 times as much court time to dispose of a case by trial as by plea." The average time to handle a drug prosecution in New York City, for example, doubled, rising from 172 days in 1973 to 351 days in 1976.

Because the law accomplished so few of its goals, cost a great deal, and otherwise created problems for the courts, it was repealed as of July 1, 1976. The findings of the evaluation are consistent with experience over many decades with mandatory sentencing laws, but such experience did not dissuade the U.S. Congress and every state—including New York—from enacting mandatory minimum-sentence laws for drug crimes during the 1980s.

(SEE ALSO: *Drug Laws, Prosecution of; Opioids and Opioid Control, History*)

BIBLIOGRAPHY

JOINT COMMITTEE ON NEW YORK DRUG LAW EVALUATION. (1977). *The nation's toughest drug law: Evaluating the New York experience.* Washington, DC: U.S. Government Printing Office.

MICHAEL TONRY

ROLLESTON REPORT OF 1926 (U.K.)
The Rolleston Report of 1926 helped to establish British policy toward OPIATES, COCAINE, and other drugs. It institutionalized a drug policy in which medical expertise and public-health considerations were given importance along with punishment and criminal penalties. The British policies were, in this sense, different from U.S. policies toward drugs that emerged during the same period and in response to similar international agreements. The historical background leading to the formation of an elite committee of British physicians, chaired by Sir Humphrey Rolleston, had four major phases.

ENDING THE COMMERCIAL OPIUM TRADE

During the nineteenth century, the British established commercial opium trading by fighting and winning two Opium Wars with China: Opium grown and sold by monopoly in British-dominated India provided a quarter of the revenue for the British gov-

ernment in India. Prepared opium (for smoking) was exported to Chinese ports by the East India Company, where British authorities collected tax revenues on it for the Chinese government. Missionaries in China and their anti-opium allies in Britain, the United States, and Canada lobbied strongly against profiting from the British-sponsored vice. They also educated the public about opium smoking and commercial opium trading.

The U.S. government stimulated the convening of several international conferences from 1909 to 1914. These conferences reached agreements that all signatory governments would enact legislation ending commercial opium trading and restricting opium and cocaine to "legitimate medical practice." The Indo-Chinese opium trade ended in 1914. These international conventions were included in the Versailles Treaty that ended World War I. "Legitimate medical practice" and appropriate controls and/or penalties were not specified in the international treaties.

OPIUM CONTROLS AND GROWTH OF THE MEDICAL PROFESSION

During the nineteenth century, opiates were the only effective way to relieve the symptoms of many physical ailments (most medicines used today, including aspirin, became available only in the twentieth century). OPIUM and its derivative MORPHINE (Britain was the world's leading manufacturer) were available in patent medicines, in alcoholic solutions, and in other commercial products. The emerging professions of pharmacist and medical physician with advanced training and specialized knowledge were anxious to differentiate themselves from a motley group of healers—chemists, herbalists, barber-dentists, patent-medicine sellers, and others. In the 1850s, such persons could provide opiates to patients since they were not then illegal, and preparations containing opiates provided substantial revenues. Opium eating and LAUDANUM (an alcoholic solution of opiates) consumption were then widespread in Britain.

British pharmacists became eager to restrict sales of opiates to qualified sellers—but only in such a way that "professional" trade would not be harmed and could be expanded. The 1868 Poisons Act restricted opiate sales to pharmacists. This act mandated the labeling of opiates and required pharmacists to keep records of purchasers. (Similar restrictions on opiate sales in the United States did not occur until the 1906 Food and Drug Act.) Pharmacists, however, could continue to sell opiates directly to customers without a prescription from a physician, and physicians could prescribe or sell opiates to patients. In the early 1880s physicians and researchers in Europe, England, and the United States almost simultaneously began to write about the opium habit and morbid cravings for opiate drugs. In 1884 physicians in England founded the Society for the Study of Inebriety, which promoted a disease model of addiction and the need for treatment.

By 1900, physicians emerged as an elite group who defined all aspects of health care and medical practice in British society; pharmacists "policed" the Poisons Act and effectively retained control of dispensing opiates and other drugs. Thus, by 1914, British pharmacists and physicians had almost a half century of experience, professional collaboration, an ongoing professional association concerned with the dispensing of opiates, and attempts to contain opiate consumption and habitual use.

PRESSURE TOWARD CRIMINAL PENALTIES

In 1914, when the international opium convention (Hague Convention) was to go into effect, several British agencies could not decide which one should take responsibility for implementing legislation and regulation of drugs. Then World War I began in August of 1914 and Sir Malcolm Delevingne, an undersecretary at the Home Office, took primary responsibility. He suggested using the War Powers Act to stop sales of cocaine and opiates to soldiers unless they were based on a prescription by a doctor that was "not to be repeated" (refilled without further prescription). Violators, however, could be fined only five pounds. Two or three cases were publicized and introduced the British public to "dope fiend" fears, but they continued to be rare.

After World War I, Delevingne argued that drug control was a police responsibility for the Home Office (where it has remained ever since). The 1920 Dangerous Drug Act was vague about two critical issues—whether doctors/pharmacists could prescribe for themselves, and whether doctors could "maintain" addicts. In 1921 and 1924, the Home Office proposed regulations that ignored the rights of professionals and imposed many complex proce-

dures. It also sought powers of search and seizure, higher fines, and longer sentences for convictions. Thus, the Home Office was making regulations that would subject doctors to criminal sanctions and circumscribe their prescribing practices—as was already happening in the United States.

APPOINTMENT OF THE ROLLESTON COMMITTEE

The Home Office needed the cooperation of the medical profession to determine the appropriateness of maintenance dosages for addicts, and it sought to determine whether gradual reduction was the appropriate treatment for addiction. The Home Office and the medical profession each recognized the legitimacy of the other's position. Both realized that a partnership was needed. Thus, these two elite groups began a collaboration to define and resolve problems and appropriate practices regarding narcotics control. All persons appointed to the committee were medical personnel representing government agencies or nongovernment physician-interest groups. The chairman, Sir Humphrey Rolleston, was president of the Royal College of Physicians and a noted exponent of the disease view of ALCOHOLISM. Another member had written the authoritative article on narcotic addiction in 1906. Police and law enforcement officials without medical training were not represented.

Committee Deliberations and Recommendations. The committee was to

> consider and advise as to the circumstances, if any, in which the supply of morphine and heroin (including preparations containing morphine and heroin) to persons suffering from addiction to these drugs, may be regarded as medically advisable and as to the precautions which it is desirable that medical practitioners administering or prescribing morphine or heroin should adopt for the avoidance of abuse, and to suggest any administrative measures that seem expedient for securing observance of these precautions.

During a year and a half of deliberations and twenty-three meetings, the committee heard evidence from thirty-four witnesses. The Home Office submitted a memorandum that structured the questions and inquiry. Witnesses represented a wide diversity of opinion, particularly regarding appropriate treatment for addicts. Prison doctors favored harsher treatment, especially abrupt withdrawal of opiates (going cold turkey). Even consultants specializing in treatment rarely agreed on points of procedure and treatment. Most witnesses and commission members accepted the disease nature of addiction.

There was wide agreement, however, that addiction to HEROIN or morphine (both opiates) was a rare phenomenon and a minor problem in BRITAIN. Most addicts were middle class and many were members of the medical profession. Relatively few criminal or lower-class addicts were then known, so criminal sanctions appeared unneeded and inappropriate. The committee report concluded that "the condition must be regarded as a manifestation of disease and not as a mere form of vicious indulgence."

From this conclusion, many recommendations followed. The most important was that some addicts might need continued administration of morphine (or other opiates) "for relief of morbid conditions intimately associated with the addiction." Thus, the committee effectively supported maintenance of an addict for long periods of time—possibly for life.

The committee also made several recommendations for administrative procedures to lessen the severity of the drug problem. Practitioners were mandated to notify the Home Office when they determined someone was addicted; but physicians could continue to provide treatment and prescribe opiates to addicts. Gradual reduction rather than abrupt withdrawal was the recommended treatment, in part to keep addicts in treatment rather than to drive them to illicit suppliers. A medical tribunal was established to promote the profession's own policing of members who became addicted. The committee also opposed banning heroin (which was a useful medication and a very small problem in Britain at the time).

LEGACY OF THE REPORT

Shortly after the Rolleston Report was completed, its recommendations were included in amendments to the Dangerous Drug Act (1926). Although this act has been amended numerous times since then, the provisions adopted from the Rolleston Report remain in effect in the 1990s. Although cocaine was included as a narcotic in this report, separate recommendations for treatment were not made. *Cannabis* (MARIJUANA) was not included in this report.

The Rolleston Report did not address the issue of illegal sales or transfers of opiates; no criminal or penal sanctions were recommended.

The *British Medical Journal* was content: The medical view of addiction as a disease needing treatment, and not a vice necessitating punishment and penal sanction, had been formally accepted as government policy. Medical professionals, rather than criminal-justice personnel, would be responsible for individual decisions about whether patients were addicts, and prescribe appropriate quantities of opiates, including on a maintenance basis. Any questions about appropriate prescribing practices and physician addiction would be handled by a committee specializing in addiction. As a result, almost no British physician has been arrested and/or tried for opiate-related violations.

The foundations of what is sometimes called the British system of drug policy had been established. From 1926 to 1960, this system worked well. Names of fewer than 1,000 addicts were forwarded to the Home Office each year, most of them medical personnel. Local practitioners could and did prescribe heroin and other opiates to their patients, including registered addicts. Some addicted patients were maintained on heroin, occasionally for years. They received their drugs from a local pharmacy. Addicts were also provided with clean needles and syringes. Drug treatment consisted almost entirely of individual physicians counseling addicted patients and providing drugs. Almost no illicit sales of opiates or cocaine occurred during these years. One staff member at the Home Office was responsible for all registrations and personally knew most of the addicts in Britain; he frequently helped addicts find doctors and/or assistance. The Home Office also covened meetings with addiction specialists to address any policy issues that arose. Thus, the British established what might be described as a system of drug control that gave due weight to medical values and public-health considerations. Most observers now agree, however, that the "system" worked because the problem was limited in size rather than that the problem was small because of the system. It worked well for half a century until the numbers of addicts increased substantially, because of drug dealing on an international scale, the widespread use of drugs during the 1960s–1980 Countercultural revolution, and the increased immigration to Britain of former colonial citizens of the crumbling empire. By the 1960s, the upsurge in heroin use and the abuse of cocaine, marijuana, and other drugs left Britain with a drug problem of both licit and illicit substances that outstripped even the British system's handling capabilities.

BIBLIOGRAPHY

BERRIDGE, V. (1980). The making of the Rolleston Report 1908–1926. *Journal of Drug Issues, Winter,* 7–28.

BRUCE D. JOHNSON

(SEE ALSO: *Britain, Drug Use in; British System of Drug-Addiction Treatment; Heroin: The British System; International Drug Supply Systems; Opioids and Opioid Control: History; Policy Alternatives: Prohibition of Drugs Pro and Con; Sweden, Drug Use in*)

RUBBING ALCOHOL   Rubbing alcohol is known as isopropyl alcohol (C_3H_8O); it is one of the more useful of the commercial alcohols, included in hand lotions and many cosmetic items as well as in antifreeze or deicer products. A 70 percent solution has more germicidal properties than does ETHANOL (drinking alcohol), so it is used in many health-care situations, both in households and in medical facilities. It is also used for massages and by athletic trainers to treat skin and muscle groups, hence the term *rubbing.* It has a drying effect on the skin and causes blood vessels to dilate; its distinctive odor is associated with doctor's offices, since it is used to clean the skin being prepared for an injection.

When rubbing alcohol is ingested either pure or added to beverages, the result is toxic—with symptoms lasting longer than those seen after drinking ethanol (alcoholic beverages), because isopropyl alcohol is slowly metabolized to acetone, another toxic substance.

(SEE ALSO: *Inhalants*)

BIBLIOGRAPHY

CSAKY, T. Z., & BARNES, B. A. (1984). *Cutting's handbook of pharmacology,* 7th ed. Norwalk, CT: Appleton-Century-Crofts.

S. E. LUKAS

RUSH   *See* Slang and Jargon.

RUTGERS CENTER OF ALCOHOL STUDIES   For all the years of its existence, the Rutgers Center of Alcohol Studies (initially founded in 1940 as the Yale Center in New Haven, Connecticut) has been centrally involved in generating significant research findings on alcohol, alcoholics, and alcoholism. Through those same years, the center's mission has also included education, service, and information dissemination to the university community of which it was a part, the nation, and the world.

The Center of Alcohol Studies was founded at Yale University by Professor E. M. JELLINEK; it was developed from the well-known Yale Laboratory of Applied Physiology, directed by Professor Howard W. Haggard, which first began to study the physiology of alcohol (ethanol) in the 1930s. In recognition of the paucity of scientific journals publishing work on alcohol and alcoholism then, faculty at the center founded *The Quarterly Journal of Studies on Alcohol* in 1940. The journal's first issue was edited by Professor Haggard; shortly thereafter, Mark Keller, longtime editor of the *Quarterly Journal,* became the journal's managing editor. Keller, who served as editor of what is now the *Journal of Studies on Alcohol* for more than 30 years as a faculty member of the Center of Alcohol Studies at both Yale and Rutgers, also became a very substantial figure in the alcohol field by virtue both of his position as longtime editor of the leading journal in the field and his many carefully wrought, penetrating, insightful talks and articles on a wide range of alcohol-related subjects.

Recognizing the absence at the time of methods and agencies for the dissemination of the practical results of research on and experience with alcohol problems, the faculty of the center founded the Summer School of Alcohol Studies (SSAS) in 1943. It was then and continues today to be oriented toward meeting the needs of persons who work directly with the problems of alcohol use and alcoholism. During its existence, the SSAS has attracted more than 14,000 students from around the world for its one- and three-week residential summer programs.

The pressing informational needs of the infant field of alcohol studies led to the development of the library of the Center of Alcohol Studies, which now possesses the most complete special collections on alcohol and alcoholism in the world, along with a complete collection of journals and books on alcohol and related subjects. The Classified Abstract Archive of the Alcohol Literature contains lengthy abstracts of scientific work from a wide range of disciplines cross-indexed in depth up to 1976; the McCarthy Collection of original scientific papers; the Ralph G. Connor Collection of Alcohol-Related Research Instruments; and several extensive, continuously updated bibliographic series.

Faculty at the Yale Center of Alcohol Studies initiated the first research program on treatment, as well as the Yale Plan on Alcoholism for Industry—a forerunner of modern EMPLOYEE ASSISTANCE PROGRAMS (EAPs). Center faculty also founded the first State Commission on Alcoholism. The research faculty at the center has continued to grow so that by the mid-1990s it comprises a substantial number of biochemists and physiologists, sociologists and psychologists, epidemiologists and preventionists—all engaged in studying an array of topics from etiology and physiology to prevention and treatment, with relevance to alcohol, alcoholics, and alcoholism.

In 1962, the Center of Alcohol Studies moved to Rutgers University, New Brunswick, New Jersey, into a building funded in part by a generous gift from R. Brinkley Smithers. From that time until he retired from Rutgers in 1975, Professor Seldon Bacon headed the Center of Alcohol Studies. A distinguished sociologist who had joined the center's faculty shortly after it was founded at Yale, Bacon played a key role for several decades in many of the most important developments in alcoholism nationally. At the Center of Alcohol Studies, he was instrumental in expanding the Yale Plan, developing the Summer School of Alcohol Studies, and nurturing the social-science research base that continues to be one of the center's major contributions.

In 1985, Mr. Smithers gave the center another extremely generous gift, permitting it to add to its building as well as to establish a prevention center and an annual prevention symposium.

(SEE ALSO: *Addiction Research Foundation of Ontario (Canada); U.S. Government Agencies*)

BIBLIOGRAPHY

MENDELSON, J. H. & MELLO N. K. (1989). Studies of alcohol: Past, present, and future. *Journal of Studies on Alcohol, 50,* 293–296.

NATHAN P. E. (1989). The Center of Alcohol Studies and the *Journal of Studies on Alcohol:* Celebrating 50 years. *Journal of Studies on Alcohol, 50,* 297–300.

NATHAN, P. E. (1987). Reports from the research centres—Rutgers: The Center of Alcohol Studies. *Journal of Addiction, 82,* 833–840.

STRAUS, ROBERT. (1993). *In memoriam:* Selden D. Bacon. *Journal of Studies on Alcohol, 54,* 130–132.

PETER E. NATHAN

S

SADD   *See* Students Against Driving Drunk.

SAFE USE OF DRUGS   *See* Prohibition.

SAMHSA   *See* U.S. Government Agencies.

SAODAP   *See* U.S. Government Agencies.

SCHIZOPHRENIA   Schizophrenia is a psychiatric illness that can be profoundly disabling and is usually chronic in nature. The cause is not known, but there appears to be a genetic predisposition. The etiology has been conceptualized in a stress/diathesis (vulnerability) model: Biological and environmental factors (e.g., drug abuse, psychosocial stresses) interact with a genetic VULNERABILITY to precipitate the illness. Several theories have been proposed to explain the observed biological abnormalities of the disorder, including overactivity of the DOPAMINE NEUROTRANSMITTER systems in the central nervous system, changes in brain structure (e.g., enlargement of the lateral cerebral ventricles) and brain function (e.g., decreased frontal lobe function [hypofrontality], as evidenced by diminished blood flow, and deficits in attention and sensory filtering). Psychological and social factors are considered important in the expression and course of the disorder.

It is likely that schizophrenia constitutes a group of disorders rather than a single entity; these disorders present with similar clinical signs and symptoms, but the etiologies, treatment responsiveness, and course of illness in each vary.

Detailed descriptions of the illness date back to the nineteenth century. Emil Kraepelin (1856–1926) used the term *dementia praecox* to describe psychiatric states with an early onset and deteriorating course. Eugen Bleuler (1857–1939) coined the term *schizophrenia* for a "splitting of the mind," in his belief that the illness was a result of the disharmony of psychological functions. The diagnosis of schizophrenia requires observation and clinical interviewing. No sign or symptom is specific for the illness, nor do any laboratory tests exist to establish the diagnosis. The DIAGNOSTIC AND STATISTICAL MANUAL *for Mental Disorders-3rd edition* contains the diagnostic guidelines of the American Psychiatric Association for schizophrenia. These include the presence of characteristic psychotic symptoms (delusions, HALLUCINATIONS, a thought disorder, inappropriate emotion); impaired work, social functioning, and self-care; and continuous signs of the illness for at least six months. The symptoms of an affected individual can change with time; therefore longitudinal follow-up is important. Furthermore, it is important to take into account the educational level, intellectual ability, and cultural affiliation of the individual when making a diagnosis. The onset of illness is usually in late adolescence or early adulthood and is generally insidious. The typical course of schizophrenia is

963

characterized by exacerbations and remissions. A gradual deterioration in functioning generally occurs that eventually reaches a plateau. However, a small proportion of persons may recover. It is estimated that 20 percent to 30 percent of affected individuals can lead somewhat normal lives whereas another 20 to 30 percent continue to experience moderate symptoms.

The prevalence rates of schizophrenia vary worldwide, but in the United States the lifetime prevalence is estimated to be 1 percent. In industrialized countries, there is a disproportionate number of schizophrenic patients in the lower socioeconomic classes.

The management of affected individuals involves hospitalization when there is an exacerbation of the illness, plus the use of medication. The mainstay of pharmacologic treatment is the class of drugs known as ANTIPSYCHOTICS. Many antipsychotics are available and they act to control the psychotic symptoms; most of them do so by blocking the actions of the neurotransmitter dopamine. After a person has recovered from an acute episode of schizophrenia, the emphasis is on practical aspects of management: living arrangements, self-care, employment, and social relationships. Education of and support made available to family members are important and can have an impact on relapse rates in the patient. Many schizophrenic patients have to remain on antipsychotic medication for prolonged periods, since the rate of relapse is high after drug discontinuation. Side effects, primarily of a neurologic nature (e.g., tardive dyskinesia), are a source of concern, but in most cases the benefits of symptom control outweigh the risks of pharmacotherapy. Making sure that the patient complies with medication use is often a problem.

(SEE ALSO: *Amphetamine; Cannabis sativa; Comorbidity and Vulnerability; Complications: Mental Disorders*)

BIBLIOGRAPHY

ANDREASEN, N. C. (1986). Schizophrenia. In A. J. Frances & R. E. Hales (Eds.), *Psychiatry update — The American Psychiatric Association annual review* (Vol. 5). Washington, DC: American Psychiatric Press.

KARNO, M., ET AL. (1989). Schizophrenia. In H. I. Kaplan & B. J. Sadock (Eds.), *Comprehensive textbook of psychiatry* (5th ed., Vol. 1). Baltimore, MD: Williams & Wilkins.

MYROSLAVA ROMACH
KAREN PARKER

SCHOOLS AND DRUGS *See* Education and Prevention.

SCID *See* Structured Clinical Interview.

SCOPOLAMINE AND ATROPINE Scopolamine (*d*-hyoscine) and atropine (*dl*-hyoscyamine) is a tropane alkaloid found in the leaves and seeds of several plant species of the family Solanaceae, including deadly nightshade (*Atropa belladonna*) and henbane (*Hyoscyamus niger*). Atropine, a major alkaloid in deadly nightshade, is also found in JIMSONWEED (*Datura stramonium*). In Europe, in centuries past, henbane was a component of so-called witches' brews or was applied as an ointment to mucous membranes. According to some folktales, the idea that witches fly on broomsticks was derived from the sensation of a flying experience after the use of such ointments.

Scopolamine and atropine have very similar actions. They act as competitive antagonists at both peripheral and central muscarinic cholinergic receptors. Scopolamine is still sometimes used clinically for the treatment of motion sickness. The compound also causes central nervous system depression, leading to drowsiness, amnesia, and fatigue. It also has some euphoric effects and abuse liability, but these are not considered to be of such magnitude to require control of the drug under the Controlled Substances Act. Atropine has fewer actions on the central nervous system than scopolamine. It is used to reduce actions at peripheral cholinergic structures—it produces decreased gastric and intestinal secretions as well as spasms and also results in pupillary dilation. It blocks the action of the vagus nerve that results in slowing of the heart. It is often used before operations to prevent unwanted reflex slowing of the heart beat.

High doses of either of these tropane alkaloids can cause confusion and delirium accompanied by decreased sweating, dry mouth, and dilated pupils.

BIBLIOGRAPHY

BROWN, J. H. (1990). Atropine, scopolamine, and related antimuscarinic drugs. In A. G. Gilman et al. (Eds.), *Goodman and Gilman's the pharmacological basis of therapeutics*, 8th ed. New York: Pergamon.

HOUGHTON, P. J., & BISSET, N. G. (1985). Drugs of ethno-origin. In D. C. Howell (Ed.), *Drugs in central nervous system disorders*. New York: Marcel Dekker.

ROBERT ZACZEK

SECOBARBITAL

Secobarbital, prescribed and sold as Seconal, is a short-acting BARBITURATE used principally as a SEDATIVE-HYPNOTIC drug but occasionally as a preanesthetic agent. It is a nonspecific central nervous system (CNS) depressant and greatly impairs the mental and/or physical abilities necessary for the safe operation of automobiles and complex machinery.

Before the introduction of the BENZODIAZEPINES, it was the drug most commonly used to treat insomnia. Prolonged or inappropriate use of secobarbital can produce TOLERANCE AND PHYSICAL DEPENDENCE. If high doses have been used, abrupt cessation can result in severe WITHDRAWAL symptoms that include convulsions. Secobarbital is more likely to be abused than benzodiazepines and appears to produce greater euphoria in certain individuals than would a comparable sedative dose of a benzodiazepine. Consequently, it is classified as a Schedule II class drug in the CONTROLLED SUBSTANCES ACT, which indicates that although it is acceptable for clinical use, it is considered to have a high abuse potential. As with other barbiturates, it should never be combined with another CNS depressant because respiratory depression can occur.

(SEE ALSO: *Abuse Liability of Drugs: Testing in Humans; Drug Interaction and the Brain; Drug Interactions and Alcohol*)

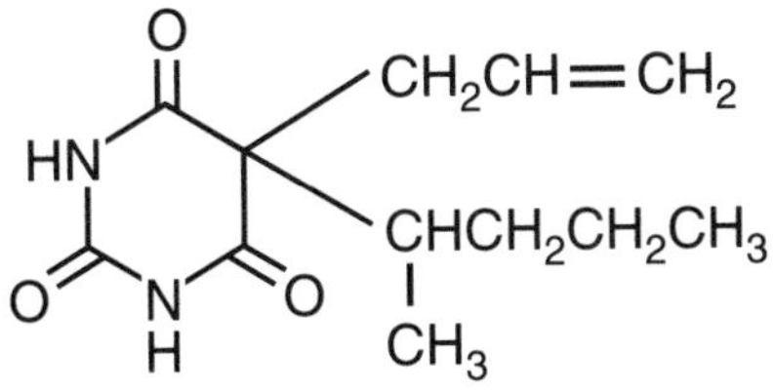

Figure 1
Secobarbital

BIBLIOGRAPHY

HARVEY, S. C. (1980). Hypnotics and sedatives. In A. G. Gilman et al. (Eds.), *Goodman and Gilman's the pharmacological basis of therapeutics*, 6th ed. New York: Macmillan.

SCOTT E. LUKAS

SECONAL   *See Secobarbital.*

SECOND GENESIS, INC.

Second Genesis, Inc., is a long-term, residential rehabilitation program for adults and teenagers with drug and/or alcohol problems. Founded in Virginia in 1969, under the direction of Dr. Sidney Shankman, Second Genesis is a nonprofit organization operating six residential THERAPEUTIC COMMUNITIES that serve Maryland, Virginia, and Washington, DC.

The Second Genesis program is often described as a school that educates people who have never learned how to feel worthy without hurting themselves and others. Through highly structured, thoroughly supervised, all-day, chemical-free treatment, Second Genesis combines the basic values of love, honesty, and responsibility with work, education, and intense group pressure to help correct the problems that prevent people from living by these values. Discovering self-respect in a supportive family-like setting, residents replace behavioral deficits and substance abuse with positive alternatives.

At Second Genesis, residents learn fundamental LIFE SKILLS that enable them to reenter their communities and live productively. An independent study by the Northern Virginia Planning District Commission Substance Abuse Advisory Council shows that the longer residents remain in treatment, the more improvement they show in education and employment status, social functioning, physical and mental health and the reduction of substance abuse and criminal activity. Five-year follow-up studies done in-house reveal that the majority of Second Genesis graduates have remained drug free.

Second Genesis has served as a drug-policy consultant to various state governments, the U.S. Senate, the U.S. House of Representatives, the White House, and a number of foreign governments and private organizations. The effectiveness of Second

Genesis has been the subject of special reports on both U.S. and British broadcasting networks—NBC, ABC, CBS, and the BBC World Service.

The program is staffed by a multidisciplinary professional team headed by Executive Director Sidney Shankman, M.D.—both an adult and a child and adolescent psychiatrist—and Deputy Executive Director Alan M. Rochlin, Ph.D.—a clinical psychologist. The staff includes mental health and vocational rehabilitation counselors (with graduate-school degrees), learning disabilities specialists, family therapists, research analysts, health education specialists, and facility counselors. Many have completed work in treatment communities like Second Genesis and have gone on for further training as mental health professionals. This diversity of specialists permits the selective application of group therapy, individual counseling, family therapy, and educational and vocational services.

The program generally accepts applicants eighteen years of age or older, male and female, without regard to race, ethnic group, or socioeconomic status.

Second Genesis estimates its 1993 per diem costs as $45/day. It is funded primarily through contracts with federal, state, and local governments for individuals unable to pay for treatment. Supplementary funding and in-kind contributions are provided by individuals, organizations, and corporate support.

(SEE ALSO: *Gateway Foundation; Marathon House; Treatment Types*)

SIDNEY SHANKMAN

SECONDHAND SMOKE *See* Tobacco Complications.

SECULAR ORGANIZATIONS FOR SOBRIETY (SOS) Secular Organizations for Sobriety (SOS National Clearinghouse, P.O. Box 5, Buffalo, New York 14215) is a self-help organization for alcohol and drug users, founded as an alternative to ALCOHOLICS ANONYMOUS (AA) and other groups based on AA. It was intended to offer help to people who are uncomfortable with the emphasis on spirituality that is a central tenet of the AA TWELVE-STEP PROGRAMS. Founded by James Christopher, SOS began with a 1985 article, "Sobriety without Supersti-

tion," describing Christopher's own path to sobriety. SOS claimed in 1991 to have an international membership of 20,000, making it the largest of the alternative groups. In 1987, it was recognized by the State of California as an alternative to AA in sentencing offenders to mandatory participation in drug rehabilitation. Members of SOS are not necessarily nonreligious; however, many do not believe in an intervening higher power who takes responsibility for their individual problems.

Unlike AA—which emphasizes that the individual is powerless over alcoholism and must look to a "higher power" for help in achieving and maintaining sobriety—SOS and other alternative organizations assert the capacity of individuals to control their own behavior. SOS stresses total abstinence, personal responsibility, and self-reliance as the means to achieve and maintain sobriety (recovery), but the organization recognizes the importance of participating in a mutually supportive group as an adjunct to recovery. Members learn that open and honest communication aids in making the appropriate life choices that are essential to recovery. SOS shares with other self-help groups the importance of anonymity and the abstention from all drugs and alcohol.

SOS consists of a nonprofit network of autonomous nonprofessional local groups dedicated solely to helping individuals with alcohol and other drug addictions. It encourages and is supportive of continued scientific inquiry into the understanding of alcoholism and drug addiction.

Among other self-help organizations that see themselves as alternatives to AA are RATIONAL RECOVERY (RR) and WOMEN FOR SOBRIETY (WFS).

(SEE ALSO: *Coerced Treatment; Disease Concept of Alcoholism and Drug Abuse; Treatment Types*)

JEROME H. JAFFE

SEDATIVE Sedative is a general term used to describe a number of drugs that decrease activity, moderate excitement, and have a calming effect. The primary use for these drugs is to reduce ANXIETY, but higher doses will usually cause sleep (a drug used primarily to cause sleep is called a *hypnotic*). Although the term *sedative* is still used, the drugs usually prescribed to produce this calming effect are BENZODIAZEPINES, which are more commonly known as antianxiety agents, or minor tranquilizers.

(SEE ALSO: *Barbiturates; Drug Types; Sedative-Hypnotic*)

BIBLIOGRAPHY

HARVEY, S. C. (1985). Hypnotics and sedatives. In A. G. Gilman et al. (Eds.), *Goodman and Gilman's the pharmacological basis of therapeutics*, 7th ed. New York: Macmillan.

SCOTT E. LUKAS

SEDATIVE-HYPNOTIC

Sedative-hypnotic drugs are used to reduce motor activity and promote relaxation, drowsiness, and sleep. The term is hyphenated because, by adjusting the dose, the same group of drugs can be used to produce mild sedation (calming, relaxation) or sleepiness. Thus, the distinction between a sedative and a hypnotic (sleeping pill) is often a matter of dose—lower doses act as sedatives and higher doses promote SLEEP.

In some people, sedative-hypnotics can produce a paradoxical state of excitement and confusion. Also, some have the potential to be abused. Very high doses of most sedative-hypnotic drugs will produce general anesthesia and can depress respiration so much that breathing must be maintained artificially or death will occur. The BENZODIAZEPINES are an exception to this in that higher doses typically produce sleep and are far less likely to severely depress respiration.

One of the first agents to be added to the list of the classic sedatives (ALCOHOL and OPIATES) was bromide, introduced in 1857 as a treatment of epilepsy. CHLORAL HYDRATE was introduced in 1869, and paraldehyde was first used in 1882. The BARBITURATES were introduced in the early 1900s and remained the dominant drugs for inducing sleep and sedation until the benzodiazepines were developed in the late 1950s and early 1960s. A number of miscellaneous nonbarbiturate sedatives (ETHCHLORVYNOL, GLUTETHIMIDE, carbromal, methylparafynol, methyprylon, METHAQUALONE) were introduced in the 1940s and 1950s and for a brief period rivaled the barbiturates in popularity, but their use declined rapidly along with the use of barbiturates. The bromides were recognized to have toxic properties, but they were still in use until the mid-twentieth century; chloral hydrate and paraldehyde were used well into the late 1970s and are still used in some places.

(SEE ALSO: *Abuse Liability of Drugs; Drug Interactions and Alcohol; Drug Types; Suicide and Substance Abuse*)

BIBLIOGRAPHY

HARVEY, S. C. (1980). Hypnotics and sedatives. In A. G. Gilman et al. (Eds.), *Goodman and Gilman's the pharmacological basis of therapeutics*, 6th ed. New York: Macmillan.

SCOTT E. LUKAS

SEDATIVES: ADVERSE CONSEQUENCES OF CHRONIC USE

Sedative drugs are also called hypnotics or SEDATIVE-HYPNOTICS. Technically, a *sedative* decreases activity and calms (tranquilizers); a *hypnotic* produces drowsiness, allowing for the onset and maintenance of a state of SLEEP similar to the natural and from which the sleeper may be easily awakened. The same drug used for sedation, for pharmacologically induced sleep, and for general systemic anesthesia may be seen to induce a continuum of central nervous system (CNS) depression. Such drugs are usually referred to, therefore, as sedative-hypnotics, and they are widely prescribed in the treatment of insomnia (sleep problems). Although some people take these drugs only occasionally and for specific sleep problems (as on long-distance flights), many more take them over months and even years as a presumed aid to nightly sleep. This occurs despite medical advice to restrict such prescriptions to periods of about two weeks.

BENZODIAZEPINES remain by far the most frequently used sedative-hypnotic drugs, although there are some new compounds with differing modes of action. The key concerns in the hypnotic use of the benzodiazepines are (1) adverse effects experienced while the patient is taking the drug, (2) possible physical and psychological DEPENDENCE on the drug, and (3) rebound insomnia and WITHDRAWAL symptoms when the patient stops taking the drug.

INSOMNIA

In laboratory studies, sleep patterns can be objectively recorded to give a clear measure of sleep and wakefulness. In clinical practice however, insomnia refers to a subjective complaint of poor quality, lack of, or disrupted sleep; it is certainly on the basis of the patient's own evaluation of sleep that most sedative-hypnotic drugs are prescribed.

A useful distinction can be drawn between primary and secondary insomnia. Primary insomnia is not related to other medical or psychiatric conditions. Secondary insomnia is associated with either physical ill health (e.g., pain, heart disease, asthma) or mental ill health (e.g., DEPRESSION).

Sleep complaints are of three main types (1) problems of getting to sleep (sleep initiation), (2) problems of staying asleep (sleep maintenance), and (3) early-morning wakening. Sleep-onset problems vary little with age; early-morning wakening is often secondary to depression; and sleep-maintenance problems show a clear and marked increase with aging. So whereas approximately 10 percent of young adults complain of serious sleep problems, this increases to 30 to 50 percent of those aged seventy or older (Morgan, 1990).

USE OF SEDATIVE-HYPNOTICS

This age-related pattern for complaints of insomnia is reflected in the pattern of use of sedative-hypnotic drugs. For example, in the United States 2.6 percent and in Britain 4 percent of adults take a benzodiazepine as a sleep inducer during a year (Mellinger, Balter, & Uhlenhuth, 1985; Dunbar et al., 1989). In the elderly, this increases to 16 percent use in a year, with 73 percent of those taking the drug regu-

larly for a year or more. Indeed, 4 percent of people older than 65 had used the drug continuously for more than a decade (Morgan et al., 1988). Across all age groups, roughly twice as many women as men take sedative-hypnotic drugs.

THE BENZODIAZEPINES

Benzodiazepines can be classified on pharmacokinetic grounds into long-acting (e.g., flurazepam, nitrazepam), medium-acting (e.g., temazepam) and short-acting (e.g., triazolam) sedative-hypnotics. Their efficacy, at least in short-term use, has been well documented. The pattern of improvement in sleep corresponds fairly closely with the pharmacokinetic properties of each drug, providing that factors of absorption and elimination are taken into account. For example, temazepam is absorbed relatively slowly and has little effect on sleep-initiation time. In contrast, triazolam is absorbed relatively rapidly, so sleep ensues more quickly.

Each sedative-hypnotic has a minimally effective dose, and the dose that is usually effective may be twice as high as the minimum. Further increases may, however, cause side effects and rebound insomnia without substantially improving sleep. In sleep-laboratory studies, many benzodiazepines are found to lose their efficacy after about two weeks of nightly use. Subjectively, however, patients often feel that their sleep is improved for longer periods than this.

TABLE 1
Commonly Prescribed Sedatives

Type of Sedative	Name	Trade Name(s)
Barbiturate	Amobarbital	Amytal
Barbiturate	Pentobarbital	Nembutal
Barbiturate	Secobarbital	Seconal
Barbiturate	Amobarbital/secobarbital	Tuinal
Barbiturate	Butabarbital	Butisol
Barbiturate	Phenobarbital	Luminal
Anxiolytic	Meprobamate	Equanil, Miltown
Anxiolytic	Hydroxyzine	Atarax
Anxiolytic	Chlordiazepoxide	Librium
Anxiolytic	Diazepam	Valium
Anxiolytic	Chlorazepane	Tranzene
Anxiolytic	Oxazepam	Serax
Anxiolytic	Alprazolam	Xanax
Anxiolytic	Triazolam	Halcion

SOURCE: U.S. Department of Health and Human Services, Washington, D.C.

ADVERSE EFFECTS

The benzodiazepines can cause dose-related "hangover" effects the next morning, including residual sedation. Thus, patients may feel drowsy, have reduced psychomotor speed, and impaired concentration. These, in turn, can adversely affect ability to function, so patients are cautioned about driving and operating machinery. The longer-acting the drug, the more pronounced are these effects. Tolerance to these sedative effects builds up to some extent over repeated use of the drug.

Age-related changes in the way that drugs are metabolized and excreted mean that benzodiazepines accumulate more in older patients and, therefore, adverse effects are more pronounced in the elderly. Lower doses are thus prescribed to elderly patients, and the longer-acting benzodiazepines are often avoided.

All benzodiazepines can impair the ability to learn and remember new information. This memory impairment is most pronounced in only a few hours after taking the drug, so when taken as a sleep aid, such effects may be much reduced by the time the person wakes the next morning. Again, the elderly are particularly prone to such effects, which compound with age-related cognitive decline. As with other adverse effects, higher doses cause greater problems.

Rarer adverse events include disinhibition and aggressive behavior, and these have been reported for some benzodiazepines (e.g., triazolam, flunitrazepam) more than others.

REBOUND INSOMNIA

This refers to the increase in insomnia that may occur when the patient stops taking the drug such that the sleep pattern is actually worse than it was before it was taken. Studies have established that rebound insomnia is generally at its worst following the shorter-acting benzodiazepines and its least following the longer-acting benzodiazepines (Roehrs et al., 1986). Rebound is clearly dose-related, so the lowest effective dose should be prescribed, with rebound effects described to warn the patient about overdosing for "faster" or "better" drug-induced sleep.

Other problems which may relate to rebound include early-morning insomnia—a worsening of sleep in the final hours of the night (Kales et al., 1983). This is found mainly with the short-acting hypnotics; it is thought to reflect clearance of the drug from receptors during the first hours of sleep, leading to rebound as the hours go by. A related syndrome is daytime anxiety, which has been seen with both short- and medium-acting benzodiazepines (Morgan & Oswald, 1982).

DEPENDENCE AND WITHDRAWAL

Some argue that rebound insomnia is itself a sign of physiological dependence on benzodiazepine hypnotics (e.g., Morgan, 1990). Others insist that dependence is shown only when withdrawal from a drug leads to symptoms other than a rebound of the original problems. Withdrawal symptoms from low doses of benzodiazepines prescribed for daytime use as anxiolytics (anxiety relievers, tranquilizers) have been described in detail (e.g., Peturrson & Lader, 1984), but less attention has been paid to benzodiazepines taken as hypnotics. In that benzodiazepine anxiolytics and hypnotics are the same pharmacologically, one would expect both of being capable of inducing dependence.

Withdrawal symptoms from hypnotics will include restlessness and anxiety. Because benzodiazepines suppress dream, or rapid-eye-movement (REM) sleep, withdrawal can be associated with excess REM sleep and bizarre dreams. The latter may break into wakefulness as paranoid ideas, delusions, and clouding of consciousness (Oswald, 1983). Such features have been noted in numerous accounts of withdrawal syndromes following the discontinuation of benzodiazepines as sleep aids.

ABUSE

Abuse refers to the nonmedical, recreational use of these drugs. The benzodiazepines have reinforcing effects that appear to be more pronounced in frequent users of other recreational drugs. For example, alcoholics and HEROIN addicts will at times use benzodiazepines to eke out their supply of first-preference drug, since ALCOHOL and heroin are also depressants.

Abuse of benzodiazepines alone is relatively rarely observed.

OTHER HYPNOTIC DRUGS

Barbiturates. These were used until the 1950s as sleeping pills but were superseded by benzodiazepines. They are now generally obsolete for this purpose.

Chloral Derivatives. These are sometimes used with elderly patients, since they are less likely to cause restlessness in confused or demented patients. These compounds can, however, cause gastric irritation and rashes.

Newer Hypnotics. These include zopiclone and zolpidem which act either atypically or selectively on benzodiazepine receptors. They are chemically distinct from benzodiazepines and from each other. Both are short-acting drugs and at normal clinical doses cause little residual (hangover) sedation. The risk of rebound insomnia or dependence with these compounds is thought to be low but not absent (Lader, 1992).

(SEE ALSO: *Accidents and Injuries from Drugs; Addiction: Concepts and Definitions; Aging, Drugs, and Alcohol; Barbiturates: Complications; Benzodiazepines: Complications; Drug Interaction and the Brain; Drug Interactions and Alcohol; Memory, Effects of Drugs on; Prescription Drug Abuse*)

BIBLIOGRAPHY

DUNBAR, G., ET AL. (1989). Patterns of benzodiazepine use in Great Britain as measured by a general population survey. *British Journal of Psychiatry, 155,* 836–841.

KALES, A., ET AL. (1983). Early morning insomnia with rapidly eliminated benzodiazepines. *Science, 220,* 95–97.

LADER, M. H. (1992). Rebound insomnia and newer hypnotics. *Psychopharmacology, 108,* 248–252.

MELLINGER, G. D., BALTER, M. B., & UHLENHUTH, E. H. (1985). Insomnia and its treatment. *Archives of General Psychiatry, 42,* 225–232.

MORGAN, K. (1990). Hypnotics in the elderly: What cause for concern? *Drugs, 10,* 688–696.

MORGAN, K., & OSWALD, I. (1982). Anxiety caused by a short-life hypnotic. *British Medical Journal, 284,* 942–943.

MORGAN, K., ET AL. (1988). Prevalence, frequency and duration of hypnotic drug use among the elderly living at home. *British Medical Journal, 296,* 601–602.

OSWALD, I. (1983). Benzodiazepines and sleep. In M. R. Trimble (Ed.), *Benzodiazepines divided: A multidisciplinary review.* New York: John Wiley.

OSWALD, I., FRENCH D., ADAM K., & GILHAM, J. (1982). Benzodiazepine hypnotics remain effective for 24 weeks, *British Medical Journal, 284,* 860–863.

PETURRSON, H., & LADER, M. (1984). *Dependence on tranquillisers.* Oxford: Oxford University Press.

ROEHRS, T. A., ET AL. (1986). Dose-determinants of rebound insomnia. *British Journal of Clinical Pharmacology, 22,* 143–147.

VALERIE CURRAN

SEIZURES, BRAIN *See* Complications: Neurological.

SEIZURES OF DRUGS The seizure of drugs is a salient consequence of a variety of U.S. enforcement programs, but particularly of interdiction. It provides evidence that the U.S. criminal-justice system is imposing costs on drug distribution. A large seizure offers the most vivid evidence that senior members of the drug trades are subject to serious risks.

Seizures from smugglers have often been used as a measure of the effectiveness of interdiction efforts. One argument suggests that the larger the quantity of drugs seized, the more smugglers have been hurt by interdiction. Others view seizures as an indicator of the quantity smuggled; this view assumes that the share of imports seized is effectively a constant. Clearly these are extreme assumptions. The quantity seized is a function of at least three factors: the quantity shipped, the relative skill of the interdictors, and the care taken by smugglers. The last element, given least attention in discussion of seizures, probably depends on the replacement cost of the drugs; if that cost goes down (e.g., because of good growing conditions in the producer country), smugglers will invest less in concealment and protection of shipments and thus the seizure rate (i.e., the share of shipments seized) is likely to rise.

Seizures of COCAINE rose throughout the 1980s, probably reflecting both the rapid increase in total shipments and the declining replacement cost of the drug; whereas in 1981, total seizures by the federal government came to less than 2 tons, that figure had risen to over 100 tons in 1990. At the same time, MARIJUANA seizures declined dramatically, from 8,800 tons in 1981 to less than 2,000 tons in 1990. This decline was primarily the result of an effective interdiction program against the marine smuggling of Colombian marijuana; therefore, Mexican and domestic U.S. producers became competitive with Colombians because neither of them were at a comparable risk of seizure.

Drugs are also seized by state and local police. The only estimate of the total gathered by these agencies was made by Godshaw, Koppel, and Pancoast (1987); they found that in 1986, nonfederal seizures of cocaine totaled 35 tons compared with 70 tons by federal agencies. If the same ratio between federal and nonfederal seizures held in 1990, total national cocaine seizures may have amounted to 150 tons, at least one-third of estimated U.S. consumption. At a replacement cost of $20,000 per kilogram, this represented a total cost to cocaine distributors of 3 billion dollars, compared with total sales of perhaps 25 billion dollars.

(SEE ALSO: *Drug Interdiction; International Drug Supply Systems; Operation Intercept; Prohibition of Drugs: Pro and Con; Source Countries for Illicit Drugs*)

BIBLIOGRAPHY

GODSHAW, J., KOPPEL, R., & PANCOAST, R. (1987). *Antidrug law enforcement efforts and their impact.*

PETER REUTER

SELF-HELP AND ANONYMOUS GROUPS *See* Treatment Types.

SENSATION AND PERCEPTION AND EFFECTS OF DRUGS
Every behavior in which an organism engages involves information from the primary senses, such as vision, hearing (audition), and touch. A number of drugs of abuse alter sensory information. Naturally occurring drugs, such as MESCALINE from the PEYOTE cactus, increase awareness of visual and auditory sensations and also produce visual illusions and HALLUCINATIONS. The PSILOCYBIN mushroom (Mexican or Magic mushroom) produces similar effects. Because of these sensory changes, mescaline and psilocybin were used since pre-Columbian times in religious ceremonies by the peoples of Mexico and the American southwest. LYSERGIC ACID DIETHYLAMIDE (LSD), which was first synthesized in the late 1930s by the Swiss chemist Albert Hoffmann, has become well known for producing intense and colorful visual sensations. People also report changes in sensory behavior with drugs that are related to LSD, such as DMT, DOM, and MDMA, also known as "ecstasy" or the "love drug." PHENCYCLIDINE (PCP) is another drug that is sometimes added to the list of drugs that alter sensory behavior; however, its sensory effects are limited to numbness in the hands and feet. The active constituent of marijuana, TETRAHYDROCANNABINOL (THC), also produces alterations in sensory behavior; however, hallucinations—such as those produced by mescaline or LSD—are less common with THC. COCAINE and AMPHETAMINE sometimes produce hallucinations and other sensory distortions, but only when they are taken for long periods of time. Hallucinations also occur in delirium tremens (DTs) and when alcohol-dependent individuals are withdrawn from alcohol.

Various names are used to describe drugs that alter sensory behavior. One term is *psychedelic*, which refers to mind-expansion or to experiencing events that go beyond normal boundaries. Another term is *psychotomimetic*, which refer to the similarities of hallucinations that occur in psychotic disorders, such as SCHIZOPHRENIA, and those produced by mescaline and LSD. The term *hallucinogenic* is slightly misleading, since not all drugs that alter sensory behavior produce hallucinations.

OBSERVATIONS IN HUMAN SUBJECTS

Most of our information about drugs and the ways in which they alter sensory behavior in people comes from individual reports (called anecdotal) rather than from well-controlled laboratory studies. People have reported vivid images, changes in perception, and hallucinations after they have taken mescaline or LSD. Synesthesias—a mixing of the senses, such as "the hearing of colors" or "the seeing of sounds"—may also occur. One of the first descriptions of LSD's effects is recounted as follows:

I was seized by a peculiar sensation. . . . Objects, as well as the shape of my associates in the laboratory, ap-

971

peared to undergo optical changes.... With my eyes closed, fantastic pictures of extraordinary plasticity and intensive color seemed to surge toward me. After two hours this state gradually wore off. [See Julien, 1988, p. 180.]

Although these sensory disturbances go away within a few hours, some people experience confusion, sensory distortions, or poor concentration for longer periods of time. In some, drug effects occur again long after the drugs have left their systems—these brief episodes are called *flashbacks*.

STUDIES IN THE LABORATORY

Since alterations in sensory behavior, such as hallucinations, cannot be observed directly, it is very difficult to examine these effects in laboratory animals. One way to investigate a drug's effect on sensory behavior is to train animals to behave differently in the presence of different types of visual or auditory stimuli. If a drug changes the animal's behavior, it is possible that these changes in behavior are due to a change in how well the animal hears or sees the stimuli. Another type of procedure examines how intense (e.g., how loud or how bright) a stimulus has to be for an organism to hear or see it. In these procedures, the intensity required to hear or see a stimulus is determined before a drug is given and then it is compared to the intensity required to hear or see the stimulus after the drug is given.

In general, drugs such as mescaline, LSD, and THC do not alter an animal's ability to tell the difference between visual or auditory stimuli; nor do they alter visual or auditory thresholds. This lack of effect in animals suggests one of two explanations. Either drugs such as LSD produce different effects in animals than they do in people, or, more likely, the procedures that are used to study alterations in sensory behavior in animals do not measure the unique ways in which drugs such as LSD alter sensory behavior.

(SEE ALSO: *Complications; Inhalants; Opiates/ Opioids; Research; Research, Animal Model*)

BIBLIOGRAPHY

JAFFE, J. H. (1990). Drug addiction and drug abuse. In A. G. Gilman et al. (Eds.), *Goodman and Gilman's the pharmacological basis of therapeutics*, 8th ed. New York: Pergamon.

JULIEN, R. M. (1988). *A primer of drug action*. New York: W. H. Freeman.

LINDA DYKSTRA

SENSATION SEEKING *See* Vulnerability As Cause of Substance Abuse.

SENTENCES FOR DRUG OFFENSES *See* Mandatory Sentencing; Shock Incarceration and Boot-Camp Prisons.

SEROTONIN Chemically named 5-hydroxy-tryptamine, this MONOAMINE transmitter is a widely distributed substance particularly prevalent in the gut, blood, platelets, and pineal gland, as well as in nine major sets of brain neurons (nerve cells). In the 1950s, chemical similarity between serotonin and the chemical HALLUCINOGEN LYSERGIC ACID DIETHYLAMIDE (LSD) focused attention on this NEUROTRANSMITTER in mental illness, a link strengthened by experimental studies in animals and humans. Neurons containing serotonin, a typical monoamine, project widely throughout the brain and spinal cord, and a large number of well-characterized serotonin-receptor subtypes mediate both direct and indirect regulation of ion channels that exist in the membranes of neurons. By regulating these channels, these serotonin RECEPTORS influence the concentration within the neuron of such ions as K^+ (potassium) and Ca^{++} (calcium) and thereby the activity of the cell.

(SEE ALSO: *Brain Structures and Drugs; Dopamine; Neurotransmission; Reward Pathways and Drugs; Serotonin-Uptake Inhibitors in Treatment of Substance Abuse*)

BIBLIOGRAPHY

COOPER, J. R., BLOOM, F. E., & ROTH, R. H. (1991). *The biochemical basis of neuropharmacology*, 6th ed. New York: Oxford University Press.

FLOYD BLOOM

SEROTONIN-UPTAKE INHIBITORS IN TREATMENT OF SUBSTANCE ABUSE

The development of effective pharmacological treatments for alcohol and drug abuse depends on our understanding of the biological mechanisms that start and maintain these behaviors. Studies in animals and humans have confirmed that SEROTONIN is one of several NEUROTRANSMITTERS that influence drug-reinforcing behaviors. Pharmacological agents that enhance central serotonergic neurotransmission—in particular, serotonin-uptake inhibitors (several of which have been marketed as antidepressants)—show considerable promise, as of the early 1990s, as effective treatments for the abuse of alcohol and some other drugs. These work by blocking the re-uptake of serotonin and thereby increase its concentration in the nerve SYNAPSE.

ALCOHOL ABUSE

In the late 1980s, serotonin-uptake inhibitors were tested in various ANIMAL models of alcoholism—including selectively bred alcohol-preferring rats given a choice between water and an alcohol solution—and showed consistent decreases in the self-administration of alcohol in a dose-dependent manner. The results of these preclinical studies led to research in human alcohol abusers. In four placebo-controlled, double-blind, randomized clinical trials, serotonin-uptake inhibitors decreased short-term (1 to 4 weeks) alcohol intake by averages of 14 to 20 percent, as compared with pretreatment. No other treatment or advice was given. The effect developed rapidly after a serotonin-uptake inhibitor was administered and disappeared rapidly after discontinuation. All subjects had had mild or moderate (not severe) alcohol dependence but no current or past depression, anxiety, other psychiatric disorder, or other substance-abuse disorder. No aversive interactions with alcohol or changes in depression or anxiety levels were observed; therefore they could not account for the effects on alcohol intake. Adverse side effects were few and mild. However, concomitant decreases in desire/urge to drink were reported by subjects during treatment with serotonin-uptake inhibitors. Therefore, experimental drinking sessions, following one or two weeks of treatment with serotonin-uptake inhibitor and placebo, were incorporated into two research studies—fluoxetine (Prozac) and citalopram, each with a placebo control—to specifically measure variations in self-re-ported desire to drink alcohol. Desire for alcohol was lower during the experimental drinking sessions after taking serotonin-uptake inhibitors than after taking placebos. In both of these studies, the effects of serotonin-uptake inhibitors on alcohol intake were also confirmed in the outpatient weeks preceding the experimental drinking sessions.

The observation that serotonin-uptake inhibitors decrease desire to drink indicates a possible mechanism of their effects on alcohol intake. In the outpatient trials, an increase in abstinent days was often the means by which alcohol intake was reduced, and similarly, in trials with animals, serotonin-uptake inhibitors decreased their number of drinking "bouts." Therefore, serotonin uptake inhibitors may, by decreasing the desire to drink, reduce the likelihood of initiating drinking. The consistency of the pharmacological effects is quite remarkable, considering the many other factors influencing drinking behavior. In an effort to enhance the pharmacological effects of serotonin-uptake inhibitors and determine their therapeutic value, a brief psychosocial intervention was combined with citalopram in a long-term (12 week) treatment research study with sixty-two mildly/moderately dependent alcoholics. Average decreases in daily alcoholic drinks from baseline were 47.9 percent during the first week of citalopram (n = 31) and only 26.1 percent during the first week of placebo (n = 31), indicating a significant improvement with citalopram. From the second to twelfth weeks of treatment, the average decreases were similar: 33.4 percent and 40.5 percent during citalopram and placebo, respectively. Craving for alcohol also decreased similarly with both citalopram and placebo. Thus, the short-term effects of citalopram are synergistic with a brief psychosocial intervention, and serotonin-uptake inhibitors seem to facilitate the initiation of reduced drinking. The true therapeutic value of serotonin-uptake inhibitors is yet to be determined, but they may be appropriate for specific applications. For example, relapse is a frequent problem among recovering alcoholics; serotonin-uptake inhibitors, by decreasing desire or urge to drink, may be particularly suitable adjuncts for relapse-prevention strategies.

COCAINE

Abuse of COCAINE increased in the 1980s; it is also common among HEROIN addicts—some who use it alone and some together with heroin. Fluoxetine decreased cocaine craving and abuse in some heroin addicts who were in a METHADONE MAINTENANCE program. These interesting results merit further study in a controlled trial.

CIGARETTE SMOKING

Cigarette smoking has not been affected by serotonin-uptake inhibitors in heavy drinkers who were not trying to reduce their smoking. Fluoxetine was found to prevent the weight gain that accompanies SMOKING CESSATION and, therefore, may be helpful in preventing relapse among exsmokers. The results of studies on the use of serotonin-uptake inhibitors in patients participating in smoking-cessation programs have not been reported yet.

PSYCHIATRIC DISORDERS

Individuals who abuse alcohol and/or drugs often have psychological or psychiatric disorders. The establishment of cause-and-effect relationships can be difficult. There is evidence that COMORBIDITY (two disease processes) adversely influences outcome in treatments of substance abuse. Some patients may self-medicate symptoms of ANXIETY or DEPRESSION with a drug of abuse, such as alcohol. Therefore, successful pharmacological treatment of the anxiety or depression may reduce the need for other drugs (the alcohol).

As antidepressants, serotonin-uptake inhibitors would be particularly suitable for treating depressed substance abusers. No research studies have been conducted, but a comparison between treatment outcomes of depressed substance abusers receiving a serotonin-uptake inhibitor and those receiving other antidepressants would be of interest.

Severe cognitive deficits (memory loss) are a frequent complication of chronic ALCOHOLISM. Low brain levels of serotonin may be a factor in this type of memory loss. Fluvoxamine, a serotonin-uptake inhibitor, improved episodic memory in patients with alcohol amnestic disorder. This might greatly facilitate success in cognitively oriented treatments for alcoholism.

CONCLUSIONS

Serotonin-uptake inhibitors decrease short-term alcohol intake and desire to drink. Their effects are synergistic with a brief psychosocial intervention for alcoholism; however, their long-term efficacy and clinical importance have not been determined. One small study indicated that a serotonin-uptake inhibitor may reduce cocaine abuse. There is currently no evidence that serotonin uptake inhibitors reduce cigarette smoking or opiate abuse.

BIBLIOGRAPHY

GORELICK, D. A. (1989). Serotonin uptake blockers and the treatment of alcoholism. In M. Galanter (Ed.), *Recent developments in alcoholism*, vol. 7. New York: Plenum.

NARANJO, C. A. & BREMNER, K. E. (1991). Recent trends in the pharmacotherapy of drug dependence. *Drugs of Today, 27*, 479–495.

NARANJO, C. A., & SELLERS, E. M. (1989). Serotonin uptake inhibitors attenuate ethanol intake in problem drinkers. In M. Galanter (Ed.), *Recent developments in alcoholism*, vol. 7. New York: Plenum.

CLAUDIO A. NARANJO
KAREN E. BREMNER

SEXUAL AND PHYSICAL ABUSE   *See* Vulnerability As Cause of Substance Abuse.

SHANGHAI OPIUM CONFERENCE   The 1909 Shanghai Opium Commission was the first multinational drug-control initiative. Through the encouragement of President Theodore Roosevelt and the organizational skills of Bishop Charles H. Brent, the United States convened this meeting of thirteen countries at Shanghai, including Great Britain, Japan, China, and Russia, to address the illegal production, trade, and use of OPIUM in China.

As a commission the participants could only recommend actions necessary to prevent opium trafficking and abuse but could not make binding international agreements. However, the participants passed resolutions urging national governments to enact measures to curb opium smoking in their countries, initiate regulation of opium use for nonmedical purposes, ban the export of opium to countries that prohibited importation, and control the manufacture and distribution of opium derivatives.

The commission was the first effective step taken by the international community to combat drug abuse. It served as a catalyst for countries to pass domestic legislation addressing drug problems within their borders. Most important, the commission united countries in an international cooperative effort to address the problem of the opium trade. The work of the commission led to the convening of the Hague Opium Conferences (1912–1914) and to the adoption of the 1912 International Opium Convention, sometimes called the Hague Opium Convention, and succeeding treaties that effectively restricted opium production and trade to legitimate purposes.

(SEE ALSO: *Asia, Drug Use in; International Drug Supply Systems; Opioids and Opioid Control: History; Psychotropic Substances Convention of 1971; Single Convention on Narcotic Drugs*)

BIBLIOGRAPHY

BEAN, P. (1974). *The social control of drugs.* New York: Wiley.

KING, R. (1992). *The drug hang-up: America's fifty-year folly.* New York: Norton.

MUSTO, D. F. (1973). *The American disease: Origins of narcotic control.* New Haven: Yale University Press.

ROBERT T. ANGAROLA

SHOCK INCARCERATION AND BOOT-CAMP PRISONS

Shock incarceration programs, frequently called boot-camp prisons, are short-term prison programs run like military basic training for young offenders—adult and youthful felons (MacKenzie & Parent, 1992). Those sentenced to them are required to arise early each day to participate in a rigorous schedule of physical training, military drill and ceremony, and hard labor. While they are in the boot camp, participants are separated from other prisoners. They are allowed few personal possessions, no televisions, and infrequent visits from relatives on the outside.

The correctional officers in the programs are referred to as drill instructors and are responsible for seeing that the inmates obey the rules and participate in all activities. When speaking to staff, inmates must refer to themselves as "this inmate" and they must proceed and follow each sentence with sir or madam as in "Sir, yes, sir." Disobedience is punished immediately using summary punishments, frequently in the form of some additional physical activity, such

as pushups or situps. More serious rule violations may result in dismissal from the program.

BOOT-CAMP PRISONS AS INTERMEDIATE SANCTIONS

The boot-camp prisons were developed during the 1980s—in part, in response to the phenomenal growth in the number of convicted offenders. Correctional jurisdictions faced severe prison overcrowding, and probation caseloads grew so large that many offenders received only nominal supervision during their time in the community. Officials searched for ways to manage the offenders. There were two options—either they were sent to prison or they were supervised in the community on probation. Neither option was entirely satisfactory for the large number of young offenders. Alternative sanctions or intermediate punishments such as intensive community supervision, house arrest, or residential-community corrections centers were proposed as solutions to the problem. These options provided more control than a sentence to probation but less than a sentence to prison. Boot-camp prisons were one relatively inexpensive alternative sanction that became particularly popular.

The first boot-camp prisons were begun in 1983, in Oklahoma and Georgia. These two programs attracted a great deal of attention and other jurisdictions soon began developing similar programs. By 1993, thirty states had opened about sixty boot-camp prisons for adult and youthful felons; additional programs were beginning in local jails and in juvenile-detention centers. Although the majority of the boot camps had male participants, some programs admitted women into the boot camps with the male offenders. Other states developed completely separate boot-camp prisons for women. The Federal Bureau of Prisons developed one boot camp for males and a separate program for females.

ENTERING AND EXITING

Since most boot camps have strict requirements about who is eligible for the camp, inmates are carefully evaluated prior to being sent there. Most programs require participants to sign an agreement saying they have volunteered. They are given information about the program and the difference between a boot-camp prison and a traditional prison. The major incentive for entering the boot camp is that the boot camp requires a shorter term in prison than a traditional prison sentence.

The first day of the boot camp involves a difficult in-take process, when the drill instructors confront the inmates. Inmates are given rapid orders about the rules of the camp, when they can speak, how they are to address the drill instructors, and how to stand at attention. The men have their heads shaved; the women receive short haircuts. This early period of time in the boot camp is physically and mentally stressful for most inmates.

The programs last from 90 to 180 days. Those dismissed prior to graduation are considered program failures. They are either sent immediately to a traditional prison to serve a longer term of incarceration or they are returned to court for resentencing.

Offenders who successfully complete the boot camp are released from prison. After graduating from the boot camp, offenders are supervised in the community for the rest of their sentence. There is usually an elaborate graduation ceremony when inmates demonstrate the military drills they have practiced. Many programs encourage family members to attend the graduation ceremony.

A DAY IN BOOT CAMP

On a typical day, the participants arise before dawn, rapidly dress, clean their living quarters, and march in cadence to an exercise area. There they will spend an hour or more doing calisthenics and running. They march back to their quarters for a quick cleanup before breakfast. As they do at every meal, they march to breakfast and stand at parade rest while waiting to be served. They stand at attention until ordered to sit and eat without conversation. Following breakfast they may work six to eight hours. This is usually hard physical labor such as cleaning state parks or public roads. They return in the late afternoon for additional physical exercise or practice in drill and ceremony. After a quick dinner, they attend rehabilitation programs until 9 P.M. when they return to their dormitories. In the short period before bedtime, they have time to be sure their shoes are shined and their clothes are clean and ready for the morning.

SIMILARITIES AND DIFFERENCES

All the boot-camp prisons incorporate the basic core components of military basic training, with physical training and hard labor. Most target young offenders convicted of nonviolent crimes such as drug, burglary, or theft offenses. Participation is limited to those who do not have an extensive past history of criminal activity.

Other than these similarities, the programs differ dramatically. Some focus only on work, military drill, and exercise. In other boot camps, offenders spend a great deal of time each day in rehabilitation programs. The camps also differ in the type of the therapeutic programming provided. Some emphasize academic education, others focus on group counseling or treatment for substance abuse.

The boot camps also differ in the ways offenders are managed after release. Some programs intensively supervise all offenders who successfully complete the boot camp; others are supervised as they would be in traditional probation caseloads. Program officials worry about the difficulty the graduates have in making the transition from the rigid structure of the boot camps to the community environment. For this reason, some boot camps developed aftercare programs to help them make the change. These aftercare programs do more than increase the surveillance over the activities of the graduates. They are designed to provide drug treatment, vocational counseling, academic education, or short-term housing to boot-camp graduates.

DRUG TREATMENT IN THE BOOT CAMPS

The earliest boot camps focused on discipline and hard work. More recently, they have begun to emphasize treatment and education. It became clear that many of the entrants were drug-involved. Realizing that the punishment alone would not effectively reduce the drug use of these offenders, corrections officials introduced drug treatment or education into the daily schedule of boot-camp activities. By the late 1980s, all the camps had some type of substance-abuse treatment or education for boot-camp inmates (MacKenzie, 1994).

As happened with other aspects of the programs, the type of treatment and the amount of time devoted to substance-abuse treatment varied greatly among programs. The 90-day Florida program included only 15 days of treatment and education; in contrast, in the New York program all offenders received 180 days of treatment. Most programs reported that drug use was monitored during community supervision; however, the schedule and frequency of this monitoring varies greatly.

New York's Therapeutic Community Boot Camps. In the boot camps that include substance-abuse treatment as a component of the in-prison phase of the program, there are large differences in the way it is delivered. The boot-camp programs, developed by the New York Department of Correctional Services, use a THERAPEUTIC-COMMUNITY model for the program. All offenders are given a similar regimen of drug treatment while they are incarcerated (New York Department of Correctional Services, 1994). Each platoon in the boot camp forms a small community. They meet daily to solve problems and to discuss their progress in the shock program. They spend over 200 hours during the six-month program in substance-abuse treatment activity. The treatment is based on the ALCOHOLICS ANONYMOUS (AA) and NARCOTIC ANONYMOUS (NA) models of abstinence and recovery. All boot camp inmates participate in the substance-abuse treatment regardless of their history of use and abuse.

Illinois's Boot Camp with Levels of Treatment. Like New York, the Illinois boot camp also targets substance abusers. However, the delivery of treatment services is very different. In Illinois, counselors at the boot camp evaluate offenders and match the education and treatment level to the identified severity level of the offender (Illinois Department of Corrections, 1992). Three different levels of treatment are provided. Inmates identified as level-one have no substance-abuse history, therefore they receive only two weeks of education. Level-two inmates are identified as probable substance abusers. They receive four weeks of treatment in addition to the drug education. The treatment consists of group therapy focusing predominately on denial and on family-support issues. Inmates identified as level-three are considered to have serious drug addictions; they receive ten weeks of education and treatment. In addition to the drug education and group therapy, they receive group sessions on substance-abuse relapse, CODEPENDENCY, behavioral differences, family addiction, and roles within the family.

Texas's Voluntary Participation Model. A third model is represented by the Texas program (MacKenzie, 1994). In the boot camp, all participants receive five weeks of drug education. During this phase, inmates may also receive individual counseling and attend TWELVE-STEP fellowship meetings. More drug treatment is available for those

who volunteer (the substance-abuse counselors in this program believe that treatment should be voluntary). These volunteers receive approximately four hours per week of treatment in the form of group therapy. The meetings are held during free time, so inmates are not released from work to attend. The group sessions focus on social values, self-worth, communication skills, self-awareness, family systems, self-esteem, and goal setting. Some inmates also receive individual counseling.

DISMISSAL RATES

As occurs in many drug-treatment programs, boot camps may have high dismissal rates. Depending upon the program, rates vary from 8 percent (Georgia in 1989) to as much as 80 percent (Wisconsin in 1993). Offenders can be dismissed from the boot camp because of misbehavior or, in some boot camps, they can voluntarily ask to leave. Those who are dismissed will either be sent to a traditional prison, where they will serve a longer sentence than the one assigned to boot camp, or they will be returned to the court for resentencing. Thus, in both cases there is the threat of a longer term in prison for those who do not complete the boot camp.

There is very little information about how drug-involved offenders do in boot camp prisons. One study of the Louisiana boot camp examined the dismissal rates of drug-involved offenders and compared these rates to offenders in the boot camp who were not identified as drug-involved (Shaw & MacKenzie, 1992). Two groups of drug-involved offenders were examined: (1) those who had a legal history of drug-involvement (an arrest or conviction for a drug offense); and (2) those who were identified as drug abusers on the basis of self-report. In this program, offenders were permitted to drop out voluntarily or they could be dismissed for misbehavior. Surprisingly, in comparison to other offenders, the drug-involved offenders were *less* likely to drop out of the program.

In another study of the Louisiana boot camp, 20 percent of the participants were identified as problem drinkers on the basis of their self-reported alcohol use and problems associated with use (Shaw & MacKenzie, 1989). The problem drinkers were no more likely to drop out of the book-camp prison than were the others.

In interviews, offenders who are near graduation from boot camp report that they are drug free and physically healthy (MacKenzie and Souryal, 1994). Unlike offenders incarcerated in conventional prisons, boot-camp participants believed that their experience had been positive and that they had changed for the better. They also reported that the reason they entered the boot camp was because they believed they would spend less time in prison—not because of the treatment or therapy offered.

PERFORMANCE DURING COMMUNITY SUPERVISION

Studies have compared the performance during community supervision of graduates from the boot-camp prisons to others who served a longer time in prison or who were sentenced to probation. In most cases, there were no significant differences between these offenders in recidivism (return) rates or in positive social activities (MacKenzie & Souryal, 1994). However, boot-camp graduates in Illinois and Louisiana had fewer revocations for new crimes. Research examining New York offenders found mixed results. Graduates had fewer new crime revocations in one study (New York Correctional Services, 1994) and fewer technical violations in another study (MacKenzie & Souryal, 1994).

All the boot-camp prisons had a military atmosphere with physical training, drill and ceremony, and hard labor. If this atmosphere alone changed offenders, we would expect all the graduates to have lower recidivism rates and better positive adjustment. The inconsistency of the results suggests that the boot-camp atmosphere alone will not successfully reduce recidivism or positively change offenders. Some other aspects of the Illinois, New York, and Louisiana programs, either with or without the boot-camp atmosphere, led to the positive impact on these offenders. After an examination of these programs, the researchers concluded that all three programs devoted a great deal of time to therapeutic activities during the boot-camp prison, a large number of entrants were dismissed, the length of time in the boot camp was longer than other boot camps, participation was voluntary, and the in-prison phase was followed by six months of intensive supervision in the community. Research as of the mid-1990s cannot separate the effect of these components from the impact of the military atmosphere. Most likely, a critical component of the boot camps for drug-involved offenders is the therapy provided

during the program and the transition and aftercare treatment provided during community supervision.

Performance of Drug-Involved Offenders. Shaw and MacKenzie (1992) studied the performance of drug-involved offenders during community supervision in Louisiana. In comparison to offenders who were not drug-involved, those who were drug-involved did poorer during community supervision. This was true of those on probation, parolees from traditional prisons, and parolees from the boot camp. The boot-camp parolees did not do better than others. During the first year of supervision, the drug-involved offenders were more likely to have a positive drug screen.

Problem drinkers who graduated from the Louisiana program were found to perform better, as measured by positive activities during community supervision (Shaw & MacKenzie, 1989). Their performance was, however, more varied—indicating that they may need more support and aftercare than other offenders.

In contrast to the Louisiana findings, research in New York indicated that those who were returned to prison were more apt to be alcoholics (New York Department of Correctional Services, 1994). In both Louisiana and New York, offenders who were convicted of drug offenses did better than self-confessed alcoholics during community supervision.

THE FUTURE OF BOOT-CAMP PRISONS

Boot-camp prisons are still controversial. People are concerned that inmates' rights will not be observed and that they are being coerced to do something that is not good for them (Morash & Rucker, 1990). These critics argue that the summary punishments and the staff yelling at offenders may be abusive for inmates; that participants may leave the boot-camp prison angry and damaged by the experience; that the military atmosphere designed to make a cohesive fighting unit may not be appropriate for these young offenders.

Advocates of the boot camp say that the program has many benefits. In their opinion, these offenders lack the discipline and accountability that are provided by the program. Furthermore, they argue, the strong relationship between the offenders and the drill instructors may be helpful to the inmates. Also, there may be some aspects of the boot camps that

are particularly beneficial for drug-involved offenders. For example, the program may educate and
may coerce offenders to stay in treatment longer
than they otherwise would. The structure, fitness
regimen, and controlled setting characteristic of the
boot camps is also a component of successful drug
treatment (Anglin & Hser, 1990).

Although controversy exists about the boot-camp
prisons, they are still a popular alternative sanction.
Advocates of the programs have been disappointed
that the research has not shown that boot camps
have a significant impact on the behavior of offenders after they are released from prison.

We do not know how appropriate these programs
are for drug-involved offenders. Boot camps differ in
aspects that will be important if the boot camps are
expected to be effective for drug-involved offenders.
One crucial component for these offenders is that of
drug treatment in the boot camps (MacKenzie, 1994;
Souryal & MacKenzie, 1994). Although the structure
and discipline may have some value, it will have to
be combined everywhere with drug treatment to be
effective in reducing the drug use and related criminal activities of drug-involved offenders.

(SEE ALSO: *Civil Commitment; Coerced Treatment;
Narcotic Addict Rehabilitation Act; Prisons and Jails;
Treatment in the Federal Prison System; Treatment
Types: An Overview*)

BIBLIOGRAPHY

ANGLIN, M. D., & HSER, Y-I. (1990). Treatment of drug
abuse. In M. Tonry & J. Q. Wilson (Eds.), *Drugs and
crime: Vol. 13, Crime and justice.* Chicago: University
of Chicago Press.

ILLINOIS DEPARTMENT OF CORRECTIONS. (1991). *Overview
of the Illinois Department of Corrections impact incarceration program.* Springfield: Author.

MACKENZIE, D. L. (1994). Shock incarceration as an alternative for drug offenders. In D. L. MacKenzie & C. D.
Uchida (Eds.), *Drugs and crime: Evaluating public policy initiatives.* Thousand Oaks, CA: Sage.

MACKENZIE, D. L., & PARENT, D. G. (1992). Boot camp prisons for young offenders. In J. M. Byrne, A. J. Lurigio,
& J. Petersilia (Eds.), *Smart sentencing: The emergence
of intermediate sanctions.* London: Sage Publications.

MACKENZIE, D. L., & SOURYAL, C. (1994). *Multi-site evaluation of shock incarceration: Executive summary.* Report to the National Institute of Justice. Washington,
DC: National Institute of Justice.

MORASH, M., & RUCKER, L. (1990). A critical look at the
ideal of boot camp as a correctional reform. *Crime and
Delinquency, 36,* 204–222.

NEW YORK STATE DEPARTMENT OF CORRECTIONAL SERVICES
AND THE NEW YORK DIVISION OF PAROLE. (1993). *The
fifth annual shock legislative report.* Albany, NY: Unpublished report by the Division of Program Planning,
Research and Evaluation and the Office of Policy Analysis and Information.

SHAW, J. W., & MACKENZIE, D. L. (1992). The one-year
community supervision performance of drug offenders
and Louisiana DOC-identified substance abusers graduating from shock incarceration. *Journal of Criminal
Justice, 20,* 501–516.

SHAW, J. W., & MACKENZIE, D. L. (1989). Shock incarceration and its impact on the lives of problem drinkers.
American Journal of Criminal Justice, 16, 63–97.

SOURYAL, C., & MACKENZIE, D. L. (1994). Shock therapy:
Can boot camps provide effective drug treatment? *Corrections Today, 56*(1), 48–54.

DORIS LAYTON MACKENZIE

SIDE EFFECTS   *See* Complications.

SINGLE CONVENTION ON NARCOTIC
DRUGS   The Single Convention on Narcotic
Drugs of 1961 is the most comprehensive international drug control agreement ever signed. It regulates the production, trade, and use of NARCOTIC
drugs, COCAINE, and cannabis (MARIJUANA).

BACKGROUND

Thirteen countries signed the first international
drug control treaty in 1912 at The Hague, Netherlands. Into the 1950s, governments entered into
eight multilateral treaties aimed at preventing the illicit trade and consumption of opium and other
drugs. Over forty years, many of the provisions had
become obsolete, had never been implemented, or
required revision as world developments presented
new challenges. The Single Convention consolidated
the existing multilateral drug-control treaties into
one agreement. Its drafters also intended to encourage governments that had not participated in earlier

drug-control agreements to join the international effort. As of November 1993, 144 governments were party to the Single Convention.

PROVISIONS OF
THE SINGLE CONVENTION

The Single Convention contains eight major provisions for the control of the production, trade, and use of drugs. All parties must establish or adjust national legislation to conform to these requirements of the convention.

Parties must require licenses for manufacturers, wholesalers, and other handlers of narcotic drugs, and they must maintain a system of permits, record keeping, reports, controls, and inspections to prevent diversion of drugs to the illicit traffic. A country that allows the domestic production of the OPIUM poppy, the COCA bush, or the *Cannabis* plant must establish a control agency to designate areas for the cultivation of these drugs and limit production to licensed growers.

Parties to the convention must prepare estimates (quotas) detailing the amount of drugs necessary to satisfy national medical and scientific needs, and they must provide these figures annually to the International Narcotics Control Board (INCB). Governments must also provide the INCB with quarterly and annual statistics on drug production, trade, and consumption. In addition, the Single Convention requires that parties maintain a system of import and export authorizations as well as import certificates so that the INCB and governments can monitor the flow of narcotics in and out of countries.

The Single Convention extends the control system over the opium poppy to the coca bush and the cannabis plant. Governments must uproot and destroy wild and illegally cultivated coca bushes and cannabis plants. Parties are furthermore required to ban opium smoking and eating, coca-leaf chewing, and cannabis smoking and ingestion. A transition period is provided to overcome any difficulties that might arise for those who use such plants or drugs in ancient rituals. Countries may reserve the right to permit the quasi-medical use of opium and coca leaves as well as the nonmedical use of cannabis.

The Single Convention encourages parties to provide assistance and treatment to drug addicts. This provision distinguishes the agreement from previous

international drug-control treaties, which focused exclusively on curbing the illicit flow of drugs.

INTERNATIONAL NARCOTICS
CONTROL BOARD AND COMMISSION
ON NARCOTIC DRUGS

Signatories to the Single Convention recognized the need for an international central monitoring and enforcement agency to oversee the production and trade of drugs. The Single Convention merged the Permanent Central Opium Board and the Drug Supervisory Board into the INCB, which serves as this central authority. The United Nations Economic and Social Council elects thirteen members to serve on the INCB.

The main responsibilities of the INCBs include limiting the cultivation, production, manufacture, and use of narcotic drugs and psychotropic substances to the amounts necessary for medical and scientific purposes, ensuring the availability of these drugs for medical purposes such as pain control. The INCB reviews estimates of opium and other drug-production figures provided by each party. These figures are formalized into production and consumption quotas. The board also analyzes information from participating countries, the United Nations, and other international organizations to ensure that there is compliance with the terms of the Single Convention. Where appropriate, it recommends that technical and financial assistance be given to those countries that may need further help. The Single Convention also provides the INCB with some direct enforcement powers, such as recommending an embargo of drug shipments to a country that is a center of drug trafficking. The INCB is more effective, however, in encouraging government to comply through confidential diplomatic initiatives than through the imposition of sanctions.

The Single Convention strengthens the role of the United Nations Commission on Narcotic Drugs (CND). The CND, which is composed of fifty governments, is the UN body that is the key information and policymaker in the drug-control area. The CND adds and deletes substances to or from the four control schedules of the convention, notifies the INCB of drug-control concerns, recommends ways to curb the illicit traffic of narcotics, and notifies nonparticipants of the actions that have been taken. It also gathers the names of the authorities that issue licenses for import and export.

DRUG SCHEDULES

In the preamble to the Single Convention, the parties recognized that "the medical use of narcotic drugs continues to be indispensable for the relief of pain and suffering and that an adequate provision must be made to ensure the availability of narcotic drugs for such purposes." In an effort to make narcotic drugs available for legitimate medical use while also curtailing drug abuse, the parties placed narcotic drugs into four schedules. Classification of a narcotic drug and the type of regulation that would be imposed on that drug substance would depend on a drug's potential for abuse as well as its medical benefit.

Schedule I is reserved for medically useful drugs exhibiting the highest potential for abuse. Examples of schedule I drugs include OPIUM, MORPHINE, and METHADONE.

Schedule II substances possess a liability for abuse that is no greater than that of CODEINE. These drugs are placed under similar controls as schedule I substances except that parties need not require prescriptions for domestic supply. Medical practitioners are not required to keep records tracking the acquisition and disposal by individuals of a controlled substance placed in schedule II. Codeine is the most commonly prescribed schedule II drug.

Drugs in schedule III are the ones intended for medical use that, as prepared, pose a negligible or nonexistent risk of abuse and a low public health risk. Schedule III drugs face substantially fewer controls than those listed in schedules I and II. Preparations of codeine and the analgesic dextropropoxyphene are two examples of drugs listed in schedule III.

To place a drug in schedules II and III governments must control the factories where these drugs are manufactured as well as the individuals involved in their manufacture, trade, distribution, and import or export. Records of the manufacture and sale of these drugs must be maintained, and limits must be imposed to ensure that they are used exclusively for medical and scientific purposes.

The special class of drugs in schedule IV exhibit strong addiction-producing properties or a high liability of abuse that cannot be offset by medical benefits or that poses too great a risk to public health to hazard using them commonly in medical practice. Drugs in this category remain subject to the same international controls that are applicable to schedule I drugs, but governments are encouraged to limit their legitimate use. Cannabis, cannabis resin, and heroin (diamorphine) are examples of schedule IV drugs. Several medical experts have questioned the appropriateness of limiting the use of diamorphine for pain control and a number of governments permit this use.

Note that these schedules or levels of control differ from those contained in the Controlled Substances Act (CSA) of the United States. For example, in this act, drugs with a high liability for abuse and no accepted medical uses are included in Schedule I. The CSA also covers all categories of drugs including sedatives, HALLUCINOGENS, and cocaine besides other stimulants, whereas the Single Convention covers only opioid drugs, cocaine, and cannabis (marijuana). Other psychoactive drugs with abuse potential are controlled under a different international treaty, the Convention on Psychotropic Substances of 1971.

The World Health Organization (WHO) is responsible for making recommendations regarding the scheduling of drugs. In evaluating the schedule of a drug, WHO considers the "degree of liability to abuse" of a substance and the "risk to public health and social welfare" that the substance in question poses or might pose. The Convention grants WHO broad discretion in interpreting these two criteria. Ultimately, the Commission on Narcotic Drugs decides, by majority vote, whether to alter or amend a schedule, thereby reserving the right to reject WHO's recommendation.

THE 1972 PROTOCOL

The 1972 Protocol Amending the Single Convention on Narcotic Drugs confers greater powers on the International Narcotics Control Board and emphasizes the prevention of drug abuse, the distribution of drug information and education, and the treatment and rehabilitation of drug addicts. It also stresses the need to balance legitimate production of narcotics for medical and scientific purposes with prevention of illicit production, manufacture, traffic, and use of these substances.

THE SIGNIFICANCE OF
THE SINGLE CONVENTION

The Single Convention has proved important in four ways. First, the aims, goals, and strategy in regard to combatting illicit drug trafficking became more focused and modernized because of its adoption. Second, the large number of participants in the Convention encourages more countries to take part in the international cooperative effort against drug abuse. Third, the placement of drugs into schedules constitutes a recognition of the differences between drug substances, and it balances the potential for abuse of the drugs with their medical benefit. The Single Convention, which openly supports the medical use of narcotics to relieve pain and suffering, states that these drugs are "indispensable" for the purpose. Narcotics with a higher potential for abuse and with a lower medical value fall subject to tighter regulation than drugs with a lower potential for abuse and a greater medical value. Fourth, the international community appreciates the need to combine strict controls of illicit drug trafficking with the treatment and rehabilitation of drug addicts. This approach, fusing strength with compassion, is now an integral part of the effort to curb the illicit production, trade, and consumption of narcotic drugs.

(SEE ALSO: *International Drug Supply Systems; Opioids and Opioid Control: History; Psychotropic Substances Convention of 1971; Shanghai Opium Conference; WHO Expert Committee on Drug Dependence*)

BIBLIOGRAPHY

BEAN, P. (1974). *The social control of drugs.* New York: Wiley.

BRUUN, K., PAN, L., & REXED, I. (1975). *The gentlemen's club: International control of drugs and alcohol.* Chicago: University of Chicago Press.

CHATTERJEE, S. K. (1981). *Legal aspects of international drug control.* The Hague: Martinus Nijhoff.

INTERNATIONAL NARCOTICS CONTROL BOARD. (1993). U.N. Publication No. E.94.X1.2. New York: United Nations.

KING, R. (1992). *The drug hang-up: America's fifty-year folly.* New York: Norton.

REXED, B., ET AL. (1984). *Guidelines for the control of narcotic and psychotropic substances.* Geneva: World Health Organization.

SINGLE CONVENTION ON NARCOTIC DRUGS. (1961). 18 U.S.T. 1407, T.I.A.S. No. 6298. March 30.

SINGLE CONVENTION ON NARCOTIC DRUGS OF 1961, AMENDMENTS. (1972). 26 U.S.T. 1439, T.I.A.S. No. 8118. March 25.

WISOTSKY, S. (1986). *Breaking the impasse in the war on drugs.* Westport, CT: Greenwood Press.

WORLD PEACE THROUGH LAW CENTER. (1973). *International Drug Control* (prepared for the Sixth World Conference of the Legal Profession, sponsored in part by the U.S. Department of Justice). Washington, DC: Author.

ROBERT T. ANGAROLA

SKID ROW   *See* Homelessness and Drugs, History of.

SKIN DAMAGE AND DRUGS   *See* Complications: Dermatological.

SLANG AND JARGON   Slang terms in the drug subculture are constantly changing, as its ethnic, social, and demographic composition changes and as new illicit drugs roll in and roll out with the tides of fashion; there are also geographical variations. Yet certain terms show a remarkable durability—for example, some of those for heroin (trademarked Heroin in Germany, 1898)—a narcotic that has been a staple street anodyne since the early 1900s. Other drug-related terms have come into the mainstream and become part of the English language, e.g., *yen, hooked, pad, spaced out, high, hip.* Many of the following words had been in use during much of the twentieth century (a few antiques of sociological or historical interest are included) and some are the product of the 1980s and 1990s (welcome to the Age of Crack). Origins, if known, are given.

a   amphetamines, a stimulant

a-bomb, bomb   LSD, a hallucinogen

acid   [a shortening of *d*-lysergic acid diethylamide; since about 1960] LSD

Adam   [originally named to connote a primordial man in a state of innocence] MDMA, a mild hallucinogen. See **ecstasy** below

amp [from *ampule*—the drug is sold in small glass ampules, which are broken open and the contents inhaled] amyl nitrite, a dilator of small blood vessels and used in medicine for angina pains; used illicitly to intensify orgasm or for the stimulation effect

amps amphetamines

angel dust [since the 1970s] phencyclidine (a brand name is Sernyl), an anesthetic used on animals but originally on humans; discontinued because of bizarre mental effects. See **PCP** below

base the pure alkaloid of cocaine that has been extracted from the salt (cocaine hydrochloride), in the form of a hard white crust or rock. See **crack** and **rock** below

beamed up [from "Beam me up, Scotty," an expression used in the television series *Star Trek*; **Scotty** is also a term for **crack** cocaine; *on a mission* means looking for crack] intoxicated by crack

beamer a **crack** addict

beast LSD

beat [from the idea of *beating*—cheating—someone] a bogus or mislabeled drug or a substance resembling a certain drug and sold as that drug (soap chips as **crack**; methamphetamine or baking soda as cocaine; catnip as marijuana; **PCP** as LSD, mescaline, or tetrahydrocannabinol (THC)—the active principle of marijuana; procaine as cocaine)

Big H heroin

blank nonpsychoactive powder sold as a drug

blast a drag of **crack** smoke from a pipe

blotter [doses of the drug are dripped on a sheet of blotter paper for sale] LSD

blow (1) to sniff a drug (2) cocaine (3) to smoke marijuana ("blow a **stick**")

blue heavens methaqualone (a sedative) pills

bone a marijuana cigarette; a **joint**

boomers hallucinogenic mushrooms containing psilocybin

bottles vials or small containers for selling **crack**

boy heroin

brown heroin from Mexico diluted with brown milk sugar (lactose), which is less pure than **China white. Also called Mexican mud**

buds [from the appearance] marijuana or sinsemilla (a hybrid variety of marijuana; see **sinse**); a quantity for sale consisting mainly of the more potent flowering tops of the marijuana plant (*Cannabis sativa*)

bump (1) a small mound of cocaine, which is sniffed. (2) cocaine

bush [from the *righteous bush*] marijuana

bust [from 1930s Harlem slang for a police raid, perhaps a shortening of *busting in*] arrest

button [from the shape of the appendages to the peyote cactus containing mescaline] peyote or San Pedro cactus

buzz, buzzed [from *buzz*, onomatopoeic equivalent of subjective feeling; the onset of the drug sometimes causes buzzing in the ears] (1) high on marijuana. (2) an inferior high from heroin

chalk [from the appearance] crystal methamphetamine or cocaine

Charlie cocaine

chasing the dragon [from a Chinese expression for inhaling fumes of heroin after heating it; the melting drug resembles a wriggling snake or dragon] (1) inhaling heroin fumes after the substance is heated on a piece of tinfoil. (2) smoking a mixture of crack and heroin

cheba marijuana

China white [from China (Indochina) **white** or **white stuff** = heroin; since the 1970s] (1) relatively pure heroin from Southeast Asia. (2) analogs of fentanyl (Sublimaze), an opioid more potent than heroin and sold on the street as **China white**

chipping, to chippy using heroin occasionally, avoiding addiction

cocoa puff [pun on the name of a chocolate-flavored breakfast cereal] a **joint,** to which cocaine has been added

coke cocaine

cola [a word play on *coke, cocaine,* and *Coca-Cola*; cocaine is derived from the coca (not the kola) plant] cocaine

cold turkey [from the gooseflesh that is part of abrupt withdrawal] by extension, ending a drug habit without medicinal or professional help, "going cold turkey"

coming down [from a **high**] losing the effects of a drug, all the way down to **crashing**

connect [from the *connection*, a drug pusher] cocaine importer or wholesaler, who fronts (consigns) cocaine to a supplier, who in turn distributes to a street retailer. See **dealing, mule, runner, steerer, touting**

cop [from British slang of the 1700s; to obtain, to steal, to buy; since the 1890s] to get or purchase illicit drugs

cop a buzz   get **high**

copping zone   an area where drugs are sold

crack   [from the crackling sound when smoked in a pipe] pebbles of cocaine **base** that are smoked

crack house   house or apartment (sometimes, an abandoned building) where **crack**-cocaine is sold and smoked on the premises 24/7 — twenty-four hours a day, seven days a week

crank   crystal methamphetamine

crank lite   [from *crank*, because of the amphetamine-like stimulant effect + *lite*, meaning lighter, as in low-alcohol beer] ephedrine, a stimulant used in nonprescription medicines as a decongestant, which is lighter than amphetamines

crash, crashing   to come all the way down from a drug **high**

cross roads   [from the scored cross on the tablets] amphetamines

crystal   [in powder form] methamphetamine or cocaine

cut   to add adulterants to a drug — extending it to make more money in selling it (some adulterants are relatively harmless, some toxic)

deadeye   blank stare produced by an overdose of phencyclidine (**PCP**) or other drug

dealing   [from *dealer*, a person who sells drugs; since the 1920s] selling drugs of all kinds

designer drugs   synthetic compounds or drug analogs that produce the effects of certain regulated drugs but have slight differences in chemical composition to evade the regulatory law; e.g., analogs of fentanyl (**China white**); analogs of amphetamine and methamphetamine such as MDA, MDMA (**ecstasy**), TMA, MMDA, MDE (**Eve**), MBDB; and toxic byproducts of the synthetic opiate meperidine (Demerol) such as MPTP and MPPP

ditch   veins on the inside of the arm at the elbow, a site for injecting heroin. See **tracks** below

do drugs   take or use illicit drugs

doobie   a marijuana cigarette; a **joint**

dope   [from Dutch *doop*, sauce (from *dopen*, to dip). In the late nineteenth century, the term came to be applied to opium, a black gum shaped into pellets and smoked in a pipe] (1) drugs (2) marijuana (3) heroin and other illicit drugs (4) intoxicating fumes of airplane fuel, glue (5) Coca-Cola

dope fiend   [opprobrious term for narcotic and il-

licit drug users since the early 1900s; the term is used ironically by drug users to defy the social stigma] drug user, drug abuser, drug addict

dosing   slipping a hallucinogenic drug into punch, brownies, etc., so that it will unwittingly be consumed by others

drag   to draw or pull on smoke from a cigarette, pipe, or other item, "to take a drag"; to convey that smoke into one's throat and lungs. See **toke** below

drop   to swallow LSD or a pill

dugie, doojee   [phonetic] heroin

dusting   (1) mixing either cocaine with tobacco in a cigarette or mixing heroin or opium with marijuana or hashish in a joint. (2) smoking **PCP**

ecstasy, extacy   [from the euphoria, heightened sensuality, intensified sexual desire attributed to the drug experience] MDMA (methylenedioxymethamphetamine), a mildly hallucinogenic drug synthesized from methamphetamine and resembling mescaline and LSD in chemical structure

eightball   an eighth of an ounce of cocaine

Emilio   [as in Emilio and Maria (Mary), from **Mary Jane**] marijuana

energize me   give me some **crack**

equalizer   pebbles of **crack**-cocaine

Eve   [variant of **Adam**, MDMA or **ecstasy**] MDE, a mild hallucinogen derived from amphetamine. *Adam* and *Eve* is a compound of MDMA + MDE = MDEA (n-ethyl-MDA or 3,4,methylene + dioxy-N-ethylamphetamine)

exing   taking **ecstasy**

fix   (1) a needed drug dose to hold off withdrawal (2) a shot of heroin. See **shoot** below

flake   [from the appearance] (1) cocaine hydrochloride (2) the sediment off a **rock** or chunk of cocaine

Flying Saucers   [trade name] hallucinogenic seeds of a variety of morning glory

freebase   [the psychoactive alkaloid, the **base,** has been *freed* or extracted from the cocaine hydrochloride] (1) crystals of pure cocaine. (2) to prepare the **base;** to smoke it

frost freak   one who inhales the fumes of Freon, a coolant gas, to get **high**

funky green luggage   a supply of marijuana in one's baggage

ganja   [from *gaja*, Hindi word for India's potent

marijuana, consisting of the flowering tops and leaves of the hemp plant, where most of the psychoactive resin is concentrated] marijuana

garbage can   drug user who takes anything, everything, combinations

ghost   LSD

girl   cocaine

gluey   one who inhales glue fumes

goofing   [from *goofballs* = barbiturates, and from *goof*, to act silly, stupidly, heedlessly] under the influence of barbiturates

grass   marijuana chopped up fine for smoking, which looks like dried grass

green   [harvested hemp leaves that are not properly cured; also, the lower leaves of the hemp plant, which contain a smaller proportion of the psychoactive resin] (1) marijuana of low potency, e.g., *Chicago green.* (2) ketamine, an anesthetic similar to phencyclidine (**PCP**) but milder in its effects, which is sprinkled on parsley or marijuana and smoked

H   heroin; also **Big H**

hash, hashish   the concentrated resin of the marijuana plant, containing a high percentage of the active principle, tetrahydrocannabinol (THC).

hash oil   liquid extracted from **hashish**, providing a more potent dose of the active principle and more easily transported in vials. It produces more sedation and deeper states of reverie than does hashish

Henry, Harry   heroin

herb   [used to connote a benign natural substance] marijuana

high   [from the sense of euphoria, being above it all, detached from unpleasant reality] intoxicated by a drug

hip   [from *laying (on) the hip*, to smoke opium—the addict lay on his side on a **pad** in an **opium den**—hence an opium user and then extended to illicit drug users. In the alienated subculture of the jazz scene of the 1930s and 1940s, using drugs was expected and made one keenly informed or *hip*—originally *hep*—until "squares" adopted the word] sophisticated, knowing, "in"; possessing taste, knowledge, awareness of the newest, and a lifestyle superior to that of conventional people

hippies   [from *hip;* diminutive of *hipster*, from *hepcat*, a 1940s term for one who was "cool," detached, self-composed, who lived for the mo-

ment; a pure hedonist who acted out sexual and aggressive drives] in the 1960s and 1970s, young people who dropped out of school or work to pursue a bucolic lifestyle. It soon centered around the taking of hallucinogenic and other mind-expanding drugs, meditation, folk and rock music, poetic perceptions, Eastern mysticism, tribalism, and love—sexual and universal (flower power). From about 1965, the term was applied by the media to a subculture of alienated baby-boomers, many of them college educated and from middle- to upper-class backgrounds. Some migrated to the Haight-Ashbury section of San Francisco, others to rural areas where they lived in back-to-nature communes. Some became antiwar, anti-establishment political revolutionaries (see **yippies**). Some became mystics and joined cults. Most returned to mainstream society, albeit affected by their experiences

hit   (1) an injection of a narcotic. (2) a **snort** of cocaine. (3) a **drag** from a crack pipe. (4) a **toke** of marijuana. (5) to adulterate (**cut**) a drug

hog   [from its original use as a veterinary anesthetic] phencyclidine (**PCP**)

horse   heroin

hot shot   a potent dose of heroin sufficient to kill; heroin laced with cyanide

ice   a crystalline methamphetamine, usually made in clandestine labs

J, jay   [from **joint**] a marijuana cigarette

jelly babies or **beans**   amphetamine pills

joint   [from *joint* as part of paraphernalia for injecting narcotics—particularly the needle; since the 1920s] a marijuana cigarette

jonesing   [after John Jones, the British physician who first described opiate withdrawal in 1700] withdrawal from addiction; by extension, craving any drug

Julio   marijuana. See **Emilio** and **Mary Jane**

junk   [from *junker*, a pusher or peddler; since the 1920s. Also possibly from a word for *opium*—a play on *junk*, a Chinese boat—which was later extended to all narcotics] heroin (which is derived from opium)

K, super K, special K   ketamine, an anesthetic similar in structure to **PCP**

keester plant   [from *keester*, rump, and *plant*, to place] drugs in a rubber container or condom concealed in the rectum

kick the gong (around)   to smoke opium (especially in a Chinese **opium den**)

kick the habit   [related to *kick it out*—to suffer withdrawal symptoms, which include muscle spasms in the legs and kicking movements from hyperactive reflexes in the spinal cord] (1) abrupt withdrawal from a drug to which one is addicted. (2) to conquer drug dependence

kind buds   potent marijuana. See **buds** above

LA coke   ketamine. See **K** above

laughing gas   nitrous oxide

lid   [from the now obsolete practice of selling a measure of marijuana in a pipe tobacco tin] an ounce of marijuana, usually sold in a plastic bag

line   (1) a thin stream of cocaine on a mirror or other smooth surface, which is sniffed through a *quill*—a rolled matchbook cover, tube, straw, or tightly rolled dollar bill, etc. (2) a measure of cocaine for sale

luding out   [from *ludes*, short for Quaaludes (a brand name for methaqualone, an addictive sedative)] taking methaqualone.

Lyle   [from *lysergic acid*] LSD

mainline   [from *main line*, a major rail route; since the 1920s] (1) the large vein in the arm; the most accessible vein. (2) *v.* to inject morphine, heroin, or cocaine into any vein

Mary Jane, MJ   marijuana

Mexican brown   marijuana from Mexico

Mexican mud   brown heroin from Mexico. See **brown** above

mind altering   the claimed mental effects of hallucinogenic drugs—altered or intensified states of perception

mind expansion   [related to *psychedelic*, mind-manifesting; a descriptive term for hallucinogenic drugs coined in the 1960s] the claimed **mind-altering** effects of hallucinogenic drugs, including greater spirituality, enhanced self-awareness, and increased sensitivity to music, art, and nature; also synesthesia—cross-sensations, such as "seeing" music or "hearing" colors

Miss Emma   morphine

monkey on one's back   desperate desire for drugs; addiction; craving

moon   [from the shape of slices of the bud of the peyote cactus] peyote

moonrock   heroin mixed with **crack** for smoking

Moroccan candy   [*majoun* (Arabic) is candy laced with **hashish**, sold in Morocco, Afghanistan] hashish. See **hash** above

mule   [animal used as a beast of burden] (1) a low-level drug smuggler from Latin America; mules often swallow a condom filled with cocaine to be delivered at a destination—a dangerous practice called *bodypacking*. (2) heroin

night train   PCP

opium den   [from *den*, an animal's lair. The term was coined by Westerners in nineteenth-century China, to have lurid connotations] a place where opium is smoked. Chinese laborers brought the practice of smoking opium to America during the gold rush of 1849 and the 1850s and the building of the transcontinental railroad

pad   [from the mats in **opium dens** on which the smokers reclined and slept. In the 1930s, Harlem apartments where marijuana was sold and smoked while reclining on couches or mattresses were called *tea pads*] (1) private place for taking drugs; a variant is **crash pad,** a place for recovering from the effects of a methamphetamine *run* (period of extended use); the user collapses (**crashes**) into an exhausted sleep. (2) by extension, since the 1950s, any dwelling place, room, apartment

PCP   [from *PeaCe Pill*] phencyclidine (brand name Sernyl), a veterinary anesthetic that induces bizarre mental states in humans

pearls   [medical nickname] amyl nitrite ampules

Persian white   fentanyl. See **China white** above

p-funk, p-dope   [*p* stands for pure] fentanyl. See **China white** above

PG   paregoric, a traditional diarrhea remedy containing opium.

piece   hashish, a form of marijuana. See **hash** above

pill popping   [from *popping* something into one's mouth] promiscuous use of amphetamine and barbiturate pills or capsules. One who does this is a *popper* and may be a **garbage can**

pit   veins on the inside of the arm at the elbow, a main site for injecting heroin and the place to look for **tracks.** See **ditch** above

pop   to inject. See **shoot** below

poppers   [the glass ampule is *popped* open and the contents inhaled] amyl nitrite ampules

pot   [from *potaguaya*, a Mexican-Indian word for marijuana] marijuana

pusher [extension from *pusher*—a person who circulates counterfeit money; since the 1920s] drug seller, drug dealer. See **dealing** above

quas, quacks [from Quaalude, a brand name of methaqualone] methaqualone pills, an addictive sedative

Raoul cocaine

reds, red birds [also called red devils, red jackets, red caps—from the color of the capsules] Seconal (a brand of secobarbital) capsules

reefer [from *grifa*, a Mexican-Spanish word for marijuana] (1) a marijuana cigarette. (2) marijuana

righteous bush marijuana plant

ringer [from the idea of "hearing bells"; *bells* is a term for **crack**] powerful effect from a **hit** of crack

roach [from its resemblance to a cockroach] the butt (end) of a marijuana cigarette

rock [from the appearance](1) large crystals or a chunk of pure cocaine hydrochloride. (2) **crack.** See **base** above

runner a messenger (often a juvenile) who delivers drugs from the seller to the buyer (not to be confused with a *drug runner*, a smuggler)

rush the quick initial onset of orgasmic sensations—of warmth, euphoria, and relaxation after injecting or inhaling heroin, cocaine, or methamphetamine

scag heroin

schoolboy (1) codeine, a derivative of opium with relatively low potency, used as a cough suppressant and analgesic. (2) morphine

Scotty crack-cocaine. See **beamed up** above

script prescription for a drug, often forged by addicts

script doctor a physician who will provide a drug prescription for a price—or one who is deceived into providing one

shake [the mixture is made by shaking the drug and the adulterant] (1) cocaine adulterated (**cut**) with a harmless substance such as mannitol. (2) loose marijuana left at the bottom of a bag that held a pressed block of marijuana.

sheet (acid) [from decorated blotter paper containing doses of the drug] LSD

shit heroin

shoot inject a drug; also *shoot up* a **fix** or a shot (usually of heroin)

shooting gallery place where heroin addicts

shoot up and share needles and other **works** (paraphernalia)

shoot the breeze inhale nitrous oxide (called **laughing gas**).

shrooming high on hallucinogenic mushrooms

Sid a play on the *s–d* sound of LSD

sinse [from *sinsemilla*, without seeds] a hybrid variety of marijuana; also called *ses*

skin popping [from **pop,** to inject] injecting heroin or any psychoactive drug subcutaneously (rather than into a vein), a practice of casual (**chippy**) users.

smack [perhaps from *shmek*, Yiddish word for sniff, whiff, pinch of snuff; since the 1910s, when heroin users sniffed the drug; in the 1920s and 1930s, some Jewish mobsters were involved in heroin trafficking] heroin

smoke marijuana

snappers [the ampule containing the drug is *snapped* open] amyl nitrite capsules

snob [from the idea of an elite—expensive—drug] cocaine.

snop marijuana

snort to sniff a drug

snow [from the appearance; also, the drug is a topical anesthetic and numbs the mucous membranes] cocaine hydrochloride.

soapers [from Sopor, the brand name of a sedative, now taken off the market] methaqualone pills

space basing or **space blasting** smoking a mixture of **crack** and phencyclidine (**PCP**)

speed (1) amphetamines (2) caffeine pills (3) diet pills

speedball [first used by GIs during the Korean War] injected mixture of heroin and cocaine.

splif a fat marijuana cigarette

spook heroin

squirrel a mixture of **PCP** and marijuana sprinkled with cocaine and smoked

stash [extension of *stash*, hobo argot for hiding place; since the 1800s (1) hiding place for drugs. (2) a supply of drugs. (3) *v.* to hide drugs

steerer member of a cocaine or heroin crew who directs people to the seller

stepped on adulterated or **cut**

stick a marijuana cigarette

street drugs drugs purchased from sellers on the street; hence, of dubious quality

strung out severely addicted

sunshine [from the type sold as an orange-colored tablet] LSD

super grass [the powder is sometimes mixed with parsley or marijuana and smoked] ketamine. See **green.**

tabs [from *tablet*, a form in which the drug is sold] LSD

tea marijuana

Thai stick potent marijuana from Thailand

thing (1) heroin. (2) *pl.* an addict's **works**—the hypodermic needle (needle and syringe)

tic [from *THC*] fake tetrahydrocannabinol

toke a **drag** on a marijuana cigarette

tooies [from Tuinal, a brand name for a preparation containing amobarbital and secobarbital] sedative capsules

toot (1) to sniff cocaine. (2) cocaine. (3) a binge, especially a drinking bout or spree (since the late 1700s)

touting (1) purchasing drugs for someone else. (2) advertising, *hawking*, drugs that one is selling

tracks a line of scabs and scars from frequent intravenous injections. See **pit** and **ditch** above

tripping [from *trip*, in the sense of a psychic "journey"] taking LSD

trips (1) LSD tablets (2) periods under the influence of various drugs, usually hallucinogens

turkey [from *turkey*, a jerk; or from a theatrical failure or flop] (1) a nonpsychoactive substance sold as a drug. (2) the seller of such phony substances

turn on take drugs, especially hallucinogens

ups, uppers amphetamines

V, Vs Valium (a brand name for diazepam, a tranquilizer) tablets

wasted [from *waste*, a street-gang term since the 1950s, meaning to kill, beat up, destroy] (1) severely addicted to the point of mental and physical depletion (2) extremely intoxicated—out of it, beyond caring

whack (1) to adulterate heroin, cocaine, or other drugs. (2) an adulterant (3) phencyclidine (**PCP**). (4) to kill

whiff [from the idea of smelling or sniffing] cocaine

white or **white stuff** heroin

white lady [from the color] cocaine

window pane [the drug is sometimes sold in a clear plastic square; also of a greater potency, providing a more intense experience and non-structured sensations—"opening a window on reality"] LSD

wired (1) extremely intoxicated by cocaine. (2) anxious and jittery from stimulants (may be related to *amped*, a play on amphetamines and amperes)

woola [phonetic spelling] a **joint** containing a mixture of marijuana and **crack**

works equipment or paraphernalia for injecting drugs

X, the X, XTC [from **ecstasy**] MDMA.

yellow jackets [from the color of the capsules] Nembutal brand of pentobarbital

yen [from English slang *yen-yen*, the opium habit, based on Cantonese *in-yan* (*in*, opium + *yan*, craving); since the 1800s] any strong craving

yippies [variant of *hippie*; the movement's official name was Youth International Party] A self-proclaimed revolutionary political movement founded by Jerry Rubin and Abbie Hoffman in 1968 to disrupt the Chicago Democratic convention and protest the Vietnam War and the politics that supported it. Yippies challenged the establishment with a Festival of Life (inviting drug-using **hippies**); this included LSD seminars, rock shows, light shows, films, marches, love-ins, put-ons, guerrilla theater, and bizarre stunts (e.g., nominating a pig named Pigasus for President). They were beaten and imprisoned by Chicago police in a rioting atmosphere and the leaders were tried as the Chicago Seven

yuppies [from *young upwardly mobile professionals*] a term coined in the 1980s for prosperous, upper-middle class people in their 30s. They were stereotyped as being dedicated to the "good life," making money, and consumerism. They retained some of the hedonistic traits of the **hippies**—drug use and a love of rock music—but they also had establishment goals, a fondness for high-tech gadgets, the fitness craze, country homes, foreign cars, and wristwatches. Politically, they were conservative, opposing government intervention in the economy and in social-welfare programs; however, they were "liberal" on some issues, such as preserving the environment, particularly that part of it adjoining their weekend homes

zenes [short for Thorazine, a brand name for chlorpromazine] tranquilizer pills

988

zombie (1) **crack** cocaine. (2) phencyclidine (**PCP**)
zooted up high on **crack**-cocaine

(SEE ALSO: *Argot; Haight Ashbury Free Clinic; Yippies*)

BIBLIOGRAPHY

EISNER, BRUCE. (1989). *Ecstasy: The MDMA story.* Berkeley: Ronin.

HURST, GEROLD, & HURST, HOLGER. (1981). *The international drug scene.* Wurzburg, Germany.

JULIEN, ROBERT M. (1992). *A primer of drug action,* 6th ed. New York: W. H. Freeman.

LINGEMAN, RICHARD. (1974). *Drugs from a to z,* 2nd ed. New York: McGraw-Hill.

MENCKEN, H. L. (1967). *The American language,* abridged with new material by Raven I. McDavid Jr. New York: Knopf.

PARTRIDGE, ERIC. (1961). *A dictionary of the underworld.* New York: Bonanza.

SEYMOUR, RICHARD B., & SMITH, DAVID E. (1987). *Guide to psychoactive drugs: An up-to-the minute reference to mind-altering substances.* New York: Harrington Park Press.

SPEARS, RICHARD A. (1986). *The slang and jargon of drugs and drink.* Metuchen, NJ: Scarecrow Press.

WENTWORTH, HAROLD, & FLEXNER, STUART BERG. (1968). *The pocket dictionary of American slang.* New York: Pocket Books.

WILLIAMS, TERRY. (1989). *The cocaine kids.* New York: Addison-Wesley.

RICHARD LINGEMAN

SLEEP, DREAMING, AND DRUGS

The use of "mind-altering" drugs and intoxicating drinks to hasten the onset of sleep and to enhance the experience of dreaming is a worldwide phenomenon and goes back to prehistory. The ancient Greeks used hallucinatory substances for religious purposes. The priestesses at Delphi, for example, chewed certain leaves while sitting in a smoke-filled chamber and going into a trance. On returning to consciousness, they would bring forth a divine prophecy. The various Dionysian cults encouraged their celebrants into ecstatic dream-like states through the use of wine and perhaps other drugs (Cohen, 1977).

The ancient Hindus imbibed a sacred drink called "soma," and MARIJUANA was used in practices of meditation. For the Arabs, HASHISH (a form of marijuana) was the substance of choice, while the Incas chewed the leaves of the COCA plant (from which COCAINE may be made). The OPIUM poppy was used in Asia, and the ancient Mexicans used a variety of powerful PSYCHOACTIVE substances, including PEYOTE, sacred mushrooms, and seeds from the Mexican MORNING GLORY plant, to enter the realm of dreams. The Australian aboriginals used the pituri, a psychoactive substance, to take them into "dream time," as they referred to it.

Belladonna and OPIATES have historically been used for the specific purpose of producing vivid dreams. The most famous illustration is the story of the English poet Samuel Taylor Coleridge (1772–1834), who allegedly wrote his most celebrated work, "Kubla Khan," during a drug-induced dream (Cohen, 1977). LYSERGIC ACID DIETHYLAMIDE (LSD) became popular in the United States and Europe during the 1960s for ostensibly facilitating higher states of consciousness and creativity. The writer John Lilley used a sensory-deprivation tank to emulate the state of sleep while taking LSD to induce creative dreaming (Cohen, 1977).

Reference to the effects of drugs and ALCOHOL on sleep and dreaming are also found in popular literature. It was a mixture made from poppies that caused Dorothy and her companions to fall into deep sleep in the *Wizard of Oz* (Baum, 1956). After ingesting a series of pills and liquids, in *Through the Looking Glass*, Alice finds herself in "Wonderland," where she has a conversation with an opium-smoking caterpillar who is sitting on a magic mushroom that alters the state of one who eats of it. After returning to the reality of her home in England, Alice realizes that she had, of course, fallen asleep and been dreaming (Carroll, 1951).

Modern study of the effects of drugs and alcohol on sleep and dreams dates to the mid-1950s. With the use of electrophysiological machines, including electroencephalograms (EEGs), electrooculograms, and electromyograms, the state of sleep most closely associated with dreaming was discovered, studied, and named REM, for the *rapid eye movements* unique to that sleep state. In humans, REM sleep recurs in approximately 90-minute cycles throughout the sleep period, resulting in 4 or 5 REM episodes per night, each lasting from 10 to 30 minutes. Adults spend about 20 to 25 percent of their sleep period in REM sleep. Abrupt, but not gradual, awakening from REM sleep is consistently associated with the recall of vivid dreaming. While the function of REM sleep is unknown, it appears to serve a necessary function. Deprivation of REM sleep by awakenings or by the administration of REM-suppressing drugs leads to a compensatory or rebound effect—specifically, a more rapid onset and a greater amount and intensity of REM sleep.

Most psychoactive substances have profound effects on sleep and particularly on REM sleep. While the effects of drugs on REM sleep are known, their effects on dreaming are being studied. Given the association of REM sleep and dreaming, one might think that REM-enhancing drugs would increase dreaming, while REM-suppressing drugs would decrease dreaming. But no data suggest such a simple relationship. After the discontinuation of REM-suppressing drugs, a REM rebound occurs, which is reported to be associated with increased and unpleasant dreams. Some have hypothesized that the visual HALLUCINATIONS experienced during discontinuation of some drugs (e.g., alcohol) is a REM rebound intruding into wakefulness. It is too simplistic to think of dreaming and REM in a one-to-one correspondence, but it is reasonable to assume that drugs affecting REM will also affect the frequency and nature of dreams.

The effects of ethanol (alcohol) on sleep are complex and somewhat paradoxical. The acute bedtime administration of ethanol to healthy, nonalcoholic volunteers shortens the latency to sleep onset and, depending on dose, may initially increase the amount of relaxed, deep slow-wave (delta-wave) sleep (Williams & Salamy, 1972). Additionally, ethanol reduces the amount of REM sleep, usually affecting the first or second REM period. An ethanol concentration in the blood of 50-milligram percent (mg%) or greater (100-mg% is legal intoxication in most states) is necessary for observing these sleep effects. The sleep effects of ethanol are observed only during the first half of an 8-hour sleep period. Ethanol is metabolized at a constant rate, and consequently the usual dose of ethanol (50–90 mg%) given in these studies is almost completely eliminated from the body after 4 or 5 hours.

Following elimination of ethanol, an apparent compensatory effect on sleep occurs. During the lat-

ter half of sleep, increased amounts of REM sleep and increased wakefulness or light sleep is found (Williams & Salamy, 1972). Within three to four nights of repeated administration of the same dose, the initial effects on sleep are lost (e.g., tolerance occurs), while the secondary disruption of sleep during the latter half of the night remains. REM sleep time and sleep latency return to their basal levels, and the effects on slow-wave sleep, when initially present, do not persist. When nightly administration of ethanol is discontinued, a REM rebound is seen. But the REM rebound after repeated nightly ethanol administration in healthy, nonalcoholic subjects is not a particularly consistent result (Vogel et al., 1990). In alcoholics, however, the REM rebound is intense and persistent (Williams & Salamy, 1972). Some believe the presence of a REM rebound is a characteristic of drugs with a high addictive potential.

MORPHINE, the opiate ANALGESIC (derived from the opium poppy), decreases the number and the duration of REM sleep episodes and delays the onset of the first REM period (Kay et al., 1969). It also increases awakenings and light sleep and suppresses slow-wave sleep. HEROIN, a semisynthetic opiate, also suppresses REM sleep and slow-wave sleep and increases wakefulness and light sleep, producing a disruption of the usual continuity of sleep. Heroin appears to be more potent than morphine in its sleep effects. The synthetic opiate, METHADONE, has similar effects on sleep and wakefulness, with a potency more comparable to that of morphine. When an opiate is administered just before the onset of sleep, the EEG pattern shows isolated bursts of delta waves on the background of a waking pattern. Animal studies have correlated these delta bursts with the behavior of head nodding (a possible physiological correlate to the street term "being on the nod"). Repeated administration of the opiates at the same dose leads to tolerance of the sleep effects of these drugs, particularly the REM sleep effects (Kay et al., 1969). The cessation of opiate use leads to a protracted REM rebound, increased REM sleep, and a shortened latency to the first REM episode.

Among the stimulants, AMPHETAMINE, when administered before sleep, delays sleep onset, increases wakefulness during the sleep period, and specifically suppresses REM sleep (Rechtschaffen & Maron, 1964). Cessation of chronic amphetamine use is associated with an increase in slow-wave sleep on the first recovery night and, on subsequent nights, with increased amounts of REM sleep and a reduced latency to the first episode of REM sleep, a REM rebound.

Cocaine also has stimulant effects on the central nervous system, and its effects on electroencephalogram readings were first studied by Berger in 1931; he was the researcher who developed the EEG (Berger, 1931). Cocaine was found to increase fast-frequency EEG activity, suggesting an alerting effect. The self-reported use of cocaine during the late afternoon and early evening is associated with reduced nocturnal sleep time. Systematic electrophysiological studies show a reduction of REM sleep (Watson et al., 1989). Cessation of chronic cocaine abuse is followed by increased sleep time and a REM rebound.

The three classic HALLUCINOGENS are LSD, MESCALINE, and PSILOCYBIN. The state experienced following use of hallucinogens is somewhat similar to dreaming. Since REM sleep is highly correlated with dreaming, scientists expected the hallucinogens to facilitate REM sleep, but LSD is the only hallucinogen that has been studied for its effects on sleep. One study done in humans showed that LSD enhanced REM sleep early in the night, although it did not alter the total amount of REM sleep for the night (Muzio et al., 1966). However, studies done in animals all indicate that LSD increases wakefulness and decreases REM sleep (Kay & Martin, 1978). The frequency changes seen in the waking EEG of animals (similar among all three hallucinogens) suggest an arousing effect. Thus the REM suppression in animals may not be a specific REM effect but rather a sleep-suppressing effect (Fairchild et al., 1979).

Another drug with hallucinogenic effects is marijuana, its active ingredient being TETRAHYDROCANNABINOL (THC). The effects of THC on the waking EEG pattern are quite distinct from the effects of the classic hallucinogens cited above (Fairchild et al., 1979). THC has sedating effects at lower doses and hallucinatory effects at higher doses. The acute administration of marijuana or THC to humans is associated with an increase in slow-wave sleep and a reduction in REM sleep (Pivik et al., 1972). When THC is administered chronically (long-term), the effects on slow-wave and REM sleep diminish, indicating the presence of tolerance. Discontinuing the use of marijuana is associated with increased wakefulness and increased REM sleep time (Feinberg et al., 1976).

Most of these drugs, which are also drugs of abuse, seem to alter sleep and specifically the amount and timing of REM sleep. Each affects chemicals in the brain that control sleep and wake and, with chronic use, some adaptation seems to occur. A characteristic REM rebound is seen on discontinuation of dependent drug use. (It may be that the ancients' experience of enhanced dreaming was the REM rebound that is typically associated with protracted drug use.) Some studies indicate that, in the former drug dependent, the occurrence and intensity of the REM rebound has been predictive of relapse to drug use. How the sleep–wake pattern changes, and specifically the REM changes associated with these drugs, contribute to abused drugs' excessive use needs further study.

ACKNOWLEDGMENTS

Supported by National Institutes of Health (NIAAA) grant no. R01 AA07147 awarded to T. Roehrs and (NHLBI) grant no. P50 HL42215 awarded to T. Roth.

(SEE ALSO: *Addiction: Concepts and Definitions; Benzodiazepines: Complications; Sedative-Hypnotics; Sedatives: Adverse Consequences of Chronic Use; Tolerance and Physical Dependence*)

BIBLIOGRAPHY

BAUM, L. F. (1956). *The wizard of Oz.* New York: Grosset and Dunlap.

BERGER, H. (1931). Über das Elektroenkephalogramm des Menschen. *Archiven Psychiat Nervenkrankheiten, 94,* 16–60.

CARROLL, L. (1951). *Alice in Wonderland.* New York: Simon and Schuster.

COHEN, D. (1977). *Dreams, visions and drugs: A search for other realities.* New York: New Viewpoints.

FAIRCHILD, M. D., ET AL. (1979). EEG effects of hallucinogens and cannabinoids using sleep-waking behavior as baseline. *Pharmacology, Biochemistry & Behavior, 12,* 99–105.

FEINBERG, I., ET AL. (1976). Effects of high dosage delta-9-tetrahydrocannabinol on sleep patterns in man. *Clinical Pharmacology Therapeutics, 17,* 458–466.

KAY, D. C., & MARTIN, W. R. (1978). LSD and tryptamine effects on sleep/wakefulness and electrocorticogram patterns in intact cats. *Psychopharmacology, 58,* 223–228.

KAY, D. C., ET AL. (1969). Morphine effects on human REM state, waking state, and NREM sleep. *Psychopharmacology, 14,* 404–416.

MUZIO, J. N., ET AL. (1966). Alterations in the nocturnal sleep cycle resulting from LSD. *Electroenceph Clin Neurophysiol, 21,* 313–324.

PIVIK, R. T., ET AL. (1972). Delta-9-tetrahydrocannabinol and synhexl: Effects on human sleep patterns. *Clinical Pharmacology Therapeutics, 13,* 426–435.

RECHTSCHAFFEN, A., & MARON, L. (1964). The effect of amphetamine on the sleep cycle. *Electroenceph Clinical Neurophysiology, 16,* 438–445.

SHEPARD, L. (1984–1985). *Encyclopedia of occultism and parapsychology,* 2nd ed. Detroit: Gayle Research.

VOGEL, G. W., ET AL. (1990). Drug effects on REM sleep and on endogenous depression. *Neuroscience and Biobehavioral Reviews, 14,* 49–63.

WATSON, R., ET AL. (1989). Cocaine use and withdrawal: The effect on sleep and mood. *Sleep Research, 18,* 83.

WILLIAMS, H., & SALAMY, A. (1972). Alcohol and sleep. In B. Kissin & H. Begleiter (Eds.), *The biology of alcoholism,* Vol. 2. New York: Plenum.

TIMOTHY ROEHRS
THOMAS ROTH

SLEEPING PILLS   This is a general term applied to a number of different drugs in pill form that help induce SLEEP. There is a wide range of such medication and many require a doctor's prescription, but some can be purchased as OVER-THE-COUNTER drugs at a pharmacy. These latter preparations generally contain an antihistamine such as chlorpheniramine maleate, which produces drowsiness.

The prescription medications are much stronger. They include barbiturates, benzodiazepines, and a number of other compounds. In general, the shorter-acting sleeping pills are used to help one relax enough to get to sleep, while the longer-acting ones are used to help prevent frequent awakenings during the night. Long-term or inappropriate use can cause TOLERANCE AND PHYSICAL DEPENDENCE.

(SEE ALSO: *Sedative-Hypnotic; Sedatives: Adverse Consequences of Chronic Use*)

BIBLIOGRAPHY

HARVEY, S. C. (1985). Hypnotics and sedatives. In A. G. Gilman et al. (Eds.), *Goodman and Gilman's the pharmacological basis of therapeutics,* 7th ed. New York: Macmillan.

SCOTT E. LUKAS

SMOKING   *See* Nicotine; Tobacco.

SMOKING CESSATION   *See* Tobacco; Treatment: Tobacco.

SMOKING CESSATION AND WEIGHT GAIN   *See* Tobacco: Smoking Cessation and Weight Gain.

SNUFF   *See* Tobacco: Smokeless.

SOBRIETY   The term *sobriety* is not defined in current medical or psychiatric literature. The term *abstinence* is found more often and is generally agreed upon as the treatment goal for severe alcoholics. Abstinence is defined as nonuse of the substance to which the person was addicted.

The term sobriety is most often used by members of ALCOHOLICS ANONYMOUS (AA) and NARCOTICS ANONYMOUS (NA). It is often preceded by the adjectives "stable" or "serene." Abstinence—the condition of being sober—is a necessary but insufficient condition for sobriety. Sobriety means something different from the *initial* abstinence so often achieved by alcoholics and other drug addicts. This initial abstinence is recognized as a time of vulnerability to RELAPSE, often referred to as a "dry drunk" or "white knuckle sobriety."

According to AA beliefs, recovery from ALCOHOLISM and other addictions calls for more than abstinence. The formerly addicted central nervous system must undergo a substantial readaptation. This means that the CRAVING, drug-seeking, dysphoria (unhappiness), and negative cognitions that characterize early abstinence must not only diminish but must also be replaced by more normal positive behavior.

This readaptation requires time and substitute activities. The activities most associated with successful readaptation are found in TREATMENT programs and in AA or NA.

Sobriety, as used by most recovering people in AA and NA, refers to abstinence plus a program of activity designed to make the abstinence comfortable and to improve functioning in relationships and in other aspects of life. The program of recovery that leads to stable sobriety usually includes (1) attending AA and/or NA meetings (2) "working" the TWELVE STEPS and continuing to use steps 10, 11, and 12 for the maintenance of sobriety; (3) working with a sponsor who acts as a mentor in maintaining sobriety; (4) belonging to a home group and engaging in service activities that help others with their sobriety and (5) other activities that enhance or support sobriety (e.g., exercise, hobbies, and psychotherapy). A program of recovery means the recognition that any activity has potential to either enhance or interfere with the recovering individual's sobriety.

There can be an almost infinite range of variability to this personal program of activities. The variability is one reason why no widely accepted definition of sobriety exists. Another is the recognition by AA and NA of the need to tolerate and accept others who are struggling with the process of recovery. A clear standardized definition would make judgment and criticism easier, thus threatening the individual's sobriety. Despite these problems, the concept of sobriety (abstinence, plus a program of activity designed to make abstinence comfortable) is a useful one for health-care professionals.

(SEE ALSO: *Addiction: Concepts and Definitions; Relapse Prevention; Treatment Types: Minnesota Model; Treatment Types: Self-Help and Anonymous Groups*)

BIBLIOGRAPHY

AMERICAN PSYCHIATRIC ASSOCIATION. (1989). *A.P.A. Task Force: Treatment of Psychiatric Disorders*. Washington, DC.

STONE, E. M. (ED.). (1988). *American psychiatric glossary*. Washington, DC: American Psychiatric Press.

WILFORD, B. B. (ED.). (1990). *Syllabus for the review course in addiction medicine*. Washington, DC: American Society of Addiction Medicine.

JOHN N. CHAPPEL

SOCIAL COSTS OF ALCOHOL AND DRUG ABUSE

Drinking, smoking, and the use of psychotropic drugs have a variety of consequences for those who partake of them, for their families and associates, and for society at large. A number of these consequences are negative. Smokers die young from heart or lung disease, drinkers get into traffic accidents and fights, drug injectors spread the HIV virus. In the context of public policymaking, where priorities must be set for the use of scarce resources, it seems important to have a measure of the overall magnitude of the social burden engendered by such consequences. One familiar approach is to express the magnitude of the problem in terms of the number of people who die each year. When we learn that there are 100,000 deaths per year in the United States from ALCOHOL abuse and perhaps four times that number from TOBACCO use, we know that the stakes are very high in devising sound policies for controlling drinking and smoking. Such statistics, compelling as they are, tell only part of the story. In addition to causing early death, substance abuse makes for a variety of consequences that reduce the quality of life, both for users and other people.

To capture this broad array of consequences in a single number, analysts have estimated various measures of social cost. The estimates are important because they figure in the political process by which federal funds are allocated to the National Institutes of Health and to other agencies that play a role in combatting substance abuse. The most prominent estimates of social costs for substance abuse have utilized the conceptual apparatus developed by a task force of the U.S. Public Health Service chaired by Dorothy Rice (Hodgson & Meiners, 1979).

Although prominent in policy debate, this cost-of-illness (COI) method has been faulted for its emphasis on production as the measure of social welfare. Economists favor a quite different approach that measures social welfare from the perspective of the consumer. The economists' preferred accounting framework is referred to in this article as the "external social-cost" approach.

THE TWO FRAMEWORKS APPLIED TO ALCOHOL ABUSE

A coherent assessment of the social costs of substance abuse requires an accounting framework that specifies criteria for judging which of the myriad effects are properly deemed to be of public concern. For example, in the case of drinking, on any one drinking occasion there may be unwanted, harmful consequences: social embarrassment, loss of reputation or affection, failure to discharge some responsibility at work or home, physical injury from an accident, victimization by a mugger or rapist, and nausea or hangover. Chronic heavy drinking may result in still other consequences, including rejection by family and friends, loss of a job or of an opportunity for promotion, progressive deterioration in physical health, and an early death. In order to capture these and other negative consequences in a single number, the list of consequences must be reviewed to determine which should be considered in establishing priorities for substance abuse policy. The consequences deemed relevant must then be quantified, translated into a standard unit of account (dollars), and summed.

The Cost-of-Illness Framework. The COI approach is concerned with measuring the loss or diversion of productive resources resulting from an illness or activity. In the case of alcohol abuse, human capital resources are lost and the gross national product reduced by the morbidity and early death suffered by some drinkers, whether as a result of injuries sustained in alcohol-related traffic accidents or violent crime or as a result of organ damage and other diseases stemming from chronic heavy drinking. The loss to society in these cases is equal to the loss of the marginal product of the victims' labor, valued at the market wage. Unpaid work at home, including housework and child care, is included in the computation, with values being assigned according to how much households pay for such services when they are performed by paid help.

The COI approach also takes account of the diversion of resources from other productive uses necessitated by alcohol abuse. Thus the costs of medical care for alcohol-related illness, treatment for ALCOHOLISM, and research on prevention and treatment are incorporated in the social-cost estimate. Similarly, the value of law-enforcement and justice resources devoted to alcohol-related crimes are included, as are the costs of replacing property damaged in traffic crashes and fires caused by drinking.

Several prominent estimates of the total costs of alcohol abuse for the United States have utilized the COI framework (Berry & Boland, 1977; Harwood et al., 1984). In 1990, Dorothy Rice and her associates produced the most complete COI study to date. They found that the economic costs to society of alcohol abuse totaled $70.3 billion in 1985, broken down as follows:

Core Costs:

Treatment	$6.3 Professional services, hospitalization
Support	$0.5 Research, training, administration
Direct TOTAL	$6.8
Morbidity	$27.4 Value of reduced productivity
Mortality	$24.0 Value of lost productivity
Indirect TOTAL	$51.4
Other Related Costs	$10.5 Law enforcement, property damage, etc.
Fetal Alcohol Syndrome (FAS)	$1.6 Long-term residential care, etc.

About three quarters of the total cost in this tabulation is the value of labor PRODUCTIVITY lost as the result of illness, injury, or early death. The human capital lost as a result of alcohol-related mortality was computed for all those who died in 1985 from causes in which intoxication or chronic heavy drinking played a role. These include traffic fatalities and deaths from liver cirrhosis, among other causes. The lost human capital was valued by estimating how much the deceased would have earned if they had lived and worked until retirement age.

The human capital lost as a result of morbidity was calculated by estimating the reduction in the productivity of the labor force resulting from alcohol dependence or abuse. Rice et al. combined two sets of estimates to arrive at this number: first, the percentage of the labor force in 1985 that was or had ever been subject to a diagnosis of alcohol dependence or abuse; and second, an estimate of the loss in earnings associated with such a diagnosis.

Critique Estimates of this sort have been challenged for two reasons. The first challenge is to the statistical methods used to generate the estimates of morbidity, mortality, and lost earnings (Cook, 1991). The second challenge is more fundamental, for it concerns the basic principles that inform the COI accounting framework.

The COI procedure estimates the cost of morbidity and mortality in terms of lost productivity, but this emphasis on production as the measure of social welfare seems misplaced. A more liberal perspective, favored by economists among others, shifts the emphasis to consumption and interprets the task of measuring social welfare in terms of aggregating individual preferences. Consumers are the best judges of their own welfare, and if sometimes they make choices that fail to maximize their productivity, that should not in itself be regarded as problematic. In this view, the choices that people make concerning how hard to work and when to retire are of little public concern. The same goes for choices that place one's own health and safety at risk. Thus in economics there is a strong presumption in favor of consumer sovereignty, the principle that the individual consumer is in the best position to define what is best for him or her, and that social welfare is enhanced by free choice within certain limits. A negative consequence is deemed to be of *public* concern only when the actions of one individual impinge negatively on the welfare of others. The basic distinction, then, is between *internal* and *external* consequences of individual decisions, where the latter impose an involuntary cost on other people.

In the case of alcohol abuse, the internal costs include those suffered by drinkers and are foreseeable as a natural consequence of their choices. A small example explains the reasoning here. Suppose a woman decides to drink heavily tonight despite knowing that she may be tired and unproductive tomorrow. By making this decision, she is indicating that for her the pleasure of partying outweighs the "morning-after" costs. If no one else is harmed by this decision, the external costs are zero. If she were to drive after drinking, however, the accounting would change. She would be risking serious injury to herself and to others on the highway. Her injury would have external costs to the extent that a third party (group insurance or Medicaid) paid her medical expenses. The risk that she might injure other people while driving is also a negative externality, to be valued at the expected loss to them. That cost, incidentally, is not limited to their lost earnings, but also includes their pain and suffering and the suffering of those who care about them.

In sum, the most fundamental challenge to the COI framework relates to its presumption that social welfare is synonymous with national product. Econ-

omists argue instead that the preferences of individuals are the proper measure of their well-being and that social welfare is the sum total of individual welfare. Some of the major costs in the COI framework, especially lost earnings, are less important in the external social-cost view, whereas a number of costs that are ignored in COI become important when the focus is on external costs.

The External Social-Cost Framework. In a recent study at the Rand Corporation, economists applied the ESC framework to alcohol abuse and other poor health habits (Manning et al., 1989, 1991). Their estimate for alcohol abuse amounted to about $30 billion in 1985, less than half the COI estimate presented above for the same year. The accounting procedures used to generate this estimate of the ESC can be briefly summarized:

1. **Earnings.** Heavy drinkers might earn less than they otherwise would have during their careers and might have their careers cut short by poor health and early death. Although the most obvious effect was a reduced standard of living, which was properly considered a private cost, a number of programs created a collective interest in the productivity of each individual. For example, those who died young saved their fellow citizens the expense of years of pension payments and medical costs. Those who retired early (perhaps because of poor health) imposed financial costs on others in the sense that their contributions to the Social Security system were reduced. Thus these collective financing arrangements had the effect of creating both external costs and benefits in relation to heavy drinking. The net effect, according to Manning et al. (1991), was negative, and equaled about 22 percent of the total external cost.

2. **Traffic Fatalities.** Manning et al. (1991) reported that about 7,400 of the 22,400 people who died in alcohol-related traffic ACCIDENTS in 1985 were "innocent," in the sense that they had not been drinking at the time. Their lives had value not because their work increased the size of the national product, but because they enjoyed life. People are willing to pay to reduce the risk of a fatal accident, and the social cost of these innocent deaths is in principle equal to the total amount the public would be willing to pay to eliminate the threat of being killed by a drunk

driver. Manning et al. (1991) employed this willingness-to-pay approach and found that nearly half of the social cost of alcohol abuse stemmed from traffic fatalities.

3. **Other Costs.** The remaining $7.2 billion in Manning et al.'s (1991) social cost estimate stemmed primarily from the burden of alcohol-related cases on the criminal justice system, and the share of collision insurance costs accounted for by the property damage caused by drunk drivers.

It appears that in several respects these estimates are incomplete. The costs of alcohol-related injuries to innocent victims are far higher than indicated by Manning et al., since they omitted the financial and personal costs of nonfatal injuries in traffic accidents (Miller & Blincoe, 1993), and also the costs of both fatal and nonfatal injuries from violent crimes perpetrated by drunks.

An even more interesting controversy has arisen over the basic perspective that informs these external social-cost estimates. Some critics reject outright the liberal doctrine that individual preferences are to be accorded primacy in the definition of social welfare and social cost. They postulate a collective interest that can somehow be defined without reference to the choices made by individuals (Beauchamp, 1980). The COI approach reflects one such definition. Other critics accept the liberal doctrine but argue about its application. A particularly difficult set of philosophical and practical issues arise in setting the boundary between internal and external costs in the context of the family. Manning et al. (1991) view the family as a unit and accept the presumption that each member of the family will internalize the concerns of the others and act accordingly. If the father is a heavy drinker or smoker, it is not because he is unaware or unconcerned about the consequences for his wife and children of his drinking or smoking, but because his enjoyment of these activities in some sense outweighs the costs to them. That presumption may seem particularly problematic in the case where the mother's substance abuse causes her baby to be born defective.

COSTS OF SMOKING AND DRUG ABUSE

Manning et al. (1989) provided an estimate of the social costs of smoking that utilized the same general approach as their estimate of drinking costs. They found that over their lifetime smokers experienced higher medical costs than they would have if they had never smoked, amounting to an average of $0.38 per pack. Since these costs were for the most part paid by insurance, government programs, or other collective sources, they included them in the external social-cost estimate. Other important external costs were the reduced contributions to the Social Security system and related programs ($0.65 per pack) resulting from the early termination of the average smoker's career, and the increased cost to group life insurance programs resulting from the reduced life expectancy of smokers ($0.11 per pack). Interestingly, these external costs were much less than the external benefits conferred by smoking. Because smokers died young, the pension payments were much less than they would have been otherwise ($1.82 per pack), and the likelihood that they would be housed in a collectively financed nursing home was also substantially reduced ($0.26 per pack). The result was that each pack of cigarettes smoked conferred a net social benefit amounting to $0.91.

Since over the course of a smoking career the social costs generally precede the benefits, the net benefit was reduced if future costs and benefits were discounted (standard practice in accounting). It turned out that with a discount rate of 5 percent, the lifetime present value of the external effects of smoking amounted to a net external cost of $0.15. Manning et al. point out that smokers more than pay this cost in the form of the state and federal excise taxes imposed on tobacco. The external effects in this calculation are all financial; they stem from private and government programs that have the effect of forcing us to pay for each other's medical care, retirement, and other benefits. Smoking, however, also causes external effects directly, since smoke pollutes the air we all breathe. The value of clean air for nonsmokers could in principle be estimated and added to the total external cost. Manning et al. chose not to do so, in part because they believe that the bulk of the costs of secondhand smoke is borne by those in the same household as smokers.

Applying the external social-cost framework to smoking and other harmfully addictive activities raises another issue. The vast majority of smokers begin their habit as adolescents, so the obvious question is whether people at that age are making well-informed decisions that take proper account of the lifetime consequences (Goodin, 1989). Adolescents tend to be as well informed about the health risks of smoking as adults, and both groups, if anything, exaggerate these risks (Viscusi, 1992). However well informed they are, most people who acquire a smoking habit nevertheless end up wishing they could quit.

In considering the social costs of illicit drug use, the illegal status of these drugs makes an enormous difference (Kleiman, 1992). The consequences of criminalizing transactions in these drugs include the bloody wars between rival drug-dealing organizations, crime by addicts seeking funds for their next fix, and the spread of disease through use of unclean needles, as well as the billions of dollars spent in law-enforcement efforts. It is difficult to generalize about these problems and estimate the social costs of drug use or abuse. Although Rice et al. (1990) offered an estimate of the costs of drug abuse (not limited to illicit drugs), their procedure mixed together some consequences of abuse with accidental poisonings and the criminal justice costs stemming from enforcing drug laws. No more specific estimates are available.

CONCLUSION

In conclusion, the effort to produce estimates of the social costs of drinking, smoking, and drug abuse is motivated by an interest in establishing a scientific basis for setting priorities in government programs. This effort has produced some useful results and a good deal of controversy surrounding the issue of what is to be counted and how. The task of estimating the social costs of substance abuse requires an accounting framework, and the choice of a framework is not a technical, scientific issue but rather a matter of political philosophy. This is surely one area where the numbers do not speak for themselves.

(SEE ALSO: *Accidents and Injuries; Complications: Overview; Economic Costs of Alcohol Abuse and Alcohol Dependence; Productivity*)

BIBLIOGRAPHY

BEAUCHAMP, D. E. (1980). *Beyond Alcoholism: Alcohol and Public Health Policy.* Philadelphia: Temple University Press.

BERRY, R. E., & BOLAND, J. P. (1977). *The economic cost of alcohol abuse.* New York: Free Press.

COOK, P. J. (1991). The social costs of drinking. In *Expert Meeting on the Negative Social Consequences of Alco-*

hol Abuse. Oslo: Norwegian Ministry of Health and Social Affairs.

GOODIN, R. E. (1989). *No smoking: The ethical issues.* Chicago: The University of Chicago Press.

HARWOOD, H. J., ET AL. (1984). *Economic costs to society of alcohol and drug abuse and mental illness: 1980.* Research Triangle Park, NC: Research Triangle Institute.

HODGSON, T., & MEINERS, M. (1979). *Guidelines for cost-of-illness studies in the public health service* (Task Force on Cost-of-Illness Studies). Bethesda, MD: Public Health Service.

KLEIMAN, M. A. R. (1992). *Against excess: Drug policy for results.* New York: Basic Books.

MANNING, W. G., ET AL. (1991). *The costs of poor health habits.* Cambridge, MA: Harvard University Press.

MANNING, W. G., ET AL. (1989). The taxes of sin: Do smokers and drinkers pay their way? *Journal of the American Medical Association, 261,* 1604–1609.

MILLER, T. R., & BLINCOE, L. J. (1993). Incidence and cost of alcohol-involved crashes. Unpublished manuscript.

RICE, D. P., ET AL. (1990). *The economic costs of alcohol and drug abuse and mental illness: 1985* (Report submitted to the Office of Financing and Coverage Policy of the Alcohol, Drug Abuse, and Mental Health Administration, U.S. Department of Health and Human Services). San Francisco: University of California, Institute for Health and Aging.

VISCUSI, W. K. (1992). *Smoking: Making the risky decision.* New York: Oxford University Press.

PHILIP J. COOK

SOCIAL MODEL *See* Disease Concept of Alcoholism and Drug Abuse.

SOCIETY OF AMERICANS FOR RECOVERY (SOAR) Society of Americans for Recovery (1317 F Street, NW, Suite 500, Washington, DC 20004; 202-347-3500) was founded by Harold E. Hughes, a former governor and senator from Iowa. It is a national grass-roots organization of concerned people whose aim is to prevent and treat dependence on alcohol and other drugs, and to educate the public about substance abuse and about its successful treatment. The organization sponsors regional conferences throughout the country and publishes a newsletter.

SOAR is particularly interested in establishing neighborhood-based treatment services with high-volume treatment capacity. The organization lobbies to fight the stigma that society places on alcoholics and addicts, and it advocates and lobbies for more and better treatment. It also encourages people to learn more about addictions and recovery and to meet others who are active in communities on behalf of substance-abuse issues.

(SEE ALSO: *Association for Medical Education and Research in Substance Abuse; Prevention Movement*)

FAITH K. JAFFE

SOLVENTS *See* Inhalants.

SOURCE COUNTRIES FOR ILLICIT DRUGS The Omnibus Drug Bill, signed into law by President Ronald W. Reagan on October 27, 1987, requires the U.S. Department of State to develop a list of all major illicit drug-producing and drug-transit countries. Inclusion on the list has an immediate effect, because 50 percent of most foreign aid allocated to each country will be withheld until a certification of improvement has been made and has survived the Congressional review process.

Major illicit drug producing country is defined in the statute as any country producing "during a fiscal year five (5) metric tons or more of OPIUM or opium derivative, 500 metric tons or more of coca, and 500 metric tons or more of MARIJUANA." (One metric ton equals 1.102 tons.)

The major source countries for HEROIN are Afghanistan, Pakistan, Iran, and Lebanon; Myanmar (formerly Burma), Thailand, and Laos; Mexico, Guatemala, and Colombia. Major source countries for COCAINE are BOLIVIA, Colombia, Peru, and Ecuador. Major source countries for marijuana are MEXICO, Belize, COLOMBIA, and Jamaica. Major source countries for HASHISH are Lebanon, Pakistan, Afghanistan, and Morocco.

Measured in U.S. dollar value, at least 80 percent of all illegal drugs consumed in the United States are of foreign origin, including all the cocaine and heroin and significant amounts of marijuana. The opium poppy flower, coca bush, and marijuana plant represent cash crops for indigenous populations—who use the proceeds of sale for subsistence, improvements in lifestyle, and/or means to procure weapons to engage in antigovernment activities. The cultivation of illicit drug crops often represents the most viable—at times the only viable—economic alternative available to otherwise impoverished farmers and political refugees.

CERTIFICATION

Chapter 8, Section 481 (h) of the Foreign Assistance Act, known as the Certification Law, links the provision of foreign aid to positive drug-control performance. The law also requires the president to certify whether major drug-producing and drug-transit countries have "cooperated fully" with the United States, or have taken adequate steps on their own, to prevent illicit drug production, drug trafficking, drug-related MONEY LAUNDERING, and drug-related corruption. Four outcomes of the certification statute deliberation are possible: (1) full and unconditional certification; (2) qualified certification for countries that would not otherwise qualify on the grounds that the national interest of the United States requires the provision of foreign assistance; (3) denial of certification; or (4) congressional disapproval of a presidential certification, which causes statutory sanctions to be imposed.

The annual International Narcotics Control Strategy Report (INCSR) is prepared by the U.S. Department of State and provides the factual basis for the president's decision on certification. The certification statute introduces the concept of variability, by using phrases such as "cooperated fully," "taken adequate steps," and "maximum achievable reductions." Judgments on a country's relative capability to perform are important factors in making certification decisions; each March, these generate spirited debate between the legislative and executive branches of the U.S. government.

(SEE ALSO: *Drug Interdiction; International Drug Supply Systems; Transit Countries for Illicit Drugs*)

BIBLIOGRAPHY

BUREAU OF INTERNATIONAL NARCOTICS MATTERS, U.S. DEPARTMENT OF STATE. (1992). *International narcotics control strategy report (INCSR)*. Washington, DC: Author. Foreign Assistance Act, as amended 1961, ch. 8, sec 481 (h). Washington, DC: U.S. Congress.

PERL, R. F. (1989). Congress and international narcotics control. *CRS Report for Congress* (October 16). Washington, DC: Library of Congress.

JAMES VAN WERT

SOUTHEAST ASIA, DRUGS AND   *See* Asia, Drug Use in; Golden Triangle; Source Countries for Illicit Drugs.

SPECIAL ACTION OFFICE FOR DRUG ABUSE PREVENTION (SAODAP)   *See* U.S. Government Agencies

SPORTS AND DRUG USE   *See* Anabolic Steroids; Lifestyles and Drug Use.

STATE DRUG PROGRAMS   *See* Appendix, Volume 4.

STEROIDS   *See* Anabolic Steroids.

STILL   Still is the colloquial term for distillery, a device used for DISTILLATION—to extract ethyl alcohol (ethanol) from various plants and food products. The simplest ones contain a cooking pot and a tightly fitted cap from which a long arm extends in a downward direction. A mash is boiled, the ethyl alcohol rises to the top and is deposited as a vapor which then condenses as it cools and passes through the arm.

(SEE ALSO: *Alcohol: History of Drinking*)

S. E. LUKAS

STIMULANTS   *See* Drug Types.

STP   *See* DOM.

STRAIGHT, INC.   *See* Appendix, Volume 4.

STREET DRUGS   *See* Slang and Jargon.

STREET VALUE   When drugs are seized by a police or interdiction agency, the significance of the seizure is often measured in terms of its street value, that is, the revenues that would be fetched if each gram were sold at the current retail price. Such measures were a routine element of police and Customs Service press releases in the United States throughout the 1980s and remained so in most other nations in the early 1990s.

The use of the term *street value* is potentially misleading when it is intended to convey the significance of the seizure as a loss to the traffickers. The price of drugs rises steeply as they move down the distribution chain from point of importation. In 1990 a gram of cocaine sold on the streets of U.S. cities for about $75; that gram (1,000 milligrams) contained approximately 700 milligrams (mg) of pure cocaine—so that the "pure gram" price was about $106. Yet when sold in 100-kilogram (kg) units at the point of import, cocaine sold for a pure-gram price of about $20. Thus it would cost drug traders $20 million to replace the 100 kilograms; that figure is the total value of payments that would have be made to growers, refiners, and smugglers in order to obtain another 100 kilograms and bring the drug to the same point in the distribution system. Valuing a 100-kg seizure at street value would imply that the government had inflicted a $106 million blow to the drug industry, more than five times as much as the true value of the loss. The extent of overstatement increases with the size of the seizure, since the price of drugs goes down as the volume increases in a given transaction.

(SEE ALSO: *Drug Interdiction; Drug Laws: Prosecution of; Seizures of Drugs*)

PETER REUTER

STRESS   Stress is best thought of as a negative emotional state—a psychophysiological experience that is both a product of the appraisal of situational and psychological factors as well as an impetus for coping (Baum, 1990). Stressors—events posing threat or challenge or otherwise demanding effort and attention for adaptation—are judged in terms of the situational variables and one's personal attributes and assets. Negative affect may ensue; and stress responses, which appear directed at the mobilization of bodily systems as a means of COPING, strengthen specific problem solving aimed at eliminating the sources of threat or demand and at reducing emotional distress (Baum, Cohen, & Hall, 1993).

(SEE ALSO: *Vulnerability As Cause of Substance Abuse*)

BIBLIOGRAPHY

BAUM, A. (1990). Stress, intrusive imagery, and chronic stress. *Health Psychology, 1,* 217–236.

BAUM, A., COHEN, L., & HALL, M. (1993). Control and intrusive memories as determinants of chronic stress. *Psychosomatic Medicine, 55,* 274–286.

LORENZO COHEN
ANDREW BAUM

STRUCTURED CLINICAL INTERVIEW FOR DSM-III-R (SCID)

This is a diagnostic interview designed for use by mental health professionals. It assesses thirty-three of the more commonly occurring psychiatric disorders described in the revised third edition of the DIAGNOSTIC AND STATISTICAL MANUAL (DSM-III-R) of the American Psychiatric Association (1987). Among these are DEPRESSION, SCHIZOPHRENIA, and the substance-use disorders. The SCID allows the experienced clinician to tailor questions to fit the patient's understanding; to ask additional questions that clarify ambiguities; to challenge inconsistencies; and to make clinical judgments about the seriousness of symptoms. The main uses of the SCID are for diagnostic evaluation, research, and the training of mental-health professionals.

The SCID is modeled on the standard clinical interview practiced by many mental-health professionals. It begins with questions about basic demographic information (e.g., age, marital status), followed by questions about the chief complaint, past episodes of psychiatric disturbance, treatment history, and current functioning. The remainder of the interview is organized into the following sections: psychotic disorders, mood disorders, substance-use disorders, anxiety disorders, somatoform disorders, eating disorders, adjustment disorder, and personality disorders. The main substance-use disorders covered in the SCID are abuse and DEPENDENCE for seven classes of PSYCHOACTIVE substances: alcohol, sedative-hypnotics-anxiolytics, *Cannabis* (MARIJUANA), STIMULANTS, OPIOIDS, COCAINE, and HALLUCINOGENS/PCP. For each substance, the interviewer determines whether the symptoms of abuse and dependence have ever been present during the subject's lifetime; whether they have been present during the last month; age when the first symptoms appeared; and the duration of symptoms during the past five years. At the end of the section on substance-use disorders, the interviewer rates the severity of dependence as mild, moderate, or severe.

The ALCOHOL (ethanol) section of the SCID begins with general questions about drinking habits. Subjects are asked whether they ever felt their drinking was excessive or problematic; whether other people ever objected to their drinking. If there is evidence that drinking was ever a problem, the interviewer identifies the most problematic or heaviest drinking period and asks a series of questions designed to measure the DSM-III-R criteria for abuse and dependence. These questions allow the interviewer to determine whether the subject has the following dependence symptoms: (1) alcohol often taken in larger amounts or over a longer period than intended; (2) persistent desire to cut down or control alcohol use; (3) a great amount of time involved in drinking-related activities; (4) alcohol used when expected to fulfill major role obligations; (5) important social, occupational, or recreational activities given up or reduced because of alcohol use; (6) persistent drinking in spite of problems; (7) marked tolerance to the effects of alcohol; (8) repeated withdrawal symptoms such as tremors, tachycardia, nausea and vomiting; and (9) drinking to relieve withdrawal symptoms.

The drug section of the SCID is conducted in a similar way.

Three versions of the instrument are available for diagnosing major psychiatric disorders, including substance use disorders: (1) the SCID-P (patient version), which is designed for use with psychiatric inpatients; (2) the SCID-OP (outpatient version), which is designed for use with psychiatric outpatients in settings in which psychotic disorders are expected to be rare; (3) the SCID-NP (nonpatient version), designed for use in studies in which the subjects are not identified as psychiatric patients (e.g., community surveys).

(SEE ALSO: *Addiction: Concepts and Definitions; Co-morbidity and Vulnerability; Complications: Mental Disorders; Disease Concept of Alcoholism and Drug Abuse; Epidemiology of Drug Abuse; International Classification of Diseases*)

BIBLIOGRAPHY

AMERICAN PSYCHIATRIC ASSOCIATION. (1987). *Diagnostic and statistical manual of mental disorders-3rd editon-revised.* Washington, DC: Author.

SPITZER, R. L., ET AL. (1992). The structured clinical interview for DSM-III-R (SCID), I. History, rationale and description. *Archives of General Psychiatry, 49*(8), 624–629.

THOMAS F. BABOR

STUDENTS AGAINST DRIVING DRUNK

(SADD) In 1981, Robert Anastis, a health educator and hockey coach in Wayland, Massachusetts, stood helplessly by as two of his students died of injuries sustained in alcohol-related traffic crashes. Anastis decided to fight back and developed a fifteen-session high school course on driving while impaired. Rather than a curriculum focusing solely on the effects of alcohol while driving, he taught strategies for preventing driving after drinking, and he emphasized the legal consequences of getting caught. In this sense, the curriculum was a significant departure from traditional driver-education approaches.

Students who took Anastis's course reacted enthusiastically and formed an organization to reduce alcohol-related traffic deaths among their peers. They called the organization Students Against Driving Drunk (SADD) in order to focus attention on the act of drunk driving, not on the drivers themselves. An anecdote related by Peggy Mann (1983) captures SADD's approach and philosophy: When a student jokingly suggested that SADD involve the governor, Anastis replied, "I believe that if you dream it, it can be done," and when the governor became the honorary chairman of SADD, its motto became "If You Dream It, It Can Be Done." Within a year, chapters had been formed throughout Massachusetts and the program was gaining national attention.

Members of the early SADD chapters had a number of goals. They sought to raise awareness of impaired driving among students through the curriculum developed by Anastis. They also sought to change norms related to impaired driving. Because they realized that most of their peers did not think of drinking and driving as wrong or risky, they reasoned that changing these norms was an important component of reducing impaired driving problems. As the students put it, they wanted to change the "drinking and driving is cool" image to another image: "Drinking and driving is dumb." Finally, students in the SADD chapters undertook to simulate discussion between high school students and their parents concerning drinking and driving. To meet this goal, they developed a "Contract for Life." The contract stipulated that a student would call a parent if he or she had been drinking or if the person responsible for driving had been drinking, and the parent, in turn, agreed to provide a ride or taxi fare.

SADD was significant in three important ways. First, it was among the earliest prevention programs to emphasize student leadership. Other programs had used peer educators or peer counselors trained and supervised by adults, but SADD chapters were run by students who planned activities and took responsibility for making them happen. Second, SADD was among the first youth programs to recognize the importance of norms in impaired-driving prevention. Earlier programs had emphasized education, attitude change, or scare tactics. Third, SADD was one of the first school-based prevention programs to venture outside the classroom. Although SADD had a curriculum, it also entailed extracurricular, community, and family involvement. In this sense, SADD was the first of the so-called comprehensive school-based prevention programs.

SADD's early growth was rapid. By the mid 1980s, there were SADD chapters in every state in the United States and chapters in Europe. SADD received considerable media attention and was the only alcohol-prevention program ever to be the subject of a nationally broadcast made-for-television movie ("Contract for Life: The Bob Anastis Story").

SADD was also controversial. Some vocal critics argued that SADD's emphasis on preventing drinking and driving implicitly condoned drinking by young people. They were particularly concerned about the Contract for Life—they argued that by insuring safe transportation, parents were communicating the message that drinking itself was not a problem. Similar charges were leveled at Safe Rides and other programs that provided sober transportation for youth. Anastis and others countered that although drinking itself *was* a problem, young people were dying from traffic crashes, not just from drinking. This debate, which resulted in the refusal by some funding agencies to allow grant money to be used to support SADD chapters, raged throughout the 1980s. It has never been fully resolved, although there is increasing recognition that reducing the *consequences* of drinking is a legitimate goal for prevention efforts.

Over the years, SADD has evolved. A junior high school program has been added, as has an emphasis on seat-belt use. The controversial Contract for Life has been de-emphasized, and the focus has shifted to alternative activities for young people. In recent years, several student safety clubs with very similar approaches to that of SADD have emerged. Members of these clubs, like SADD members, encourage students reaching out to other students to reduce highway deaths.

As is the case with many widespread, visible prevention efforts, few data can be summoned to show whether or not SADD is effective in reducing drinking and driving among youth. In the mid-1980s, the NATIONAL INSTITUTE ON ALCOHOL ABUSE AND ALCOHOLISM (NIAAA) funded a study of SADD in two western states (1994). It looked at changes in driving by impaired youth as a result of SADD and also examined the extent to which SADD might cause increases in drinking as its critics had claimed. The study found no evidence that SADD had either reduced impaired driving or had increased drinking. The SADD programs studied were not, however, particularly well executed; they may not have been representative of what well-implemented programs might accomplish. Thus, the question as to whether SADD or the student safety clubs based on SADD can decrease driving by young people while impaired remains unanswered.

(SEE ALSO: *Accidents and Injuries from Alcohol; Dramshop Liability Laws; Drunk Driving; Mothers Against Drunk Driving; Prevention Movement*)

BIBLIOGRAPHY

KLITZNER, M., ET AL. (1994). A quasi-experimental evaluation of Students Against Driving Drunk. *American Journal of Alcohol and Drug Abuse, 20*(1), 57–74.

MANN, P. (1993). *Arrive alive: How to keep drunk and pot-high drivers off the highway.* New York: Woodmere Press.

MICHAEL KLITZNER

SUBSTANCE ABUSE *See* Addiction: Concepts and Definitions.

SUBSTANCE ABUSE AND AIDS "AIDS" stands for *acquired immunodeficiency syndrome*: AIDS is a serious, life-threatening disease that results from severe damage to part of the body's cellular immune system—the defense system against opportunistic infections and some cancers. The disease is acquired (as opposed to genetic or hereditary) and presents a myriad of clinical manifestations (syndrome) that result from severe damage to the immune system. AIDS was first identified in 1981, among homosexual men in California and New York, and among illicit injected-drug abusers in New York City. Between 1981 and 1994, the numbers and types of AIDS patients increased rapidly; it was diagnosed in millions of persons throughout the world. In the United States alone, more than 350,000 cases and 200,000 deaths had been reported as of the end of 1993.

By 1992, injecting drug abusers accounted for 20 percent of cases among men, 50 percent of cases among women, and about 55 percent of pediatric cases—the children of mothers who are either injecting drug abusers or the sexual partners of male injecting drug abusers. AIDS has been diagnosed among injectors of various illicit substances, including OPIATES, COCAINE, AMPHETAMINES, and ANABOLIC STEROIDS. AIDS has also been reported among non-injecting drug abusers, such as alcoholics, cocaine "snorters," and crack (cocaine) smokers, who have been infected through sexual contact. An epidemic like AIDS that spans the continents is appropriately called a pandemic.

CAUSE

AIDS is caused by a viral infection. In the United States, the virus is called HIV (for *human immunodeficiency virus*); it is one of a group of viruses called retroviruses (so-called because it can make DNA copies of its RNA—the reverse of which is what typically occurs in animal cells). In 1983, French researchers discovered the virus, which they had linked to an outbreak of enlarged lymph nodes (one early sign of HIV infection) that had been reported among French male homosexuals. The French named it the lymphadenopathy-associated virus (LAV). In 1984, U.S. researchers isolated HIV from AIDS patients and named it human T-lymphotropic virus type III (HTLV-III). U.S. investigators found a way to grow HIV in laboratories in large amounts, which led to the development of laboratory tests that detect HIV infection.

HIV gradually destroys certain white blood cells called T-helper lymphocytes or CD4+ cells. The loss of these cells results in the body's inability to control microbial organisms that the normal immune system controls easily. These infections are called opportunistic because they take advantage of damage to part of the immune system. A few select cancers are also frequently diagnosed, such as Kaposi's sarcoma, a cancer of blood vessels, which appears as purplish spots on the skin or mucous membrane.

The sharing of needles contaminated with HIV for injecting drugs of abuse may lead to infection with HIV—but drug abuse may also act as a cofactor with HIV, affecting the development of AIDS. A cofactor in AIDS is a non-HIV-related influence operating in conjunction with HIV to affect the cause of the disease. For example, HIV-infected individuals who continue to inject drugs and/or continue tobacco use may not survive as long as those who do not abuse those substances. The abuse of nitrite INHALANTS ("poppers") among HIV-infected homosexual men may promote the development of Kaposi's sarcoma.

SIGNS AND SYMPTOMS

The natural history of HIV disease and the time intervals between clinical events vary greatly from individual to individual. The general course, however, is one of exposure to HIV, which leads to infection. Within a few weeks or months of infection, laboratory evidence of infection can be detected. Antibodies to HIV are found in the blood and indicate that infection has occurred. Some patients develop flulike symptoms or peripheral nerve abnormalities that are self-limited. Most patients have no symp-

toms during this period. Over the ensuing years (1 to 15 or more years), laboratory evidence of a decreasing number of helper T-lymphocytes can be measured. As the helper T-lymphocyte count decreases, patients are more likely to develop signs and symptoms such as enlarged lymph glands, fatigue, unexplained fever, weight loss, diarrhea, and night sweats. At about the same time or later, patients develop opportunistic infections or cancers. The diagnosis of one of the opportunistic infections or cancers indicates that the patient has developed AIDS. *Pneumocystis carinii* pneumonia is the most common opportunistic infection among AIDS patients. It is a parasitic infection of the lung. Tuberculosis is another serious infection that has become increasingly common because of the AIDS pandemic. Other opportunistic infections affect the lungs and other parts of the body, especially the brain or central nervous system and the gastrointestinal tract. Kaposi's sarcoma is the most common cancer among AIDS patients. Kaposi's sarcoma usually arises in the skin and looks like a bruise or an area of bruises, but it grows and spreads.

DIAGNOSIS AND TREATMENT

Infection with HIV can be diagnosed with a blood test measuring antibodies to the virus. Antibodies are proteins produced by certain white blood cells in response to infection. The HIV antibody test became widely available in 1985. It was immediately used in the United States and other countries to screen blood and blood plasma donations. The use of this test by blood banks has greatly reduced the chances of contracting infection from transfusions.

AIDS is the final stage of HIV infection, diagnosed when an infected individual develops an opportunistic infection or a cancer indicative of immune dysfunction. By 1993, AIDS could be diagnosed when a very low number of helper T-lymphocytes were present in an HIV-infected person's blood.

Although a cure for AIDS has not been discovered, three drugs—AZT (zidovudine), ddC (zalcitabine), and ddI (didanosine)—have been approved by the U.S. Food and Drug Administration (FDA) and they prolong the relatively symptom-free incubation period before AIDS occurs. Antimicrobial agents are used to counter many of the more common AIDS-defining opportunistic infections, and prophylactic use of antibiotics for *Pneumocystis carinii* pneumonia has been effective. These measures have served to prolong the lives of people with AIDS.

HIV TRANSMISSION

HIV can be transmitted from person to person in three ways: (1) by contact with infected blood or blood components; (2) through intimate sexual contact; and (3) from an infected pregnant mother to her fetus. Drug abusers commonly become infected by sharing needles, syringes, and other injecting paraphernalia; injecting substances—such as heroin, cocaine, and amphetamines—after an HIV-infected person uses the needle and syringe causes direct inoculation of HIV. Using any paraphernalia contaminated with blood (even in quantities too small to see) can result in HIV or hepatitis B virus transmission. Sexual contact is a common route of transmission from drug abusers to their sex partners (who can transmit the virus to other sex partners, other drug abusers, or to unborn children). Health-care workers have also been exposed to HIV through unprotected or accidental direct contact with blood of infected patients in health-care settings.

We do not know how many individuals are HIV infected worldwide. The World Health Organization estimates more than 10 million infections worldwide; the U.S. Public Health Service estimates 1 million infected in the United States. Numerous HIV surveys have been conducted among injecting drug abusers in several parts of the world. As those currently HIV infected progress to AIDS, the health-care systems and social fabric of many nations will be severely challenged.

HIV does not appear to be contagious in other settings. No known cases of AIDS have been linked to transmission in nonsexual social or household situations, through air, food, or water.

PREVENTION AMONG DRUG ABUSERS

Because no reliable cure or vaccine for HIV infection exists now (or probably in the near future), the hope for slowing the spread of HIV infection is through education and behavior-changing strategies. Among injecting drug abusers, the most effective way to avoid HIV infection is to stop sharing infected needles, or, better yet, stop injecting drugs, and to avoid sexual contact with individuals who may be HIV-infected. Former drug abusers in drug-abuse treatment have been consistently found to have lower HIV infection rates than those "on the streets." Methadone maintenance therapy has been shown to be an effective therapy for opiate addicts and has decreased HIV transmission among compliant patients. The National Institute on Drug Abuse (NIDA) continues to conduct research on innovative treatment for drug abuse. The use of HIV antibody tests, counseling about HIV infection, and partner notification projects in drug-abuse treatment programs have met with limited success so far.

Some investigators recommend that injecting drug abusers employ "safer" needles and syringes. One approach to reduce HIV transmission among injecting drug abusers is to educate addicts about cleaning needles and syringes between each use. Mechanical cleansing to remove any visible evidence of blood or other debris in the paraphernalia is followed by rinsing with a disinfectant. Of the various disinfectants tested, household bleach appears to be the most effective against HIV. Another approach has been the establishment of needle/syringe exchange programs. Rigorous studies of the effects of such programs on (1) HIV transmission and (2) the recruitment of "new" injectors of drugs will help to show how useful this strategy is.

(SEE ALSO: *Alcohol and AIDS; Complications: Route of Administration; Injecting Drug Users and HIV; Needle and Syringe Exchanges and HIV/AIDS; Sweden, Drug Use in*)

BIBLIOGRAPHY

BATTJES, R. J., & PICKENS, R. W. (1988). Needle sharing among intravenous drug abusers: National and international perspectives. *NIDA research monograph no. 80.* Washington, DC: U.S. Government Printing Office.

HAHN, R. A., ET AL. (1989). Prevalence of HIV infection among intravenous drug users in the United States. *Journal of the American Medical Association, 261,* 2677–2684.

HAVERKOS, H. W. (1989). AIDS update: Prevalence, prevention, and medical management. *Journal of Psychoactive Drugs, 21,* 365–370.

MASUR, H., ET AL. (1981). An outbreak of community-acquired *Pneumocystis carinii* pneumonia: Initial manifestations of cellular immune dysfunction. *New England Journal of Medicine, 305,* 1431–1436.

SELWYN, P. A. (1989). Issues in the clinical management

of intravenous drug users with HIV infection. *AIDS,* 3(suppl. 1), S201–S208.

HARRY W. HAVERKOS
D. PETER DROTMAN

SUBSTANCE ABUSE AND MENTAL HEALTH SERVICES ADMINISTRATION (SAMHSA)   *See* U.S. Government Agencies.

SUDDEN INFANT DEATH SYNDROME (SIDS)   *See* Fetus: Effects of Drugs on; Tobacco: Medical Complications.

SUICIDE AND SUBSTANCE ABUSE

With 29,000 annual victims, SUICIDE is the eighth leading cause of death in the United States. Alcohol and illicit drugs are involved in about 50 percent of all suicide attempts. About 25 percent of completed suicides occur among alcoholics and drug abusers. Substance abuse among young adults is largely responsible for the increased suicide rates under age thirty.

The relationship between substance abuse and suicidal behavior has been more extensively studied for alcoholism than for drug abuse. To evaluate this relationship, it is helpful to understand the statistical association between ALCOHOL and drug abuse and suicide, to learn which substance abusers are at particular risk to attempt or commit suicide, and to appreciate how this knowledge may be used to prevent suicide.

SUBSTANCE ABUSE INCREASES SUICIDE RISK

Suicides are not random; each occurs in a particular context. The association between specific psychiatric syndromes—such as DEPRESSION or abuse of alcohol or drugs—and suicidal behavior has been studied by epidemiologists using both retrospective and prospective methods. Since interviews with suicide completers are impossible, retrospective reviews of the circumstances predating suicides have been conducted. By using interviews of relatives and others familiar with the suicide victim, together with study of medical records, suicide notes, and coroner reports, each suicide case is subjected to a "psychologic autopsy." Factors that distinguish successful suicide cases from suicide attempters and substance abusers who have never attempted suicide are compared in the hope that differences in these factors may identify those at particular risk of attempted or completed suicide. A limitation of retrospective studies is termed *recall bias:* informants may provide information about the suicide victim that is distorted by their attempt to explain the suicide event. Although written records and use of standardized methods to collect diagnostic information can reduce this bias, prospective studies are more reliable. Prospective studies in the general population are not feasible, because suicide is rare, occurring in only about 1 in 10,000 annually; however, about 10 percent of suicide attempters, 15 percent of depressed people, and 3 percent of alcoholics eventually commit suicide. By prospective study of such high-risk groups, additional risk factors can be identified during a follow-up period.

Although most heavy drinkers are not alcoholic, heavy drinking in young adulthood is associated with suicide in middle adulthood. A prospective study of Swedish military conscripts found that those who drank more than twenty drinks weekly had three times the death rate, prior to age forty, of light drinkers. Most of these premature deaths were due to suicide or accidents. Those who develop alcohol dependence or abuse are, together with drug abusers, at increased risk of death from accidents, liver disease, pancreatitis, respiratory disease, and other illnesses; however, suicide is among the most significant causes of death in both male and female substance abusers. U.S. and Swedish prospective studies, for example, found that alcoholism increased the risk of suicide fourfold in men and twentyfold in women.

Next to depression, alcoholism and drug abuse are the psychiatric conditions most strongly associated with suicide attempts. In the U.S. Epidemiologic Catchment Area (ECA) Study conducted in the 1980s, the risk of suicide attempts was increased forty-onefold by depression and eighteenfold by alcoholism. While COCAINE users had increased rates of SUICIDE ATTEMPTS, users of MARIJUANA, SEDATIVE-HYPNOTICS, and AMPHETAMINES did not.

Among completed suicides, the proportion who were alcoholics or drug abusers is large: Prior to 1980, ALCOHOLISM accounted for about 20 to 35 percent, and drug abuse for less than 5 percent, of suicides in a variety of countries. In the San Diego Suicide Study, conducted in the early 1980s, well over 50 percent of 274 consecutive suicides had alcoholism or drug abuse or dependence. Much of the increase in young-adult suicide rates since the 1960s is attributable to alcoholism and drug abuse or dependence.

RISK FACTORS FOR
SUICIDE ATTEMPTS

Alcoholics and drug abusers frequently threaten to kill themselves. Many, particularly women and young adults, actually attempt it. Among alcoholics studied in the ECA communities, 32.5 percent had attempted suicide during a period of active alcoholism. About 15 to 25 percent of alcoholics in treatment programs report having previously attempted suicide. In a group of treated opiate addicts, 17 percent had attempted suicide. This represents at least a fivefold increased frequency of suicide attempts compared to those among nonsubstance abusers.

Although only about 10 percent of substance abusers who attempt suicide will die in a subsequent attempt, most substance abusers who commit suicide have attempted suicide at least once before. Thus, a review of the risks of suicide attempts may guide the identification of those substance abusers at risk of suicidal death. The risk of attempting suicide by an alcoholic or drug abuser is increased by coexisting depression, ANTISOCIAL PERSONALITY disorder (ASP), and a history of parental alcoholism.

Even among people who do not abuse alcohol or drugs, major depression increases the risk of attempting suicide. Major depression is itself 50 percent more common among alcoholics than nonalcoholics: it was found among 5 percent of male and 19 percent of female alcoholics living in the five ECA communities. Depressive feelings (but not necessarily the syndrome of major depression) often motivate alcoholics and drug addicts to enter a treatment program. Typically 20 to 40 percent of alcoholics in such programs have had a period of major depression during their lifetime. While many people drink alcohol or use drugs such as cocaine to reduce feelings of depression, experiments show that consumption produces an initial state of euphoria, followed within a few hours by anxiety, depression, and enhanced suicide ideas. Retrospective studies have found that depressive symptoms are more common among alcoholics who have made a suicide attempt.

Several studies have found that alcoholism in a parent is associated with suicide attempts among alcoholics. In addition, antisocial personality disorder (ASP) and drug abuse, which commonly occur in genetically predisposed males who develop alcoholism early in life, are associated with suicide attempts. Many clinicians have noted the repetitive high-risk behaviors of intravenous drug addicts, who often are quite aware that they may acquire infection or die by overdose with each injection. Overdoses occur more commonly among HEROIN addicts who have attempted suicide than among those who have not. Highly impulsive and aggressive alcoholics or drug abusers with ASP may be a subgroup at elevated risk of attempting suicide. Transient but intense dysphoria (feeling unwell or unhappy), though not of sufficient scope or duration to meet criteria for major depression, may nonetheless increase this group's risk of attempting suicide.

Prospective studies have found that depression, anxiety, and histories of violence and legal problems were predictive of suicide attempts in previously nonsuicidal drug addicts. Retrospective studies of alcoholics and drug addicts have found that poor social supports, occupational losses, personal losses such as divorce, and other family problems increase their risk of making a suicide attempt.

RISK FACTORS FOR
COMPLETED SUICIDE

Although in the general population there is considerable overlap between those who attempt suicide and those who complete suicide, substantial differences exist between these groups. For example, women are three times more likely than men to attempt suicide, while men are three times more likely to commit suicide. Despite these differences, suicide attempters are at higher risk of completed suicide. What, then, are the risk factors for completed suicide in substance abusers?

Depression. Depressed people, particularly men, typically kill themselves in young adulthood. Among pure alcoholics, over 90 percent of suicides occur among men. In contrast to depressives, alcoholic men typically commit suicide in their fifth and sixth decades; usually this follows about twenty years of alcoholism. Men with depression, but not those with alcoholism, continue to be at elevated suicide risk beyond age sixty. Drug abuse shortens the interval preceding suicide: in the San Diego Suicide Study, drug addicts committed suicide after an average of only nine years of heavy use. They typically did so in young adulthood. This suggests that factors other than alcoholism may shorten the suicide risk period in this group. About three of four alcoholic suicides communicate their suicidal intent prior to their deaths. Thus, middle-aged male alcoholics and young polysubstance abusers, especially those who talk of suicide, are at high risk of suicide.

Long-term Use. Ongoing substance use makes suicide more likely. Nearly all alcoholic suicides occur among active drinkers, and alcohol consumption often immediately precedes the suicide. The abstinent alcoholic is only partly protected from suicide, however, for 3 percent of suicides among alcoholics occur among those who are abstinent. It is likely that impulsiveness and transient or syndromal depression contribute to these suicides.

Psychiatric Conditions. Coexisting psychiatric conditions, particularly depression, play an important and perhaps crucial role in the suicide of alcoholics and drug abusers. The vast majority of suicide victims have depressive symptoms at the time of their death. Concurrent depression is the leading factor in at least 50 percent of suicides among alcoholics and drug abusers. SCHIZOPHRENIA, mania, and ASP are also associated with suicide in substance abusers.

Timing. What determines the timing of suicide among substance abusers? Substance abusers often accumulate interpersonal problems throughout their drinking or drug-use careers, but one-third of those who commit suicide sustain a major interpersonal disruption (such as separation or divorce) within the six weeks preceding their deaths. They often are unemployed, living alone, and unsupported by family and friends at the time of this final and most severe disruption. In contrast, only 3 percent of nonalcoholics with depression suffer such a loss in the period before they commit suicide. Beyond psychiatric diagnoses, the strongest indicator of suicide risk in substance abusers is such an interpersonal loss. Beyond these actual losses, anticipated losses, such as impending legal, financial, or physical demise may also increase the risk of suicide among substance abusers. Among alcoholics, those who develop serious medical problems, such as liver disease, pancreatitis, or peptic ulcers, are also at higher risk of suicide.

Summary. Which of these risk factors is the most important, and how do they interact to affect the risk of suicide? To partly answer these questions, Murphy and colleagues studied 173 white male alcoholics, 67 of whom committed suicide. After adjusting for age, the most potent risk factor for suicide was (1) current drinking, followed by (2) major depression, (3) suicidal thoughts, (4) poor social support, (5) living alone, and (6) unemployment. All suicide cases had at least one, and 69 percent had at least four, of these six risk factors. These factors act cumulatively to increase the risk of suicide in male alcoholics significantly. Their relative roles in other groups of substance abusers have not been reported.

CLINICAL FEATURES

Substance abusers who commit suicide often see a physician or are psychiatrically hospitalized in the months prior to their deaths. Those who talk of suicide may be ambivalent about their wish to die. They may thus be amenable to clinical interventions such as detoxification, substance-abuse rehabilitation, or psychiatric hospitalization. Conversely, those who take special precautions against discovery during a prior suicide attempt are much more likely to die in a subsequent suicide attempt.

Feelings of hopelessness are common in depression. While suicide attempters who are depressed and who report hopelessness are more likely to die of suicide, hopelessness is not a particular risk for completion of suicide among alcoholics. This may occur because substance abusers are motivated to commit suicide less by persistent hopelessness and more by impulsive anger, dysphoria, or feelings of isolation or abandonment.

PREVENTION

Prediction of those who will complete suicide remains poor in individual cases, even among high-risk groups such as substance abusers. Despite their high prevalence, alcoholism and drug abuse often go unrecognized by physicians and other health-care professionals. Recognition of alcohol and drug use disorders and of risk factors such as major depression that increase the risk of suicide may assist clinicians with preventive interventions. The substance abuser with active suicide plans or a recent suicide attempt may need hospitalization, detoxification, and/or rehabilitation designed to foster abstinence from alcohol and drugs of abuse. Firearms should be removed from the homes of substance abusers with active suicide ideation, especially adolescents and young adults. Treatments designed to enhance social supports and foster abstinence from alcohol and drugs, together with those directed at resolution of major depression, often reduce the risk of suicide.

(SEE ALSO: *Accidents and Injuries; Comorbidity and Vulnerability; Complications: Mental Disorders; Epidemiology; Social Costs of Alcohol and Drug Abuse*)

BIBLIOGRAPHY

Acta Psychiatrica Scandinavica 81, 565–570.
ALLEBECK, P., & ALLGULANDER, C. (1990). Suicide among young men: Psychiatric illness, deviant behaviour and substance abuse.
FOWLER, R. C., RICH, C. L., & YOUNG, D. (1986). San Diego suicide study, II: Substance abuse in young cases. *Archives of General Psychiatry, 43*, 962–965.
HESSELBROCK, ET AL. (1988). Suicide attempts and alcoholism. *Journal of the Study of Alcohol, 49*, 436–442.
MOSCIDKI, E. K. ET AL. (1992). Suicide attempts in the epidemiologic catchment area study. *Yale Journal of Biological Medicine, 61*, 259–268.
MURPHY, G. E. (1992). *Suicide in alcoholism*. New York: Oxford University Press.
ROSEN, D. H. (1976). The serious suicide attempt: Five year follow-up study of 886 patients. *Journal of the American Medical Association, 235*, 2105–2109.

MICHAEL J. BOHN

SUICIDE ATTEMPT Suicide is the act of taking one's own life. When such an act is undertaken but does not result in death, it is called a suicide attempt. Some authors have categorized attempted suicides according to the underlying motivation, with a gradation from a deliberate will to die (failed suicide) to a wish for temporary respite to a manipulative act where the intention is to survive.

A recent epidemiologic survey in the United States found that 3 percent of respondents acknowledged having attempted suicide at some time in their lives. Predominant clinical features of those attempting suicide included female gender, lower socioeconomic status, disrupted marital status, and a psychiatric diagnosis. Between 13 and 35 percent of individuals attempting suicide will repeat the act within 2 years. Factors associated with eventual suicide include age greater than 45 years, male gender, being divorced or widowed, unemployment, chronic physical illness, and psychiatric disorders—including ALCOHOLISM and previous suicide attempts. These are associated factors, however, and none have been found to be clearly predictive of suicide.

The acute precipitant of a suicide attempt is often a life change, particularly interpersonal stress, but individuals who attempt suicide are also more likely to have chronic social problems, such as marital discord, unemployment, and, among ADOLESCENTS, problems with schoolwork, parents, and peers. The psychiatric disorders most commonly associated with a suicide attempt are personality disorders, DEPRESSION, and alcohol and drug abuse. Many of the people who attempt suicide have had recent contact with their family physician, often because of symptoms arising from their social and interpersonal problems. The majority of suicide attempts are impulsive and involve drug overdoses. Many psychiatrists believe that any patient who has made a suicide attempt should be hospitalized, but admission rates for such patients vary greatly. Hospitalization provides a secure environment, the opportunity to mobilize social support, initiate treatment with medications for psychiatric illness, and make plans for longer term management of the patient (e.g., treatment for alcohol problems).

(SEE ALSO: *Barbiturates; Barbiturates: Complications; Comorbidity and Vulnerability; Complications: Mental Disorders; Overdose; Suicide and Substance Abuse; Suicide Gesture*)

BIBLIOGRAPHY

ROY, A. (1989). Suicide. In H. I. Kaplan & B. J. Sadock (Eds.), *Comprehensive textbook of psychiatry* (5th ed., Vol. 2). Baltimore, MD: Williams & Wilkins.
ROY, A. (ED.). (1986). *Suicide*. Baltimore, MD: Williams & Wilkins.

MYROSLAVA ROMACH
KAREN PARKER

SUICIDE GESTURE Often used disparagingly, the term *suicide gesture* refers to acts of self-harm that the clinician believes were not intended to end in death but were deliberately planned and carried out for the purpose of interpersonal gain.

They are seen as manipulative behaviors in which there is minimal threat to the individual's survival. Usually, they are of limited lethality and include behaviors such as cutting and overdosing. Although these "gestures" are often repeated and can elicit intense animosity in clinical staff when the individual seeks help, they are just as often signals of distress and should be viewed as no less authentic than more lethal acts. They usually indicate an inability to communicate emotional distress in a more socially acceptable way and are often seen in individuals with poor self-esteem and COPING skills. Frequently, the psychosocial history of these people entails deprivation, disorganization, and severe interpersonal difficulties. Individuals showing repetitive suicidal behavior often suffer from severe personality disorders and are at high risk for eventual SUICIDE.

(SEE ALSO: *Barbiturates; Barbiturates: Complications; Comorbidity and Vulnerability; Overdose; Suicide Attempt; Suicide and Substance Abuse*)

BIBLIOGRAPHY

ROY, A. (1989). Suicide. In H. I. Kaplan & B. J. Sadock (Eds.), *Comprehensive textbook of psychiatry* (Vol. 2). Baltimore, MD: Williams & Wilkins.

MYROSLAVA ROMACH
KAREN PARKER

SURGEON GENERAL, REPORT OF THE *See* Tobacco: Medical Complications; Treatment: Tobacco.

SWEDEN, DRUG USE IN Sweden is roughly the size of California—or twice that of the United Kingdom. Sweden's capital city, Stockholm, has a population of about 1.3 million, and the country as a whole has some 8.8 million inhabitants. The first well-documented example of drug abuse in Sweden arose during the 1940s, when the technique of injecting AMPHETAMINE began to spread among criminal elements and bohemians in Stockholm. This form of intravenous (IV) drug abuse quickly spread to other major towns and cities and also to the neighboring countries of Finland, Norway, and Denmark. In 1944, central nervous system (CNS) stimulants were subjected to the same strict prescrip-

tion control regulations as narcotic drugs in general. In Sweden, CNS stimulants were formally scheduled as narcotics in 1958. Indeed, the classification of CNS stimulants as psychotropic substances in the international convention of 1971 was largely a result of Sweden's efforts.

MARIJUANA (*Cannabis* leaves) enjoyed a briefly popularity around 1954, when the habit of smoking a "joint" was started by American jazz musicians who were performing in Sweden. HASHISH (*Cannabis* resin) was introduced in the early 1960s and became popular among young people as the habit of smoking "pot" (marijuana) emerged along with the youth rebellion. In the 1990s, the domestic growing of *Cannabis* plants started on a small scale.

The intravenous use of heroin stems from the mid-1970s, and this mode of drug abuse quickly attracted attention from the news media when several overdose deaths were reported. COCAINE was introduced into Sweden in the late 1970s, but on a small scale.

LEGISLATION

In Sweden, the term *narcotic drugs* refers to all pharmaceutical substances controlled under the provisions of the Narcotic Drugs Act (1968) and listed on the Narcotic Drug Schedules issued by the Swedish Medical Products Agency. These schedules contain all internationally controlled substances and some additional substances, such as KHAT (leaves and branches from *Catha edulis*). The use of Schedule I drugs (*Cannabis*, LSD, HEROIN, MDMA, khat, etc.) is prohibited, even for medical purposes.

Narcotic offenses in Sweden fall into three classes:

1. Petty offenses involving possession of small amounts of the drug punishable with a fine or imprisonment for a maximum of six months.
2. Narcotic offenses, which might entail selling ("pushing") drugs on the streets, carry a maximum of three years imprisonment.
3. Grave (serious) narcotic offenses, such as the import of large amounts of illicit drugs or the production and sale of narcotics. These offenses are punishable by imprisonment for two to ten years.

Compulsory (coercive) treatment of drug abusers is possible under the 1988 law for "Treatment of Alcoholics and Drug Misusers." Young offenders may be subjected to compulsory treatment under the Care of Young Persons Act of 1990. The decision to invoke this treatment for young drug abusers is made by the county administrative courts. METHADONE MAINTENANCE treatment for opiate addicts, using very strict admission criteria, is currently available at three university hospital clinics—at Stockholm, Uppsala, and Malmö-Lund.

Doping compounds, such as ANABOLIC STEROIDS, are regulated under the Doping Compounds Act of 1992. These substances cannot be imported, produced, traded, or possessed without special permits; however, consumption/use of anabolic steroids is not a punishable offense at the present time.

CURRENT SITUATION AND TRENDS

Since the 1970s, hashish has been the most widespread of the illicit drugs used in Sweden; it often serves as the starting point, or gateway, into abuse of other drugs. During the screening of job applicants in 1986, as many as 4 percent had traces of TETRA-HYDROCANNABINOL (THC) in their urine. An estimated 50,000 people regularly smoke hashish in Sweden as of the mid-1990s.

Amphetamine, which is relatively easily obtained throughout the country, is the most popular drug of abuse for intravenous use; about 10,000 people are currently using this CNS stimulant. Injection of heroin seems to be mainly concentrated in the southern and central metropolitan areas, where some 2,000 to 3,000 are known to indulge in this form of drug abuse. The abuse of cocaine is only seen within jet-set circles in the major cities. The smoking of CRACK-cocaine is still uncommon in Sweden. HALLUCINO-GENS (such as LSD and MDMA) are used to some extent by adolescents who follow the "rave" culture. As of the mid-1990s, plant hallucinogens such as PSILOCYBIN are rarely encountered, and the police have not seized PHENCYCLIDINE (PCP) or "ice" (crystallized METHAMPHETAMINE) or phentanyl (e.g., fentanyl, sufentanil) opioids.

Increased immigration into Sweden during the 1980s brought the development of new subpopulations of drug users, with use patterns derived from their home drug cultures. These included the smoking of opium and heroin, which is common to the Middle East, or the chewing of khat from East Africa. The relaxing of border controls with the Eastern bloc led to new smuggling routes for drugs into Sweden—hashish from Russia and amphetamine from Poland.

According to figures obtained from the Stockholm Remand Prisons, human immunodeficiency virus (HIV) infection rates in the early 1990s were approximately 30 percent among IV abusers of heroin and 5 percent among IV abusers of amphetamine. About 600 individuals are apprehended each year in Sweden on suspicion of driving under the influence of drugs. The most common drug encountered in suspected drugged drivers is amphetamine, followed by *Cannabis* and then various SEDATIVE-HYPNOTIC prescription drugs belonging to the BENZODIAZEPINE family.

SHIFTS IN CONTROL POLICY

Sweden has experienced dramatic shifts in public policy concerning the control of illicit drugs. In 1965, after a turbulent media campaign, the medical authorities were obliged to allow certain doctors to prescribe what were illicit drugs to registered addicts for their personal use, as part of the so-called legal prescription experiment. Over a two-year period, about 4 million doses of amphetamine and 600,000 doses of morphine had been distributed to a total of only 150 addicts. The project rapidly became unmanageable; it was stopped as the IV drug habit began to spread widely and several fatal overdoses were reported. During the final twelve months of the project, the prevalence of IV drug use among the arrestee population in Stockholm had doubled.

In 1969, a nationwide police offensive against all sorts of drug-related crime brought about a dramatic decrease in drug abuse in Sweden. The tendency among public prosecutors to dismiss petty drug offenses during the 1970s led to an escalation in drug abuse once again. Since 1980, all drug offenses have been either referred to the courts for trial or, if the suspects plead guilty to petty offenses, they are fined directly. In the late 1980s, the police began a new strategy against drug abuse, by focusing more attention on all kinds of drug activity on the streets—with the aim of decreasing the demand for drugs.

The fight against drug abuse in Sweden grew progressively stricter between 1983 and 1993. In 1988, the taking of illicit drugs was made a punishable offense. Since July 1, 1993, the police have been allowed to order chemical analyses of body fluids for evidence that a suspect has been taking illicit drugs.

(SEE ALSO: *Amphetamine Epidemics; Britain, Drug Use In; Drug Testing and Analysis; Italy, Drug Use in;*

Netherlands, Drug Use in; Prohibition of Drugs: Pro and Con)

BIBLIOGRAPHY

BEJEROT, N. (1975). Drug abuse and drug policy. *Acta Psychiatrica Scandinavica*, Supplement 256.

BEJEROT, N. (1970). *Addiction and society.* Springfield, IL: Charles Thomas.

SWEDISH COUNCIL FOR INFORMATION ON ALCOHOL AND OTHER DRUGS. (1991). *Trends in alcohol and drug abuse in Sweden*, Report 91. Stockholm: CAN.

SWEDISH NATIONAL CRIME PREVENTION BOARD. (1990). *Current Swedish legislation on narcotics and psychotropic substances, 2.* Stockholm: Allmänna Förlaget.

SWEDISH NATIONAL POLICE BOARD. (1992). *Narcotic drugs, laws, facts, arguments.* Stockholm: Allmänna Förlaget.

JONAS HARTELIUS
A. W. JONES

SYNANON   Founded in 1958 by Charles E. Dederich, Synanon pioneered a breakthrough approach to the treatment of drug dependence. Using some of the approaches he had personally experienced in ALCOHOLICS ANONYMOUS, a mixture of self-reliance and Buddhist philosophies, and his own bombastic and confrontational interpersonal style, Dederich shaped a self-help organization that grew from a small storefront in Santa Monica, California, to over 2,000 members in multiple residential settings across the United States by the early 1970s.

While originally successful with long-term HEROIN addicts, the methodologies and operating activities developed and refined by Synanon became the precursor for the drug-free THERAPEUTIC COMMUNITY approach. This strategy has proven significantly effective for both ADOLESCENTS and adults, regardless of the types of drug they use.

The salient ingredients that Synanon pioneered, along with the ones they incorporated into a systematic approach, remain fundamentally intact in drug-free therapeutic communities in the United States and elsewhere, though the approach has been refined and complemented over time. These fundamental ingredients fall into four major categories: (1) behavior management and behavior shaping, (2) emotional and psychological life, (3) ethical and intellectual development, and (4) work and vocational life. Within each of these categories, elaborate

sets of techniques use deliberate but artful dissonance and confrontation as major tools for changing behavior.

Synanon programs were residential and originally proceeded in three stages, lasting from two to four years for the early members. These stages included (1) live-in, work-in, (2) live-in, work-out, and (3) live-out, work-out (while still maintaining affiliation). As Synanon grew in membership, Dederich's charismatic and autocratic leadership moved the organization away from a concern with restoring people to society—and focused instead on the development of an alternative intentional community, one in which people could remain indefinitely.

Some of Synanon's techniques—such as the intense expressions of raw emotion and the colorful discipline practices incorporating signs or shaved heads—gained national media attention. Public controversy over these techniques obscured the rest of the treatment approach—for example, environmental stimuli, slogans, books, and community rituals. All of Synanon's activities took place in the context of the community as the healing force. The emphasis was that each member was a part of that healing force and that, as such, each member was needed and valued. All members had the potential to become leaders of the community (the treatment environment), with rewards of status and privilege, and a critical lesson for others—you can succeed, you can make a difference.

After a highly recognized and acclaimed ten-year success with reformed drug users, and as a result of a series of money-generating businesses that produced millions of dollars as well as the acquisition of major real estate throughout California, the large Synanon membership enjoyed great affluence. It possessed its own schools, planes, fleets of autos and trucks, and, according to some reports, overseas bank accounts. As they became increasingly insulated, without accountability to the outside, Synanon members began to consider their process a religion. By the mid-1970s, many forms of arbitrary and total control over members appeared—such as forced weight reduction and mass exercise programs. Violent and invasive forms of control were also imposed, in the form of coerced vasectomies, abortions, marital separations, and separation of children from parents.

Furthermore, by the late 1970s, a new population began to be referred to Synanon—tough youths from various correctional, youth-authority, and probation departments. These adolescents did not easily conform to Synanon's rigid rules, so acts of violence became a tool of attempted control over them. The whole system also began to have increasingly violent interactions with outsiders—including intimidation and actual physical assaults. The organization, so lauded in the press during its early years, became an object of national criticism. Then Dederich reversed his earlier position of shunning chemicals and began to drink. In 1978, he was indicted for conspiracy to commit murder, and the court instructed him to vacate leadership. Synanon became faced with IRS violations and other court problems that still plague the organization. Eventually, Dederich's alcoholism intensified and he experienced a series of small strokes that left him impaired but still loved by a small cadre of members who live in the foothills east of Fresno, California. Their number has fallen from over 2,000 to less than 100. Synanon ceased its drug-treatment programs in the 1980s and is no longer involved in any human-service business.

(SEE ALSO: *Amity, Inc; Daytop Village; Gateway Foundation; Marathon House; Phoenix House; Second Genesis, Inc; Treatment, History of*)

DAVID DEITCH
ROBIN SOLIT

SYNAPSE, BRAIN The term synapse is from the Greek word *synaptein*, for "juncture" or "fasten together," by way of the Latin *synapsis*. It refers to the specialized junction found between nerve cells. It was conceived by the British pioneer neurophysiologist Sir Charles Sherrington (1857–1952) to describe the then-novel microscopic observations that the "end-feet" of one neuron physically contacted, in an intimate manner, other NEURONS to which it was structurally connected. A similar point of connection between peripheral nerves and their targets is usually referred to as a *junction*.

Synapses in the brain (see Figures 1 and 2) are morphologically typed by several features (1) a dilation of the presynaptic terminal (nerve ending) that contains accumulations of synaptic vesicles in various sizes, shapes, and chemical reactivities; (2) mitochondria; (3) a specialized zone of modified thickness and electron opacity in the presynaptic

membrane, in which a presynaptic grid is perforated to provide maximum access of transmitter-containing vesicles to the presumptive sites of transmitter release; and (4) a specialized zone of altered thickness and opacity in the postsynaptic membrane termed the *active zone* and believed to be the site of initial response.

The synaptic vesicles have been shown to contain the NEUROTRANSMITTERS by a series of extensive analyses of meticuously purified vesicles. The vesicles differ in their protein content and may include the transmitter's synthetic enzymes, as well as the transporters that can concentrate the transmitter within the vesicles. For MONOAMINE neurons, the vesicles also contain specific proteins (named for their sites of discovery in the adrenal medulla as *chromogranins* but now termed more generally *secretogranins*. These are assumed to facilitate storage and release. Superficially, synapses with a thinner postsynaptic specialization, of about the same thickness as that at the presynaptic membrane (hence termed *symmetrical*), are often inhibitory; those with a thickened postsynaptic membrane (*asymmetrical*) are often excitatory.

Monoaminergic synapses, however, are often asymmetrical, as are those for peptide-containing neurons that do not obey these simple physiological categorizations. Synapses can also be discriminated on the basis of the pairs of neuronal structures that come together at this site of functional transmission. Most typical is the *axo-dendritic* synapse in which the axon of the presynaptic neuron contacts either the smooth or spiny surface of the dendrite of the postsynaptic neuron. A second common form is the *axo-somatic* synapse in which the presynaptic axon contacts the surface of the post-synaptic neuron's cell body (or somata). Less frequently observed are axo-axonic relationships in which one axon contacts a second axon-terminal that is in its own axo-dendritic relationship; such triads of axo-axo-dendritic synapses are found most frequently in spinal cord and certain midbrain structures, in which channels of information flow are necessarily highly constrained. Most rarely, junctions between cell bodies (somato-somatic) and dendrites (dendro-dendritic) have also been described.

The nature of the proteins that provide for the thickened appearances of the active zones by electron microscopy are not completely known, but they include the postsynaptic RECEPTORS and associated

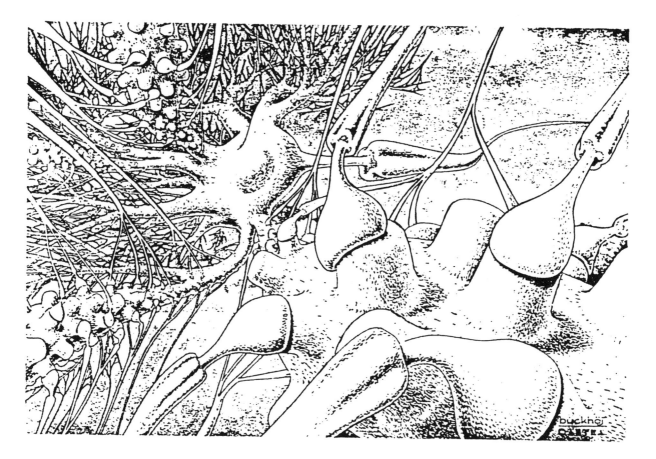

Figure 1

Synapse. The nerve ending from one neuron forms a junction, the synapse, with another neuron (the postsynaptic neuron). The synaptic junction is actually a small space, sometimes called the synaptic cleft. Neurotransmitter molecules are synthesized by enzymes in the nerve terminal, stored in vesicles, and released into the synaptic cleft when an electrical impulse invades the nerve terminal. The electrical impulse originates in the neuronal cell body and travels down the axon. The released neurotransmitter combines with receptors on postsynaptic neurons, which are then activated. To terminate neurotransmission, transporters remove the neurotransmitter from the synaptic cleft by pumping it back into the nerve terminal that released it.

SOURCE: Figures 1 and 2 have been modified from Figure 1, in M. J. Kuhar's "Introduction to Neurotransmitters and Neuroreceptors," in *Quantitative Imaging,* edited by J. J. Frost and H. N. Wagner. Raven Press, New York, 1990.

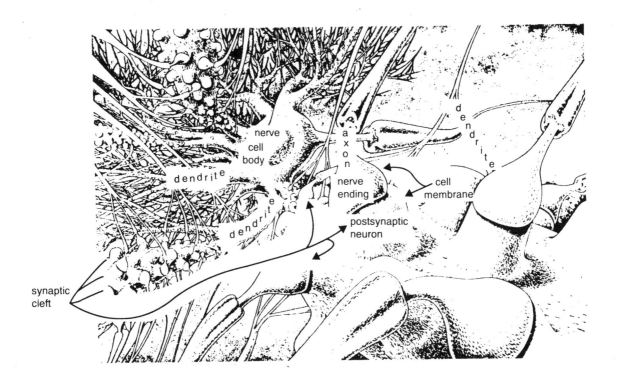

Figure 2

Neuronal Network. Synapses can be seen here with their narrow synaptic clefts, only 20 micrometers wide, across which a nerve impulse is transmitted from one neuron to the next. Hundreds of thousands of nerve endings may form synapses on the cell body and dendrites of a single neuron. As an electrical impulse reaches the synaptic cleft, it cannot be transmitted because of a discontinuation in the cell membrane. To bridge this cleft, another type of transmission, a chemical transmission, begins, mediated by a chemical compound—the transmitter substance or a neurotransmitter.

molecules that can transduce the signals from the activate receptors, as well as those molecules that serve to concentrate the receptors in such locations.

(SEE ALSO: *Brain Structures and Drugs; Neurotransmission; Reward Pathways and Drugs*)

BIBLIOGRAPHY

BLOOM, F. E. (1990). Neurohumoral transmission in the central nervous system. In A. G. Gilman et al. (Eds.), *Goodman and Gilman's the pharmacological basis of therapeutics*, 8th ed. Pergamon.

COOPER, J. C., BLOOM, F. E., & ROTH, R. H. (1991). *The biochemical basis of neuropharmacology*, 6th ed. New York: Oxford University Press.

FLOYD BLOOM

SYRINGE EXCHANGE AND AIDS *See* Needle and Syringe Exchanges and HIV/AIDS.

T

TASC   *See* Treatment Alternatives to Street Crime.

TAX LAWS AND ALCOHOL   The first internal revenue measure adopted by the U.S. Congress, in 1790, was an excise tax on domestic whiskey; a subsequent increase in that tax from nine to twenty-five cents per gallon led to an armed insurrection in the summer of 1794, the so-called Whiskey Rebellion of western Pennsylvania. It was put down by the militia of Pennsylvania and other states.

This matter of the appropriate level for alcoholic beverage taxes has remained contentious to this day; although there is consensus that alcoholic beverages should be subject to higher taxes than other commodities, there is substantial disagreement on the appropriate level for such taxes. The principal impetus for raising tax rates has always been the quest for increased government revenue, and that remains true today. Since the 1970s, however, increasing attention has been paid to the public-health benefits of alcohol taxes, as research demonstrated that raising the excise tax rates, and hence the prices of alcoholic beverages, has had a tendency to reduce traffic fatalities and other costly consequences of alcohol abuse.

HISTORY

Alcoholic beverage taxes were a major source of revenues for the federal government through much of U.S. history. As recently as 1907 this source accounted for 80 percent of federal internal-tax collections but only for about 10 percent in 1941. Currently the federal excise taxes and import duties continue to have a considerable effect on the prices of alcoholic beverages, but they figure very lightly (less than 1%) in overall federal tax collections.

Because federal excise taxes are set in dollar terms per unit of liquid, rather than as a percentage of the price, inflation gradually erodes the real value of these taxes. For example, although between 1951 and 1991 Congress increased the tax per fifth of 80-proof spirits by 29 percent (to $2.16), the overall level of consumer prices increased by more than 400 percent during this same period. The result is that the *real* value of the federal liquor tax in the early 1990s became one quarter of its 1951 value. A dramatic reduction in the average price of whiskey and other spirits, relative to the prices of other commodities, has been the inevitable result.

The states also impose special excise taxes on alcoholic beverages, as do some local governments. In addition, alcoholic beverages are generally subject to state and local sales taxes. The relative importance of these tax collections in state budgets differs widely, but in the mid-1990s it amounted to less than 10 percent of revenues.

TAX EFFECTS

When a legislature raises the excise tax rates on alcoholic beverages, the resulting cost to distributors is passed along to consumers in the form of higher prices. As is true for other commodities, the sales of alcoholic beverages tend to fall when prices increase. This is not to say that price is all that matters. Indeed, the steady decline in sales and consumption of alcohol during the 1980s cannot be explained by increased prices, since the prices of alcoholic beverages remained more or less constant (in real terms) during this relatively affluent period. The downward trend in consumption presumably resulted from the aging of the population and increasing public concern with healthy lifestyles, among other factors. Per capita sales and consumption of alcohol are nevertheless inversely related to alcohol beverage prices, and if Congress had increased federal excise taxes substantially during the 1980s, sales would have declined still more rapidly than they did.

Although they differ somewhat, a number of published estimates of the price elasticity of demand for beer, wine, and liquor tend to confirm that price is one of the important variables influencing sales. One recent review of these estimates concluded that the price elasticity for liquor is approximately -1.0; this implies that, other things being equal, a percentage increase in the average price of liquor will result in an equal percentage reduction in the quantity of liquor sold. Beer and wine sales tend to be somewhat less responsive to price, with estimated price elasticities in the neighborhood of -0.5 (Leung & Phelps, 1993).

These results do not in themselves imply that a general price increase for alcoholic beverages will reduce consumption of ethyl alcohol (ethanol), the intoxicating substance in all these beverages. In the face of higher prices, consumers can switch to cheaper brands, substitute at-home consumption for drinking at bars, or attempt to make their own beer or wine. Research suggests that these substitutions are nevertheless not large enough to negate the price effect. Ethanol consumption does tend to fall nationwide in response to a general increase in the price of alcoholic beverages.

Given the fact that higher alcohol excise taxes increase prices and reduce ethanol consumption, there remains the vital question as to whether alcohol taxes are effective instruments in preventing alco-

hol-related harms. The most costly harms are associated either with the *acute* (immediate) effects of inebriation—injuries stemming from accidents and violent crime—or the *chronic* (long-term) deterioration in health and productivity that tends to result from heavy drinking over a period of years.

Considerable evidence points to an incidence of such harms that is sensitive to the prices of alcoholic beverages. For the acute effects, Cook (1981) studied 39 instances in which states increased their liquor tax between 1960 and 1975 and found strong evidence that traffic fatalities in those states fell as a result. Coate and Grossman (1988), using survey data, found that the prevalence of frequent drinking by youths was highly responsive to state beer taxes; Saffer and Grossman (1987), using panel data on state traffic fatality rates, found a fairly large negative effect of the beer tax for each of the youth age groups (i.e., 15–17, 18–20, and 21–24). Cook and Moore (1993), also using panel data on states, found a close link between per capita ethanol consumption and violent crime rates, along with some evidence that an increase in the beer tax helped suppress rape and robbery. There is also evidence of a link between alcohol prices and the prevalence of chronic heavy drinking. Cook and Tauchen (1982) demonstrated that changes in state liquor taxes had a statistically discernible effect on the rate of mortality from cirrhosis of the liver. Since a large percentage of deaths due to liver cirrhosis result from many years of heavy drinking, this finding suggests that people given to chronic heavy drinking are quite responsive to the price of alcohol. This conclusion is supported by evidence from clinical experiments and other sources (Vuchinich & Tucker, 1988).

Thus there is indeed evidence that alcohol taxes are an effective instrument for preventing alcohol-related harms, and the claim that alcohol taxes promote the public health has been an increasingly important consideration in the public debate over raising federal and state alcohol taxes.

FAIRNESS

Although alcohol taxes reduce consumption and save some lives that would otherwise be lost to alcohol-related accidents, the question remains as to whether they are fair. Fairness is indeed largely in the eye of the beholder (or taxpayer); nevertheless, several standards are commonly used as bases for judging the fairness of a tax. Two of the major ones are that a tax should fall equally on households that are in some sense equally situated, and that it should not be regressive.

If equals are to be treated equally, is it fair that alcohol taxes force drinkers to pay more taxes than are paid by nondrinkers of similar incomes? The bulk of all alcohol taxes are paid by the small minority of people who drink heavily; about 50 percent of all alcohol consumption is accounted for by just 6 to 7 percent of the adult population. One response is that it *is* fair for drinkers to pay more, because drinking imposes costs on others. A recent estimate suggests that drinkers impose an average cost on others that amounts to about 25 cents per drink (Manning et al., 1990). Thus if the alcohol tax is considered a sort of "user fee," whereby the drinker pays in proportion to the amount of alcohol consumed, then it may seem fair.

The other concern, that alcohol taxes may be regressive, refers to the idea that on the average, wealthier households spend a smaller fraction of their income on alcohol taxes than do poorer households, even if they buy large quantities of alcoholic beverages. The evidence on this point is not clear, although it is often taken as self-evident in political debates over raising beer taxes (Sammartino, 1990; Cook & Moore, 1993).

Another debated issue is that of uniform taxation. A can of beer, a glass of wine, and a shot of spirits all contain approximately the same amount of ethanol, but they are taxed quite differently; the federal excise tax on a shot of spirits exceeds the tax on a can of beer by a factor of two—and on a glass of wine by a factor of three. If special taxes on alcoholic beverages are ultimately justified by the fact that such beverages are intoxicating, then these disparities are difficult to explain. Part of the explanation may be the widespread belief that spirits are in some sense more intoxicating than beer or wine, and hence more subject to abuse, whereas beer is the "drink of moderation" and wine the "drink of connoisseurs." Much of the evidence works against this view, however. Beer consumption may indeed be more costly to society (per drink) than spirits, because of the demographics of beverage choice—young men, as a group, consume most of their ethanol in the form of beer and are involved in by far the highest incidence of alcohol-related traffic accidents and violent crimes.

(SEE ALSO: *Accidents and Injuries from Alcohol; Alcohol; Legal Regulation of Drugs and Alcohol; Prevention*)

BIBLIOGRAPHY

COATE, D., & GROSSMAN, M. (1988). Effects of alcoholic beverage prices and legal drinking ages on youth alcohol use. *Journal of Law Economics, 31,* 145–171.

COOK, P. J. (1981). The effect of liquor taxes on drinking, cirrhosis, and auto fatalities. In M. H. Moore & D. R. Gerstein (Eds.), *Alcohol and public policy: Beyond the shadow of prohibition.* Washington, DC: National Academy of Sciences.

COOK, P. J., & MOORE, M. J. (1993). Economic perspectives on alcohol-related violence. In S. E. Martin (Ed.), *Alcohol-related violence: Interdisciplinary perspectives and research directions.* NIH Publication no. 93-3496. Rockville, MD: National Institute on Alcoholism and Alcohol Abuse.

COOK, P. J., & MOORE, M. J. (1993). Taxation of alcoholic beverages. In M. Hilton & G. Bloss (Eds.), *Economic research on the prevention of alcohol-related problems.* NIH Publication no. 93-3513. Rockville, MD: National Institute on Alcoholism and Alcohol Abuse.

COOK, P. J., & TAUCHEN, G. (1982). The effect of liquor taxes on heavy drinking. *Bell Journal of Economics, Autumn,* 379–390.

GROSSMAN, M. (1989). Health benefits of increases in alcohol and cigarette taxes. *British Journal of Addiction, 84,* 1193–1204.

HU, T. Y. (1950). *The liquor tax in the United States 1791–1947.* New York: Columbia University Press.

LEUNG, S. F., & PHELPS, C. (1993). The demand for alcoholic beverages. In M. Hilton & G. Bloss (Eds.), *Economic research on the prevention of alcohol-related problems.* NIH Publication no. 93-3513. Rockville, MD: National Institute on Alcoholism and Alcohol Abuse.

LYON, A. B., & SCHWAB, R. M. (1991). Consumption taxes in a life-cycle framework: Are sin taxes regressive? (Working Paper No. 3932). Cambridge, MA: National Bureau of Economic Research.

MANNING, W. G., ET AL. (1991). *The costs of poor health habits.* Cambridge, MA: Harvard University Press.

MOORE, M. H., & GERSTEIN, D. R. (EDS.). (1981). *Alcohol and public policy: Beyond the Shadow of Prohibition.* Washington, DC: National Academy Press.

PHELPS, C. E. (1988). Death and taxes: An opportunity for substitution. *Journal of Health Economics, 7,* 1–24.

POGUE, T. F., & SGONTZ, L. G. (1989). Taxing to control social costs: The case of alcohol. *American Economic Review, 79,* 235–243.

SAFFER, H., & GROSSMAN, M. (1987). Beer taxes, the legal drinking age, and youth motor vehicle fatalities. *Journal of Legal Studies, 16,* 351–374.

SAMMARTINO, F. (1990). *Federal taxation of tobacco, alcoholic beverages and motor fuels.* Congressional Budget Office Report. Washington, DC: U.S. Government Printing Office.

VUCHINICH, R. E., & TUCKER, J. A. (1988). Contributions from behavioral theories of choice to an analysis of alcohol abuse. *Journal of Abnormal Psychology, 97*(2), 181–195.

PHILIP J. COOK

TEA   Tea is the most widely consumed beverage in the world, except for water, and provides over 40 percent of the world's dietary CAFFEINE. In the United States, caffeine from tea accounts for about 17 percent of caffeine consumed; per capita caffeine consumption from tea is about 35 milligrams per day, which is a little over one-third of the daily caffeine provided by coffee beverages. Tea consumption in the United Kingdom is substantially higher, averaging 320 milligrams per capita per day and accounting for 72 percent of the United Kingdom's caffeine consumption.

Although tea contains a large number of chemical compounds, the relatively high content of polyphenols and caffeine is responsible for tea's pharmacological effects. The primary psychoactive component of tea is caffeine. Tea also contains two compounds that are structurally related to caffeine, theophylline and THEOBROMINE, however, these compounds are found in relatively insignificant amounts. On average, a 6-ounce (177-milliliter) cup of leaf or bag tea contains about 48 milligrams of caffeine, a little less than half the caffeine in the same amount of ground roasted coffee, and only slightly more than the amount found in 12 ounces of a typical COLA soft drink. Six ounces of instant tea contain 36 milligrams caffeine, on average. Individual servings of tea contain amounts of caffeine that can affect mood and performance of adult humans.

Although the term *tea* has been used to refer to extracts from a large number of plants, only teas derived from leaves of *Camellia sinensis* plants are of special interest here, because they contain caffeine.

Figure 1
Tea

The term *tea* has come to be used especially for extracts of *Camellia sinensis* and that restricted usage is maintained in this entry.

Consumption of *Camellia sinensis* was first documented in *China* (where tea is called *cha* or *chai*) in 350 A.D., although there is some suggestion that the Chinese consumed tea as early as 2700 B.C. Tea was introduced to Japan around 600 A.D. but did not become widely used there until the 1400s. Through the China trade, tea became available in England in the 1600s, where it became the national drink. Tea was introduced into the American colonies around 1650 but in 1773 became a symbol of British rule. Americans protested the British tax on tea by raiding ships anchored in Boston Harbor and dumping boxes of tea into the water. This event, referred to as the Boston Tea Party, along with other similar protests that followed, became important in shifting the predominant caffeinated beverage in North America from tea to coffee.

India, China, and Sri Lanka are the major producers and exporters of tea—producing about 60 percent of the world's tea and providing about 55 percent of world tea exports. The United Kingdom, the United States, and Pakistan are the leading importers of tea.

Two types of tea, black and green tea, account for almost all of the tea consumed in the world. Black tea makes up over 75 percent of the world's tea; green tea accounts for about 22 percent. The method by which tea is manufactured determines whether black or green tea is produced. Black tea is dark brown in color and is produced by promoting oxidation of a key tea constituent. Green tea is yellow-green in color and is produced by preventing such oxidation, a less processed tea. Oolong tea, a less common type, is partially oxidized and is intermediate in appearance to that of black and green tea. Flavored teas were originally prepared by adding a range of fruits, flowers, and other plant substances to the tea prior to final packaging, although artificial flavors are often added today.

(SEE ALSO: *Chocolate; Plants, Drugs from*)

BIBLIOGRAPHY

BARONE, J. J., & ROBERTS, H. (1984). Human consumption of caffeine. In P. B. Dews (Ed.), *Caffeine.* New York: Springer-Verlag.
SPILLER, G. A. (ED.). (1984). *The methylxanthine beverages and foods: Chemistry, consumption, and health effects.* New York: Alan R. Liss.

KENNETH SILVERMAN
ROLAND R. GRIFFITHS

TEMPERANCE MOVEMENT Many temperance movements and societies emerged in the United States during the nineteenth century. These movements began in the early 1800s and gained ascendancy during the mid-to-late 1800s, culminating in the Prohibition Movement, the Prohibition Amendment (Article 18) to the U.S. Constitution in 1919, and the start of Prohibition in 1920. Gusfield (1986), an eminent scholar of the temperance movement, has argued that the term *temperance* is not appropriate, because the broad reformist ideology of the movement focused mainly on abstinence—not moderation—in the intake of alcoholic beverages. Blocker (1989) observed that the many temperance movements that emerged in the United States represented men and women from varying ethnic, religious, social, economic, and political groups who selected out temperance as the solution to what they perceived as problems in their own lives and in those of others. By the end of the nineteenth century, the temperance movement had evolved through several phases, and the strategies used by the proponents changed from persuasive efforts to moderate the intake of alcoholic beverages to more coercive strategies, even laws, to bring about the control of all drinking.

EARLY PHASE: 1800–1840

In colonial America and during the early 1800s, alcoholic beverages (brewed, fermented, and distilled) were a staple of the American diet, were often homemade, and were viewed as "the good creature of God." Among the colonists, the drinking of alcoholic beverages was integrated with social norms; all social groups and ages drank alcoholic beverages, and the consumption rate was very high. Alcohol was also traded, sold, and given to Native Americans, who had no long history of daily drinking, with almost immediate negative consequences for these peoples.

By 1840, a revolution in American social attitudes had occurred, in which alcohol came to be seen as "the root of all evil" and the cause of the major problems of the early republic, such as the crime, poverty, immorality, and insanity of the Jacksonian era (Tyrell, 1979). Temperance was advocated as the ideal solution for these problems by such people as Anthony Benezet, a popular Quaker reformer; Thomas Jefferson; and Dr. Benjamin Rush, the surgeon general of the Continental Army and a signer of the Declaration of Independence. Temperance-reform organizations, such as the American Temperance Society, emerged, committed to the eradication of these social problems.

The American Temperance Society (ATS), founded in Boston in 1826 as the American Society for the Promotion of Temperance, was the first national (as opposed to local) temperance organization. It had its roots in the processes of industrialization and the commercialization of agriculture. The people who developed the movement were committed to hastening the processes of economic and social change. These processes involved the educating of Americans to value sobriety and industry, in order to create the conditions for the development of an industrial-commercial society. The movement was supported by entrepreneurs who needed a disciplined and sober work force to help create the economic change necessary for the material improvement of the young republic.

During the so-called Great Awakening the evangelical clergy as well as that of other U.S. Protestant groups supported temperance as a means of promoting the morality needed for building a "Christian nation," through social and economic progress. According to Gusfield, these groups helped to place the issue of drinking on the public and political agenda, providing their personnel as authorities on the cognitive aspects of drinking and becoming the legitimate source of public policies on drinking. Also, in the early 1820s and 1830s, small-scale farmers and rural groups were active in promoting the temperance movement; they saw temperance as a way to promote social progress in a time of transition from a rural to an urban-industrial order, from small-scale farming to entrepreneurial forms of agriculture.

By 1836, the American Temperance Society had become an abstinence society, and ideas about problems associated with alcohol had begun to change—inebriety or habitual drunkenness was being called a disease. The ideology of the movement placed the source of alcohol addiction in the substance itself—alcohol was inherently addicting—a finding supported by research conducted by Rush, who in 1785 wrote *Inquiry into the Effects of Ardent Spirits upon the Human Body and Mind* (approximately 200,000 copies were published between 1800 and 1840). Blocker (1989) observed that the general focus of the American Temperance Society was on persuading the already temperate to become abstinent, rather than persuading drunkards to reform their drinking behavior. According to Gusfield (1986), abstinence became a symbol that enabled society to distinguish the industrious, steady American worker from other people—which resulted in the movement becoming democratized instead of associated only with the New England upper classes. Attempts to reform and save drunkards was the focus of another temperance movement, the Washingtonians.

MIDDLE PHASE: 1840–1860

Where well-to-do groups and Protestant evangelical clergy dominated the early phase of temperance reform, the middle phase included the efforts of artisans and women of the lower and lower-middle classes, who promoted self-help groups among largely working-class drunkards trying to give up drinking (Tyrell, 1979). These artisans organized into the Washingtonian societies (named for George Washington), dedicated to helping working-class drunkards who were trying to reform.

In 1840, the (first) Washingtonian Temperance Society was established in Baltimore. Members took a pledge against the use of all alcoholic beverages and attempted to convert drunkards to the pledge of teetotalism (c. 1834, derived from *total* + *total* =

abstinence). By the end of 1841, Washingtonian societies were active in Baltimore, Boston, New York, and other areas throughout the North. These groups were not socially homogeneous. Tyrell (1979) observed that the relationships between the old organizations and the new societies culminated in various struggles for control over the Washingtonian societies, with fragmentation of these groups occurring.

Washingtonian members who wanted respect from the middle-class temperance reformers, including the evangelical reformers, elected to remain with the mainstream temperance movement. The wage earners and reformed drunkards remained in their own societies, and they opposed early efforts at legal coercion—for example, the passage of the Maine Law of 1851. Gusfield (1986) has interpreted support for this law as a reaction against the drinking practices of the Irish and German immigrants to the United States between 1845 and 1855. He argued that temperance reform in this period represented a "symbolic crusade" to impose existing cultural values on immigrant groups. Tyrell interpreted the Maine Law as a way for middle-class reformers to control and reform the laboring poor. From 1851 on, many local laws were passed that attempted to limit the consumption of alcohol; however, throughout the remainder of the century, these statutes were repealed, liberalized, or unenforced.

LATE PHASE: 1860–1920

The Civil War, World War I, and the rapid demographic changes that accompanied immigration during this period contributed to the support of abstinence during the last phase of the temperance movements. Urban areas were expanding, factory towns were a reality, and there was an increase in the socializing at the end of the workday as well as at the end of the workweek; consequently there was an increase in the production and consumption of alcoholic beverages. Several temperance societies that emerged during this period included the active participation of women and children—since wives and children were often neglected or abused by drunken husbands and fathers. Irish-American Catholics formed the Catholic Total Abstention Union in 1872; the WOMEN'S CHRISTIAN TEMPERANCE UNION (WCTU) was formed in 1874; and the Anti-Saloon League (ASL) emerged in 1896. These societies were able to mobilize tremendous support for abstinence, rather than mere moderation in the intake of alcoholic beverages. At this time, the ideology of the temperance movements centered upon the evil effects of all alcohol, espousing the view that alcohol had become the central problem in American life and that abstinence was the only solution for this problem.

The WCTU was founded in Cleveland in 1874 and emerged as the first mainstream organization in which women and children were systematically involved in the temperance movement. Annie Wittenmeyer, Frances Willard, and Carrie Nation provided this temperance-reform movement with creative and dynamic leadership. The WCTU—a crusade to shut down saloons and promote morality—took a radical stance, criticizing American institutions by aligning itself with the feminist movement, the Populist party, and Christian Socialism. Gusfield (1986) argues that, although, under the leadership of Frances Willard (1879–1898), the WCTU was unsuccessful in establishing these alliances, it did achieve the following: It united the Populist and more conservative wings of the movement and it united the political forces of "conservatism, progressivism, and radicalism in the same movement." In addition, the WCTU provided backing for Prohibitionist candidates, including workers for their campaigns as well as audiences to listen to their positions on alcohol use. The WCTU still exists, based in Evanston, Illinois, and lists about 100,000 members as of 1990.

By the late 1800s, coercive reform became the dominant theme of the temperance movement. In 1893, the ASL of Ohio was organized by Howard H. Russell, a Congregational minister and temperance activist. In 1895, this group combined with a similar group in the District of Columbia, establishing a national society in 1896. By the end of the 1800s, the ASL, which represented a skillful political leadership resource for the Prohibition movement, mobilized tremendous support for abstinence instead of just temperance. In 1896, the movement began to separate itself from a number of economic and social reforms, concentrating on the struggle of traditional rural Protestant society against developing urban systems and industrialization.

Part of the success of the ASL was its determination to remain a single-issue (prohibition) pressure group that cut across all political party lines; the ASL also maintained a strong relationship with the Protestant clergy. It always put its own issue first but worked peacefully with the major political parties and especially with legislators (Blocker, 1989). By

1912, local prohibition laws had been passed to render most of the South legally dry.

In 1917, a major event boosted the cause of national prohibition. The United States entered into World War I, which prompted the ASL to push for the suspension of the industrial distilling of alcohol (ethanol). Very shortly after the U.S. entry into the war, the selling of liquor near military bases and to servicemen in uniform was prohibited (Blocker, 1989). By 1918, the Eighteenth Amendment to the U.S. Constitution had been proposed and the ASL had pushed prohibition through 33 state legislatures. Consequently, the Volstead Act—called Prohibition—was ratified on January 16, 1919. It went into effect one year later, on January 16, 1920, prohibiting the manufacture, sale, or transportation of alcoholic beverages.

CONCLUSION

Where the temperance movement was a middle-class reform movement, because it articulated the theme of self-control that was central to the middle-class ideology of the nineteenth century, some members of the working class also supported reform (Blocker, 1989). An ideology of ABSTINENCE became a rallying point for middle-class people who saw the rich as greedy, the working class as increasingly restless, and the poor as uneducated immigrants. Thus, they felt the need to restore a coherent moral order, especially after the upheaval of the Civil War and the ensuing period of industrial greed. At this time, the United States was undergoing economic expansion and deepening division along class lines. Other reform groups, such as the Progressive political party, joined the prohibitionists in their commitment to rid cities of saloons so that the United States could move toward becoming a virtuous and moral republic. At the end of the nineteenth century, Americans seemed to be more receptive to moral than scientific arguments for temperance reform and abstinence from alcohol.

Members of the temperance movements were concerned not only with changing the behavior of other social classes and groups but also about changing themselves (Levine, 1978). They were concerned that the pernicious effects of alcohol were also destroying the lives of Protestant middle-class people. While some of these reform groups were not complete supporters of an abstinence ideology, they were concerned with rebuilding a national community and promoting the common welfare. Abstinence became the governing ideology of the many diverse groups that had mobilized to promote a new social order.

As more scholars turn their attention to the study of the temperance era and the various temperance movements and societies, additional knowledge and interpretations will continue to be published. The bibliography that follows provides examples of some new interpretations of this period.

(SEE ALSO: *Alcohol; Prohibition: Pro and Con; Treatment*)

BIBLIOGRAPHY

BLOCKER, J. S., JR. (1989). *American temperance movements: Cycles of reform.* Boston: Twayne Publishers.

BLUMBERG, L. U., WITH PITTMAN, W. L. (1991). *Beware the first drink! The Washingtonian temperance movement and Alcoholics Anonymous.* Seattle, WA: Glenn Abbey Books.

BORDIN, R. (1981). *Women and temperance: The quest for power and liberty, 1873–1900.* Philadelphia: Temple University Press.

CLARK, N. (1976). *Deliver us from evil.* New York: Norton.

Dictionary of American temperance biography. (1984). Westport, CT: Greenwood Press.

EPSTEIN, B. (1981). *The politics of domesticity: Women, evangelism and temperance in nineteenth-century America.* Middletown, CT. Wesleyan University Press.

GUSFIELD, J. R. (1986). *Symbolic crusade: Status politics and the American temperance movement,* 2nd ed. Urbana, IL: University of Illinois Press.

HOFSTADER, R. (1955). *The age of reform.* New York: Vintage.

LENDER, M., & HOUSTON, J. K. (1982). *Drinking in America: A history.* New York: Free Press.

LEVINE, H. (1978). The discovery of addiction: Changing conceptions of habitual drunkenness in America. *Journal of Studies on Alcohol,* 39, 143–174.

RORABAUGH, W. (1979). *The alcoholic republic: An American tradition.* New York: Oxford University Press.

TYRELL, I. R. (1979). *Sobering up: From temperance to prohibition in antebellum America, 1800–1860.* Westport, CT: Greenwood Press.

PHYLLIS A. LANGTON

TEMPOSIL   *See* Calcium Carbimide.

TERRORISM AND DRUGS

The term *narcoterrorism* has entered the popular lexicon as a shorthand to refer to the complex relationship between the illicit drug trade and terrorism. The term, however, has often been used interchangeably to refer to two distinct aspects of this issue.

EXPLOITING THE DRUG TRADE

Narcoterrorism refers, first, to the activities of a number of guerrilla groups worldwide. These groups engage in terrorism and insurgency and also exploit the drug trade for financial gain. In most cases this exploitation involves rural-based guerrillas. Guerrillas and the drug trade (especially cultivation and processing) both tend to thrive in rugged, remote areas where government control is weak and where a nationally integrated economic infrastructure is lacking.

Rural-based guerrillas make money primarily by extorting "war taxes" from growers and traffickers. Thus the relationship between guerrillas, on the one hand, and the growers and the traffickers, on the other, is frequently rooted in coercion and conflict.

Nevertheless, guerrillas, growers, and traffickers sometimes cooperate in a marriage of convenience. The degree of government pressure exerted in an area can at times act as a unifying factor. Local family and/or personal relationships in a drug region can bring guerrillas, growers, and traffickers together, at least for periods of time.

A number of guerrilla groups have used both coercion and cooperation to exploit the drug trade. Examples include the following: The Revolutionary Armed Forces of Colombia (FARC), the country's largest and oldest insurgent group, and Colombia's National Liberation Army (ELN); Peru's Sendero Luminoso (Shining Path) and the Revolutionary Movement Tupac Amaru (MRTA); and the Kurdish Workers' Party (PKK) in the Middle East.

In addition to or apart from "taxation" and "protection" arrangements, various groups themselves have been directly involved in the drug trade:

In COLOMBIA, the FARC controls its own coca fields and processing laboratories for COCAINE. FARC may have some drug distribution networks, although evidence for this is fragmentary.

In Southeast Asia's GOLDEN TRIANGLE of Thailand, Burma, and Laos, guerrillas have long been actively involved in every stage of the OPIUM/HEROIN pipeline. They have frequently devolved into warlord trafficking organizations and dominate the drug business in the area.

- Some guerrillas in the South Asian subcontinent (the Indian peninsula of Bangladesh, Bhutan, Nepal, Pakistan, Sikkim, and India), such as the Tamil Tigers (LTTE) and the Sikhs, have used expatriate communities abroad to smuggle heroin.
- Lebanon's Hizballah reportedly smuggles drugs as a result of a *fatwah* (an Islamic religious decree). In 1987, the police uncovered narcotics in a Hizballah terrorist arms cache near Paris, France.

USING THE TACTICS OF TERROR

The second aspect covered under the rubric of *narcoterrorism* has been the drug traffickers' use of the tactics of political terrorism—such as the car bomb, kidnapping, and selective assassination—to undermine the resolve of various governments at the highest levels to fight the drug trade.

Traffickers usually use members of their own organization to carry out such attacks. Sometimes, however, traffickers have subcontracted to guerrillas. In late 1990, Colombia's Pablo Escobar used the ELN to help conduct kidnappings to pressure the Colombian government into negotiating with him.

Colombia has been hardest hit by the traffickers' use of terrorist tactics. Escobar's Medellín trafficking group was responsible for a string of vicious attacks in the 1980s and early 1990s. Among the victims and targets were a justice minister, an attorney general, Supreme Court justices, the editor of a leading newspaper, several presidential candidates, a commercial airliner, and the headquarters of Colombia's equivalent of the FBI.

Escobar scored a major victory by using narcoterrorism along with bribery to ensure the banning of extradition between Colombia and the United States in 1991. With the aid of corrupt officials, Escobar escaped from a jail in 1992 and continued to carry out sporadic attacks until he was killed by Colombian authorities in December 1993. The use of terrorist tactics by traffickers in Colombia subsided in the wake of Escobar's demise.

Italy too has suffered from drug violence. During the 1980s and early 1990s, the Sicilian Mafia retaliated for government crackdowns by killing a number of the country's leading prosecutors and law enforcement officers—often with car bombs, in spectacular fashion.

IMPLICATIONS

Narcoterrorism in both its incarnations challenges government efforts to control political violence, organized crime, and the drug trade.

Although involvement in the drug trade may sometimes decrease the revolutionary fervor of a guerrila group, the ability to derive income from this lucrative source strengthens the resources and capabilities of the groups to oppose the central government either as subversives or as a criminal element. Whether or not the guerrillas obtain the funding through coercion or cooperation with growers and traffickers, the result is usually a more formidable foe. Most observers, for example, believe that exploitation of the drug trade is the chief source of funding for Peru's Sendero Luminoso. The group remains the Western Hemisphere's most ruthless insurgency despite significant setbacks, such as the 1992 capture of leader Abimael Guzmán. Guerrilla groups such as Colombia's FARC and ELN reportedly use their drug-related profits to purchase weapons on the black market, sometimes from drug traffickers who also frequently deal in contraband arms. In general, the presence of guerrillas with an economic stake in the survival of the drug trade makes counternarcotics efforts an even more risky undertaking.

The willingness and ability of drug barons in some countries to use the tactics of terrorism adds a dangerous dimension to the threat posed by the drug trade. In Colombia, Escobar's narcoterrorism further weakened the country's judiciary—a mainstay of any democracy. His violence and the related scandal surrounding the lenient terms of his incarceration and his 1992 prison escape weakened the presidency of César Gaviria. In other countries, such as Mexico and some of the newly independent states of the former Soviet Union, there is growing concern about the volatile mix of drugs, violence, and organized crime.

(SEE ALSO: *Crop-Control Policies; International Drug Supply Systems*)

MARK S. STEINITZ

TERRY & PELLENS STUDY In a time when the use of many drugs is illegal in the United States and the public is inundated with information on such drug use, it is probably surprising that this set of circumstances is a historically recent phenomenon. Throughout most of the history of the United States, the manufacture, possession, and use of most drugs now considered addictive were legal, and very little was known about these drugs, their use or abuse.

Other than ALCOHOL (through the TEMPERANCE MOVEMENT), the drug that first captured the attention of policymakers and medical and public-health sciences was OPIUM. An interest in the addiction to opiates in the United States can be found as far back as 1877, when Dr. Marshall conducted a study of the number of opiate addicts in Michigan. However, this and the handful of similar efforts at epidemiological research conducted through 1920 were plagued with methodological problems. Generally these studies were conducted by sending short questionnaires to physicians or pharmacists who, at that time, legally supplied people with OPIUM and opium-based products. These physicians or druggists were simply asked to report the number of opium addicts they saw in their communities. All these studies were done in only one city, county, or state—with one exception. The exception was a study done by the U.S. Department of the Treasury, in an attempt to provide direct estimates of the number of opium-addicted people in the nation. Unfortunately, none of these studies would come close to meeting the requirements of sampling or of measures taken that would be required today.

A very important step forward in the study of drug addiction or dependency in general, and opiate addiction in particular, took place in a now classic study done for the Committee on Drug Addictions of the Bureau of Social Hygiene, in cooperation with the U.S. Public Health Service, by Charles E. Terry and Mildred Pellens from 1923 to 1924 (Terry & Pellens, 1924, 1927, 1928). This study was groundbreaking in several ways. First, rather than sending questionnaires to physicians and pharmacists, only about 30 percent of whom had responded in any of the previous studies, Terry and Pellens used field study techniques—their staff went to the sites of data collection. Second, rather than relying on self-reports, Terry and Pellens took advantage of official records that physicians, dentists, veterinarians, institutions, and laboratories were required to keep for all opium distribution, as mandated by the HARRISON NARCOTIC ACT of 1914. Third, and perhaps most important, Terry and Pellens conducted their study in six sites across the United States: Sioux City, Iowa; Montgomery, Alabama; Tacoma, Washington; Gary, Indiana; Elmira, New York; and El Paso, Texas. Although no known precedent existed for such a research strategy, they selected these six cities on the basis of racial characteristics, occupations, geographic region, and other social demographic factors, so that in aggregate these six sites could represent the United States as a whole.

As a consequence of these efforts, Terry and Pellens not only attempted to collect data more accurately but also produced the first study of the EPIDEMIOLOGY of drug addiction or dependence that tried to take into account social and demographic factors that, now as then, affect the number and distribution of people who are addicted to or dependent upon chemical substances. Their book, *The Opium Problem*, which contains chapters on the history of the problem, theories of its etiology, and contemporary treatments, is considered a classic in the field.

(SEE ALSO: *Epidemiology of Drug Abuse; High School Senior Survey; National Household Survey on Drug Abuse; Treatment*)

BIBLIOGRAPHY

TERRY, C. E., & PELLENS, M. (1928). *The opium problem.* New York: Bureau of Social Hygiene.

TERRY, C. E., & PELLENS, M. (1927). *A further study and report on the use of narcotics under the provisions of federal law in six communities in the United States of America, for the period July 1st, 1923 to June 30th, 1924.* New York: Bureau of Social Hygiene.

TERRY, C. E., & PELLENS, M. (1924). *Preliminary report on studies of the use of narcotics under the provisions of federal law in six communities in the United States of America, for the period July 1st, 1923 to June 30th, 1924.* New York: Bureau of Social Hygiene.

ERIC O. JOHNSON

TETRAHYDROCANNABINOL (THC)

This is a chemical found in the HEMP plant, CANNABIS SATIVA, that causes the PSYCHOACTIVE effects in the drugs MARIJUANA, BHANG, HASHISH, and GANJA. It is one of the three natural cannabinoids—chemical constituents of *Cannabis*—the other two being cannabinol (CBN) and cannabidiol (CBD).

Although some agent like THC was long considered to cause the mental effects of *Cannabis*, it was not until the mid-1960s that THC's exact chemical structure and synthesis were determined. A slightly different chemical, synhexyl, had been thought to mimic the action of THC, and this proved to be the case when the two materials were compared. It now appears that a variety of THC homologs (cannabinoids that differ chemically only in minor ways from each other) are found in various types of *Cannabis*, differing only in potency.

MECHANISM OF ACTION

For more than thirty years, the discovery of the mechanism of action of cannabinoids had eluded the best researchers. The problem seems finally to have been resolved by the detection of specific cannabinoid-binding sites (RECEPTORS) in the brain. Synthesized cannabinoid agonists, which show high potency and enantiospecificity (mirror imaging) in behavioral assays, were used to characterize cannabinoid-receptor binding.

One of these synthetic cannabinoids, radiolabeled CP55,940, is 10 to 100 times more potent than THC. The receptor was identified by attaching a radioactive element to CP55,940. It attached to the binding sites in the brain and the radioactivity allowed researchers to see just where the binding sites were. The question of whether these binding sites were really the receptors that were responsible for the psychoactive effects of THC was answered by comparing the potencies of a number of natural and synthetic cannabinoids with their relative potencies in several biological assays. The high correlations between binding receptors and biological effects indicated that the receptors were the same ones that mediate behavioral and pharmacological effects of cannabinoids—including the human subjective experience. In the central nervous system, these receptors are most dense in the outflow nuclei of the basal ganglia, the substantia nigra pars reticulata, and the globus pallidus, as well as the hippocampus and the cerebellum. Generally, high densities in the forebrain and cerebellum implicate roles for cannabinoids in cognition and movement, while sparse densities in brainstem areas controlling cardiovas-

cular and respiratory functions may explain why high doses of THC are rarely if ever lethal (Herkenham et al., 1990).

A further step in unraveling the mechanism of THC's action has been the cloning of the cannabinoid receptor.

METABOLISM AND KINETICS

Once taken, THC is extensively metabolized by the body and new metabolites are still being discovered. One metabolite, 11-hydroxy-THC, is actually more active than THC itself; however, it is not nearly as abundant, and one must assume that the major part of the activity of *Cannabis* derives from THC (Agurell et al., 1986). The high lipid (fat) solubility of the drug leads to extensive sequestration (uptake and storage) in the lipid compartments of the body (the body fat and the gray matter of the brain). Metabolites may be released from these fatty stores and then excreted out of the body for as long as a week after a single dose. Whether accumulation in the body of unchanged THC can occur remains questionable.

The pattern of plasma concentrations of THC after smoking marijuana parallels that following intravenous administration. That the drug should be so rapidly absorbed is an indication of the efficiency of the lung as a trap for the drug. THC is quickly redistributed into other tissues so that plasma concentrations decline over the course of three hours to negligible amounts. The usual symptoms of intoxication are almost completely gone by that time.

THC is absorbed slowly and unreliably from the gut after oral administration, with THC plasma levels of most subjects peaking between one and two hours after ingestion. Peak concentrations are also considerably lower than those following smoking.

DRUG-TESTING AND FORENSIC ISSUES

Plasma concentrations of THC do not correlate very well with levels of psychomotor impairment (motor incoordination, mental diminution). During the early stages of intoxication, plasma levels are very high while impairment is just beginning. After a rapid distributive phase, plasma concentrations begin to decline, becoming quite low by three to four hours after smoking. During the period from one to four hours, some correlation can be found between plasma THC concentrations and impairment. The absolute levels associated with impairment are highly variable. One estimate is that only levels of 25 mg/ml or higher would be definitely associated with impairment in every case (Hollister et al., 1981). Such levels would be attained only within one to two hours of smoking.

Urine testing is hardly useful for determining impairment, since metabolites of THC are detectable for days to weeks following brief exposures to the drug. Unless urine concentrations were extremely high, it would be difficult, if not impossible, to make judgments regarding impairment.

MENTAL EFFECTS

The psychoactivity of various cannabinoids and their metabolites has been systematically studied. The delta-8 isomer of THC has about 75 percent the activity of the delta-9 isomer (whether this isomer occurs in nature is questionable). Chemical variants of THC (made by modifying the molecule through addition of side chains)—such as tetrahydrocannabivarin, which has a three-carbon side chain—show approximately 25 percent the activity of THC. Delta-6a-THC, which has the double bond in a slightly different position, is only about 30 percent as active as THC; the same is true for its six-carbon side-chain homolog, synhexly. Among THC metabolites, 11-hydroxy-THC is somewhat more active, having about 120 percent of the activity of THC. The 8a- and 8b-hydroxy-THC metabolites have activity 20 to 25 percent of THC. When given orally in substantial doses, neither of the natural cannabinoids—cannabinol (CBN) nor cannabidiol (CBD)—show activity. An intravenous infusion of 18 milligrams of CBN showed definite activity, although only 10 percent that of THC itself. In general, psychoactivity is found only in cannabinoid compounds with THC-like structures.

(SEE ALSO: *Drug Metabolism; Drug Testing and Analysis; Pharmacokinetics*)

BIBLIOGRAPHY

AGURELL, S., ET AL. (1986). Pharmacokinetics and metabolites of delta-1-tetrahydrocannabinol and other cannabinoids with emphasis on man. *Pharmacological Review, 38,* 21–34.

HERKENHAM, M., ET AL. (1990). Cannabinoid receptor localization in the brain. *Proceedings of the National Academy of Science, 87,* 1932–1936.

HOLLISTER, L. E., ET AL. (1981). Do plasma concentrations of delta-9-tetrahydrocannabinol reflect the degree of intoxication? *Journal of Clinical Pharmacology, 21,* 1715–1755.

LEO E. HOLLISTER

THC *See* Tetrahydrocannabinol.

THEOBROMINE This ALKALOID belongs to the class of drugs called methylxanthines; it is similar to theophylline and to CAFFEINE. Theobromine (3,7-dimethylxanthine), however, is somewhat weaker than these two compounds and currently has almost no practical use in medicine.

Theobromine is found in the seeds of the plant *Theobroma cacao,* which is the well-known source of CHOCOLATE and cocoa. The cacao seeds have caffeine too (as does TEA, which contains small amounts of theobromine and theophylline); caffeine has powerful stimulant effects on the brain, whereas theobromine has very little (although popular articles alleged for years that theobromine makes one feel "happy"). High doses of theobromine can, however, affect several physiological functions in the body, such as increasing the formation of urine in the kidney.

BIBLIOGRAPHY

RALL, T. W. (1990). Drugs used in the treatment of asthma: The methylxanthines, cromolyn sodium, and other agents. In A. G. Gilman et al. (Eds.), *Goodman and Gilman's the pharmacological basis of therapeutics,* 8th ed. New York: Pergamon.

MICHAEL J. KUHAR

THERAPEUTIC COMMUNITIES *See* Treatment Types: Therapeutic Communities.

TOBACCO: DEPENDENCE In the United States as of the mid-1990s, there are about 45 million cigarette smokers—representing 25 percent of the adult population. Another 5 percent (men) use SMOKELESS TOBACCO (chewing tobacco or snuff). Most (70–80%) say they would like to quit. Unfortunately, they are dependent on (addicted to) NICOTINE, an alkaloid that makes it difficult to stop using tobacco. Most of them will have to try to quit several times before they are successful. Both the direct effects of nicotine on the body and behavioral associations with those effects learned over the years of tobacco use keep people going back for more even when they want to quit.

The role of nicotine in tobacco use is complex. Nicotine acts on the body *directly* to produce effects such as pleasure, arousal, enhanced vigilance, relief of anxiety, reduced hunger, and body-weight reduction. It may also reverse the WITHDRAWAL symptoms that occur in a nicotine-dependent person trying to quit—when nicotine levels in the body fall. These symptoms include anxiety, irritability, difficulty concentrating, restlessness, hunger, depression, sleep disturbance, and craving for tobacco. When this happens, the use of nicotine (whether tobacco or nicotine-containing medications) usually make people feel better by reversing the unpleasant withdrawal symptoms.

Nicotine also acts *indirectly*, through a learning process that occurs when the direct effects of nicotine occur repeatedly in the presence of certain features of the environment. As a result of the learning process, called conditioning, formerly insignificant environmental factors become cues for the direct actions of nicotine. These factors can become either pleasurable in themselves or they can serve as a triggering mechanism for lighting up a cigarette. For example, the taste, smell, and feel of tobacco often evoke a neutral response and sometimes repugnance in a nonsmoker. After years of experiencing the direct effects of nicotine in the presence of tobacco, however, a smoker finds the sensory aspects of tobacco pleasurable.

The indirect or conditioned effects of nicotine are responsible for much more complicated learning than the learning associated with nicotine's direct effects. Conditioning is also the process whereby the situations in which people often smoke—such as after a meal, with a cup of coffee, with an alcoholic beverage, while doing a task at work, while talking on the phone, or with friends who also smoke—become in themselves powerful cues for the urge to smoke. When people stop using tobacco, therefore, the direct effects of nicotine are not the only plea-

TABLE 1

Causes of Death and Their Established Epidemiological Association with Cigarette Smoking

Category	Cause of Death
A	Cancer of lung
	Chronic obstructive lung disease (includes emphysema)
	Peripheral vascular disease
	Cancer of larynx
	Cancer of oral cavity (pharynx)
	Cancer of esophagus
B	Stroke
	Coronary heart disease
	Cancer of bladder
	Cancer of kidney
	Cancer of pancreas
	Aortic aneurysm
	Perinatal mortality
C	Cancer of cervix, uterus
	Cancer of stomach
	Gastric ulcer
	Duodenal ulcer
	Pneumonia
	Cancer of the liver
	Sudden infant death syndrome
D	Alcoholism
	Cirrhosis of liver
	Poisoning
	Suicide
E	Endometrial cancer
	Parkinson's disease
	Ulcerative colitis

A = diseases for which a direct causal association has been firmly established and smoking is considered the major, single contributor to excess mortality from the disease; B = diseases for which a direct causal association has been firmly established, but smoking is but one of several causes; C = diseases for which an increased risk (association) has been epidemiologically demonstrated, but the exact nature of that association has not been firmly established; D = diseases for which excess mortality in smokers has been observed, but association is attributed to confounding; E = diseases for which smokers have lower death rates than nonsmokers.

SOURCE: U.S. Department of Health and Human Services, Washington, D.C.

sures they must give up. They must also learn to forgo the indirect effects of nicotine: those experiences that, through learning, have become either pleasurable in themselves or a cue to smoke.

MOTIVATION FOR QUITTING

Most Americans who use tobacco would like to quit, and the reasons for wanting to quit vary. The most common include (1) a concern for one's health; (2) a concern for the health of one's family and friends (this may entail concern about the harmful effects on children of secondhand smoke or concern about setting a bad example for them); (3) social pressure; (4) and economic factors (cigarettes are expensive).

STAGES OF QUITTING

Successful quitting of tobacco use usually occurs as a process over time, a series of mental stages or steps that the smoker goes through in quitting:

1. *Precontemplation.* The person is smoking and is not motivated to stop smoking during this stage.
2. *Contemplation.* The person is still using tobacco and is motivated to quit but has not settled on a quit date that is within one month.
3. *Action.* The person has a stop date and a plan that either was already implemented or will be implemented within one month.
4. *Maintenance.* The person has discontinued the regular, daily use of tobacco for a minimum of one month.

RELAPSE

Most tobacco users who try to quit agree with Mark Twain, who said, "To cease smoking is the easiest thing I ever did; I ought to know because I've done it is a thousand times." People who are addicted to tobacco and who try to quit are able to do so for a brief period of time, but most relapse; that is, they resume smoking. For example, 66 percent of smokers who try to quit on their own or with minimal outside help relapse within 2 days, 90 percent relapse within 3 months, and 95 percent to 97 percent relapse within 1 year of quitting.

The key to successful smoking cessation is an understanding of what triggers relapse, and what strategies are effective in preventing relapse.

Some of the most important triggers for lighting up a cigarette are withdrawal symptoms, environmental cues acquired through learned associations, and emotional upset. Relapse is promoted by such common withdrawal symptoms as difficulty concentrating, irritability, and weight gain. Environmental cues to relapse include the presence of other smokers—such as a spouse, friends, or coworkers who smoke—and occasions when alcoholic beverages are consumed. Emotional upset and DEPRESSION are also commonly reported cues for lighting up.

MANAGING URGES TO SMOKE

A smoker who contemplates quitting often thinks that smoking cessation is a simple matter of refraining from smoking during a period of nicotine withdrawal. Urges to smoke are powerful, however, and occur long after the period of nicotine withdrawal has ended. Tobacco users must not only not smoke but must, in fact, learn a new, tobacco-free lifestyle. Some learn on their own; others seek professional help.

Key aspects of learning a tobacco-free lifestyle include anticipating and managing withdrawal symptoms and environmental triggers for smoking. The environment might be managed to minimize smoking triggers by, for example, (1) sitting in nonsmoking sections of restaurants; (2) removing ashtrays from one's home and office; (3) leaving the table as soon as possible after meals and engaging in other activities such as talking, walking, or doing the dishes; (4) avoiding (at least temporarily) situations that trigger smoking, such as drinking alcohol or coffee when smokers are around and going to places, parties, or bars where people smoke; (5) actively seeking social support for smoking cessation. The encouragement of a husband or wife or of friends and other people who have quit or are quitting also makes it easier to quit.

Smokers who enjoy handling cigarettes or having something in their mouths need to substitute something for these smoking-related behaviors. They may chew gum, toothpicks, sunflower seeds, or something similar; munch food or low-calorie snacks; exercise to take up time they might otherwise spend smoking and to reduce any weight gain; snap, roll, or twist rubber bands on their wrist.

What people think about while quitting is an important factor in relapse. They need to teach themselves to maintain thoughts that may be useful in overcoming urges to smoke. Instead of thinking about the expected pleasures of a cigarette, the would-be quitter can substitute a stream of thoughts about the risks of smoking, the benefits of not smoking, the commitment to not smoking, the pleasures of an anticipated reward for not smoking, or the day's next activity.

Stress management is also important for successful quitting. Smokers soon recognize that giving up smoking is a substantial stress in itself. They can resort to some strategies that may reduce stress, such as meditation, relaxation, and physical exercises.

Other aspects of self-management during smoking cessation include setting realistic goals and some sensible rewards for behavior that leads to reducing tobacco use. Some days a realistic goal is a short-term one and involves just getting through each urge to smoke without succumbing. The smoker who is quitting can use any of the already mentioned substitution or distraction strategies while remembering that urges to smoke are likely to continue to come and go for some time.

Rewarding oneself for meeting even the short-term goals is important. Rewards for not using tobacco can include new clothes, a new book, time to develop a new hobby, or anything else the former smoker might enjoy. Many rewards can be paid for from money saved by not buying tobacco.

INDEPENDENT QUITTING

Most smokers quit smoking without professional help. People who quit on their own can benefit by (1) clearly identifying the reasons they want to quit (i.e., health, cost of cigarettes, etc.); (2) anticipating potential barriers to or problems with quitting and how to manage them; (3) setting a firm quit date and on that date removing all cigarettes and ashtrays from the home or office. In addition, any friends or family members who smoke should be asked not to offer cigarettes.

Persistence in trying to quit almost always works. Smoking a cigarette in the course of trying to quit should not become the end of the smoking-cessation effort. Most smokers try to quit several times before they are successful.

Many aids are available to tobacco users who quit on their own. Smoking-cessation program guides and motivational and educational tapes—audiotapes and videotapes—may be obtained from physicians, hospitals, or organizations such as the American Lung Association, the American Cancer Society, or the

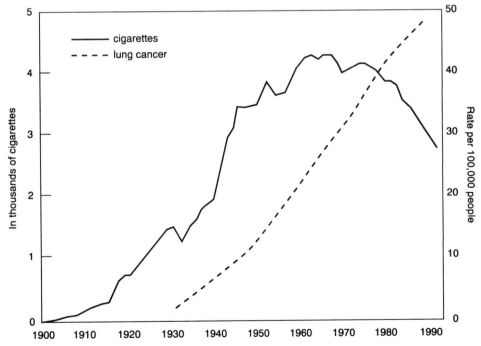

Figure 1
Cigarette-Smoking and Lung-Cancer Rates

American Heart Association, or they may be found in bookstores and libraries.

ASSISTED QUITTING

Smoking-Cessation Programs. These programs are available to help smokers in most communities. The programs usually involve attending meetings made up of small groups of quitting smokers who discuss their reasons for not smoking, their problems with quitting, and how they manage these problems. Participants in the programs can pick up practical skills in managing their smoking-cessation attempts and also obtain social support for their efforts. The cessation programs are offered by public-health organizations such as the American Lung Association and the American Cancer Society, and also by private companies such as Smokestoppers and Smokenders.

Physician- and Clinic-Assisted Quitting. Many physicians' offices and some hospital clinics offer assistance in smoking cessation. The clinics are particularly useful for people who have medical problems that need to be treated at the same time, for people who have tried before and failed to quit, or for people who may benefit from taking nicotine-replacement medications. Smokers can turn to these health-care facilities for advice on how to quit and for self-help material as well as for support and information during the different stages of quitting.

Nicotine-Replacement Medications: Nicotine Chewing Gum. Nicotine-replacement therapy can serve to reduce the severity of the nicotine-withdrawal symptoms of a smoker trying to learn to live without cigarettes. The nicotine-replacement medications are particularly useful with more seriously addicted smokers, but they are not a simple cure; rather, they must be used as part of a program of learning to live a tobacco-free life-style. Currently, two nicotine-replacement products are marketed in the United States: NICOTINE CHEWING GUM (also called Nicorette) and nicotine patches (also called transdermal NICOTINE DELIVERY SYSTEMS). Recent research has shown that nicotine replacement increases by about twofold the likelihood of a person's successful quitting smoking.

Nicotine chewing gum contains nicotine (bound to a resin, a chemical substance that binds other chemicals) and sodium bicarbonate. The sodium bicarbonate is necessary for keeping the saliva at an alkaline (basic) pH, which in turn is necessary for allowing nicotine to cross the lining of the mouth.

The gum is available in strengths of 2 and 4 milligrams (mg), although the dose actually delivered to the chewer is 1 mg and 2 mg, respectively. Nicotine is absorbed from the gum gradually over 20 to 30 minutes, in the course of which levels of nicotine similar to those seen after smoking a cigarette are produced in the blood. The gum is meant to be chewed intermittently, to allow time for the nicotine in the saliva to be absorbed. One should not chew the gum while drinking coffee, fruit juice, or cola drinks, because these beverages, by making the mouth more acidic, reduce the absorption of nicotine from the gum.

Smokers are instructed to quit smoking and then to chew the gum regularly throughout the day, and also whenever they have the urge to smoke a cigarette. For maximum efficacy, nicotine gum should not be chewed within ten minutes of drinking any beverage. Most people need to chew eight to ten pieces per day to obtain optimal benefits. Usually they chew the gum for three to six months but need to chew fewer pieces during the last month or two.

Side effects from chewing nicotine gum may include fatigue and soreness of the jaw, loosening of dental fillings, and occasionally nausea, indigestion, gas, or hiccups, particularly if one has chewed the gum so rapidly as to swallow nicotine-rich saliva.

Some tobacco users are concerned about the hazards of taking in nicotine, but the hazards of nicotine-replacement therapy are much less than those associated with smoking. In the first place, the amount of nicotine ingested in replacement therapies is less than that taken in from cigarettes. In the second place, nicotine-replacement medications do not expose smokers to the other hazards of cigarette smoke. The smoke contains, in addition to nicotine, carbon monoxide, tar, cyanide, and a number of other toxic substances. On balance, using nicotine gum is much safer than smoking cigarettes.

Nicotine Patches. To make it easier to stop smoking, researchers developed patches that administer nicotine without the side effects of nicotine chewing gum. Patches deliver nicotine in its un-ionized (uncharged) chemical form, thereby allowing the drug to pass through the skin readily. Various patches deliver different doses and are applied to the skin once a day, for times that range from sixteen or twenty-four hours. Four patches are available (as of 1994) in the United States: *Habitrol* (Ciba-Geigy),

Nicoderm (Marion-Merrell Dow), *Nicotrol* (McNeil), and *Prostep* (Lederle).

The patches deliver nicotine doses that are equivalent to smoking fifteen to twenty cigarettes (a pack) per day. Higher-dose patches are used during the initial three months of quitting, and lower-dose patches are available for subsequent tapering. Smokers who want to quit are instructed to first stop smoking and then to apply the patch daily. People who use the patch are warned not to smoke because concurrent smoking can easily make for excessive nicotine levels in the body.

The usually minor side effects from nicotine patches may include itching or burning over the patch site, which usually subsides within an hour, and local redness and mild swelling. Some people experience a sense of stimulation and, occasionally, insomnia; with sleep may come vivid dreams. These effects tend to occur during the first few days of patch use but not thereafter.

Other Therapies. A number of other treatments are available or have been used in the past to aid in smoking cessation. Although the effectiveness of these treatments has not been established by medical research, some individuals may benefit from them. None of these treatments, however, can magically cure smokers of their tobacco addiction without the commitment and effort that are usually required to quit.

HYPNOSIS has been widely used to increase a smoker's motivation or commitment to stop. While under hypnosis, the smoker receives suggestions, such as "smoking is a poison to your body," "you need your body to live," "you owe your body respect and protection." This treatment probably works best in combination with the previously discussed behavioral modification programs.

Treatment with ACUPUNCTURE, also widely advertised as a smoking-cessation technique, involves the placement of needles or staples in various parts of the body, most commonly the ears. As with hypnotism, there are no impartial scientific studies demonstrating the efficacy of acupuncture, but individuals report success in quitting by using it.

A variety of medications have also been depicted as aids in smoking cessation, but thus far, no evidence clearly shows that they are effective. Among these are prescription medications such as CLONIDINE (a blood pressure–lowering drug), tranquilizers, and ANTIDEPRESSANTS. Some smokers use these medications in the hope of reducing their nicotine-

withdrawal symptoms, particularly the mood disturbances.

OVER-THE-COUNTER (OTC) medications have been available in pharmacies without a physician's prescription, including lobeline and silver acetate. Lobeline, a chemical similar to nicotine but with less psychoactivity, has been recently removed from the market by the Food and Drug Administration. Lobeline has been available in prescriptions such as CigArrest, Bantron, and Nikoban. Silver acetate, available in a chewing gum, mouthwash, mouth spray and lozenges, acts as a deterrent. Tobacco smoke combines with the silver in the mouth to precipitate silver sulfide, which has an unpleasant taste. The unpleasant taste presumably decreases the incidence of smoking.

TREATMENT OF SMOKELESS TOBACCO ADDICTION

Most of the therapies described above have been developed and tested for treatment of cigarette smokers. Much less is known about effective treatment for smokeless tobacco (i.e., snuff or chewing tobacco) addiction. The general behavioral approach is similar to that for cigarette smoking, although the specific learned associations and cues are naturally somewhat different. Self-help materials are available from a variety of sources in the United States. Some strategies include the use of alternative activities, such as chewing gum, hard candy, sunflower seeds, nuts, toothpicks, or beef jerky. Formal treatment programs are also available in some parts of the country. Nicotine gum or patches may be helpful for users of smokeless tobacco who failed in their previous attempts (unaided by medication) to stop using tobacco.

(SEE ALSO: *Addiction: Concepts and Definitions; Relapse Prevention; Tobacco Smoking Cessation and Weight Gain; Treatment*)

NEAL L. BENOWITZ
ALICE B. FREDERICKS

TOBACCO: HISTORY OF *Tobacco* generally refers to the leaves and other parts of certain South American plants that were domesticated and used by Native Americans for the alkaloid NICOTINE. Tobacco plants are a species of the genus *Nicotiana*, belonging to the Solanaceae (nightshade) family; this also includes potatoes, tomatoes, eggplants, belladonna, and petunias. Including plants used for tobacco, there are sixty-four *Nicotiana* species. The two widely cultivated for use as tobacco are *Nicotiana tabacum* and *Nicotiana rustica*, the latter of which contains the higher levels of nicotine.

Nicotiana tabacum is however the major source of commercial tobacco, although it has been hybridized with other *Nicotiana* species, with resultant alteration in chemical composition. *Nicotiana tabacum* is a broad-leaf plant that grows from 3 to 10 feet (1–3 m) tall and produces 10 to 20 leaves radiating from a central stalk. *Nicotiana rustica*, also known as Indian tobacco, was first cultivated by Native Americans and was probably the tobacco offered to Columbus. The word *tobacco* comes into English (c. 1565) from the Spanish word *tabaco*, probably from the Taino word for the roll of leaves containing *N. rustica* that the American natives of the Antilles smoked.

HISTORY OF TOBACCO USE

Tobacco was introduced to Europeans by Native Americans at the time of Columbus's exploration of the New World (1492–1506). The first written records of tobacco use date from this time, but there is archaeological evidence for tobacco's wide use in the Americas as early as A.D. 600–900. Native Americans considered tobacco a plant used in ritual as it was sacred; it was used for social, fertility, and spiritual purposes. For example, tobacco was used for seasonal ceremonies, for sealing friendships, preparing for war, predicting good weather or good fishing, planting, courting, consulting spirits, and preparing magical cures. The desired effects of tobacco were a trance state, achieved by using the leaves in various ways, including smoking, chewing, snuffing, drinking (tobacco juice or tea), licking, and administering enemas.

Acute nicotine poisoning was a central aspect of the practice of shamanism in many parts of South America. South American shamans would smoke or ingest tobacco to the point of producing a nicotine-mediated trance or coma. The dose of nicotine could be titrated to produce a coma state resembling death but from which the shaman would recover. Recovery from apparent death enhanced the perception of the shaman's magical powers.

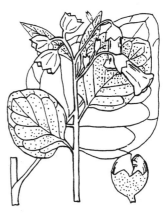

Figure 1
Wild Tobacco

In 1492 Columbus encountered natives in Hispaniola smoking tobacco in the form of large cigars. Enticed by the sacred and special regard in which they held tobacco, Columbus's crew experimented with tobacco smoking and soon became enthusiasts. Tobacco was brought back to Europe and, within a few decades, its use spread. People smoked it in the form of cigars and pipe and used it as snuff or chewing tobacco. Within forty years of Columbus's arrival, Spaniards were cultivating tobacco in the West Indies. Tobacco use then became widespread in Europe and in Spain and Portugal's American colonies by the late 1500s.

In 1570 the tobacco plant had been named *nicotiana* after Jean Nicot, the French ambassador to Portugal who introduced tobacco to France for medicinal use. Tobacco was said to be useful in the prevention of plague and as a cure for headache, asthma, gout, ulcers, scabies, labor pains, and even cancer. In the late 1500s Sir Walter Raleigh popularized the smoking of tobacco for "pleasure" in the court of Queen Elizabeth (reigned 1558–1603); from there it spread to other parts of England.

James I of England (reigned 1603–1625)—who succeeded Queen Elizabeth—was strongly opposed to tobacco use and wrote the first major antitobacco treatise, entitled "Counterblast to Tobacco," in 1604. King James described tobacco as "a custome loathsome to the eye, hateful to the nose, harmful to the brain, dangerous to the lungs, and in the black stinking fume thereof nearest resembling the horrible stygian smoke of the pit that is bottomless." Despite James's opposition, however, tobacco use flourished. Eventually even James lessened his opposition to tobacco, because of the lucrative income from its taxation.

During the 1600s, tobacco use had spread throughout Europe, Russia, China, Japan, and the west coast of Africa. Over the centuries, draconian penalties for tobacco use were occasionally promulgated. For example, Murad the Cruel of Turkey (1623–1640) ordered that tobacco users be beheaded, quartered, and/or hanged. Nevertheless, smoking persisted. In the American colonies, tobacco became the most important export crop and was instrumental in the economic survival of the colonies.

By the nineteenth century, tobacco production was a mainstay of American capitalism. Most tobacco was smoked as cigars or in pipes, or used as snuff. Cigarettes were hand rolled. A skillful worker could roll four cigarettes per minute. Cigarette smokers were primarily boys or women, and smoking was a behavior confined to the lower socioeconomic class. The invention of the cigarette rolling machine by James Bonsack in 1881 made tobacco use inexpensive and convenient. Bonsack went into business with W. B. Duke and Sons in Durham, North Carolina. Together they improved the machine until by April 30, 1884, the device could roll 120,000 cigarettes per day.

The phenomenal surge in cigarette smoking in the twentieth century is attributable to efficient manufacture and well-aimed marketing strategies that made smoking attractive and broadened the user base. Tobacco companies became financially powerful. W. B. Duke and Sons formed the American Tobacco Company, which was disbanded in 1911, because of antitrust concerns, into four companies: Liggett & Myers, R. J. Reynolds, P. Lorillard, and American Tobacco. At present, these four companies plus Brown & Williamson and Philip Morris manufacture most of the cigarettes in the United States. U.S. Tobacco is the major producer of snuff and chewing tobacco. These corporate giants are now multinational companies with subsidiaries in industries such as food and beverages, merchandising, hotels, entertainment, mining, and many other areas.

Just as cigarettes were becoming widely available and affordable, tobacco manufacturers strongly promoted their use. Massive advertising campaigns, government issue of cigarettes to soldiers during the world wars, glamorization of cigarettes in motion pictures, and the gradual incorporation of women into the smoking market increased the popularity of cigarette smoking in the United States and around the world. Smoking rates peaked in the United States for men in 1955, with 50 percent of men smoking, and in 1966 for women, with 32 percent of women smoking. As a result of clever marketing by the cigarette companies, smoking at that time was considered to be sophisticated, glamorous, individualistic, and even healthful.

While there had been occasional reports on the health hazards of cigarette smoking from the time of King James, the first large-scale studies documenting the link between cigarette smoking and cancer appeared in 1952 (Doll & Hill) and 1956 (Wynder & coworkers). Subsequently, hundreds of studies have shown that cigarette smoking accounts for 30 percent of cancers—including some cancers of the lung, mouth, throat, esophagus, bladder, and kidney, as well as some leukemia; and that it is the cause of some heart and vascular disease and stroke, emphysema and chronic obstructive lung disease, and other health problems. In 1962 the Royal College of Physicians in the United Kingdom, and in 1964 the U.S. surgeon general issued reports on smoking and health, indicating that cigarette smoking most probably caused some lung cancers and other health problems. These reports mark the beginning of modern public-health efforts to control tobacco use.

Subsequent landmarks in tobacco control in the U.S. include the following:

- 1965—Federal Cigarette Labeling and Advertising Act (PL89-92) required health warnings on cigarette packages and an annual report to Congress on the health consequences of smoking.
- 1969—Public Health Cigarette Smoking Act (PL91-222) strengthened health warnings on cigarette packs, prohibited cigarette advertising on television and radio.
- 1973—Little Cigar Act (PL93-109) extended the broadcast ban on cigarette advertising to little cigars.
- 1984—Comprehensive Smoking Education Act (PL98-474) required rotation of four specific health warnings and mandated that the cigarette industry provide a list of cigarette additives.
- 1986—Comprehensive Smokeless Tobacco Health Education Act (PL99-252) required three rotating health warnings on SMOKELESS TOBACCO packages and advertisements and a list of additives and nic-

otine content in smokeless tobacco products, prohibited smokeless tobacco advertising on television and radio, and mandated reports to Congress on smokeless tobacco and a public information campaign on the health hazards of smokeless tobacco.

The four warnings currently rotated among cigarette packs are the following:

1. Surgeon General's Warning: Smoking Causes Lung Cancer, Heart Disease, Emphysema, and May Complicate Pregnancy
2. Surgeon General's Warning: Quitting Smoking Now Greatly Reduces Serious Risks to Your Health
3. Surgeon General's Warning: Smoking by Pregnant Women May Result in Fetal Injury, Premature Birth, and Low Birth-Weight
4. Surgeon General's Warning: Cigarette Smoking Contains Carbon Monoxide.

The three smokeless tobacco warnings that are rotated are these:

1. Warning: This Product May Cause Mouth Cancer
2. Warning: This Product May Cause Gum Disease and Tooth Loss
3. Warning: This Product Is Not a Safe Alternative to Cigarettes

As a consequence of education and other public-health activities, tobacco use has begun to decline in the United States. Currently, 25 percent of Americans, about 43 million people, smoke. About 45 million former smokers have quit. Unfortunately, adult smoking rates have been declining very slowly in recent years because adolescents are taking up smoking at undiminishing rates and grow up to become addicted adult smokers.

NATURE OF TOBACCO PRODUCTS

In the United States, tobacco is consumed in five forms: cigarettes, cigars, pipe tobacco, snuff (oral and nasal), and chewing tobacco. Ninety-five percent of American tobacco consumption is in the form of cigarettes. Cigarettes are composed of shredded tobacco leaves and other parts of tobacco plants, with flavorings and other ingredients added, encased in paper, and often attached to a filter through which the smoke is puffed. American cigarettes are made almost exclusively of blond tobacco, while some Eu-

ropean and Middle Eastern cigarettes are made of dark tobacco. Cigars compose about 1.5 percent of the U.S. tobacco market. Cigars contain dark tobacco leaves that are rolled and covered with a wrapper that may be tobacco leaf or reconstituted tobacco sheet.

Tobacco for pipes and in the form of smokeless tobaccos represent the smallest segment of the market. Pipe tobacco consists of a blend of a variety of leaf types and is often heavily flavored to give particular aromas and tastes. Four types of smokeless tobacco are manufactured in the United States—(1) oral snuff and (2) loose-leaf, (3) plug, (4) twist or roll chewing tobacco. In other countries, particularly the United Kingdom, dry snuff is also widely used. Loose-leaf chewing tobacco consists of tobacco leaf that has been heavily treated with licorice and sugars. Plug tobacco is produced from leaves that are immersed in a mixture of licorice and sugar and then pressed into a plug (usually a square). Twist or roll tobacco is made from leaves flavored and twisted to resemble a rope.

Moist snuff consists of fine particles of tobacco that contain considerable moisture; many are treated with flavorings such as wintergreen or mint. The oral use of snuff, called "snuff dipping," means placing a pinch of the tobacco between the cheek or lips and the gum, or beneath the tongue. Dry snuff is powdered tobacco that contains flavor and aroma additives. Dry snuff is meant to be snorted or taken up into the nose (producing the sneezing seen in costume films about the eighteenth and nineteenth centuries).

PRODUCTION OF TOBACCO PRODUCTS

The four major types of tobacco leaf used in the manufacturing of American tobacco products are (1) flue-cured tobacco (also called bright leaf or Virginia tobacco), (2) light air-cured tobacco (burley and Maryland varieties), (3) dark tobaccos, and (4) Oriental tobaccos. Flue-cured tobacco is grown in relatively sandy soils from Virginia to Florida, with its production center in North Carolina. Flue-curing refers to the process of hanging mature tobacco leaves in a dark barn with heat applied (at one time via a flue from a furnace) to the leaves as they ferment and dry.

Both the type of tobacco cultivated and the manufacturing process varies around the world. The burley and Maryland tobacco types, used in light air-cured tobacco, are produced mostly in Kentucky, Tennessee, and western North Carolina. Air-curing involves hanging tobacco in a shaded barn without heat, typically for six to eight weeks. Dark tobaccos resemble the tobaccos originally used by Native Americans. They are air-cured and often fermented. Dark tobaccos are widely used in French and Spanish cigarettes, in the bidi smoked in India, and in clove cigarettes smoked in Indonesia. Small-leaf Oriental tobacco, which is sun-cured and fermented during storage, is valued for its aroma. It is produced primarily in Turkey, Greece, Bulgaria, and Russia.

Reconstituted tobacco sheets as well as tobacco leaves are important components of cigarettes and sometimes cigars. The sheet consists of tobacco stems and scraps of leaves (once wasted by-products of cigarette manufacturing) chopped up and extracted with water. The tobacco extract, which contains nicotine and other alkaloids, is concentrated and held for reapplication. After the tobacco is chopped fine and made into a slurry, it is formed into a paperlike web or sheet, which is then dried. The dried sheet is impregnated with the concentrated tobacco extract, then cut into shreds for use in tobacco products.

Cigarettes are made by blending different varieties of tobacco leaf and reconstituted tobacco sheet. The typical American blended cigarette contains 30 to 70 percent flue-cured Virginia, 15 to 45 percent burley, 1 to 5 percent Maryland, and 5 to 15 percent Oriental. Cigarettes smoked in the United Kingdom are primarily flue-cured tobacco, and cigarettes smoked in France, Spain, and North Africa are primarily dark air-cured tobacco.

Taste modifiers are added to the tobacco leaves in two steps. The first step is the application before shredding of a casing solution, containing sugars and glycerine, to tobacco strips. Sugars are added to overcome the bitterness and to make the smoke smooth, and glycerine keeps the tobacco moist. The second step is the addition of flavorings after the tobacco is shredded. Flavorings are sprayed onto the tobacco, which produce the aroma considered a pleasing feature of the package and the smoke. Cigarette flavorings include menthol and others. Smokeless tobacco—that is, snuff and chewing tobacco—is commonly impregnated with flavorings such as mint, wintergreen, or licorice.

Other components of the cigarette are the wrapping paper, the filter plug, and the overwrap paper that joins the filter plug to the tobacco column. Filters are usually made of cellulose acetate filaments, which trap some of the particles in the smoke.

THE BURNING CIGARETTE AND COMPONENTS OF TOBACCO SMOKE

When a smoker puffs on a cigarette, air is drawn into the lit end of the cigarette, resulting in combustion at about 1650° (900° Celsius). Combustion is incomplete and results in a tarry vapor, water, and gases. The products of combustion are examined by dividing them into particulate and gaseous materials. The particulates form during combustion, when nicotine vaporizes, cools, and condenses with other combustion products. The smoke that is taken in by the user during a puff is called mainstream smoke. Sidestream smoke is generated between puffs as the cigarette smolders; 75 percent of the smoke is emitted as sidestream smoke. Sidestream smoke wafts into the air and is often called environmental smoke.

Examining the particulate and gaseous components of cigarette smoke, scientists have found, in addition to tars and nicotine, several thousand different chemicals. Many of these chemicals are known to be toxic or hazardous to people. Some of the tobacco toxins include carbon monoxide, cyanide gas, and nicotine, as well as a number of chemicals that may contribute to the development of cancer. Cancer-causing substances include polycyclic aromatic hydrocarbons, nitrosoamines, benzene, formaldehyde, vinyl chloride, metals such as arsenic and chromium, and radioactive elements such as polonium-210.

CIGARETTE YIELDS

In recent years tobacco manufacturers have attempted to develop and promote cigarettes that would be perceived by smokers as less harmful, even though the producers have never acknowledged that cigarette smoking is hazardous. They report their tobacco products as low in nicotine and tar yields and promote these attributes in print advertisements and on billboards. However, lower yields do not necessarily mean that the tobacco product delivers less nicotine and tar to the smoker.

Advertised yields of nicotine and tar are measured by machines, not by examining the levels of these substances in smokers. Unfortunately, the machine does not accurately reflect the smoking behavior of the consumer. The device consists of a mechanized syringe that draws 35-milliliter (ml) puffs of smoke over 2 seconds, once every minute, until the cigarette is burned to a specific distance above the filter overwrap. The smoke is drawn through a paper filter that collects the particulates. The particulates are assayed for nicotine and tar. Tar is a dark, tarry substance similar to that which collects on chimneys and on the tops of barbecue grills. It remains in smoke aerosole after the nicotine and water are removed. Tar contains many cancer-causing chemicals, particularly polycyclic aromatic hydrocarbons. The part of the smoke that passes through the filter is gaseous. The gaseous material is typically analyzed for carbon monoxide levels, although many other components are also toxic.

Yields of tar and nicotine as measured by machines have decreased considerably since the mid-1950s. However, the tobacco itself and the amount of nicotine in the tobacco remain the same for high-yield and low-yield cigarettes. The decreased yields have been produced by engineering cigarettes mechanically. Thus low-yield cigarettes typically use filtration, reduction in the weight of tobacco, faster-burning paper that accelerates the rate of burning, and dilution to produce lower yields of tar and nicotine on the smoking machine.

Cigarette filters remove some but not all of the particulates. Typically, filters remove 20 to 60 percent of tar and nicotine. Filters do not reduce the smoker's exposure to gases. Reducing the amount of tobacco in cigarettes by making the cigarettes narrower (i.e., slims) or by using expanded or puffed tobacco (lighter cigarettes) means that less tobacco is burned, so the resultant yield is reduced. The rate of burning can be controlled by the use of wrapper papers that burn faster. With faster-burning paper, the cigarette on the smoking machine burns to a specified length in a shorter time. As a result, fewer puffs can be taken by the machine during the testing period, less mainstream smoke is inhaled, and more sidestream smoke is released into the air. Since the machine measures the content of inhaled or mainstream smoke from the puffs, the total yield of the mainstream smoke from low-yield cigarettes will be less because less smoke is captured.

Finally, yields are reduced by diluting cigarette smoke with room air. Cigarette manufacturers use more porous wrapping papers and/or place ventilation holes in the filter or overwrap papers. The resulting cigarette produces lower yields on the smoking machine because fresh air enters the mainstream smoke when the cigarette is puffed.

Unfortunately, a smoker who puffs a cigarette consumes much more tar and nicotine than the smoking machine tests indicate. Humans smoke differently than do machines, taking more puffs, puffing more deeply, and unintentionally blocking ventilation holes. In reality, smokers of low-yield cigarettes tend to have the same exposure to nicotine and tar as smokers of high-yield cigarettes. Thus, no health benefit is realized from smoking cigarettes marketed as lower yield.

(SEE ALSO: *Advertising and Tobacco Use*; *Nicotine Delivery Systems for Smoking Cessation*)

NEIL L. BENOWITZ
ALICE B. FREDERICKS

TOBACCO: INDUSTRY The tobacco industry is made up of the complex of primary suppliers, manufacturers, distributors (both wholesale and retail), advertising agencies, and media outlets that produce, promote, sell tobacco products, as well as the law, public relations, and lobbying firms that work to protect these products from stringent public-health regulation and control. The industry evolved in the late nineteenth and early twentieth century from many, relatively small enterprises that produced tobacco products for puffing, snuffing, and chewing. The products of these small firms delivered nicotine to the nasal and oral mucosa. With the evolution and refinement of the cigarette, the industry developed first into a monopoly and then into an oligopoly in which a handful of major producers made this more sophisticated nicotine delivery system: a device that delivers nicotine by inhalation to the lungs and thence rapidly to the brain. Although its popularity is declining in the United States, cigarette use is increasing worldwide at over 2 percent per year, especially in much of Asia, Eastern Europe, and the former Soviet Union. An integrated system of suppliers, manufacturers, marketers, and sales outlets is constantly evolving to supply this vast and growing market. Sophisticated legal and lobbying enterprises manage to protect this industry from the sort of regulation advocated by a number of public health groups—regulations that governments routinely impose on far less toxic products.

PRIVATE ENTERPRISE VERSUS STATE MONOPOLY

Tobacco (nicotiana) is a plant of the nightshade family (genus *Nicotiana*) and is native to the Americas; it was a major commodity of commerce in colonial times. Cigar tobaccos were key exports from the Spanish and Portuguese colonies of the Caribbean and South America, while tobaccos for snuff, pipe, and chew were the economic mainstays of the English colonies in Virginia, Maryland, and the Carolinas. Whereas most of Europe (and the rest of the world) established state-run monopolies for tobacco distribution, private enterprise was the vehicle of tobacco commerce in Great Britain (and eventually in the United States). The state monopolies provided both a popular product for the populace and revenue for the national treasury—but private enterprise, which always paid excise tax in Great Britain, was more resourceful in expanding the market. This phenomenon has been exploited in the twentieth century and is especially apparent in the 1990s, with the remaining state monopolies becoming privatized and adopting the marketing techniques of the by-now enormous transnational tobacco companies, often actually merging with them.

FROM COTTAGE INDUSTRY TO MONOPOLY TO OLIGOPOLY

Relatively expensive, hand-rolled cigarettes became popular novelties in the United States and Europe in the mid-nineteenth century. The novelty came to dominate the industry over a period of forty years, from the mid-1880s to the mid-1920s, when, for the first time, more tobacco in the United States was used for cigarettes than for chewing tobacco.

A number of changes in the nineteenth century laid the groundwork for the cigarette's commercial success. The development of flue-cured tobacco and air-dried burley tobacco—easily processed into tobaccos for smoking (where the smoke might be inhaled) were major factors (Slade, 1993). Cigarette-making machines—first used commercially in 1883 by the American Tobacco Company—the development of safe matches, and an extensive railroad network to transport centrally manufactured cigarettes throughout the United States were among the other key factors responsible for this product's success.

Duke of Durham, North Carolina. These elements were successfully harnessed by Benjamin Newton (Buck) Duke, head of the American Tobacco Company. A working cigarette-making machine had been invented in 1881 by James Bonsack in response to a contest held by the cigarette maker Alan & Ginter of Richmond, Virginia (Smith, 1990). But the contest sponsors decided against using the invention since they did not know how to sell as many cigarettes as the machine was capable of making. Duke, however, realized that the low prices made possible by mass production, together with advertising to stimulate demand, would create a large enough market to absorb the vastly expanded production. He obtained favorable terms for using the machine in exchange for technical assistance in perfecting it. The machine Duke put on line in 1883 produced 120,000 cigarettes per day, the equivalent of 60 expert hand rollers. Duke's competitors had to pay more for Bonsack machines than he had, and Duke engaged in price wars to further weaken other manufacturers. Gradually, he bought out his competitors and monopolized the U.S. cigarette industry. By 1890, Duke controlled the cigarette market, and by 1910, just before his monopoly was broken, he controlled more than 80 percent of all tobacco products manufactured in the United States, except for cigars (Robert, 1952).

Seeking further growth, Duke began to expand his cigarette business overseas (Robert, 1952). By 1900, a third of America's domestic production was being sent to Asia, and company factories were operating in Canada, Australia, Germany, and Japan. In 1901, Duke purchased a cigarette factory in Liverpool, England. Alarmed British manufacturers, seeking to avoid the fate of their U.S. compatriots, banded together as the Imperial Tobacco Company. The resulting trade war between American and Imperial ended in a truce. American was given exclusive trading rights in the United States and Cuba, and Great Britain became Imperial's exclusive territory. A new company, jointly controlled by both giants, was to sell cigarettes to the rest of the world. This modest sinecure was the birthright the parent companies gave the British-American Tobacco Company (BAT).

Antitrust Litigation. In 1907, the U.S. government filed an antitrust case against the American Tobacco Company. The result of this litigation was the dissolution of the trust four years later into a number of successor companies, some of which retain major roles in the U.S. cigarette market. These companies were the American Tobacco Company, the R.J. Reynolds Tobacco Company, Liggett & Myers, and P. Lorillard.

Once it had emerged from the confines of the trust, R.J. Reynolds, which had never before made cigarettes, developed and introduced Camel, a novel brand, in 1913 (Tilley, 1985). Camel was the first brand to combine air-dried burley, which had previously been important in chewing-tobacco products, with the then-conventional cigarette tobaccos—the flue-cured and Turkish (Oriental) varieties (Slade, 1993). Camel featured a coherent, national advertising campaign from N.W. Ayer that relied entirely on mass-media outlets in magazines and on billboards instead of on package-based promotions such as cigarette cards, coupons, and premiums. The legacy of this startling departure from the conventional cigarette-marketing techniques of the time is captured by the sly legend that still graces each pack of twenty unfiltered Camels sold in the United States: "Don't look for premiums or coupons, as the cost of the tobaccos blended in CAMEL Cigarettes prohibits the use of them."

The other thing that distinguished Camel from its competitors was its price. While the leading brands of the time, such as Fatima, sold for fifteen cents per pack of twenty, a pack of Camel sold for a dime. In short order, Camel overwhelmed the competition and ushered in a dramatic expansion of the domestic cigarette market. American Tobacco copied the Camel formula with Lucky Strike, and Liggett & Myers followed with its copycat product Chesterfield. Cigarette cards, premiums, and coupons were abandoned in favor of the mass media, and prices fell. Cigarette use, then only rising slowly, began an unprecedented increase. This growth continued virtually unabated for forty years or so, until finally slowed and eventually reversed by alarms that lung cancer and other major diseases could be caused by cigarettes (Fiore et al., 1993).

Only two firms that had no roots in the tobacco trust have played major roles in the U.S. cigarette market (Sobel, 1978). After Buck Duke's death in 1929, BAT purchased the Brown & Williamson Tobacco Company in Louisville, Kentucky. BAT gradually built this company into a major cigarette producer. For decades, its Kool brand dominated the menthol category, and during the 1930s and 1940s, its Wings brand gained market share by undercutting the prices of the majors. Brown & Williamson continues to offer a full range of cigarettes for the U.S. market. It also produces cigarettes for export to many of BAT's international markets.

The other upstart company was Philip Morris, which began its U.S. operations as a specialty cigarette maker in New York in the first quarter of the century. In addition to its standard brand called Philip Morris, it produced Marlboro—a cigarette for "ladies." The company expanded in the 1930s with a low-priced brand (Paul Jones) and a clever pricing scheme for Philip Morris English Blend (Robert, 1952; Sobel, 1978). It suggested a retail price for the latter slightly above that for the major brands, but it gave retailers a larger margin, thus encouraging prominent display of the brand in stores. In the mid-1950s, Philip Morris gave Marlboro a filter and had the Leo Burnett advertising agency remake its image entirely to one of rugged masculine outdoor daring on horseback. (The entire sweep of Marlboro advertising is included in the special advertising collection of the American Museum of National History in Washington, D.C.) By the mid-1970s, Marlboro was the leading U.S. cigarette and by the 1990s, thanks to the strength of Marlboro's appeal to teens and young adults, Philip Morris overtook R.J. Reynolds to become the nation's largest tobacco-product manufacturer.

Smokeless Tobacco. Moist snuff and chewing tobacco have enjoyed a 1980s and 1990s resurgence in popularity—this is based on the successful efforts of U.S. Tobacco (UST). It sells oral tobacco (e.g., Skoal Bandits, Skoal, Copenhagen) to adolescents and preadolescents (Denny, 1993). Oral tobacco is the only category of tobacco product whose consumption has increased in recent years in the United States. This increase is attributable to UST's innovative marketing of moist snuff to adolescent boys, and to imitation products from other manufacturers. Although UST envisions a global market for snuff, the World Health Organization has declared that countries in which oral tobacco is not a traditional product should ban it. A number of countries—including Australia, New Zealand, Hong Kong, and the European Community—have taken this step, often defying intense pressure from the U.S. government when doing so.

TABLE 1
Leading U.S. Tobacco Companies, 1992

Company	Home Office	Major Brands
CIGARETTES:		
Philip Morris	New York	Marlboro
		Basic
		Benson & Hedges
		Merit
		Virginia Slims
RJR/Nabisco	New York	Winston
		Camel
		Salem
American Brands	New York	Carlton
		Lucky Strike
British American Tobacco (BAT)	London	Kool
Loews Corporation	New York	Newport
		Kent
Brooke Group	Miami	Generics
MOIST SNUFF:		
UST	Greenwich, CT	Copenhagen
		Skoal Bandits
		Skoal Classic

Table 1 lists the major tobacco-product manufacturers in the United States, the location of their corporate headquarters, and the major tobacco brands they market.

INNOVATION

The tobacco industry adapts to changing circumstances in many ways. Product innovation is a key strategy. Since the early 1950s, the major changes in cigarette design have come in response to public-health concerns that cigarettes constitute a leading cause of illness and death (McGinnis, 1993; Slade, 1993). Most of these innovations have been variations on filters and so-called low-tar designs. Ballyhooed with multibillion-dollar advertising budgets, these innovations propped up cigarette consumption over the years despite the complete absence of demonstrated benefit at the time they were introduced. Years of study (and as many years of unregulated sale) have only produced evidence for decidedly marginal benefits, yet the innovations have become

firmly established. These supposed advances have been criticized by some as being nothing more than public relations gimmicks in the face of and in mocking response to profound public-health problems.

The cigarette companies continue to invent novel ways to deliver nicotine to the brain. Electronic devices, smokes with charcoal fuel elements, and tiny aerosol cans are but some of the gimmicks the companies have patented to facilitate the inhalation of nicotine. Despite these efforts, the industry remains dependent on smoking, with variations of the tobacco-filled cigarette the mainstay of its business for the foreseeable future.

INTERNATIONAL EXPANSION

Cigarette smoking has been declining in the United States, Canada, and Western Europe. Since the 1960s, however, the biggest cigarette manufacturers (BAT, Philip Morris, RJR/Nabisco, and, recently, Japan Tobacco Incorporated) have steadily increased their business in international markets (Taylor, 1984). This expansion has been accompanied by the weakening and dissolution of both national private and state-owned tobacco companies. The process got under way in Latin America in the 1960s, spread to eastern Asia in the late 1980s, and developed into a frenzy of deal making in Eastern Europe and the republics of the former Soviet Union in the early 1990s (Shepherd, 1985; Sesser, 1993).

Shepherd has described the process whereby a transnational corporation moves toward dominating a formerly self-contained market through product innovation, smuggling, aggressive advertising, and pricing policies. The result is a larger market for tobacco products than existed previously and a corporate management that is better able to oppose public-health efforts at regulation and control. Although cigarette consumption is down in the United States, Canada, and Western Europe, it is rapidly growing in most of the world—especially the so-called third world. The transnational companies have positioned themselves to both fuel and profit from this trend.

DIVERSIFICATION

The giant cigarette makers have invested their tobacco profits in other enterprises for more than twenty years, ranging from soft drinks and cookies to office products, insurance, and real estate. This process has resulted in the ownership by tobacco companies of some widely known consumer-product companies, including Kraft and Nabisco. Although

the parent tobacco companies pretend that this phenomenon makes them somehow less involved in tobacco (none now have the word "tobacco" in their corporate name), a thoughtful examination of these businesses reveals the following:

Tobacco products remain by far the most profitable sector of each of these conglomerates; and tobacco products are always responsible for most of the company profits (see Tables 2 and 3).

Not one of these companies has backed away from any available opportunity to sell tobacco products. Indeed, the strongest companies continue to invest in domestic and overseas ventures that have as their goal the expansion of tobacco consumption.

These companies make ready use of nontobacco subsidiaries to support their tobacco businesses. For example, RJR/Nabisco fired the ad agency that did their Oreo Cookie advertising after that agency also produced ads promoting an airline offering smoke-free flights. Philip Morris has used one of its Kraft-General Foods warehouses for its coupon-redemption program for the Marlboro Adventure Team.

Tobacco companies do not diversify to get out of the tobacco business. They diversify because tobacco has given them profits, the acquisitions seem sound investments, and the resulting product mix complements the core business in some manner.

TABLE 2
1992 Overall Earnings for Six U.S.-Based Tobacco-Product Manufacturers (in Millions of U.S. Dollars)

Company	Tobacco Revenues	Nontobacco Revenues	Tobacco as % of Revenue	Tobacco Income	Nontobacco Income (or Loss)	Tobacco as % of Income
Philip Morris	25,677	33,454	43	7,203	3,757	66
RJR/Nabisco	9,027	6,707	57	2,687	947	74
American Brands	8,157	6,467	56	1,091	757	59
Loews Corp. (Lorillard)	2,185	11,506	16	915	(1,217)	233
Brooke Group (Liggett & Myers	606	114	84	53	(59)	211
UST	884	163	84	509	14	97
Totals:	46,536	58,411	44	12,458	4,199	75

SOURCES: Corporate annual reports.

TABLE 3
Profitability of Selling Tobacco Products Compared to Selling Other Goods and Services, 1992

Company	Gross Profit Margin on Tobacco Product Sales	Gross Profit (or Loss) Margin on Sales Other than Tobacco
Philip Morris	28%	11%
RJR/Nabisco	30	14
American Brands	13	12
Loews Corp (Lorillard)	42	(11)
Brooke Group (Liggett & Myers)	9	(52)
UST	58	9
Overall:	27	7

SOURCES: Corporate annual reports.

PRICE WARS

Price competition has long been part of the tobacco industry strategy. It was the major tool for the achievement of monopoly power in the 1880s and was a key element in the early twentieth-century dominance of the market by Camel. In the 1930s, price competition, made possible by overly aggressive price increases by the majors, contributed to the emergence and growth of Brown & Williamson and Philip Morris (Sobel, 1978). From the end of World War II (1945) until 1980, however, price competition was virtually absent from the U.S. cigarette market.

In 1980, tiny Liggett & Myers, a firm that had become too small to enjoy oligopolistic profits, broke ranks with its fellows by introducing generic cigarettes. The strategy was made possible by the pattern of price increases in the industry—increases that had exceeded the rate of inflation for years. Brown & Williamson soon followed suit with its own generic brands, and within a few years every cigarette manufacturer had a multitiered pricing structure, with the heavily advertised, standard brands at the top. Prices for the major brands continued to rise steeply, far faster than inflation, through early 1993. Customers who might have stopped smoking because of high prices were kept in the market by the increasingly available lower priced offerings. By early 1993, however, investment analysts had become concerned because lower priced brands accounted for more than 25 percent of all cigarette purchases—with attendant threats to profits—and Philip Morris had become alarmed by the market share losses sustained by its cash cow, Marlboro, to less than 25 percent of all cigarettes sold.

Philip Morris had a number of key strengths that gave it a flexibility not possessed by its competitors, including market leadership, an absence of corporate debt, and a strong youth market for Marlboro. Its principal competitor, RJR/Nabisco, had an enormous corporate debt—and although Camel had been making inroads into Marlboro's youth market, it was still far from the dominant cigarette. These factors led Philip Morris to cut prices substantially (while mounting the most elaborate promotional campaign ever seen in the industry). The competition was forced to follow suit with lower prices. Marlboro's brand share surged; the threat to profitability from lower priced brands subsided; and the competition was left somewhat weakened.

LOBBYING AND PUBLIC RELATIONS

In 1915, the U.S. tobacco industry formed the Tobacco Merchants Association (TMA) to lobby against the anticigarette laws that had become a problem for the industry in a number of states (Robert, 1952). These laws came about as a result of the efforts of antitobacco advocates, including Henry Ford and Thomas Edison. The TMA accomplished its objectives: By 1930, the state prohibitions on cigarettes had been diminished to easily ignored prohibitions that only barred the sale of cigarettes to minors.

In the 1950s, the industry faced a more substantial challenge—proof that cigarettes caused lung cancer. In addition to putting cosmetic filters on the product and making outrageous claims for their benefit (P. Lorillard trumpeted its asbestos-filtered Kent as "the greatest health protection in cigarette history"), the industry developed a sophisticated public relations and lobbying capability (Wagner, 1971). The public relations firm of Hill & Knowlton organized the Tobacco Institute to meet the industry's public relations and lobbying needs. The cigarette makers also formed the Tobacco Industry Research Committee (later reorganized and renamed the Council for Tobacco Research) to create the pretense that the industry was conscientiously involved in biomedical research to get to the bottom of the smoking and health question (Freedman & Cohen, 1993). Despite the research, such as it was, and in the face of growing evidence of harm from a variety of other quarters, the smoking epidemic continued unabated.

The Tobacco Institute, in alliance with the various branches of the industry, has stood as a bulwark against public-health activities for a generation. The Council for Tobacco Research has funded studies of marginal importance for public relations gain while operating a Special Projects branch for the benefit of tobacco-product liability defense. In these and other ways, the tobacco industry has insulated itself from significant regulation and from acceptance of any responsibility for the harm its products cause. Similar organizations exist to protect the interests of oral-tobacco manufacturers.

OWNERSHIP

The major tobacco-product manufacturers are publicly owned and traded corporations. As such, they are owned by their investors. Major institutions, including banks, insurance companies, and pension funds, hold the majority of shares in the tobacco industry.

SUMMARY AND CONCLUSION

The tobacco industry is a powerful oligopoly of product manufacturers in alliance with a network of suppliers and associated service organizations. Although its products form the leading cause of preventable death, it defends itself against appropriate regulation by extensive legal, public relations, and lobbying efforts. The industry is understandably driven by an interest in making money. It has never acted out of a primary concern for the health of its customers or the health of those around them. For a variety of reasons, including clever intervention by the industry, government has utterly failed to provide the sort of regulatory control expected when it comes to something as addicting and toxic as nicotine-containing tobacco products.

(SEE ALSO: *Advertising and Tobacco Use; Nicotine*)

BIBLIOGRAPHY

BROOKS, J. E. (1949). *The mighty leaf.* New York: Little, Brown.

DENNY, J. (1993). The king of snuff. *Common Cause Magazine, 19*(2), 20–27.

FIORE, M. C., NEWCOMBE, P., & MCBRIDE, P. (1993). Natural history and epidemiology of tobacco use and addiction. In C. T. Orleans & J. Slade (Eds.), *Nicotine addiction: Principles and management.* New York: Oxford University Press.

FREEDMAN, A. M., & COHEN, L. P. (1993, February 11). How cigarette makers keep health question "open" year after year. *Wall Street Journal,* p. A-1.

MCGINNIS, J. M., & FOEGE, W. H. (1993). Actual causes of death in the United States. *Journal of the American Medical Association, 270*(18), 2207–2212.

ROBERT, J. C. (1952). *The story of tobacco in America.* New York: Alfred A. Knopf.

SESSER, S. (1993, September 13). Opium war redux. *The New Yorker, 69*(29), 78–89.

SHEPHERD, P. L. (1985). Transnational corporations and the international cigarette industry. In R. S. Newfarmer (Ed.), *Profits, progress and poverty.* Notre Dame, IN: University of Notre Dame Press.

SLADE, J. (1993). Nicotine delivery devices. In C. T. Orleans & J. Slade (Eds.), *Nicotine addiction: Principles and management.* New York: Oxford University Press.

SMITH, J. W. (1990). *Smoke signals.* Richmond, VA: The Valentine Museum.

SOBEL, R. (1978). *They satisfy.* Garden City, NY: Anchor Press/Doubleday.

TAYLOR, P. (1984). *The smoke ring: Tobacco, money, and multi-national politics.* New York: Pantheon.

TILLEY, N. M. (1985). *The R.J. Reynolds Tobacco Company.* Chapel Hill: University of North Carolina Press.
WAGNER, S. (1971). *Cigarette country.* New York: Praeger.

JOHN SLADE

TOBACCO: MEDICAL COMPLICATIONS

The idea that smoking tobacco is injurious to the body is not of recent origin. King James I of England, in his classic "*Counterblaste to Tobacco,*" written in 1604, outlined a number of beliefs about tobacco's ill effects on health and urged his subjects to avoid it. He called smoking a "filthie noveltie ... A custome lothsome to the eye, hatefull to the nose, harmefull to the braine, dangerous to the Lungs. ..." Opinions on the possible benefits and health damage caused by use of tobacco varied over the next 300 years. Some nineteenth-century arguments that tobacco use injured health were linked to moral arguments against its use rather than to what today would be considered medical evidence.

In 1926 Sir Humphrey Rolleston of Cambridge University (the same ROLLESTON who headed the committee on the use of opioids) addressed the Harrogate Medical Society on the subject of medical aspects of tobacco and the possible toxic effects of nicotine. He drew few firm conclusions. Only a few health problems were clearly linked to tobacco. These included some irritation of the throat and upper air passages by furfural, pyridine derivatives, ammonia, and carbon monoxide, which he ascribed to combustion of vegetable material and "not, like NICOTINE, in any way special to tobacco." He did mention tobacco amblyopia, a disorder of the optic nerve leading to blindness, now thought to be a rare complication. Among the heart disorders Rolleston mentioned were extrasystoles (irregular heartbeats) and angina (pain due to insufficient blood reaching the heart). He noted that nicotine constricted coronary arteries but suggested that people who suffered from extrasystoles might consider giving up coffee and tea before tobacco. He observed that cigarette smoking could cause arterial spasms, noting that it was linked to obliterative diseases of the large arteries among young Jews living in London's East End. Rolleston believed that cancers of the lip and oral cavity observed in smokers were probably caused by syphilis and therefore not firmly linked to smoking. He devoted only a few lines to smoking's adverse effects on the respiratory tract, observing that smoking was responsible for "causing cough, hoarseness, bronchial catarrh, and so emphysema of the lungs." In general, Rolleston observed that considering "the large number of heavy smokers, the comparative rarity of undoubted lesions due to smoking is remarkable." He concluded that "to regard tobacco as a drug of addiction may be all very well in a humorous sense, but it is hardly accurate."

But even as Rolleston was lecturing, researchers were looking at the evidence suggesting that smoking was responsible for the increasing number of lung cancer cases, a rare disease in the nineteenth century. Within thirty years there would be a growing consensus among the medical scientific community that tobacco smoking was the principal cause of lung cancer, causally related to other cancers, and a major contributor to cardiovascular diseases, peripheral artery disease, and chronic obstructive lung disease (emphysema). Yet from the 1920s to the 1960s, cigarette smoking gained almost universal social acceptance. Using doctors and nurses and health-related slogans ("not a cough in a carload") in their advertisements, cigarette manufacturers implied that cigarette smoking was without health risk. By the 1960s the majority of adult males were smokers, with more than 70 percent in some age groups.

The turning point in the public's perception of the adverse consequences of tobacco smoking came with the publication of the *Report of the Royal College of Physicians* in England in 1962 and the *Report of the Surgeon General* in the United States in 1964. These two reports documented the experimental, epidemiological, and pathological evidence linking tobacco smoking to a variety of diseases, the most notable of which were lung cancer, illness and death from heart disease, and chronic bronchitis and other lung disorders. Many more reports on the health consequences of smoking followed these two pivotal publications. Since 1969 the Office of Smoking and Health of the U.S. Public Health Service has coordinated the annual publication of a Surgeon General's Report on the health consequences of smoking, with several of the reports focusing on specific topics. In approaching such major reviews of specific health consequences of smoking, the Office of Smoking and Health assigns recognized experts to review and summarize all the existing scientific literature on the topic and then draw some conclusions from it. Some of the special topics that have been considered are health consequences of smoking for women (1980), the changing cigarette (the implications for health of low tar/nicotine cigarettes and filters) (1981), chronic obstructive lung disease (1984), cancer and chronic lung disease in the workplace (1985), and nicotine addiction (1988). The 1972 report was the first to explore the health consequences of involuntary smoking (passive or secondhand smoking).

The 1979 and 1989 reports were overall reviews of the field, marking the fifteenth and twenty-fifth anniversaries of the landmark 1969 report produced when Dr. Luther Terry was surgeon general. The 1979 report described tobacco smoking as "the largest preventable cause of death in America." It noted that statisticians were able to identify the following as deaths related to smoking: 80,000 each year from lung cancer; 22,000 from other cancers; up to 225,000 from cardiovascular disease; and more than 119,000 from chronic pulmonary disease.

Briefly summarized below are the main findings in these voluminous reports on the adverse health consequences of smoking or using SMOKELESS TOBACCO.

Cancer. Tobacco smoking has been shown to be the major cause of lung cancer in both men and women. The increased risk for lung cancer is directly related to the amount smoked. The risk of death from lung cancer is about twenty times greater for men who smoke two packs a day than for those who have never smoked. It is about ten times higher for those who smoke one-half to one pack a day. Depth of inhalation also influences risk of disease. Tobacco smoking is synergistic (produces a multiplier effect) with the effects of other carcinogenic risks, such as exposure to radon or asbestos. Smoking is also synergistic with alcohol in causing cancers of the oral cavity, larynx, and esophagus.

Cardiovascular Disease. Smoking is one of three major causes of coronary heart disease (CHD); risk of death from CHD is 70 percent higher for men who smoke, with a similar effect for women. The risk due to smoking increases if there are risk factors present such as hypertension and elevated cholesterol levels. Smoking increases risk for stroke. For example, women who smoke twenty-five cigarettes or more per day have a risk for stroke almost four times higher than nonsmokers. Smoking also increases the risk of atherosclerosis (formation of plaques) in the peripheral arteries and the aorta. In peripheral arteries this condition can lead to insuf-

ficient oxygen reaching the muscles; in the aorta it can lead to a rupture that is usually fatal.

Lung Disease. The link between tobacco smoking and chronic obstructive lung disease (COPD) was noted in the 1964 Surgeon General's Report. COPD includes three related disorders: chronic mucous hypersecretion that causes cough and phlegm production; airway thickening and obstruction of expiratory airflow; and emphysema—abnormal dilation of air sacs and destruction of walls of the alveoli. Compared to nonsmokers, male smokers are three times more likely and female smokers are twice as likely to have a persistent cough.

Other Disorders. These include peptic ulcers, upper respiratory infections, osteoporosis, and cancers of the pancreas, bladder, and esophagus.

The toxic properties and carcinogenic effects of tobacco smoke and its constituents have been studied in the laboratory using animals. The evidence linking tobacco use to death and disease in humans, however, relies heavily on epidemiological studies comparing the rates of various diseases as they occur in smokers versus nonsmokers, in light versus heavy smokers, and in continuing versus former smokers. The level of certainty that links tobacco use to a particular disease varies. Shopland and Burns (1993) have grouped diseases according to their established epidemiological association with cigarette smoking in five categories. These are outlined below.

Category A. Diseases for which a direct causal association has been firmly established and smoking is considered the major single contributor to excess mortality from the disease: cancers of the lung, larynx, pharynx (oral cavity), and esophagus; chronic obstructive lung disease, including emphysema; peripheral vascular disease

Category B. Diseases for which a direct causal association has been firmly established but for which smoking is only one of several causes: stroke; coronary heart disease; cancers of the bladder and pancreas; aortic aneurysm; perinatal mortality

Category C. Diseases for which an increased risk (association) has been demonstrated but a risk whose exact nature has not been firmly established: cancers of the cervix, uterus, stomach, and liver; gastric and duodenal ulcers; pneumonia; sudden infant death syndrome

Category D. Diseases for which excess mortality in smokers has been observed but for which this observation is attributed to confounding variables (other factors that are commonly found among smokers):

alcoholism; cirrhosis of the liver; poisoning; suicide *Category E.* Diseases for which smokers have lower death rates than nonsmokers: endometrial cancer; Parkinson's disease; ulcerative colitis

The effects of tobacco use are not limited to specific diseases that lead to death. Tobacco use can stimulate enzymes in the liver, and this stimulation can result in alterations in the way various medications are metabolized. This alteration in metabolism can mean that the levels of medications in the body will not be high enough to be optimally therapeutic.

The overall increased mortality from smoking varies with the amount smoked. For those who smoke two or more packs of cigarettes per day, it is about double that of nonsmokers; for those who smoke less, it is about 1.7 times higher than for nonsmokers. The risk for various diseases can be powerfully affected by cessation, but not all risks decline at the same rate. Cardiovascular disease risk decreases markedly within a year of quitting smoking; risks of cancer decline more slowly, with some elevated risk still evident ten years after cessation. By ten to fifteen years after quitting, overall mortality of former smokers is not much higher than that of nonsmokers. Increased mortality rates are not as marked for pipe and cigar smokers, but they are still substantially elevated. The mortality risk for users of smokeless tobacco comes primarily from cancers of the oral cavity and throat.

The adverse effects of passive inhalation (secondhand smoke) are not considered here except in connection with the higher incidence of respiratory illness among the infants of mothers who smoke. But there is no question that there are differences in composition of mainstream smoke (the smoke inhaled by the smoker), sidestream smoke (produced by tobacco burning between puffs), and environmental smoke (the mixture of exhaled mainstream and sidestream smoke). Sidestream smoke is produced at lower combustion temperatures and has higher concentrations of carbon monoxide and organic constituents believed to be carcinogenic.

WOMEN AND SMOKING

Women who smoke tobacco have the same risks for adverse effects as men. The early impression that women suffered fewer adverse effects from smoking was really due to lower levels of exposure (fewer women smokers and a tendency of women smokers

to smoke less heavily.) As has been written more than once, women who smoke like men die like men. In 1986 deaths due to lung cancer among women exceeded deaths from breast cancer, becoming the leading cause of cancer death for women. Some women are at special risk. It has been documented that while the use of oral contraceptives alone does not constitute a serious health risk, the combination of oral contraceptives and cigarette smoking raises substantially the risk of cardiovascular disease, including subarachnoid hemorrhage (bleeding inside the skull).

Women who smoke have higher infertility rates than those who do not and are also more likely to have menstrual irregularities. Nicotine crosses the placenta. Because it constricts blood vessels, a decreased amount of oxygen is delivered to the fetus. In addition, smoking elevates the amount of carbon monoxide in the mother's blood so that it carries less oxygen to the fetus. Women who smoke during pregnancy have higher rates of premature detachment of the placenta (abruptio placentae), premature rupture of membranes, and preterm delivery. The greater the amount of tobacco smoked during the pregnancy, the higher the frequency of spontaneous abortion and fetal death. In the United States smoking has been associated with a 20 percent increase in preterm births among women who smoked a pack a day or more compared with those who did not smoke.

There is no consensus on whether smoking increases the probability of congenital malformations. However, it is well established that babies born to women who smoke during pregnancy weigh on average about seven ounces less than those born to nonsmokers. Apgar scores, a composite of measurements of the breathing, skin color, and reflexes of infants taken at one and five minutes after delivery, are lower for babies of women who smoked during pregnancy. Women who stop smoking early in pregnancy increase their likelihood of having normal deliveries and normal-birth-weight babies. Interestingly, epidemiological data suggest that passive smoke exposure during pregnancy (e.g., living with a smoker) can adversely affect birth weight of the baby. Infants born to mothers who smoke are far more likely to die before their first birthday, primarily as a result of respiratory complications and sudden infant death syndrome. Children of mothers who smoke seem in general more likely to suffer from colds, asthma, bronchitis, pneumonia, and other respiratory problems.

Efforts to educate the public about the health consequences of smoking, including smoking-prevention programs directed at young people and encouragement of smokers to quit, have led to a reduction in the prevalence of smoking in the United States and in several European countries since the mid-1960s. In general, white males in higher socioeconomic groups have lowered their smoking rate more than women and members of ethnic and racial minorities and lower socioeconomic groups. By the early 1960s lung cancer deaths among African-American men exceeded those among white men; by 1990 it was 30 percent higher. The lung cancer rate among both African-American and white women was virtually the same, reflecting similar smoking patterns.

In contrast to the general decline of smoking in the West, the prevalence of smoking may actually be increasing in developing and newly industrialized countries where, even among medical students, cigarette smoking retains a cachet of sophistication and affluence.

(SEE ALSO: *Advertising and Tobacco Use; Complications; Nicotine; Treatment: Tobacco*)

BIBLIOGRAPHY

COOK, P. C., PETERSEN, R. C., & MOORE, D. T. (1994). *Alcohol, tobacco, and other drugs may harm the unborn.* Rockville, MD: U.S. Department of Health and Human Services, Public Health Service.

CORTI, E. (1932). *A history of smoking.* Translated by P. England. New York: Harcourt, Brace.

GRITZ, E. (1980). Problems related to the use of tobacco by women. In O. J. Kalant (Ed.), *Alcohol and drug problems in women.* New York: Plenum.

ROLLESTON, H. (1926). Medical aspects of tobacco. *Lancet,* May 22.

SHOPLAND, D. R., & BURNS, D. M. (1993). Medical and public health implications of tobacco addiction. In C. T. Orleans & J. Slade (Eds.), *Nicotine addiction: Principles and management.* New York: Oxford University Press.

U.S. DEPARTMENT OF HEALTH, EDUCATION AND WELFARE. (1979). *Smoking and health: A report of the surgeon general.* DHEW Publication no. (PHS) 79-50066. Washington, DC: U.S. Government Printing Office.

JEROME H. JAFFE
DONALD R. SHOPLAND

TOBACCO: SMOKELESS   Since tobacco is a plant native to the New World, Native Americans were the first to use it. In addition to smoking it, they used it in smokeless forms—mainly chewing it, making teas and drinks from it, even using the ash in rituals that ranged from South America to Central America and the Caribbean to North America. It was used along with many other plants for both ritual and medicinal purposes.

The use of tobacco was brought to Europe by Columbus and other explorers, where it was taken up for recreation in both the smoked form (cigars and pipes) and the smokeless. Smokeless tobacco (ST) became popular in British society in the practice called sniffing, but British colonists in the Americas preferred to chew tobacco or use snuff. In the 1800s, chewing tobacco was widespread in the United States; its use decreased, however, when the spitting that resulted (into spittoons or cuspidors or wherever the spit fell) was linked to the spread of tuberculosis, one of the most dreaded and fatal of diseases. Consequently, the twentieth century had declining sales of chewing tobacco until about 1970; those who were using it were, for the most part, backwoodsy and rural, or baseball players.

In the twentieth century, there have primarily been two types of ST: (1) snuff, the type one dips by placing it between the cheek and gum, or (2) chewing tobacco, the type one chews and places in the cheek area. Snuff is a cured, ground tobacco that comes in three forms: (1) fine-cut tobacco, (2) moist snuff, or (3) dry snuff (Glover et al., 1988; Christen et al., 1982; Christen & Glover, 1987). Fine-cut tobacco and moist snuff are used by placing a pinch between the cheek and gum or lower lip and gum. Dry snuff may be used by inhaling a pinch through each nostril or by placing a pinch between the cheek and the gum or the lower lip and the gum. Chewing tobacco is also produced in three forms: (1) loose-leaf tobacco; (2) plug tobacco; or (3) twist chewing tobacco (Christen et al., 1982; Penn, 1902; Christen & Glover, 1987; Voges, 1984; U.S. Department of Agriculture, 1969; Smokeless Tobacco Council, 1984). All three forms are used by placing a "chaw" in the cheek and periodically chewing.

In the 1970s, the use of ST surged in the United States, with smokers showing a preference for moist snuff. In the early 1990s, the reported numbers of U.S. users ranged from 6 million to 22 million (Smight, 1981; Harper, 1980). It is increasingly evident that youngsters and adolescents are using ST products much more than they did in the recent past. This resurgence of popularity has been attributed to innovative advertising campaigns by tobacco companies that use sports superstars, cowboy celebrities, and entertainers to promote their products. These campaigns represent an attempt to get past or erase the old, unsanitary image of the habit, replacing it with a manly or macho image (Christen et al., 1982; Shelton, 1982; Glover, Christen, & Henderson, 1981, 1982).

A spokesperson for the tobacco industry described the average consumer of smokeless tobacco as someone who is generally between the ages of eighteen to thirty but is most likely to be between eighteen and twenty-four (Maxwell, 1980). Youngsters, however (as young as ten to twelve years of age), also appear to be influenced by the trend to using smokeless tobacco (Hunter et al., 1986; Marty, McDermott, & Williams, 1986; Schroeder et al., 1987; Christen, 1980).

NICOTINE, a dependence-producing drug found in ST, is the same drug that is found in smoking tobacco. Cigarette smokers inhale smoke containing nicotine into their lungs, and the nicotine is then transported into the bloodstream. ST users absorb nicotine directly through the lining of their mouths. Each time smokers smoke a cigarette, they absorb approximately 1 milligram of nicotine into their system. By comparison, people who use chewing tobacco receive approximately 4.5 milligrams of nicotine per chaw, and people who use snuff receive approximately 3.6 milligrams of nicotine per pinch (Benowitz, 1988).

ST is sometimes viewed as a safe alternative to cigarettes, but it is not. ST is directly related to a variety of health problems: bad breath, abrasion of teeth, gum recession, periodontal bone loss, tooth loss, leukoplakia, nicotine dependency, and various forms of oral cancer (Christen, 1985; Schroeder, Chen, & Kuthy, 1985). There are indications that smokeless tobacco also plays a role in cardiovascular alterations and neuromuscular toxicity (Schroeder & Chen, 1985; Squires et al., 1984).

Survey data as of the mid-1980s indicate that it is predominantly males who use smokeless tobacco. In a large national survey of smokeless tobacco use in college, Glover and colleagues reported that about 22 percent of collegiate males were users of smokeless tobacco, whereas only 2 percent of collegiate females used it (Glover et al., 1986). In a study of 5,078 students from 67 high schools throughout the state of Massachusetts, 16 percent of males and 2 percent of females reported using it "once or twice." Eight and 4 percent of the males studied reported using it "several times" and "very often," respectively (McCarty & Krakow, 1985).

The increasing numbers of individuals who use ST demonstrate that there is a need for cessation programs. Few formal cessation programs have been developed, however, but these few need to be rigorously evaluated.

(SEE ALSO: *Adolescents and Drug Use*; *Advertising and Tobacco Use*)

BIBLIOGRAPHY

BENOWITZ, N. L. (1988). Nicotine and smokeless tobacco. *CA: A Cancer Journal for Clinicians, 38*(4), 244–247.

CHRISTEN, A. G. (1985). The four most common alterations of the teeth, periodontium and oral soft tissues absorbed in smokeless tobacco users: A literature review. *Journal of the Indiana Dental Association, 64*, 15–18.

CHRISTEN, A. G. (1980). The case against smokeless tobacco: Five facts for the health professional to consider. *Journal of the American Dental Association, 101*, 464–469.

CHRISTEN, A. G., & GLOVER, E. D. (1987). History of smokeless tobacco use in the United States. *Health Education, 18*(3), 6–11, 13.

CHRISTEN, A. G., ET AL. (1982). Smokeless tobacco: The folklore and social history of snuffing, sneezing, dipping and chewing. *Journal of the American Dental Association, 105*, 821–829.

GLOVER, E. D., CHRISTEN, A. G., & HENDERSON, A. H. (1982). Smokeless tobacco and the adolescent male. *Journal of Early Adolescence, 2*, 1–13.

GLOVER, E. D., CHRISTEN, A. G., & HENDERSON, A. H. (1981). Just a pinch between the cheek and gum. *Journal of School Health, 51*, 415–418.

GLOVER, E. D., ET AL. (1988). An interpretative review of smokeless tobacco research in the United States: Part 1. *Journal of Drug Education, 10*, 285–309.

GLOVER, E. D., ET AL. (1986). Smokeless tobacco use trends among college students in the United States. *World Smoking and Health, 11*(1), 4–9.

GLOVER, E. D., ET AL. (1984). Smokeless tobacco research: An interdisciplinary approach. *Health Values, 8*, 21–25.

HARPER, S. (1980). In tobacco, where there's smokeless fire. *Advertising Age, 51*, 85.

HUNTER, S. M., ET AL. (1986). Longitudinal patterns of cigarette smoking and smokeless tobacco use in adolescents: The Bogalusa heart study. *American Journal of Public Health, 76*, 193–195.

MARTY, P. J., MCDERMOTT, R. J., & WILLIAMS, T. (1986). Patterns of smokeless tobacco use in a population of high school students. *American Journal of Public Health, 76*, 190–192.

MAXWELL, J. C., JR. (1980). Maxwell manufactured products report: Chewing snuff is growth segment. *Tobacco Reporter, 107*, 32–33.

MCCARTY, D., & KRAKOW, M. (1985, January 28). *More than "just a pinch": The use of smokeless tobacco among Massachusetts students.* Report by the Massachusetts Department of Public Health. Boston: Division of Drug Rehabilitation.

PENN, W. A. (1902). *The soverane herbe: A history of tobacco.* New York: Grant Richards Co.

SCHROEDER, K. L., & CHEN, M. S., JR. (1985). Smokeless tobacco and blood pressure. *New England Journal of Medicine, 312*, 919.

SCHROEDER, K. L., CHEN, M. S., JR., & KUTHY, R. A. (1985). Smokeless tobacco: The new thing to chew on. *Ohio Dental Journal, 59*, 11–14.

SCHROEDER, K. L., ET AL. (1987). Bimodal initiation of smokeless tobacco usage: Implications for cancer education. *Journal of Cancer Education, 2*(1), 1–7.

SHELTON, A. (1982). Smokeless sales continue to climb. *Tobacco Reporter, 109*, 42–44.

SMIGHT, T. A. (1981). A man's chew. *Nutshell, 43,*

SMOKELESS TOBACCO COUNCIL. (1984). *Smokeless tobacco.* Peekskill, NY: Author.

SQUIRES, W. G., ET AL. (1984). Hemodynamic effects of oral smokeless tobacco in dogs and young adults. *Preventive Medicine, 13*, 195–206.

U.S. DEPARTMENT OF AGRICULTURE. (1969). *Tobacco in the United States* (Miscellaneous Publication 867). Washington, DC: Author.

VOGES, E. (1984). *Tobacco encyclopedia.* Mainz: Germany Tobacco International.

ELBERT GLOVER
PENNY N. GLOVER

TOBACCO: SMOKING CESSATION AND WEIGHT GAIN

Smokers weigh less than nonsmokers, and when smokers quit, they gain an average of 5 pounds (2.3 kg). Research indicates that not only is there more concern about postcessation weight gain in women than in men, but women actually tend to gain more weight than men. Data suggest that fear of postcessation weight gain may be a predictor of relapse, but that those who gain the most weight after cessation are least likely to relapse to smoking.

The mechanisms underlying the gender difference in postcessation weight gain are not known. The mechanisms contributing to postcessation weight gain across the genders have been evaluated, however, with three potential factors receiving the most attention—physical activity, energy expenditure, and dietary intake.

PHYSICAL ACTIVITY

There is little evidence that changes in physical activity play a role in postcessation weight gain. It does appear, however, that energy expenditure is increased by NICOTINE, which may contribute to the lower weight of smokers versus nonsmokers. This thermogenic role of nicotine is further supported by the findings that, despite their weight differences, smokers and nonsmokers have similar caloric intake. When energy expenditure has been assessed after smoking cessation, however, the evidence implicating decreased energy expenditure in postcessation weight gain is mixed. Some studies have found no changes in energy expenditure after cessation, while others have found decreases of 5 to 10 percent. Thus, while decreased energy expenditure may occur after cessation, the decreases do not appear to be large enough to account for most of the postcessation weight gain.

EATING

Increased eating is a recognized effect of smoking cessation, and data suggest that food intake increases from 200 to 300 kilocalories per day after cessation. Some animal and human data also suggest that increased intake of sweet foods accounts for a large proportion of the increased caloric intake. Although increased eating after cessation has not always been observed, of the three mechanisms, it has been observed most consistently. Increased eating, then, combined with the elimination of the thermogenic effects of nicotine result in increased weight after smoking cessation.

Behavioral and pharmacologic methods for preventing or suppressing postcessation weight gain have been evaluated. Dietary interventions have been largely ineffective for preventing postcessation weight gain in healthy populations; one study found that those assigned to a dietary intervention were more likely to relapse than those in the dietary control group. Physical exercise has been evaluated in only one study. While suggesting a potentially positive effect, the study had a very small sample size. Thus, the efficacy of behavioral methods for preventing postcessation weight gain has not been established and requires further study.

NICOTINE REPLACEMENT

Pharmacologic methods for preventing short-term postcessation weight gain have shown more promise than behavioral methods. Studies evaluating nicotine polacrilex (a replacement for nicotine) have found that it can suppress postcessation weight gain, but weight increased when this form of therapy was discontinued. Two large clinical trials evaluated the efficacy of a transdermal nicotine patch and did not find a difference in weight gain between the nicotine replacement and placebo groups; however, both assessed weight change after the nicotine patch use had been discontinued. Weight change while actually using the patch was not reported.

Finally, phenylpropanolamine and *d*-fenfluramine have shown some short-term efficacy in suppressing postcessation weight gain, but their effects on smoking cessation and long-term weight suppression have yet to be established.

SUMMARY

In sum, smoking cessation results in weight gain in most quitters, especially in women—primarily because of changes in caloric intake and, to a lesser extent, because of changes in energy expenditure. Females are most concerned about postcessation weight gain, and concern about weight gain appears to play a role in relapse to smoking. The efficacy of diet and exercise for preventing weight gain after cessation has not been established, while some pharmacologic methods have shown some efficacy.

(SEE ALSO: *Nicotine Delivery Systems for Smoking Cessation; Nicotine Gum; Serotonin Uptake Inhibitors in Treatment of Substance Abuse*)

BIBLIOGRAPHY

GROSS, J., STITZER, M. L., & MALDONADO, J. (1989). Nicotine replacement: Effects on post-cessation weight gain. *Journal of Consulting and Clinical Psychology, 57*(1), 87–92.

HALL, S. M., ET AL. (1989). Changes in food intake and activity after quitting smoking. *Journal of Consulting and Clinical Psychology, 57*(1), 81–86.

HALL, S. M., GINSBERG, D., & JONES, J. T. (1986). Smoking cessation and weight gain. *Journal of Consulting and Clinical Psychology, 57,* 81–86.

HOFSTETTER, A., ET AL. (1986). Increased 24 hour energy expenditure in cigarette smokers. *New England Journal of Medicine, 314*(2), 79–82.

LEISCHOW, S. J., & STITZER, M. L. (1991). Effects of smoking cessation on caloric intake and weight gain in an inpatient unit. *Psychopharmacology, 104,* 522–526.

MARCUS, B. H., ET AL. (1991). Usefulness of physical exercise for maintaining smoking cessation in women. *American Journal of Cardiology, 68*(4), 406–407.

TONNESON, P., NORREGAARD, J., & SIMONSEN, K. (1991). A double-blind trial of a 16 hour transdermal nicotine patch in smoking cessation. *New England Journal of Medicine, 325*(5), 311–315.

U.S. DEPARTMENT OF HEALTH AND HUMAN SERVICES (1990). *The health benefits of smoking cessation: A report of the surgeon general.* DHHS Pub. no. (CDC)90–8416. Washington, DC: U.S. Government Printing Office.

WILLIAMSON, D. F., ET AL. (1991). Smoking cessation and severity of weight gain in a national cohort. *New England Journal of Medicine, 324*(11), 739–745.

SCOTT J. LEISCHOW

TOBACCO: WITHDRAWAL (ABSTINENCE) SYNDROME The cessation of using cigarettes, chewing tobacco, or other nicotine-containing products causes anxiety, craving, depression, difficulty concentrating, impatience, irritability, restlessness, and insomnia. Cessation causes physiologic changes as well, such as decreased heart rate and a decrease in mental efficiency, which can be measured by objective tests. Some people with a history of depression may experience heightened symptoms of depression with nicotine cessation. Tobacco abstinence also produces weight gain, by increasing hunger and food intake and by decreasing resting metabolism.

About 50 percent of smokers have significant WITHDRAWAL symptoms when they stop abruptly. Gradually reducing one's smoking appears to decrease withdrawal symptoms but whether this increases the chances of permanently stopping remains unclear.

Nicotine withdrawal peaks two to four days after stopping, and most symptoms last three to four weeks—although hunger and craving can persist for at least six months. Most withdrawal effects are due to the absence of NICOTINE, as evidenced by the fact that slowly provided nicotine via chewing gum or by use of a skin patch reduces most symptoms. As with all drug withdrawal states, the severity of nicotine withdrawal varies widely among individuals. Causes for these differences in tobacco withdrawal among smokers are unknown. The severity of withdrawal does not appear to differ between men and women and, surprisingly, is not well-linked to the number of cigarettes smoked per day.

Tobacco withdrawal differs from other drug-withdrawal states in several ways. First, cessation of tobacco does not produce certain physical signs and symptoms seen with other drugs—such as tremor, headache, dizziness, sweating, and so on. Second, other than hunger and tobacco craving, there is no protracted or prolonged withdrawal syndrome (those symptoms that last six months or more). Third, in other drug-withdrawal syndromes (e.g., OPIOIDS or BENZODIAZEPINES), a medication that blocks the effects of the drug (such as the antagonist NALOXONE) precipitates the withdrawal syndrome. This does not occur with tobacco withdrawal. Fourth, in other drug-withdrawal syndromes, cues or settings that have been reliably associated with drug withdrawal can come to produce withdrawal symptoms on their own, through a conditioning process. Although this seems likely with cigarettes, it has not been shown thus far.

Over 50 percent of those who try to stop smoking relapse back to smoking within the first week. If this suggests that unpleasant withdrawal symptoms may be driving abstinent smokers back to smoking, scientific studies have not shown that tobacco withdrawal plays a major role in why smokers so often begin again; however, those who are treated with NICOTINE GUM or nicotine skin patches are less likely to experience withdrawal and are also less likely to relapse.

In summary, cessation of tobacco produces a well-defined set of symptoms—a syndrome—that appears to be caused by nicotine deprivation. In a rigorous scientific sense, however, the role of such symptoms in causing relapse back to smoking remains unclear.

(SEE ALSO: *Smoking Cessation and Weight Gain; Withdrawal: Nicotine*)

BIBLIOGRAPHY

HENNINGFELD, J. E. (1985). *Nicotine: An old-fashioned addiction.* New York: Chelsea House.

JOHN R. HUGHES

TOLERANCE   *See* Addiction: Concepts and Definitions; Tolerance and Physical Dependence.

TOLERANCE AND PHYSICAL DEPENDENCE

Tolerance and physical dependence are common *consequences* of drug self-administration. For those interested in understanding and modifying alcohol and drug abuse and the problems they cause, the greatest importance of tolerance and physical dependence is in the contribution they make as *determinants* of drug self-administration. An alcoholic, for example, can appear normal at BLOOD ALCOHOL CONCENTRATIONS (BAC) that would prostrate a social drinker. What role, if any, does tolerance play in paving the way to an escalation in drug use and in the medical and psychological problems caused by heavy drug use? In addition to being highly tolerant, alcoholics will also be physically dependent on their drug—alcohol. What evidence is there to support the common assumption that physical dependence is a critical factor in maintaining drug self-administration?

Such questions are best answered in the context of a general theory of how drug consumption is regulated. A useful starting point is the proposition that behavior is motivated by its consequences. Where tolerance is concerned, the important consequences of drugs are only those that depend on pharmacological effects. The pharmacological consequences that determine self-administration can be grouped according to whether they promote or restrain drug use. Rewarding consequences are those that increase the likelihood of drug use. Drugs may make a person feel alert, powerful, confident, relaxed, friendly, sexy, or talkative. They may alleviate ANXIETY, DEPRESSION, and physical PAIN. In cases where physical dependence develops, the alleviation of withdrawal symptoms may become an acquired reward for drug use. All these consequences and more have been hypothesized and evaluated as promoters of drug use.

Although people may initiate and maintain an episode of drug use in the pursuit of rewarding consequences, they may end it because drugs also have aversive pharmacological consequences at higher doses. These effects should also be taken into account as restraints on self-administration. Many restraining consequences of drug use can be suggested, ranging from unwanted dysphoria (a state of unease) to frank physical illness.

In summary, a simple regulatory theory asserts that "reward" drives drug use and "aversion" restrains it. If there is tolerance to the rewarding or aversive effects of drugs, it is clear how tolerance might determine drug use. A reduction in the rewarding effectiveness of a given dose would require an increased dose to obtain the same degree of reward. Similarly, tolerance to aversive effects of a drug might mean a much larger dose could be taken before the restraining aversive effect occurred.

Unfortunately, there is remarkably little scientific evidence for the common view that tolerance to the rewarding effects occurs. The common and plausible view that tolerance results in a loss of rewarding effectiveness is based mainly on anecdotal evidence. In contrast, there is ample scientific evidence of substantial tolerance to drug effects that could be viewed as restraints on the motivation to self-administer.

Physical dependence as a promoter of self-administration can be dealt with briefly. The earliest theories of dependence assumed that the avoidance of withdrawal was the most compelling motivation for persistent drug use. The experimental evidence for this view is strongest in the case of opiates—and weak to nonexistent for other drugs, including alcohol.

Tolerance can be characterized as a facilitator of consumption and its consequences, independent of the underlying reasons for drug use. If I become able to drink a lot more before I get sleepy or dizzy, my capacity to drink is increased regardless of why I do so. If the ability of tissue to resist damage does not increase with the body's capacity to resist the drug effects that regulate consumption, tolerance becomes an important determinant of medical and other problems.

(SEE ALSO: *Addiction: Concepts and Definitions; Causes of Substance Abuse; Research, Animal Model; Withdrawal*)

BIBLIOGRAPHY

CAPPELL, H. (1981). Tolerance to ethanol and treatment of its abuse: Some fundamental issues. *Addictive Behaviors, 3,* 197–204.
CAPPELL, H., & LeBLANC, A. E. (1983). The relationship of tolerance and physical dependence to alcohol abuse and alcohol problems. In B. Kissin and H. Begleiter (Eds.), *The biology of alcoholism.* Vol. 7, *The pathogenesis of alcoholism: Biological factors.* New York: Plenum.
CAPPELL, H., & LeBLANC, A. E. (1981). Tolerance and physical dependence: Do they play a role in alcohol and drug self-administration? In Y. Israel et al. (Eds.), *Research advances in alcohol and drug problems.* New York: Plenum.

HOWARD D. CAPPELL

TOLUENE *See* Inhalants.

TOPS *See* Treatment Outcome Prospective Study.

TOUGHLOVE The generic term *toughlove* (or tough love) describes a style of caring applied in diverse interpersonal contexts whereby one person or group reasserts power over another for whom he or she is responsible. Claire Kowalski was the first person to use the term in published material, in 1976, to differentiate a respectful means of caring for elderly people that preserves self-mastery from a smothering style that promotes dependence. Since that first use, others have found the term useful. The Association of the Relatives and Friends of the Mentally Ill endorses the concept (Roberts, 1985). In its most common use today, the term describes the means by which parents of abusive, delinquent, or drug-abusing children can regain parental control. Toughlove is also the name of a SELF-HELP program for these parents and their children.

Toughlove, the self-help program, was developed by Phyllis and David York in 1980. They found that rescuing their daughter, who engaged in highly destructive behavior, did more harm than good. Instead, they permitted natural and logical consequences to correct their daughter's behavior while they sought emotional support from their friends. They wrote and published *Toughlove* (1980) and founded an organization called the Toughlove Support Network (which is described in their later book, 1984). The network's mission is to promote what they view as a mode of intervention for individuals, families, and communities.

According to the Toughlove philosophy, parents are the ones with the dominant power in a family. Children misbehave when parents fail to assert themselves or to take responsibility for their role as parents, but when parents' expectations are stated clearly, a child will no longer control the family. Parents are urged to describe the behavior they expect from their children. Speculation about the causes of child misbehavior is discouraged. Parents do not need to understand why their child misbehaves. Instead, they must act in coalition with other parents to assert control of themselves and their home environment.

Toughlove parents are taught not to feel guilty about their child's misbehavior, because children are responsible for their own actions. A Toughlove parent of a destructive child might say: "We have had enough. We are not rescuing you from the trouble you have caused. We love you enough to say no." Proponents of Toughlove believe that drug and alcohol abuse is the most important causative factor in the disruptive behavior among teens. Once parents suspect drug and alcohol abuse, it is important that they investigate by questioning their child's friends, school officials, other family members, and anyone else their child meets frequently. When parents find drug and alcohol abuse, they must require abstinence. Strict discipline and limit setting are seen as the only means of enabling children to behave and to have a chance of regaining control of their lives.

Parents must confront their child about the drug and alcohol abuse and stipulate the behavior they expect. Toughlove recommends that they require the child to stop using drugs and seek treatment if needed. If a child refuses to comply, he or she is to be ejected from the home. Many uncooperative children are sent to live with another Toughlove family until they are serious about meeting their own parents' stipulations. Children who refuse to live with another Toughlove family are out on their own until they agree to their parents' rules.

To gain help in maintaining firmness and setting appropriate rules, parents attend a support group consisting of other parents who endorse the Toughlove principles. Toughlove support groups are organized by the parents without any professional leadership. Besides providing support for parents, Toughlove groups evaluate the effectiveness of treatment programs and the effectiveness of professionals who treat children for alcohol and drug abuse.

Hollihan and Riley (1987) used qualitative research methods to study a Toughlove parent group. They found that several themes characterized group sessions and defined the Toughlove program experience for parents. First, the lay-led group emphasized that old-fashioned values are superior to those inherent in today's method of raising children. Second, members regarded child-development professionals as advocates for modern child-raising methods that blame parents for child misbehavior. Third, they described the Toughlove group as their island of support within a pro-child social environment made up of the police, educators, social workers, and the courts. Last, the group provided successful models of rule setting by parents and enforcement of strict discipline—including as a final resort forcing a child to leave home. The group presented a persuasive and comforting rationale for the use of strict discipline that addressed the needs of parents who were experiencing great stress and feelings of failure (Hollihan & Riley, 1987).

Toughlove has been criticized as being simplistic and heavy-handed. According to Hollihan and Riley (1987), parents in the group they observed who did not believe their child was abusing drugs or alcohol were nevertheless instructed in how to document such abuse. Other possible causes of their child's misbehavior were ignored, because the Toughlove solution is supposed to apply in all situations. The tactic of throwing an unruly child out of the house is especially controversial. Although most children go to live with other Toughlove families, some are forced to leave with nowhere to go and can become homeless, a predator or a victim, or a threat to themselves and others. For example, John Hinckley, who attempted to kill President Ronald W. Reagan in 1982, had been cast out of his home by parents who endorsed Toughlove and who later warned other parents to be cautious in disciplining their children.

Neither the Toughlove program nor the style of caring identified with it has been evaluated. On the one hand, there is anecdotal evidence from parents to vouch for it. On the other, as illustrated by the Hinckley family, Toughlove solutions can make matters worse. At present, we do not know whether the positive or the negative is the more common outcome, or whether positive outcomes result from factors having nothing to do with Toughlove.

(SEE ALSO: *Adolescents and Drug Use; Parents Movement; Prevention Movement*)

BIBLIOGRAPHY

HOLLIHAN, T., & RILEY, P. (1987). The rhetorical power of a compelling story: A critique of a "Toughlove" parental support group. *Communication Quarterly, 35,* 13–25.

KLUG, W. (1990). *A preliminary investigation of Toughlove: Assertiveness and support in a parents' self-help group.* Paper presented at the Annual Convention of the American Psychological Association, Boston.

KOWALSKI, C. (1976). Smother love vs. tough love. *Social Work, 21,* 319–321.

LAWTON, M. (1982). Group psychotherapy with alcoholics: Special techniques. *Journal of Studies on Alcohol, 43,* 1276–1278.

NEMY, E. (1982). For problem teenagers: love, toughness. *New York Times,* April 26, p. B12.

ROBERTS, A. (1985). A.R.A.F.M.I.: Association of the Relatives and Friends of the Mentally Ill. *Mental Health in Australia, 1,* 37–39.

WOHL, L. (1982). The parent training game—from Toughlove to perfect manners. *Ms.,* May, pp. 40–44.

YORK, P., & YORK, D. (1980). *Toughlove.* Sellersville, PA: Community Service Foundation.

YORK, P., YORK, D., & WACHTEL, T. (1984). *Toughlove solutions.* Garden City, NY: Doubleday.

GREGORY W. BROCK
ELLEN BURKE

TOXICITY *See* Complications; Poison Control Centers, Appendix I, Volume 4.

TRANQUILIZERS *See* Benzodiazepines.

TRANSIT COUNTRIES FOR ILLICIT DRUGS

Transit countries are those through which drug shipments travel to reach local dealers and users. U.S. strategy to deal with the cocaine problem, for example, might best be described as a series of concentric circles around the source and trafficking countries of the Andes, through (1) the surrounding countries in South America into (2) the transit countries of MEXICO, Central America, and the Caribbean, to (3) the major consumer countries. In the 1990s, the United States has similar objectives for dealing with both source and transit countries— namely, to strengthen their governments' political will and capability; to increase their effectiveness in terms of military and law-enforcement activities; and to help inflict significant damage on drug-trafficking organizations.

President George H. Bush's National Drug Control Strategy of January 1990 called for detailed implementation plans for expanded drug-control activities on a regional and country-specific basis. The strategy emphasizes the major choke points at either end of the international chain: The three source countries of COLOMBIA, Peru, and BOLIVIA at one end and the primary transit countries of Mexico and the Bahamas at the other end. In addition to the source countries, in South America only Ecuador, Venezuela, and Brazil have the potential for profitable cultivation of COCA leaf, but the U.S. government believes that only small-scale cultivation and involvement in drug-transit activities exist in these countries. Consequently, only modest drug-control assistance has been made available to them—largely in the form of training, technical assistance, and commodities—to encourage them to take their own actions against high-value elements, such as money flows and essential and precursor chemicals. Brazil and Venezuela, for example, manufacture essential chemicals, used in COCAINE production.

As more success is achieved against cocaine source countries, and as pressure builds against trafficking through Mexico and the Bahamas, drug traffickers disperse their growing and processing operations and develop new smuggling routes. Diversionary routes, deliveries, and source areas are already being established, therefore.

INTERMEDIATE COUNTRIES

The president's drug control strategy discusses "intermediate" transit countries in the Caribbean and South America, where traffickers are expanding and where only narrow windows of opportunity exist for drug interdiction. Notable exceptions are Panama, where major shipments of South American cocaine already transit, and Guatemala, where OPIUM production has increased significantly and the trafficking infrastructure already exists for shipping major amounts of South American cocaine into North America. The small Caribbean states lack resources to perform adequate law enforcement; air drops of drugs to waiting boats have become common, because no Caribbean nation has a marine or security force capable of completely controlling territorial waters.

INTELLIGENCE NETWORKS

The key to successful drug control in the surrounding and transit countries lies in U.S. ability to develop and use effective intelligence networks. At the same time, efforts must continue that strengthen both the judicial systems and the law-enforcement procedures that act on the intelligence. The intended result will be improved interdiction of drugs and the essential chemicals needed to manufacture those drugs; effective forfeiture of assets through money-laundering investigations; and enhanced legal sanctions through more competent law enforcement and judicial activities.

STRATEGY SUCCESS

The success of the U.S. strategy for potential source and transit countries is predicated on building long-term institutions in these countries that work with the United States. To be successful, U.S. agencies must expand their efforts in the Pacific and the Caribbean to (1) collect and process intelligence; (2) help the transit countries develop their own intelligence collection, sharing, and dissemination capabilities; (3) help these countries take action on their own to apprehend traffickers and seize drug shipments and (4) direct bilateral and multilateral efforts against drug trafficking MONEY LAUNDERING, asset forfeiture, chemical diversion, and drug shipments.

(SEE ALSO: *Crop Control Policies; Drug Interdiction; International Drug Supply Systems; U.S. Government*)

BIBLIOGRAPHY

BUREAU OF INTERNATIONAL NARCOTICS MATTERS, U.S. DEPARTMENT OF STATE. (1992). *International narcotics control strategy report (INCSR)*. Washington, DC: Author.

JAMES VAN WERT

TREASURY, U.S. DEPARTMENT OF *See* U.S. Government: Agencies in Drug Law Enforcement and Supply Control.

TREATMENT *See* Treatment/Treatment Types.

TREATMENT ALTERNATIVES TO STREET CRIME (TASC) This is a program designed to divert drug-involved offenders into appropriate community-based treatment programs by linking the legal sanctions of the criminal-justice system to treatment for drug problems. The program now serves as a court diversion mechanism or as a supplement to probation or other justice-system sanctions and procedures. Created by President Richard M. Nixon's SPECIAL ACTION OFFICE FOR DRUG ABUSE PREVENTION (SAODAP) and funded by the Law Enforcement Assistance Administration (LEAA) and the National Institute of Mental Health (NIMH), TASC was an attempt to find a way to break the relationship between drug use and crimes committed to support the cost of obtaining illegal drugs. The idea for the initial TASC programs derived from an analysis of the criminal-justice system indicating that many drug-addicted arrestees were released on bail while awaiting trial and were likely to continue to commit crimes. Although there were provisions for supervision of drug-dependent offenders after conviction (on probation) or after release from prison (parole), no such mechanisms were in place to provide supervision of those awaiting trial. Yet, if arrestees could be directed to treatment, success in treatment could be taken into consideration at time of trial.

The first TASC programs, in Wilmington, Delaware, and Philadelphia, Pennsylvania, became operational in 1972. TASC currently operates in more than 100 jurisdictions in 28 of the U.S. states and territories. In the mid-1990s, TASC programs receive support from the U.S. Department of Justice through the Bureau of Justice Assistance (BJA) Criminal Justice Block Grants to state and local governments. LEAA was discontinued in 1982. Many TASC programs have expanded their base of support so that state and federal funding is supplemented by private donations and grants or client fees.

TASC programs initially focused on pretrial diversion of first offenders. The original TASC model was structured around three goals: (1) eliminating or reducing the drug use and criminal behavior of drug-using offenders; (2) shifting offenders from a system based on deterrence and punishment to one that, in addition, fostered treatment and rehabilitation; and (3) diverting drug-involved offenders to community-based facilities so as to limit criminal labeling and also to avoid the learning of criminal behavior that occurs in prisons. These goals were based on the assumption that treatment intervention had a better chance of success with first-offenders, since they had not yet been labeled as criminals. It also reflected community concerns that serious or dangerous offenders who might otherwise be incarcerated would instead be released. In practice, it turned out that most first-time *drug* arrests were not necessarily *first* arrests, so the program was quickly expanded to reach all drug-involved offenders that the courts were willing to divert into treatment.

TASC procedures determine a drug-dependent offender's eligibility for intervention, and they include assessment of the offender's risk to the community, severity of drug dependence, and appropriateness for treatment placement. After an individual is referred to a treatment program, TASC case-managment services monitor that individual's compliance with the conditions of the treatment and rehabilitation regime, including expectations for abstinence, employment, and improved personal and social functioning. Progress is reported to the referring justice-system agency. Clients who violate the conditions of their justice mandate—TASC "contract" (or treatment agreement)—are usually returned to the justice sys-

tem, where the legal process interrupted by TASC diversion goes forward.

Specific "critical program elements" define the parameters of a well-described national TASC model. These have been carefully worked out by The National Consortium of TASC Programs (NCTP) (444 North Capitol Street, NW, Suite 642, Washington, DC 20001) (Phone: 202/783-6868; FAX: 202/783-2704). These critical elements provide the structure for the linkages between the criminal-justice and treatment systems. This model makes it possible to easily replicate TASC programs anywhere in the United States, including urban, suburban or rural settings, and is easily adaptable to specific population needs. NCTP provides technical assistance for implementation of the model program, training for program development, systems coordination, program assessment, development and dissemination of materials (such as model policies, procedures, protocols, etc.), training in the use of the "critical elements," internships, and accreditation of TASC programs.

Many of the states have expanded the TASC model to provide a wide array of adjunct services to a wide variety of participants in TASC programs. Illinois TASC, for example, founded in 1976 by Melody Heaps, uses the name Treatment Alternatives for Special Clients (TASC, Inc.) in order to better describe the scope of its programs. The program provides case management and a comprehensive array of services throughout Illinois for men, women, and adolescents who have a variety of social, welfare, and health-related needs. Populations served include youth in the child-welfare system, AIDS-affected clients, DUI (drunk-driving) offenders, juvenile offenders, students, welfare recipients, offenders sentenced to home confinement, youth in community-based programs and those in the child-welfare system, Supplemental Security Income (SSI) recipients, pretrial arrestees, and Cook County Jail inmates. For each special population targeted and served, appropriate interventions and services have been devised, such as a school intervention program, a gang intervention program, and youth services for substance-abusing students and adolescents. Adult criminal-justice services include monitoring of offenders in home confinement using technologies such as electronic monitoring and drug testing; a jail project providing screening and assessment, orientation, intensive therapeutic-community counseling, transition counseling, and aftercare planning and management. Illinois TASC is the sole agency providing substance-abuse assessment and recommendations for the Illinois courts. As well as providing offender case-management services, it offers training for judges, state's attorneys, public defenders, criminal-justice planners, and federal and state probation and parole staffs.

TASC programs play an important role in reducing the growing rates of drug-related street crime and alleviating court backlogs. They have been effective in identifying drug-involved offenders in need of treatment, assessing the nature and extent of their drug use and their specific treatment needs, and referring them to treatment. TASC clients have been found to remain in treatment longer and to have better posttreatment success. In addition, as an adjunct to parole and work release, the programs have the potential to help ease prison overcrowding. TASC also effectively fulfills its original purpose of linking the criminal-justice and treatment systems by providing client identification and monitoring services for the courts, probation departments, and other segments of the criminal-justice system.

(SEE ALSO: *California Civil Commitment Program; Civil Commitment; Coerced Treatment for Substance Offenders; Crime and Drugs; Narcotic Addict Rehabilitation Act*)

BIBLIOGRAPHY

INCIARDI, J. A., & MCBRIDE, D. C. (1991). *Treatment alternatives to street crime: History, experiences, and issues.* DHHS Publication no. (ADM) 91-1749. Rockville, MD: U.S. Department of Health and Human Services, Public Health Service, Alcohol, Drug Abuse, and Mental Health Administration, National Institute on Drug Abuse.

MORGAN, J. (1992). Treatment alternatives to street crime. *State ADM Reports* no. 15 (June), Intergovernmental Health Policy Project. Washington, DC: George Washington University.

JEROME H. JAFFE
FAITH K. JAFFE

TREATMENT CENTERS DIRECTORY

See Appendix, Volume 4.

TREATMENT FUNDING AND SERVICE DELIVERY

No single accepted method or setting exists for the treatment of substance abuse—alcohol and other drug-abuse disorders. Treatment is offered in specialty units of general and psychiatric hospitals, residential facilities, halfway houses, outpatient clinics, mental-health centers, jails and prisons, and the offices of private practitioners.

In the United States during the 1970s and 1980s, drug abusers were commonly treated in programs distinct from those serving alcoholics. By the 1990s, the two treatment systems were merged; in 1991, of the estimated 11,000 substance-abuse treatment programs in the United States, 79 percent reported that they served both drug and alcohol abusers. Some 88 percent were enrolled in outpatient programs. Another 10 percent were in residential facilities. Only 2 percent were hospital inpatients.

The cost of treatment varied greatly depending on setting. In the early 1990s, hospital inpatient care was the most expensive on a daily basis ($300–600/day), but it was usually of short duration (30 days or less). Treatment in nonhospital residential programs was less expensive ($50–60/day), but it commonly lasted longer (a few months to 2 years). Programs that did not require the individual to live in a specialized facility were the least expensive, both on a daily basis ($5–15/day) and over a full course of treatment.

PRIVATE HEALTH INSURANCE

The availability of private health-insurance coverage for substance-abuse treatment grew in the 1980s. By 1990, better than 90 percent of health-insurance plans had explicit coverage for drug treatment. Individuals with such private insurance have a greater range of treatment providers from which to choose than those who are indigent and have only government-funded programs at their disposal. Programs that mainly rely on insurance reimbursement, however, tend to be more expensive than those that receive the bulk of their support from government sources.

U.S. GOVERNMENT FINANCING

In the U.S. general health-care system, 68 percent of the cost of services is borne by the individual, insurance company, or other private third-party payer.

For substance-abuse or mental-health care, in contrast, the government supplies 63 percent of the funds for substance-abuse treatment. After the private sector, which provides 37 percent of the funds, the states traditionally have been the major source of treatment support (31%), followed by the federal government (24%), and then county and local agencies (8%). States often finance treatment by reimbursing providers through public-welfare programs or through grants or contracts. Some states transfer funds to county and local governments, which, in turn, purchase services from providers. Another financing mechanism is Medicaid, a combined state and federal program that pays medical bills for low-income persons. Under Medicaid, states can pay for substance-abuse care in inpatient general hospitals, clinics, outpatient hospital and rehabilitation services, and in group homes with sixteen or fewer beds.

A federal program that pays the health-care costs of persons 65 years of age or older, or those who are disabled, is Medicare. This primarily covers inpatient hospital treatment of alcohol or drug abuse, as well as some medically necessary services in outpatient settings. The primary federal mechanism for paying for alcohol and drug treatment is the Substance Abuse Block Grant, administered by the Department of Health and Human Services. Funds from the block grant are distributed to the states (and territories) using a formula that takes the characteristics of the state's population into account. In fiscal year 1994, Congress appropriated approximately 1.3 billion dollars for the Substance Abuse Block Grant. The federal government also makes grants to individual treatment providers to support innovative treatment approaches, improve the quality of treatment, or to ensure services for underserved or special populations.

(SEE ALSO: *Treatment; U.S. Government Agencies*)

BIBLIOGRAPHY

HEALTH INSURANCE ASSOCIATION OF AMERICA. (1991). *Source book of health insurance data.* Washington, D.C.: Author.

INSTITUTE OF MEDICINE. (1991). *Treating drug problems,* vol. 1. Washington, D.C.: National Academy Press.

INSTITUTE OF MEDICINE. (1990). *Broadening the base of treatment for alcohol problems.* Washington, D.C.: National Academy Press.

SUBSTANCE ABUSE AND MENTAL HEALTH SERVICES ADMINISTRATION. (1992). *Highlights from the 1991 National Drug and Alcoholism Treatment Unit Survey (NDATUS).* Rockville, MD: Author.

SALVATORE DI MENZA

TREATMENT, HISTORY OF, IN THE UNITED STATES

The history of the treatment of alcohol and other drug problems is often assumed to be a straightforward story of progress—moralism, neglect, and brutality were displaced by scientific knowledge, medical activism, and professional civility; a view that the addict exercised free will in choosing to use drugs was succeeded by an understanding of how a "disease" or "disorder" could overrule the capacity to choose.

This assumption is historically incorrect. First, it neglects the coexistence and mutual influence of views emphasizing free will or social or biological determinism. While one view may have enjoyed greater influence at a given time, its competitors have never been vanquished. No generation has any more solved the puzzle of addiction than it has resolved the related enigmas of the relationship between mind and body, choice and compulsion. Second, it is equally incorrect to associate condemnation and neglect with the free-will position or kindness and activism with the determinist perspective. The truth is more complicated.

As various studies have demonstrated, there is a tenacious American folk wisdom about addiction. Simply put, it goes as follows: While addicts experience a compulsion to take a drug, this develops as the result of repeated bad choices that are socially influenced; further, addicts can rid themselves of compulsion only by developing self-discipline, perhaps with some skilled influence in the form of treatment. Thus, in our culture, and despite the modern message that "addiction is a disease like hypertension or diabetes," addicts are understood to be both sick and immoral, blameless and culpable, free and determined. In the popular mind, and among treatment professionals, addicts are ambiguous characters.

The history of treatment in the United States reflects this cultural dilemma. Cultures limit the range of possible responses to a problem, and because they tend to change very slowly in fundamental ways, to the extent that an important problem recurs or remains unsolved, the range of possible responses will be explored repeatedly as new generations search for fresh insights and effective methods of intervention. At various times, treatment has embraced exhortation and coercion, sermons and miracle drugs, democratic mutual aid, and autocratic professional prerogative—often simultaneously.

THE PREMODERN ERA

Modernity has different meanings with respect to the treatment of habitual drunkenness and drug addiction. In the case of habitual drunkenness, the modern era is traceable to the birth of ALCOHOLICS ANONYMOUS (AA) in 1935. In the case of drug addiction, delineating historic periods is more difficult, but we will mark the modern era by the introduction of methadone maintenance (for heroin dependence) in 1965 and passage of the federal NARCOTIC ADDICT REHABILITATION ACT (NARA) in 1966.

We should also clarify our choices of terminology. The terms *alcoholism* and *alcoholic* date from the middle of the nineteenth century, but they did not come into common professional use until the early twentieth century and were not embedded in the American vernacular until after the rapid growth of AA during the 1940s. The more common professional terms in the premodern era were *inebriety* and *inebriate*, but as these often were used to refer to a heterogeneous group now called "substance abusers," we will use the durable term *drunkard* when writing about this era. Similarly, the term *drug addict* was not in common use until the early 1900s; before this time habitual users of drugs were known as "morphinists," "cocainists," or sometimes, "dope fiends." In order to speak generally and to avoid pejorative (if historically accurate) terminology, we will use drug addict, and we will use addict and addiction when speaking of both habitual drunkards and drug addicts.

THE TREATMENT OF HABITUAL DRUNKARDS

The Tradition of Mutual Aid. The organized, specialized effort to help habitual drunkards began with the Washington Total Abstinence Movement in 1842. This Washingtonian Movement stands at the head of a tradition of mutual aid that developed throughout the 1800s in close connection to American Protestantism, particularly its evangelical expressions. The Salvation Army, which traces its American incarnation to the mid-1870s, is also in this line, and so is AA and the many other "Anonymous" fellowships it inspired.

Washingtonian societies were dedicated to sobering up hard drinkers, usually (but not always) men. The societies intended to foster a solidarity based on shared experience with suffering that transcended profound social divisions. (They were neutral on the divisive question of prohibition.) Although some famous teetotalers like Abraham Lincoln were members, the societies included the disreputable, the unlettered, and sometimes nonwhites and women as equals. Their motives were couched in terms of Christian charity, economic self-improvement, and democratic principles.

The hallmark of mutual aid is the banding together of people in similar circumstances to help one another. (The popular term "self-help" is thus misleading.) The Washingtonians and their successors did not invent the methods by which they fostered solidarity and mutual support. However, in adapting the voluntary association to the reform of drunkards, the Washingtonians introduced new elements.

Owing its provenance to the revival meeting, the most striking and controversial (some found it distasteful) Washingtonian innovation was the confession of drunkards before their peers, and sometimes before a general audience. We are familiar with its contemporary form: "I am Jim B. and I am an [alcoholic, drug addict, etc.]"; but the practice dates from Washingtonian "experience lectures," forums for the telling of "drunkard's tales," stories of degradation, struggle, and redemption through sobriety. These introduced the drunkard's tortured inner life to the polite public. "You all know me and what I used to be," Salvation Army lecturers often began.

Some Washingtonian societies also established temporary homes, or refuges, for drunkards. These were places where drunkards could live for a short time while they sobered up and were introduced to the Washingtonian fellowship, whose members found them jobs and other necessities. A century later, AA would reinvent this institution (the recovery home) as part of its twelfth-step work—the commitment to help other drunks.

Although not continuous with these early refuges, beginning in Boston (1857), San Francisco (1859), and Chicago (1863), a number of formal inebriate homes were established to treat drunkards in the Washingtonian tradition. Typically, these were small institutions (fewer than 50 beds), operated as private charities, sometimes under religious or temperance auspices. They relied on the voluntary cooperation of their residents and used temperance fellowship as a form of what we now call aftercare. They were located in urban environments and did not isolate their residents from community life. Although they often were superintended by physicians, residence rarely exceeded three weeks and medical treatment was considered important only in managing withdrawal symptoms or DELIRIUM TREMENS (DTs). The terms *disease* and "vice," *cure* and "reformation" were used interchangeably, and sober outcomes were attributed to the influences of family, friends, and the fellowship, not to medical intervention. Inebriate homes practiced a profoundly social (and sometimes spiritual) form of treatment based on the belief that the human capacity for transformation was never extinguished, no matter how "despotic" the "appetite" for alcohol.

For those in the Washingtonian line, the source of such optimism was their belief in the presence of an immortal God in the human mind. The mind, they believed, was distinct from the brain and other corruptible flesh and was formed in God's image. By the mid-1800s, the image of God was far more benign and rational than the often wrathful, finally inscrutable deity of even the early 1700s. This gradual change in the conception of God owed much to the spread of the market as arbiter of economic affairs and social relations. The rigorous logic of the market reordered economics from the academy to the workshop. In its train, a disciplined, optimistic rationalism—and the ideas of moral progress and human perfectibility—suffused popular culture and theology.

At the same time, another form of rationalism, that of natural science, was pervading popular discourse and causing tumult in seminary and pulpit. Science did not overthrow religion so much as assume a place alongside it. For believers, scientific order was a wonder of the divine plan. The natural "laws of health," as various rules of disciplined self-denial were known, were signals of divine intent, of God's ideas about right living. The drunkard was therefore both sinful and sick, having contracted the disease as the result of moral transgression. (A common analogy of the time was to syphilis; today, some religious leaders speak similarly of AIDS.) Thus, while Washingtonians and their successors spoke of addiction as a disease—by which they meant an organically based compulsion—they also employed clerical images, for they believed in the power of the divinely inspired human mind to choose the rational good (total abstinence from alcohol) and to thus achieve health. In the Washingtonian tradition, the languages of morality and disease became assimilated, and remain so in the many contemporary Anonymous fellowships' claim that addiction is in part a "spiritual disease."

Although the Washingtonian Movement as such was defunct by 1850, Washingtonianism was extremely influential until about 1865. The tradition did not disappear, but in the decades following the Civil War (1861–1865), profound changes in American culture and society, and related changes in the temperance movement, blunted Washingtonian influence and gave new prominence to a competing philosophy of treatment and its attendant practices and institutional embodiment. The philosophy was that of biological determinism, or "somaticism," and its institutional expression was the "inebriate asylum."

The Asylum Tradition. In 1810, Benjamin Rush, a Philadelphia physician, signer of the Declaration of Independence, and first formulator of a disease theory of addiction (though not the inventor of the idea), proposed "sober houses" for drunkards. However, Samuel Woodward, a Massachusetts insane asylum superintendent and temperance orator was the father of institutional treatment based on a somatic explanation of habitual drunkenness. In a tract written in 1835, Woodward contributed two critical ideas to what would become the inebriate asylum movement of the nineteenth and early twentieth centuries. The first was that drunkards could not be treated successfully on a voluntary basis. The second, which flowed from the first, was that they needed legal restraint in a "well-conducted institution"—by which Woodward meant something like the insane asylum that he superintended.

The line of thinking staked out by Rush and Woodward had no institutional realization until an inebriate asylum subsidized by the State of New York opened in Binghamton in 1864. Another was opened in Kings County, New York, in 1869. In subsequent decades, pursuant to arduous promotion by the

American Association for the Cure of Inebriates (AACI, founded in 1870), public inebriate asylums opened in Massachusetts (1893), Iowa (1904), and Minnesota (1908). Other jurisdictions chartered inebriate asylums but never built them (Texas and Washington, D.C.), and in California an inebriate asylum chartered in 1888 was converted to an insane asylum before the facility opened in 1893. Indeed, Binghamton was converted to an insane asylum in 1879. By the advent of Prohibition in the United States in 1920, all public inebriate asylums had been closed or converted to other use.

The inebriate-asylum movement spawned dozens of private sanitaria that treated well-to-do drunkards and, by the 1890s, drug addicts. However, judged by its manifestation in brick and mortar, the movement for public treatment was a failure. For two related reasons, the AACI was notably unsuccessful in converting legislatures to its cause. First, its physician members never could produce a strictly medical "cure" for addiction. Although its theorist-practitioners developed rigorously somatic explanations of addiction that dispensed with will power, spirituality, and the therapeutic necessity of fellowship, they relied on recuperation by bed rest, a healthy diet, and therapeutic baths (hydrotherapy), followed by the discipline of useful labor. This regime was highly structured (military analogies were popular) and medically supervised, and was set in a context of prolonged legal restraint (involuntary commitment). However, there was nothing particularly innovative or medical about this approach. Its methods already were the staples of lunatic asylums (called mental hospitals in most states after about 1900), almshouses, and county jails, institutions that managed huge numbers of habitual drunkards and, after the 1880s, drug addicts. Second, the inebriate asylum was an ambitious undertaking: like the insane asylum, it was to accommodate several hundred patients on a sequestered rural estate. Few legislatures could be persuaded that such costly new institutions were worth the price. In a word, the inebriate asylum was viewed as redundant.

The ideology of the inebriate-asylum movement—its adherents' view of the world—was shaped by two profound, contemporaneous developments in American culture and society: (1) the rising esteem and secularism of science and (2) the growing disorder and complexity of American society after the Civil War. The movement reflected the grand aspirations of Gilded Age science, whose practical applications were transforming American life: railroads and streetcars, the telephone, gas and electrical lighting—all attested to the power of science and human ingenuity. It was a time when "scientific" understanding became the basis for professional standing, not only for medicine, but for all manner of professional groups, from proto-social workers to plumbers. The metaphor of disease, and the optimistic message implicit in its use—that all defects could be cured—became popular among forward thinkers. In the most widely read book of its time, the utopian novel, *Looking Backward* (1888) by Edward Bellamy, the author characterized all sorts of misconduct as disease, and his near-perfect world of the year 2000 cured its rare wayward citizens in public hospitals.

If Washingtonians assimilated the languages of morality and disease, the rising generation of inebriate-asylum enthusiasts radically separated them, and often reduced human volition to a by-product of neurology. In the United States and Europe, they initiated research on the biology (and later, the genetics) of addiction. Primitive by today's standards, it nonetheless established a robust tradition of inquiry that remains lively.

The inebriate-asylum movement appealed to American aspirations to create a better world through science, but it also addressed growing fears of social disorder. The extent of such disorder should not be exaggerated, however; pre-industrial America was more disorderly than nostalgic chroniclers have made it seem, and urbanization and industrialization were less chaotic than critics sometimes contend. On the whole, though, life after the Civil War was more complex, more anonymous, and less certain.

Immigration from abroad was an important fuel for such change and promoted the (American) nativist fears that accompanied it. In the 1830s, free Americans were overwhelmingly Anglo-Saxon in origin and Protestant in belief. By the 1880s, this was changing dramatically. Burgeoning northern and western cities were becoming testing grounds for the promise and limits of diversity—indeed, for explanations of diversity. Amid glaring inequality of wealth and opportunity, cultural conflicts often were played out around practices of consciousness alteration. Protestant, native-born Americans (including African Americans) were remarkably abstemious (a notable success of the Protestant-driven temperance movement); the mostly Roman Catholic Italians and French were daily wine drinkers; Poles, Germans, and some Scandinavians drank large quantities of beer (some on Sunday—in public beer gardens).

Of Irish Catholics, who had a large temperance movement of their own but also a penchant for drunkenness (what is known as a "bi-modal distribution" of drinking habits), a California temperance editor wrote in 1883: "They are by far the worst and meanest material in which to store whisky." Native Americans had been introduced to alcohol by traders and government agents from colonial times, so "firewater" became a factor in the westward movement and the ensuing Indian Wars. The "idolatrous" (non-Judeo-Christian) Chinese introduced opium smoking to America, a practice that crossed the color line during the 1870s and became popular among young white men and women during the 1880s. Then from 1900 to 1920, Hindus and Mexicans became associated with *Cannabis* (MARIJUANA) use in the West and Southwest. In the South, African-American men frequently were accused of the riotous use of CO-CAINE, with subsequent designs on white women.

The increasingly diverse backgrounds of the U.S. population became a source of conflict and disorder; the roller-coaster ride of industrial capitalism was another. The United States experienced two prolonged economic depressions (then called "panics") between the Civil War and the turn of the century—from 1873 to 1878 and from 1893 to 1898. In between, a short but sharp slump during the mid-1880s took its toll on stability. During these years, the noun "tramp" entered the American language; the country experienced its first pronounced labor violence and political bombings (dynamite being an 1860s product of scientific ingenuity); and in the spring of 1894, "armies of the unemployed" converged on Washington, D.C., from all over the country.

This era of mounting diversity and instability was marked by a failing faith in exhortation (verbal appeal) as a method to achieve social regulation and by a concomitant exaltation of coercive means (force). Although never abandoning altogether its sympathy for drunkards, the temperance movement made securing prohibitionist measures its primary objective. Although never withdrawing its support from surviving Washingtonian institutions, temperance adherents simultaneously supported the more stringent regime promoted by inebriate asylum enthusiasts, some of whom believed that an orderly, peaceful society required the lifetime detention of incurable addicts. Indeed, the temperance movement helped to popularize theories which purported

to demonstrate a biological basis for the failure of certain racial and ethnic groups to live up to the abstemious standard of so-called native stock—or to benefit from treatment. In the name of "prevention," such views justified not only prohibition laws but statutes which in a few states permitted the forced sterilization of addicts.

In sum, the legacy of the inebriate asylum movement was the biologically based approach to understanding addiction, the corollary claim that addiction is the special province of medicine and physicians, the notion that successful treatment requires legal coercion, and the assertion that treatment is both a responsibility of government and a commodity to be sold on the market. These ideas endure as part of the complex intellectual, professional, and political fabric of treatment.

The Tradition of Mental Hygiene. The mental-hygiene movement, customarily dated from the 1908 publication of Clifford Beers' *A Mind That Found Itself*, represented a departure from the somatic tradition of thought about mental disorder and addiction. At the same time, it did not appeal to spiritual explanations nor did it dwell on will power. Rather, mental hygienists employed a socio-biological determinism: While addiction could be the result of hereditary biological defect, and could be incurable, its origins were mainly familial and social, and if the condition was addressed early on, could be arrested. Mental hygienists stressed the important roles of family, friends, and occupation in creating a salubrious environment for an addict's continuing sobriety. Mental hygiene did not speak the language of mutual aid, but it was similarly environmental in outlook. This was the beginning of what later would be called community mental health, and its point of view virtually defines what we understand to be "modern" about treatment and the biopsychosocial perspective.

The environmentalism of mental hygiene challenged the rationale of the asylum model of treatment. Mental hygienists criticized the asylum's lack of connection with community life and its reliance on involuntary treatment, claiming that only voluntary access to free or inexpensive care would attract patients in the early stages of drinking or drug-taking careers. The history of the Massachusetts Hospital for Dipsomaniacs and Inebriates (1893–1920) illustrates well the influence of mental-hygiene philosophy and practice. Between 1893 and 1907, the hospital was run on the asylum model. After a complete reorganization in 1908, it followed a mental-hygiene course: most of its admissions were legally voluntary; the hospital established a statewide network of outpatient clinics; it worked closely with local charities, probation offices, employers, and the families of patients. Known finally as Norfolk State Hospital, it was a preview of what treatment was to become, beginning in the 1940s.

Even so, Norfolk created on its campus a "farm" for the long-term detention of "incurables." The mental-hygiene movement modified the emphasis of the asylum tradition but did not entirely abandon its practices. Indeed, under the banner of mental hygiene, between 1910 and 1925, many local governments across the United States established "farms" to segregate repeated public drunkenness offenders and drug addicts. Some of these persisted until the 1960s, and some have been reopened in recent years to accommodate homeless people with alcohol and drug problems. As discussed below, the asylum tradition remained particularly important in the treatment of drug addicts.

TOBACCO

Although tobacco use is now widely considered in the United States to be a problem akin to drug dependence, for most of the twentieth century it was not treated as such by either the medical or criminal-justice establishment. However, nineteenth-century temperance groups saw tobacco use as another form of inebriety. As far back as the 1890s, advertisements for patent medicines claimed to help people break the tobacco habit. In the great temperance upsurge of the early twentieth century, more than twenty states passed tobacco prohibition laws, but most of these were quickly repealed. Public concern with habitual tobacco use declined dramatically from the 1920s through the 1950s, and cigarette smoking (over smokeless tobacco, pipes, or cigars) became normative behavior among men and grew steadily among women. This situation changed abruptly with the publication of the 1964 *Report of the U.S. Surgeon General* that linked cigarette smoking to cancer. Since then, increasing attention has been paid to the tobacco habit, or TOBACCO DEPENDENCE, and to treatment for it. Treatment approaches are at least as varied as those described here for alcohol and other drugs. Pharmacological treatments, such as nicotine chewing gum and skin patches have been

used, as have acupuncture, hypnosis, mutual aid, aversive electric shock, and other techniques. While many people advocate that government or private insurance should pay for treatment of this addiction, to date there have been no suggestions that tobacco addicts should be treated on a compulsory basis, although the places where it is legal to smoke have been diminishing.

THE TREATMENT OF DRUG ADDICTS

Although the San Francisco Home for the Care of the Inebriate (1859–1898) treated a few opium addicts as early as 1862, Washingtonian institutions mainly treated drunkards. Similarly, although a few reborn drug addicts were among the legions of the Salvation Army and other urban missions by 1900, they were vastly outnumbered by reformed drunkards. Until the organization of NARCOTICS ANONYMOUS (NA) in 1953, there was no large or well-defined group of addicts involved in the practices of mutual aid, and there were a variety of reasons for this.

Drug addiction was not a matter of widespread concern until after the Washingtonian philosophy had been eclipsed by the asylum model of treatment. Further, drug addicts were quickly perceived to be more exotic and ominous than habitual drunkards. While there were many people addicted to morphine as a result of ill-advised medical treatment or attempts at self-treatment during the late 1800s, this more or less respectable population declined after the turn of the century as physicians and pharmacists reformed their dispensing practices and new laws required the disclosure of the content of patent medicines and nostrums. At the same time, a growing number of urban young people began to experiment with drugs, especially smoking opium, morphine, and cocaine. By 1910, drug addiction was popularly associated with petty thieves, dissipated actors, gamblers, prostitutes, and other nightlife aficionados, and with racial minorities and dissolute youth. Unlike habitual drunkards, drug addicts never were caricatured as boisterous and occasionally obstreperous nuisances or buffoons; especially after 1900, they usually were portrayed as dangerous predators and corrupters of society, alternating between drug-induced torpor (in the case of opiates) or hyperactivity and hallucination (in the case of cocaine) and a craving that propelled them on relentless and unscrupulous searches for drugs and the means to buy them.

The "criminal taint" of drug addiction, and the widespread view that most addicts were incurable and would do anything to alleviate withdrawal symptoms, provided a powerful rationale for their prolonged confinement under strict conditions. Even the mental hygienists at Norfolk State Hospital had no expectation that addicts would remain sober and favored incarcerating them in the Massachusetts State Farm at Bridgewater, a correctional facility. Indeed, state hospitals were generally more opposed to admitting addicts than habitual drunkards, preferring to have them incarcerated in jails. Even more than drunkards, addicts disturbed the routine and good order of state hospitals, in no small part because they were, as a group, considerably younger and less conventional than other hospital patients. They pursued sexual liaisons in violation of institutional rules against fraternization; they smuggled drugs into the hospitals; and once through withdrawal, they escaped in droves.

Nor were jails and prisons anxious to take in addicts, mainly because of the problem of smuggling. By the late 1880s, opium was a customary (though illicit) medium of exchange at San Quentin Prison in California, and it was routinely available in the big county jails of the United States at the turn of the century. As state laws against the sale or possession of opiates and cocaine proliferated in the 1890s, and as they began to be more strictly worded and enforced after 1910, county jails and state prisons faced a major problem of internal order. This intensified with the implementation of the federal HARRISON NARCOTICS ACT (passed in 1914 to take effect in March 1915), particularly after a U.S. Supreme Court decision in 1919 made it illegal for physicians to prescribe opiates for the purpose of maintaining an addict's habit. The vast majority of drug offenders, even those arrested by federal agents, were prosecuted under state drug and vagrancy laws and sent to state and county lockups. The resulting crisis led jailers to support two related treatment strategies.

The first of these was the creation of special institutions for drug addicts. Thus the county farms mentioned earlier in this essay were created, or laws were passed to allow addicts to be committed to existing state or county hospitals with wards designated for this purpose. Mendocino State Hospital in California, Worcester State Hospital in Massachusetts,

Norwich State Hospital in Connecticut, and Philadelphia General Hospital, to name a few, treated significant numbers of addicts in the 1910s and 1920s. Later, California (1928) and Washington (1935) opened state-sponsored variations on the jail farm, though under the auspices of their state hospital systems.

The growing number of addict-prisoners in the federal system also led to their segregation, first at Leavenworth, Kansas (mainly), and then at two narcotic hospitals opened at Lexington, Kentucky (1935) and Fort Worth, Texas (1938). Operated by the U.S. Public Health Service, these hospitals were in fact more like jails, although they were authorized to admit voluntary patients of "good character" whose applications were approved by the U.S. Surgeon General. Initially, these patients were kept involuntarily once they had been admitted, but a federal district court ruling in 1936 affirmed that voluntary patients could leave after giving notice. Before they were closed in the 1970s, the two facilities had admitted more than 60,000 individuals comprising over 100,000 admissions.

Jailers were also an important part of local political coalitions in support of a short-lived and controversial treatment strategy of the early 1920s—drug dispensaries for registered addicts. At least forty-four such clinics were established nationwide, most in late 1919 or early 1920, following the Supreme Court's antimaintenance ruling.

In principle, these were not to be maintenance clinics. Addicts initially were to receive their customary dosages of morphine (occasionally heroin, and very rarely, smoking opium), and were then to be "reduced" over a short time to whatever dosage prevented withdrawal. At this point, abstinence was to be achieved.

In practice, few of the clinics worked this way. Many clinic operators believed that their primary aim was to mitigate drug peddling by supplying addicts through medical channels. This implied a maintenance strategy at odds with the Supreme Court's interpretation of the Harrison Act and with some earlier state laws forbidding maintenance (in California and Massachusetts, for example). Further, most clinic operators agreed with the American Medical Association (AMA) that dispensaries could only work effectively within the law if prolonged institutional treatment was available once the addict's dosage had been reduced to the brink of withdrawal. In the absence of such institutional capacity, reduction was useless, and so clinic doctors rarely bothered. The Prohibition Unit of the U.S. Department of the Treasury (which enforced the Harrison Act), state boards of pharmacy (which typically enforced state drug laws), and local medical societies and law enforcement agencies regarded the clinics as stop-gaps, valuable only until adequate public hospitals could be opened.

In the midst of the inflation following World War I, localities looked to the states to finance such institutions and states looked to the federal government, particularly the U.S. Public Health Service, which had operated hospitals for merchant mariners since 1792. But legislation to create a federal treatment program failed to pass and the states were thrown on their own resources. The Prohibition Unit, convinced that the clinics were doing more harm than good, moved to close them, threatening dispensing physicians with prosecution. The clinics closed rapidly. The last one, at Shreveport, Louisiana, closed in 1923. Addicts were consigned to their customary ports of call in jails, prisons, or for the fortunate few, private sanitaria.

The controversy over maintenance did not disappear, however, particularly on the West Coast, where efforts to loosen its prohibition in the states of California and Washington continued until the United States entered World War II (1941). Further, both federal and state governments permitted the maintenance of a small number of addicts, usually of middle age or older, suffering from severe pain related to a terminal illness or an incurable condition. However, the period from 1923 through 1965 was generally characterized by the strict enforcement of increasingly severe laws against drug possession and sales, by relentless opposition to maintenance, and by treatment that was essentially in the asylum tradition, supplemented by the mental-hygiene innovation of supervised probation. In 1961, California passed legislation permitting the compulsory treatment of drug addicts (including marijuana users) and established the CALIFORNIA CIVIL ADDICT PROGRAM within the Department of Corrections. From 1962 to 1964, more than 1,000 people were committed to a 7-year period of supervision, which typically involved an initial year of residential treatment in a facility surrounded by barbed wire to discourage premature departure. In 1964, New York passed similar legislation but assigned its implementation to a special commission rather than to the

Department of Corrections. As in California, New York's residential treatment facilities were "secure." As late as 1966, the federal NARCOTIC ADDICT REHABILITATION ACT (NARA), in most respects a piece of "modern" legislation, nonetheless provided for the compulsory treatment of addicts and made the hospitals at Lexington and Fort Worth into the institutional bases of the NARA program.

THE MODERN ERA

The modern history of alcohol and drug treatment has been shaped by the therapeutic pluralism descended from the mutual-aid, asylum, and mental-hygiene traditions; the growing prestige of clinical and basic medical research; the coexistence of public and private sectors of treatment; and an increasingly complex field of interorganizational relationships involving several layers of government and substantial fragmentation within each layer.

ALCOHOLISM TREATMENT

The influence of ALCOHOLICS ANONYMOUS can hardly be exaggerated. Whatever its therapeutic success—a point of warm debate among scholars—AA has profoundly affected the treatment of people now regularly known as alcoholics. AA's impact has been both ideological and institutional; that is, its promotion of "disease theory" within the mutual-aid tradition has changed how recent generations think about excessive or problem-causing alcohol consumption and treatment methods, and the penetration of policymaking bodies and treatment institutions by people recovering from alcoholism has shaped the funding and practices of treatment.

AA's impact was facilitated by the growing influence of the mental-hygiene movement during the 1920s and 1930s, for AA provided the critical therapeutic bridge between the segregating institution and the community at large. This was recognized quickly by men like Clinton Duffy, the great "reform" warden of San Quentin, who encouraged the establishment of AA groups in his prison in 1942. Much early twelfth-step work was done in U.S. county jails. Harvard psychiatrist Robert Fleming opined in 1944 that the prolonged institutionalization of alcoholics was no longer necessary; a week's medical care in a general hospital followed by community-based psychotherapy and AA participation was his new prescription. The growth of AA permitted the first substantial stirrings of community care since the Washingtonian Movement.

During the early 1960s, some state hospitals, particularly in Minnesota, incorporated recovering alcoholics and the principles of AA into their treatment programs. What became known as the Minnesota model of short-term inpatient care (usually 28 days) and subsequent AA fellowship and recovery-home living spread slowly but discernibly among private treatment providers such as the HAZELDEN Foundation, also in Minnesota, and the Mary Lind Foundation in Los Angeles. Across the country, local councils on alcoholism, dominated by people recovering from alcoholism and encouraged by the NATIONAL COUNCIL ON ALCOHOLISM AND DRUG DEPENDENCE and the National Institute of Mental Health (NIMH, created in 1946, was an ardent promoter of community psychiatry), began to press states and localities for outpatient clinics, diversion of alcoholics from jail, and other methods consistent with the traditions of mutual aid and mental hygiene. Even so, treatment resources for alcoholics did not expand dramatically. A survey in 1967 found only 130 outpatient clinics and only 100 halfway houses and recovery homes dedicated to serving alcoholics. Alcoholics continued to be barred from most hospital emergency rooms.

All this advocacy and organizing activity was propelled by the concept of "alcoholism as a disease," a proposition given its most systematic modern exposition by E. M. Jellinek in *The Disease Concept of Alcoholism* (1960). Jellinek was more provisional in his use of the term than most of his readers appreciated, but he understood the important strategic value of such a claim. In the first instance, the language of disease challenged the legal and correctional system's jurisdiction over alcoholics; in addition, it provided a rationale for the increased availability of services for alcoholics within established medical facilities and under the aegis of public health. Jellinek was widely read in the literature of the earlier inebriate asylum movement, and although he disparaged its science he understood and sympathized with its aims. He fully understood that whatever its equivocal status as scientific truth, the assertion that alcoholism is a disease carries important implications for treatment policies.

Several important court decisions in the 1960s endorsed the view that alcoholism was a disease; and, in 1967, a presidential commission on law enforcement concluded that it was both ineffective and inhumane to handle public drunkenness offenders within the criminal-justice system and recommended creating a network of detoxification centers instead. In 1970, Congress passed the Comprehensive Alcohol Abuse and Alcoholism Prevention, Treatment and Rehabilitation Act (the "Hughes Act"). Senator Harold Hughes, a former governor of Iowa, was a recovering alcoholic. A persuasive speaker, Hughes became the conscience of the Congress in developing support for a more humane and decent response to people with alcoholism and related problems. He was supported in these efforts by Senator Harrison Williams, Congressman Paul Rogers, and several advocacy groups led by the National Council on Alcoholism and the North American Association on Alcohol Problems. While Hughes's early efforts had been supported by President Lyndon Johnson and Assistant to the President Joseph Califano, it was President Richard M. Nixon who signed the legislation establishing the NATIONAL INSTITUTE ON ALCOHOL ABUSE AND ALCOHOLISM (NIAAA). This legislation made federal funds available for the first time specifically for alcoholism-treatment programs.

The Hughes Act accomplished three goals of the modern alcoholism treatment movement. First, it effectively redefined alcoholism as a primary disorder, not a symptom of mental illness. Second, and based on this distinction, it created the federal agency—NIAAA—that would not be dominated by the mental-health establishment competing for the same resources. Finally, and of great practical importance, the Hughes Act established two major grant programs in support of treatment. One authorized NIAAA to make competitive awards (grants and contracts) directly to public and nonprofit agencies; the other was a formula-grant program, which allocated money to states based on a formula accounting for per capita income, population, and demonstrated need.

NIAAA aggressively sought state adoption of the model Uniform Alcoholism and Intoxication Treatment Act, first drafted in 1971 by the National Conference of Commissioners on Uniform State Laws. Section 1 of the Uniform Act, as it was known, stated that "intoxicated persons may not be subject to criminal prosecution because of their consumption of alcoholic beverages but rather should be afforded a continuum of treatment." By 1980, thirty states had adopted some version of the Uniform Act, thereby decriminalizing public drunkenness.

The thrust of federal and state grant making was to create an effective system of community-based alcoholism treatment services. This occurred in tandem with the deinstitutionalization process that was rapidly depopulating state mental hospitals. Although we customarily think of deinstitutionalization as affecting only the mentally ill, in fact it had an important impact on alcoholics. In 1960, a decade before deinstitutionalization began in earnest, thirty-six states had provisions specifically for the involuntary hospitalization of "alcoholics," "habitual drunkards," and "inebriates." In addition, many states had voluntary-admission statutes. By the mid-1970s, however, these laws were history. Prepared or not, local communities had to provide.

The alcoholism-treatment field was not static during the 1980s. The federal "block grant system," stringent drunk-driving laws, and the rise of EMPLOYEE ASSISTANCE PROGRAMS (EAPs) and insurance coverage for treatment, all important developments, will be discussed following a description of the modern era of drug treatment.

DRUG TREATMENT

Even by the late 1950s, the tough law, anti-maintenance consensus of an earlier era of drug control and treatment was breaking down. A joint report of the American Bar Association and the American Medical Association in 1958, finally published in 1961, cautiously favored outpatient treatment and limited opioid maintenance as alternatives to "threats of jail or prison sentences." In 1962, appealing to disease theory, the U.S. Supreme Court struck down a California statute that made drug addiction *per se* a crime. Medical treatment, not the "cruel and unusual punishment" of incarceration, was the Court's *desideratum.* In 1963, the President's Advisory Commission on Narcotic Drug Abuse made substantially similar recommendations.

It was the experimental success of METHADONE MAINTENANCE that finally altered the discussion of opioid maintenance. Methadone, a synthesized drug with opioid properties, was invented by German pharmacologists during World War II and had been used at the U.S. PUBLIC HEALTH SERVICE HOSPITAL at Lexington to block addicts' withdrawal symptoms. In 1963 and 1964, with the support of the prestigious Rockefeller University, medical researchers Vincent

Dole and Marie Nyswander began to study its wider use in the treatment of heroin addiction. Their research proceeded despite opposition by the federal Bureau of Narcotics, and was first published in 1965. The remarkable changes they observed in their patients soon were replicated by other scholars. Methadone maintenance attracted considerable notoriety and generated new enthusiasm for maintenance as a strategy of treatment.

Methadone maintenance did not become widespread overnight, however, and it has never been without controversy. The most fundamental criticism of maintenance has always been that it presumes "incurability," encourages users to continue to rely on a narcotic medication, and thereby undermines abstinence-based approaches. During the 1960s, and especially during the 1970s, when methadone maintenance programs expanded dramatically, this criticism came mainly from two sources: (1) abstinence-based programs run by recovering addicts more or less in the mutual-aid tradition and (2) minority poverty activists who saw in methadone a palliative strategy to treat what they saw as a symptom of economic deprivation without addressing its causes.

Opposition from those working in the mutual-aid tradition came chiefly from veterans of THERAPEUTIC COMMUNITIES inspired by SYNANON (established in Southern California in 1958) and DAYTOP VILLAGE (opened in New York City in 1964). While most therapeutic communities saw addiction primarily as a result of characterological deficits and immaturity, some drew financial support from the Office of Economic Opportunity (OEO), the short-lived, principal arm of the War on Poverty, and relied on an analysis of heroin addiction that located its social sources in adaptations to poverty. This was an important theme of much scholarship on addiction during and after the late 1950s. In this analysis, still vital today, no form of treatment is effective without job and community development to support aftercare and prevent relapse. Descending from the mental-hygiene tradition, this view provided a rationale for great skepticism about any narrow medical approach that was proclaimed as a "solution" rather than as a first step. There was (and remains) no inherent contradiction between maintenance and antipoverty strategies, and many workers in antipoverty programs embraced methadone as a viable and useful treatment. But many did not, and the result was an uneasy pluralism in drug-treatment approaches. In

1966, when New York City launched a major expansion of treatment for drug addiction, it chose to make drug-free therapeutic communities the centerpieces of its effort.

The middle to late 1960s were marked by a modest expansion of publicly supported programs for drug addiction, characterized by competition among a variety of distinct and sometimes incompatible treatment philosophies: therapeutic communities; methadone maintenance programs; compulsory treatment with prolonged residential components; twelve-step programs; overtly religious programs; and a number of traditional mental-health approaches offering detoxification followed by supportive psychotherapies.

Despite the variety of approaches, accessibility to voluntary treatment remained limited throughout the 1960s. In 1968, NIMH undertook a survey to identify every private or public program focused on the treatment of drug addiction in the United States; it located only 183. Most of these were in New York, California, Illinois, Massachusetts, Connecticut, and New Jersey. Of these, 77 percent had been open for less than 5 years. Only the federal hospitals at Lexington and Fort Worth had been in operation for 20 years or more.

In addition to establishing the federal civil commitment program, the Narcotic Addict Rehabilitation Act of 1966 authorized NIMH to make grants to establish community-based treatment programs. The first of these were awarded in 1968; they provided federal support for therapeutic communities and methadone maintenance. This expansion of treatment capacity was also notable for its attention to problems associated with a variety of drugs. It came at a time of sharp increase in marijuana use among middle-class youth, an epidemic of amphetamine use, growing experimentation with LSD, and media preoccupation with the counterculture, or the "youth revolt." Thus, the political urge to provide treatment was fueled by two enduring concerns of Americans—unconventional and disorderly behavior by young people and minority group members; and the connection between drug use and crime. Anything that might work was tried.

The administration of President Richard M. Nixon took office in 1969 and made the connection between drugs and crime a priority, concentrating first on law enforcement, federal legislation (the CONTROLLED SUBSTANCES ACT of 1970), and a reorganization of federal enforcement agencies. In 1970, while the administration was beginning to consider the role of treatment in its overall strategy, heroin use among service personnel in Vietnam captured media attention. In response, on June 17, 1971, Nixon declared a War on Drugs and created, by executive order, the SPECIAL ACTION OFFICE FOR DRUG ABUSE PREVENTION (SAODAP) within the executive office of the president. He appointed as director Dr. Jerome H. Jaffe, a psychiatrist and pharmacologist from the University of Chicago and the director of the Illinois Drug Abuse Programs. SAODAP was the first in a two-decade series of differently named White House special offices concerned with the drug problem; Jaffe was the first in a series of so-called Drug Czars (though the title might most appropriately fit Harry ANSLINGER, autocratic boss of the Bureau of Narcotics for over 30 years.)

The creation of SAODAP marked the federal government's first commitment to make treatment widely available. Indeed, SAODAP's goal was to make treatment so available that addicts could not say they committed crimes to get drugs because they could not get treatment. Over the next several years, a variety of community-based programs were initiated and/or expanded. The major modalities were drug-free outpatient programs, methadone maintenance, and therapeutic communities. SAODAP deliberately de-emphasized hospital-based programs, allowing the civil commitment program under NARA to wither away. Even so, the need to expand treatment for the Veterans Administration (VA) resulted in funding VA hospitals to use their beds for both detoxification and rehabilitation. SAODAP fully supported methadone maintenance, regarded as experimental by NIMH and federal law-enforcement agencies, and became a focal point of controversy as it presided over the dramatic growth of methadone programs beginning in the early 1970s. Treatment within the military also was legitimized as an alternative to court martial.

SAODAP was given a legislative basis in 1972. The same legislation, the Drug Office and Treatment Act, also created a formula grant program for drug treatment comparable in intent to that for alcoholism treatment. The legislation required the production of a written National Strategy, and authorized establishment of the NATIONAL INSTITUTE ON DRUG ABUSE (NIDA), analogous to NIAAA. Like NIAAA, NIDA was lodged within NIMH.

During its first two years, SAODAP directed an unprecedented expansion of treatment. In early 1971 there were 36 federally funded treatment programs in the U.S. By January 1972 there were 235, and by January 1973, almost 400. For a brief, three-year period, the federal resources allocated to treatment, prevention, and research exceeded those allocated to law enforcement, actually comprising two-thirds of the drug resources in the 1973 federal budget.

In 1973, Dr. Robert Dupont, also a psychiatrist, succeeded Jaffe at SAODAP. Dupont had established and directed a treatment program in Washington, D.C., and had extensive experience with methadone treatment. He extended the work of SAODAP and then provided for continuity of policy when he became the first director of NIDA.

During the administration of President Gerald R. Ford (August 1974–January 1977), the sense of urgency about drug problems declined. This was not due to indifference; it reflected a belief that the metaphor of war was not appropriate to a problem that might be controlled but was unlikely ever to be eliminated. The recent lesson of Vietnam—that wars must be quickly won to be popular—was not lost on Ford's advisors. Thus, Ford did not appoint a Drug Czar, leaving coordination of drug activities to a unit within the Office of Management and Budget. There were no sharp changes in policy, but the treatment budget was substantially reduced from the highwater mark of the Nixon era.

The administration of President Jimmy Carter heightened the expectations of those interested in expanded and improved treatment. One of Carter's close advisors, Dr. Peter Bourne, was a psychiatrist who had established treatment programs in Georgia and who had worked briefly in SAODAP during the Nixon administration. Bourne enjoyed more White House influence than any previous presidential advisor on drug issues. However, Bourne resigned in July 1978, and in the wake of his resignation, drug issues resumed their low profile. Resources for treatment from 1978 to 1980 were stagnant despite an unprecedented inflation rate.

Measured in 1976 dollars, the level of federal support for treatment was cut almost in half between 1976 and 1982. The Ford, Carter, and Reagan administrations all presided over this decline. At the same time, as the result of the impact of inflation on the cost of state and local government, these juris-

dictions also curtailed their support, thus aggravating the impact of federal reductions.

However, the Reagan administration was ideologically different from its predecessors—it was characterized by considerable skepticism about federal activism in general and about the efficacy of drug treatment in particular. While it increased resources for law enforcement and supply control, and it introduced a stringent policy of ZERO TOLERANCE that filled American prisons and newly popular (though hardly innovative) therapeutic boot camps with drug offenders, the Reagan administration downplayed treatment in favor of prevention—especially First Lady Nancy Reagan's "Just Say No" campaign and the president's public advocacy of widespread drug testing of employees in industry and government. The 1980 reorganization of the federal block grant program that supported both alcohol and drug treatment combined these funds into an Alcohol, Drug Abuse and Mental Health Services (ADMS) block grant and turned these funds over to the states. In the process, overall funding was reduced from 625 million to 428 million dollars and federal oversight was virtually abandoned. After 1984, federal regulation required that a certain percentage of these funds be spent on prevention rather than treatment. The Institute of Medicine estimated that the proportion of the ADMS block funds available to support drug treatment fell from 256 million dollars in 1980 to 93 million dollars in 1986—and this estimate did not account for inflation.

In spite of the Reagan administration's lack of interest in drug treatment, congressional interest was rekindled. It was apparent by 1984 that HIV was being transmitted among drug injectors and by drug injectors to others, especially their female partners and their fetal young. Crack, an extremely potent and inexpensive form of smokable cocaine, was being aggressively marketed in areas of concentrated poverty, although it took the deaths of several prominent athletes, particularly Len Bias, a first-round draft choice of the Boston Celtics, to pique concern with the growing use of cocaine. Prodded by Congress, the second Reagan administration, in its closing years, did increase funding for both research and treatment. However, according to the Institute of Medicine, these increases did not compensate for the effects of previous budget cuts and inflationary erosion. Adjusted for inflation, public funding for drug treatment in 1989 (the last Reagan budget) was substantially below the level of 1972–1974, the opening years of Nixon's War on Drugs.

Even so, the Reagan administration retained its emphasis on law enforcement and prevention. To better focus on prevention, in 1987 it created the Office for Substance Abuse Prevention (OSAP), placing it within the Alcohol, Drug Abuse, and Mental Health Administration (ADAMHA). Most prevention activities carried out by the National Institute on Drug Abuse (NIDA) were transferred to OSAP, the first director of which was Dr. Elaine Johnson.

In 1989, President George H. Bush reinvigorated the position of drug czar when he appointed Dr. William Bennett, former secretary of education in the Reagan administration, to head his new White House drug policy office, the Office of National Drug Control Policy (ONDCP). ONDCP was charged with coordinating demand-side (prevention and treatment) and supply-side (law enforcement) matters relating to drugs. There were increases in resources for treatment—and even more substantial increases in law-enforcement efforts. Although Bennett had recruited a noted drug-abuse scholar, Dr. Herbert Kleber, as his deputy for demand-side activity, the ONDCP chief and his staff remained skeptical about the value of treatment, continuing the decade-long policy of emphasizing prevention and law enforcement.

Later in 1989, much of the authority and funding for drug treatment was transferred from NIDA to another new agency created within ADAMHA, the Office for Treatment Improvement (OTI). Dr. Beny J. Primm, a major figure in drug treatment, was recruited to organize OTI and to be its first director. OTI was given responsibility for oversight of the block (formula) grant for drug and alcohol treatment and prevention and was given new authority and budget resources to make grants for treatment-demonstration projects.

In 1992, Congress decided that the placement of OTI and OSAP within ADAMHA, which also housed NIDA, NIAAA, and NIMH, was leading to conflicts between the missions of research and those of treatment and prevention. In still another reorganization, the three research institutes—NIDA, NIAAA, and NIMH—were transferred to the National Institutes of Health (NIH), and the remaining service functions were incorporated into a new agency, the SUBSTANCE ABUSE AND MENTAL HEALTH SERVICES ADMINISTRATION (SAMHSA). SAMHSA was composed of three centers: the Center for Substance Abuse Prevention (CSAP), consisting primarily of the former OSAP; the Center for Substance Abuse Treatment (CSAT), consisting primarily of the former OTI; and the Center for Mental Health Services (CMHS), consisting of the service-demonstration grant projects that were formerly within NIMH.

Succeeding President Bush in 1992, President Bill Clinton appointed Dr. Lee Brown as his Drug Czar. Brown, a criminologist by academic training, had been a police chief in New York and Texas. Although there were some signs within the administration that drug treatment was understood to be an important part of attacking persistent joblessness and welfare dependency, the early Clinton budgets made only slight shifts in resource allocation. Further, as Clinton's health-care reform, welfare reform, and crime and employment strategies became hostage to management of the national budget deficit and partisan politics, no major initiatives specifically on drug treatment were introduced during the first two years he was in office. Some provisions for more treatment within the criminal-justice system were part of the original crime bill. As a result of the recession of the early 1990s, and faced with the necessity of accommodating in their jails and prisons huge numbers of drug offenders incarcerated on mandatory sentences, states and counties have also failed to restore the support an earlier era provided for treatment. In some cases they have retrenched considerably.

A TWO-TIERED SYSTEM

Beginning in the 1970s, and promoted by NIAAA, NIDA, and a few insurance industry leaders like The Travelers, health insurance policies began to provide coverage for the treatment of alcohol and drug dependence. Sometimes this was the result of labor negotiations; sometimes it was the result of state insurance commission mandates for its inclusion. In response to the availability of support, private hospitals (both nonprofit and for-profit) expanded their treatment capacities dramatically. There had been no such growth in the private-treatment sector since the boom of the inebriate asylum era.

Commonly, treatment programs within the private sector were based on the Minnesota model, emphasizing twelve-step principles and employing recovering people. Such programs typically consisted of a brief period of inpatient detoxification followed by several weeks of inpatient rehabilitation. Twenty-eight days was such a common duration of inpatient care that the programs often were referred to as 28-day programs. The post-hospital phase of treatment usually consisted of participation in AA, Narcotics Anonymous, or Cocaine Anonymous.

Such programs—often called chemical-dependency programs because they admitted people with drug *and* alcohol problems—catered almost exclusively to those with health insurance. (In many instances, they represented important profit centers for medical institutions needing to subsidize financial losses from other services, like emergency rooms.) Those without insurance either had no access to treatment or made use of the network of publicly supported programs—a network that became increasingly thin during the 1980s and increasingly under pressure to find sources of funds other than public grants and contracts and payments from medical programs for the indigent (such as Medicaid). Sliding fee scales became more commonly used, and in some places scarce public treatment slots were absorbed by fee-paying drinking drivers mandated to treatment by stricter penalties for drunk driving and more systematic enforcement of such laws.

The growth of the private sector was spurred as well by EMPLOYEE ASSISTANCE PROGRAMS (EAPs), efforts to intervene in alcohol and/or drug problems at places of employment. This strategy goes back at least to the Washingtonian movement, but formal EAPs date from the 1940s. Their ranks swelled during the 1970s and 1980s. Generally, EAPs referred people with more serious alcohol and drug problems to formal—usually private—treatment programs, which were paid primarily by fees derived from third-party payers, such as insurance companies, who in turn derived their funds from policies paid for or subsidized by employers. The sharply rising cost to employers of providing alcohol and drug treatment was a major factor in the rise of managed care, which was aimed initially at controlling the cost of mental health and alcohol and drug treatment. The major mechanism by which the managed-care industry addressed the cost of treatment was to challenge the practice of using several weeks of inpatient care as the initial phase of treatment for alcohol and drug dependence. In practice, treatment providers were told that inpatient treatment beyond a few days could not be justified and would not be paid for under the insurance policy.

The success of managed care in reducing costs by constraining the use of inpatient treatment resulted in a dramatic growth of managed-care organizations and an equally significant contraction and restructuring of the private alcohol and drug treatment system. By the early 1990s, a number of states had obtained federal permission to use managed-care approaches to contain the costs of treatment for individuals covered by federal programs like Medicaid. The future of funding for treatment, the various public grant and contract programs notwithstanding, is inseparable from the broader national debate on the financing of health care.

In 1990, the Institute of Medicine described U.S. treatment arrangements as a two-tiered system, comprised of public and private sectors, in which the private sector served 40 percent of the patients but garnered 60 percent of total treatment expenditures. While the ratio of patients to revenues cannot be known for earlier eras, this two-tiered structure is a creature of the nineteenth century, when treatment was established both as a public good and a commodity. Barring some revolution in the organization of U.S. health care, this is unlikely to change soon. What remains to be seen is what the balance of public and private treatment will be, what innovations or reinventions will be born of financial necessity, or as the result of homeless addicts and a groaning correctional system. History allows us to predict the likely questions, but it is not a very reliable guide to specific answers.

(SEE ALSO: *Disease Concept of Alcoholism and Drug Abuse; Temperance Movement; Treatment Types; U.S. Government: Drug Policy Offices in the Executive Office of the President; U.S. Government: The Organization of U.S. Drug Policy*)

BIBLIOGRAPHY

BAUMOHL, J. (1993). Inebriate institutions in North America, 1840–1920. In C. Warsh (Ed.), *Drink in Canada: Historical essays.* Montreal: McGill-Queens University Press.

BAUMOHL, J. (1986). On asylums, homes, and moral treatment: The case of the San Francisco Home for the Care of the Inebriate, 1859–1870. *Contemporary Drug Problems, 13,* 395–445.

BAUMOHL, J., & ROOM, R. (1987). Inebriety, doctors, and the state: Alcohol treatment institutions before 1940.

In M. Galanter (Ed.), *Recent developments in alcoholism*, vol. 5. New York: Plenum.

BAUMOHL, J., & TRACY, S. (1994). Building systems to manage inebriates: The divergent paths of California and Massachusetts, 1891–1920. *Contemporary Drug Problems, 21,*

BESTEMAN, K. J. (1991). Federal leadership in building the national drug treatment system. In D. R. Gerstein & H. J. Harwood (Eds.), *Treating drug problems*, vol. 2. Committee for the Substance Abuse Coverage Study, Division of Health Care Services, Institute of Medicine. Washington, DC: National Academy Press.

COURTWRIGHT, D. T. (1991). A century of American narcotic policy. In D. R. Gerstein & H. J. Harwood (Eds.), *Treating drug problems*, vol. 2. Committee for the Substance Abuse Coverage Study, Division of Health Care Services, Institute of Medicine. Washington, DC: National Academy Press.

COURTWRIGHT, D. T. (1982). *Dark paradise: Opiate addiction in America before 1940.* Cambridge: Harvard University Press.

COURTWRIGHT, D., JOSEPH, H., & DES JARLAIS, C. (1989). *Addicts who survived: An oral history of narcotic use in America, 1923–1965.* Knoxville: University of Tennessee Press.

GERSTEIN, D. R., & HARWOOD, H. J. (EDS.). (1990). *Treating drug problems.* Committee for the Substance Abuse Coverage Study, Division of Health Care Services, Institute of Medicine. Washington, DC: National Academy Press.

INSTITUTE OF MEDICINE. (1990). *Broadening the base of treatment for alcohol problems.* Report of a Study by a Committee of the Institute of Medicine, Division of Mental Health and Behavioral Medicine. Washington, DC: National Academy Press.

JAFFE, J. H. (1979). The swinging pendulum: The treatment of drug users in America. In R. I. Dupont, A. Goldstein, & J. O'Donnell (Eds.), *Handbook of drug abuse.* Washington, DC: U.S. Government Printing Office.

MUSTO, D. F. (1986). *The American disease: Origins of narcotic control.* New York: Oxford University Press.

RUBINGTON, E. (1991). The chronic drunkenness offender: Before and after decriminalization. In D. J. Pittman & H. R. White (Eds.), *Society, culture, and drinking patterns reexamined.* New Brunswick, NJ: Rutgers Center for Alcohol Studies.

TICE, P. (1992). *Altered states: Alcohol and other drugs in America.* Rochester, NY: The Strong Museum.

JIM BAUMOHL
JEROME H. JAFFE

TREATMENT IN THE FEDERAL PRISON SYSTEM The federal prison system of the United States has made repeated efforts to treat drug-abusing prisoners. The issue was first raised in 1928 by the chairman of the Judiciary Committee of the U.S. House of Representatives. He reported that the three then-existing federal penitentiaries—Atlanta, Leavenworth, and McNeil Island—held 7,598 prisoners, 1,559 of whom were "drug addicts." To deal with these prisoners he called for a "broad and constructive program in combatting the drug evil," and he recommended the establishment of special federal "narcotics farms" for the "individualized treatment" of drug-abusing prisoners. He hoped that there would become institutions that "will reduce and also prevent crime . . . and greatly alleviate the suffering of those who have become addicted."

In 1930, the U.S. Bureau of Prisons (BOP) was established to handle the burgeoning population of federal prisoners, caused mainly by the enforcement of PROHIBITION. The BOP's first directorate was eager to launch special programs for drug-abusing prisoners, but many in Congress and elsewhere believed that prisons should have little or no direct role in treating drug-abusing offenders. A compromise was struck. The U.S. Public Health Service (USPHS) was authorized to establish and administer two hospitals that would offer state-of-the-art drug-abuse treatment, and the BOP was permitted to freely assign addict prisoners to the facilities. The first USPHS HOSPITAL opened in 1935 at Lexington, Kentucky; the second was opened in 1938 at Fort Worth, Texas.

REHABILITATION EFFORTS

In the 1960s, a broad consensus emerged that prisons should do whatever possible to rehabilitate drug-abusing inmates. In 1966, Congress passed the NARCOTIC ADDICT REHABILITATION ACT (NARA), which, among other initiatives, ordered in-prison and aftercare treatment for narcotic addicts who had been convicted of violating federal laws. Between 1968 and 1970, the BOP established NARA-mandated drug-treatment units within five of its prisons. In the 1970s, the BOP assumed direct control over both USPHS hospitals and began to develop an extensive network of programs for the treatment of drug-abusing prisoners throughout the system. In

1979, the BOP required the development of NARA-standard drug-treatment programs in all its prisons, publishing it *Drug Abuse Incare Manual*. In 1985, the BOP established a task force to evaluate the state of drug-abuse treatment programs within federal prisons. The review found that administrative problems had hampered the BOP's drug-treatment efforts. In response, in 1986, the position of chemical-abuse coordinator was established within each prison, and in 1988, the position of national drug-abuse coordinator was created to oversee drug-abuse treatment efforts throughout the federal prison system.

At the end of 1990, the BOP held some 59,000 prisoners. About 54 percent of federal prisoners were serving sentences for drug-related crimes. At the time of their admission, 47 percent of federal prisoners were classified as having moderate to serious drug-abuse problems. Under the BOP's classification scheme, a moderate problem designation indicates that the inmate's use of drugs or alcohol had negatively affected at least one "major life area"—school, health, family, financial, or legal status—in the two-year period prior to arrest.

In 1991, the BOP's drug-education program was required for all inmates with any history of drug abuse or drug-related crime. By the end of 1992, an estimated 12,000 to 15,000 federal inmates completed drug-education programs. Counseling services—ALCOHOLICS ANONYMOUS (AA), NARCOTICS ANONYMOUS (NA), group therapy, stress management, prerelease planning—were available on an ongoing basis at most federal prisons, and the BOP planned to make them available to inmate volunteers at all institutions at any time during their incarceration.

Transitional drug-abuse treatment services were being developed throughout the BOP. The administration of these services were divided into two six-month components, each of which included individual and family counseling, assistance in identifying and obtaining employment, and random urine testing. The first component was provided in the BOP's community corrections centers; the second component was provided as post-release aftercare, in conjunction with the Probation Division of the Administrative Office of the U.S. Courts.

To assess the effectiveness of its current multidimensional drug-abuse treatment efforts, the BOP has begun a major evaluation of these programs that will analyze data on both in-prison adjustment and postrelease behavior for up to five years after release.

(SEE ALSO: *Coerced Treatment for Substance Offenders; Prisons and Jails, Drug Treatment in*)

BIBLIOGRAPHY

DiIULIO, J. J., JR. (1992). *Barbed-wire bueaucracy: Leadership, administration, and culture in the Federal Bureau of Prisons.* New York: Oxford University Press.

KEVE, P. W. (1990). *Prisons and the American conscience: A history of U.S. federal corrections.* Carbondale, IL: Southern Illinois University Press.

U.S. DEPARTMENT OF JUSTICE, FEDERAL BUREAU OF PRISONS. (1991). *State of the bureau 1990: Effectively managing crowded institutions.* Washington, DC: Author.

U.S. DEPARTMENT OF JUSTICE, FEDERAL BUREAU OF PRISONS, OFFICE OF RESEARCH AND DEVELOPMENT. (1990). *Proposal for the evaluation of the Federal Bureau of Prisons drug abuse treatment programs.* Washington, DC: Author.

WALLACE, S., ET AL. (1991). Drug treatment. *Federal Prisons Journal, 2* (3), 32–40.

JOHN J. DiIULIO, JR.

TREATMENT OUTCOME PROSPECTIVE STUDY (TOPS) This is a prospective clinical, epidemiological study of clients who entered drug-abuse treatment programs from 1979 to 1981. During the course of TOPS, 11,182 clients were interviewed at admission to drug-abuse treatment by program researchers hired to work in assigned clinics and professionally trained and supervised by Research Triangle Institute (RTI) field staff. The interviews at admission covered demographics, history of drug use, treatment, arrest and employment behavior in the year prior to treatment, and status upon admission to treatment. The study was sponsored by the NATIONAL INSTITUTE ON DRUG ABUSE (NIDA) and by the RTI. The study population included 4,184 clients from 12 outpatient methadone programs, 2,891 clients from 14 residential programs, and 2,914 clients from 11 outpatient drug-free programs in 10 cities. Interviews with questions on behavior, services received, and satisfaction were collected by the program researchers every three months while clients remained in treatment. The self-report data were supplemented with data abstracted from the clinical and medical records of all clients selected for the follow-up, and questionnaires describing the treatment philosophy, structure, practice, and process were completed by counselors and program directors.

The follow-up data included interviews 1 and 2 years after treatment with 1,130 clients who were admitted in 1979; follow-ups 90 days and 1 year after treatment of 2,300 clients who entered treatment in 1980; and follow-ups 3 to 5 years after treatment of 1,000 clients who entered programs in 1981. Professional field interviewers hired, trained, and supervised by RTI field staff were able to locate and interview between 70 and 80 percent of the clients selected for these interviews.

TOPS has resulted in a substantial body of important knowledge about drug-abuse treatment and treatment effectiveness. The client populations of outpatient METHADONE PROGRAMS, long-term residential programs, and outpatient drug-free programs who participated in TOPS differed on many sociodemographic and background characteristics. The residential clients were significantly more likely to report multiple use of drugs, more drug-related problems, suicidal thoughts and attempts, heavy drinking, predatory crimes, and less full-time employment compared to the methadone clients. Outpatient drug-free clients were more likely than methadone clients to report drug-related problems, suicidal thoughts or attempts, predatory crimes, and heavy drinking, but they were less likely than residential clients to use multiple drugs. These results demonstrated that each type of program served very different, important segments of the drug-abusing population. The high rates of self-referrals to methadone (48%) and criminal-justice referrals to residential and outpatient drug-free treatment (31%) suggest differences in clients' motivations for seeking treatment and, consequently, differences in retention, services received, and outcomes.

The drug-abuse patterns reveal the differential concentration of types of drug abusers across the major categories. Clients on methadone were primarily (52%) traditional heroin users who used only cocaine, marijuana, and alcohol, in addition to heroin. One in five of these clients, however, used heroin and other narcotics, as well as a variety of nonnarcotic drugs. The remaining quarter of clients on methadone were classified as former daily users who had histories of regular use but did not use heroin on a weekly or daily basis in the year before treatment. Residential clients had diverse patterns of use, and the majority of outpatient drug-free clients were users of alcohol and marijuana (36%) or single nonnarcotics users (22%).

Symptoms of depression are very commonly reported by clients entering drug-abuse treatment programs. Overall, about 60 percent of TOPS clients reported at least one of three symptoms of depression at intake: nearly 75 percent of the women under 21 years of age reported one or more symptoms of depression. Other results suggest that the duration of regular drug use and the number of prior treatment episodes are important indicators of the effectiveness of any single treatment episode; clients with lengthy drug-abuse or drug-treatment histories have poorer prognoses.

Clients who have come into treatment by way of the criminal justice system do as well or better than other clients in drug-abuse treatment. Formal or informal mechanisms of the criminal justice system appear to refer individuals who had not previously been treated and many who were not yet heavily involved in drug use. Involvement with the criminal justice system also helps retain clients in treatment up to an estimated six to seven additional weeks. Drug abuse treatment programs vary in the nature and intensity of the treatment services provided, the types of therapists and therapies provided, the average length of stay, and the inclusion or exclusion of aftercare.

The study of the treatment process in TOPS programs focused on many important aspects of the structure, nature, duration, and intensity of drug-abuse treatment. Descriptions of aspects of the treatment process were developed from clients' self-reports of needs for treatment services, services received, and satisfaction, combined with abstractions of clinical and medical records and descriptions of programs by counselors and directors. The outpatient methadone and outpatient drug-free treatment programs had budgets per slot of approximately 2,000 dollars per year. Therapeutic communities had an average expenditure of 6,135 dollars per bed.

The number of available services (medical, psychological, family, legal, educational, vocational, and financial services) varied during the years 1979 to 1981. Fewer services appeared to be available in the later years of the study. The proportion of clients in residential treatment programs who received family, educational, and vocational services decreased noticeably during the three-year period. During this same period, the clients' demands for services increased. Programs in TOPS appeared to focus on the client's primary drug of abuse rather than addressing the client's multiple drug use, drug-related problems, and social and economic functioning. Low-dose methadone (69% of the clients admitted were initially treated with less than 30 mg of oral methadone daily) was the most common pattern of methadone treatment in the programs participating in TOPS.

In TOPS, multiple measures of treatment outcome were necessary to describe changes in the client's ability to function in society after treatment. In general, clients who remained in treatment at least three months had more positive posttreatment outcomes, but the major changes in behavior were seen only in those who remained in treatment for more than twelve months. Analyses of the TOPS data show that the posttreatment rate of daily heroin, cocaine, and psychotherapeutic-agent use among clients who spent at least three months in treatment was half that of the pretreatment rate. The posttreatment rates of weekly or more frequent use for clients who stayed in treatment at least three months were 10 to 15 percent lower than the rates for shorter-term clients. The results showed that time spent in treatment was among the most important predictors of most treatment outcomes. Stays of one year or more in residential or methadone treatment, or continuing maintenance with methadone, produced significant decreases in the odds of a client using heroin in the follow-up period. Clients in TOPS also reported a substantial decrease in depression symptoms during the years after treatment.

Analyses of the effects of treatment on behavior have focused on reductions in predatory crime and the costs associated with crime. The assessment of the benefit/cost ratio indicates that substantial benefits are obtained in reductions of crime-related costs regardless of the measures used within the year after treatment. Reducing transmission of the AIDS virus would increase the benefit portion of benefit/cost ratio even more.

(SEE ALSO: *Drug Abuse Treatment Outcome Study; Treatment Alternatives to Street Crime; Treatment Types*)

ROBERT HUBBARD

TREATMENT/TREATMENT TYPES

The following series of articles provides the reader with brief descriptions of some of the diverse ways that people with substance-related problems can be helped. It is organized into two subsections. *Treatment* consists of summaries of the common ways that problems relating to specific substances are currently treated. Different approaches are described for alcohol, cocaine, heroin, polydrug abuse, and tobacco. *Treatment Types* presents descriptions of distinct interventions that are applicable to dependence on a variety of drugs.

In practice, many treatment programs are hybrids, incorporating features from several distinct treatment modalities and adapting them to specific needs having to do with age, gender, ethnic, racial, and socioeconomic factors, provider preference, and the economic realities that govern delivery of treatment.

Neither of the sections is exhaustive. A variety of substance dependence interventions employed in other countries and by certain ethnic groups in the United States (such as sweat lodges among some Native American tribes) are not covered. Nevertheless, the entries included here should allow the reader to become reasonably familiar with what is considered mainstream treatment in the United States in the 1990s.

TREATMENT   This section contains summaries of the common ways that problems relating to specific substances are currently treated. It is organized first by drug and then by treatment approach. Different approaches are described for *Alcohol, Cocaine, Heroin, Polydrug Abuse,* and *Tobacco*. The reader should also see the entries for each of these topics and the entries for *Barbiturates, Inhalants,* and *Nicotine* under their individual headings, and the section below entitled *Treatment Types.*

This section contains the following articles:

Alcohol, An Overview;
Alcohol, Pharmacotherapy;
Alcohol, Psychological Approaches;
Cocaine, An Overview;
Cocaine, Pharmacotherapy;
Cocaine, Psychological Approaches;
Heroin, Pharmacotherapy;
Heroin, Psychological Approaches;
Polydrug Abuse, An Overview;

Polydrug Abuse, Pharmacotherapy;
Tobacco, An Overview;
Tobacco, Pharmacotherapy;
Tobacco, Psychological Approaches.

Alcohol, An Overview Alcohol abuse and ALCOHOLISM are serious problems. *Alcohol abuse* refers to heavy, problematic drinking by nondependent persons, while *alcoholism* suggests TOLERANCE, PHYSICAL DEPENDENCE, and impaired control of drinking. There are an estimated 9 million alcohol-dependent persons and 6 million alcohol abusers in the United States (Williams et al., 1989).

Problems that arise from misuse of alcohol vary widely, but they often include the following areas: financial, legal, family, employment, social, and medical. Medical complications include alcoholic liver disease, gastritis, pancreatitis, organic brain syndrome, and the FETAL ALCOHOL SYNDROME (FAS). It is estimated that more than 100,000 alcohol-related deaths occurred in the United States in 1987 (Centers for Disease Control, 1990). The most common alcohol-related death is a motor vehicle fatality.

Despite the complex nature of alcohol abuse and dependence, research has burgeoned over the past decade and has deepened our understanding of the causes, prevention, and remediation of alcohol abuse and alcoholism. Here, we briefly review assessment of alcohol problems, detoxification, and treatment.

ALCOHOLISM ASSESSMENT

To appropriately assign an individual to treatment, his or her condition must be accurately evaluated. Management of alcoholism may be seen as involving a five-stage sequential process: screening, diagnosis, triage, treatment planning, and treatment-outcome monitoring. Specific procedures exist to help inform clinical decisions at each of these stages (Allen, 1991). Screening tests help determine whether a drinking problem might exist. If this seems likely, formal and more lengthy diagnostic procedures are performed to specify the nature of the problem. If the diagnosis of alcoholism is established, determination of the type of treatment setting and intensity of care needed for detoxifying and treating the patient must be made next. Treatment planning can then be initiated to establish rehabili-

tation goals and strategies appropriate to the patient. Finally, outcome is monitored to determine if further treatment is needed or if a different treatment approach is advisable.

DETOXIFICATION

When an alcohol-dependent person abruptly stops drinking, physiological symptoms may occur. This cluster of symptoms is termed *alcohol withdrawal*, and symptoms can range from relatively mild discomfort to life-threatening problems. Mild symptoms include sweating, tachycardia (rapid heartbeat), hypertension, tremors, anorexia, sleeplessness, agitation, and anxiety. More serious consequences involve seizures and, rarely, DELIRIUM TREMENS (DTs), characterized by agitation, hyperactivity of the autonomic nervous system, disorientation, confusion, and auditory or visual hallucinations. It has been postulated that as the number of untreated withdrawal episodes increases, the potential for more serious symptoms in subsequent withdrawals may also escalate. This phenomenon is known as *kindling* (Brown, Anton, Malcolm & Ballenger, 1988).

Treatment of alcohol withdrawal includes both pharmacological and nonpharmacological interventions. It is generally believed that if the withdrawal symptoms are mild to moderate, no medications are needed. Instruments such as the Clinical Institute Withdrawal Assessment Scale (Foy, March & Drinkwater, 1988) have recently been developed to gauge severity of withdrawal symptoms. Nonpharmacological techniques used to treat milder forms of alcohol withdrawal include efforts to reduce anxiety and to provide emotional reassurance. Patients in withdrawal should receive the B vitamin thiamine so as to prevent the occurrence of the WERNICKE–Korsakoff syndrome, a serious neurological complication of alcoholism.

If the symptoms are more severe, however, drugs should be prescribed. The most commonly used medications to treat withdrawal have been BENZODIAZEPINES. The benzodiazepines have been demonstrated in randomized clinical trials to reduce the occurrence of seizures and other serious withdrawal symptoms. They have a wide margin of safety. Side effects, however, include transient memory impairment, drowsiness, lethargy, and motor impairment. Benzodiazepines must be tapered down and then stopped after the patient is no longer suffering from withdrawal because patients can develop dependence on them. In addition, the physiological effects of benzodiazepines are synergistic or additive with those of alcohol—hence, it is important that patients not drink while taking them. Other medications to treat withdrawal include beta-adrenergic blockers, alpha-2 adrenergic agonists, calcium channel blockers, and anticonvulsant agents such as carbamazepine; however, the first two categories of drugs do not prevent seizures and, therefore, are less useful than benzodiazepines. Recent research suggests that carbamazepine may be an effective alternative to benzodiazepines, while calcium channel blockers are still in early stages of research.

TREATMENT OPTIONS

After screening, diagnosing, and detoxifying a patient, the clinical staff has numerous options for short- and long-term treatment. While a more detailed review of these interventions can be found in Hester and Miller (1989), the techniques can be categorized as follows:

Alcoholics Anonymous. Since the 1940s, ALCOHOLICS ANONYMOUS (AA) has been an important component of alcoholism rehabilitation, and many recovered alcoholics are convinced that AA was essential for their recovery. As a means of achieving and maintaining SOBRIETY, AA consists of regular meetings utilizing fellowship, mutual support for sobriety, open discussions, and a program known as the TWELVE STEPS. The effectiveness of AA has not been established by randomized clinical trials, largely because the organization was developed outside the scientific mainstream. A well-designed study by Walsh, Hingson, and their colleagues (1991) was, however, done in the setting of an EMPLOYEE ASSISTANCE PROGRAM (EAP). Employees seeking or referred for treatment were randomly assigned to inpatient treatment with AA as a component, AA alone, or self-choice of treatment. All three treatment conditions resulted in equal improvement in job performance; however, inpatient treatment did better than AA or self-choice in terms of several aspects of drinking behavior. Inpatient treatment was particularly valuable for those employees who were abusing both alcohol and COCAINE. Other self-help groups that do not use the twelve-step program (e.g., RATIONAL RECOVERY) also exist.

Minnesota Model. The MINNESOTA MODEL is so named because it originated in several alcoholism programs in Minnesota and is the most common type of inpatient treatment for alcoholism in the United States. It stresses complete abstinence and employs methods such as group and individual therapy, alcohol education, family counseling, and required attendance at AA meetings. The staff in these programs are usually a mixture of professional individuals and recovering alcoholics. The evidence for its effectiveness is limited. The study by Walsh et al. (1991) supports the idea that these programs are effective. Studies on health-care utilization costs before and after treatment for alcoholism also add evidence that these programs are effective. When this general program is used to treat drug problems other than alcoholism, it is often referred to as a chemical-dependency program.

Group Psychotherapy. Group psychotherapy is widely used in the treatment of alcoholics. The many types of group psychotherapy employ supportive, cognitive, psychoanalytic, or confrontational techniques. Also, group psychotherapy is often used in conjunction with other approaches, such as AA and pharmacologic adjuncts to treatments.

Individual Psychotherapy. Individual psychotherapy attempts to probe possible underlying reasons for problem drinking and subsequently strives to guide the patient in working through emotional difficulties. Some of the cognitive and behavioral approaches described below can also be considered forms of psychotherapy. Similar to group psychotherapy, individual psychotherapy is often combined with other treatment activities. Despite the widespread use of group and individual psychotherapy, the scientific evidence supporting their efficacy as isolated treatments is limited.

Family and Marital Therapy. This type of therapy involves the problem drinker, spouse, and sometimes other family members. Over the past several years, research interest has heightened in determining the contribution of family and marital factors in aiding the patient to sustain recovery. Generally, family and marital therapy seeks to enhance communication, problem-solving, and positive reinforcement skills.

Social-Skills Training. Social-skills training includes techniques for improving communication skills, forming and maintaining interpersonal relationships, resisting peer pressure for drinking, and becoming more assertive. Research on its effectiveness has been encouraging.

Relapse Prevention. RELAPSE PREVENTION is a behavioral approach that deals with teaching the patient to successfully cope with environmental situations that may serve as high-risk drinking stimuli. Relapse prevention is important in alcoholism treatment, since many patients who are successfully detoxified and stabilized tend to revert to drinking. While relapse prevention is widely used, the evidence of its effectiveness is again limited, albeit promising.

Stress Management. Stress-management techniques may be employed to reduce emotional discomfort, which may contribute to drinking behavior. Specific techniques include deep-muscle relaxation, biofeedback, systematic desensitization, and cognitive and behavioral strategies to cope with stress-inducing stimuli.

Pharmacotherapy. Since the 1950s, DISULFIRAM (Antabuse) has been the most widely used medication in the treatment of alcoholism. Patients on disulfiram are deterred from drinking because to do so would cause physical discomfort, including headaches, flushing, and rapid heartbeat. A major problem in using disulfiram is lack of patient compliance. Several techniques have been developed to enhance compliance, including establishing a contract with the client or significant other on disulfiram administration, offering positive and negative incentives for taking the medication, and using implants.

In addition to disulfiram, recent advances have been made in the development of medications that directly curb desire to drink. The most promising include serotonergic agents and opioid antagonists—these agents act on brain mechanisms that are believed to be related directly to drinking.

Aversive Therapy. This type of therapy attempts to establish a conditioned avoidance response to alcohol. Drinking is paired with unpleasant experiences, such as electric shock, nausea, vomiting, or imagined unpleasant consequences. The underlying rationale of AVERSION conditioning is that patients will be less likely to drink if they associated alcohol consumption with immediate negative consequences. Good evidence that this approach is effective is lacking, because of the absence of randomized clinical trials evaluating aversive therapy. Some programs using it report very high levels of abstinence, however, in the months following in-hospital treatment.

Patient-Treatment Matching. A newer strategy in alcoholism treatment attempts to match particular types of treatments to relevant patient characteristics, rather than assigning all patients to similar treatments. Common patient-matching variables include the patient's collateral psychopathology, degree of alcohol involvement, and personality and motivational characteristics. Approximately forty studies, although based on small numbers of patients, have supported the concept that patient-treatment matching improves treatment outcome.

Community-Reinforcement Approach. The community-reinforcement approach (CRA) is a broad-spectrum treatment approach that focuses on positive reinforcers for abstinence in the patient's natural environment. Specific techniques include adding improvements to the patient's employment conditions, marital relationships, problem-solving skills, social skills, and stress management—and different components of the program are chosen for the individual, depending on his or her life problems. The initial studies of CRA are encouraging.

CONCLUSIONS

Advances in treatment research have led to a variety of treatment interventions. The alcoholism-treatment community must become better able to assist the recovery of alcoholics and alcohol abusers. Advances in assessment technology have helped identify patient needs more clearly; this subsequently enables the clinician to provide a treatment regime tailored to the needs of the patient. An important future direction for alcoholism-treatment research is to discover how to more precisely match patients with specific types of treatment interventions. Also, development of new medications to directly reduce drinking behavior will have a major impact. Future treatments will likely combine pharmacologic interventions with behavioral and psychosocial therapies to further improve treatment outcome.

(SEE ALSO: *Accidents and Injuries from Alcohol; Complications; Treatment, History of; Treatment Types*)

BIBLIOGRAPHY

ALLEN, J. P. (1991). The interrelationship of alcoholism assessment and treatment. *Alcohol Health and Research World, 15,* 178–185.

BROWN, M. E., ET AL. (1988). Alcohol detoxification and withdrawal seizures: Clinical support for a kindling hypothesis. *Biological Psychiatry, 23,* 507–514.

CENTERS FOR DISEASE CONTROL. (1990). Alcohol-related mortality and years of potential life lost—United States, 1987. *Morbidity and Mortality Weekly Report, 39*(11), 173–175.

FOY, A., MARCH, S., & DRINKWATER, V. (1988). Use of an objective clinical scale in the assessment and management of alcohol withdrawal in a large general hospital. *Alcohol: Clinical and Experimental Research, 12*(3), 360–364.

HESTER, R. K., & MILLER, W. R. (EDS.). (1989). *Handbook of alcoholism treatment approaches: Effective alternatives.* New York: Pergamon.

WALSH, D. C., ET AL. (1991). A randomized trial of treatment options for alcohol-abusing workers. *New England Journal of Medicine, 325,* 775–782.

WILLIAMS, G. D., ET AL. (1989). Epidemiologic Bulletin no. 23: Population projections using DSM-III criteria. *Alcohol Health and Research World, 13*(4), 366–370.

RICHARD K. FULLER
JOHN P. ALLEN
RAYE Z. LITTEN

Alcohol, Pharmacotherapy Research on pharmacotherapy for ALCOHOLISM is rapidly expanding. Currently, the most widely used medication for the treatment of alcoholism is DISULFIRAM. Disulfiram (Antabuse) does not act to reduce the CRAVING for ALCOHOL or ameliorate the euphorigenic (feeling of well-being) effect of alcohol. However, recent advances in understanding the biological bases for drinking behavior have opened up exciting new possibilities for the long-term management of alcohol (ethanol) abuse and dependence. A variety of new drugs are being developed and may soon become available to the treatment community. In addition, further research on existing medications is leading to more effective approaches for alcoholism treatment. "Anticraving" medications and medications that reduce the "high" from drinking alcohol may be particularly useful in recovering alcoholics who are prone to relapse. Medications originally developed to treat DEPRESSION and ANXIETY have also exhibited potential for managing drinking behavior in specific subgroups of alcoholics.

The focus here is on four categories of medications that are either currently available or appear promising for the treatment of alcoholism and alcohol abuse. These include the following: alcohol-sensitizing agents; agents that directly attenuate drinking behavior; agents to improve cognition in patients with alcohol-induced impairments; and agents to treat psychiatric problems concomitant to alcoholism. The most promising medications within each of the above categories are examined, addressing their stage of development, clinical efficacy, potential side effects, and future research. For more detailed information on this topic, see the review articles by Liskow and Goodwin (1987) and Litten and Allen (1991). First, we describe briefly the methodology used to conduct clinical pharmacotherapy studies.

CONTROLLED CLINICAL TRIALS

The method used to determine medication efficacy is called the controlled clinical trial. The key components of clinical trials include the following: control groups; random assignments of eligible subjects to medication or to control groups; use of placebos (identical-appearing but inactive medications) for the control group—unless a standard effective medication is available to serve as the comparison; assurance that neither the patients receiving the drug nor the physicians administering/prescribing know whether they are getting the active medication or the placebo (called double-blind); methods that validly and reproducibly measure the response to the medication; methods to monitor whether subjects take the medication; and procedures to follow all the patients who entered the study for the duration of the clinical trial. After the data are collected, they must be analyzed by using the appropriate statistical tests.

Randomizing. It is important to randomize eligible patients to the treatment and placebo groups, because this assures that the two groups are comparable except for the medications being prescribed. If some method other than randomization is used to assign patients to treatments, it is likely that the groups will differ in important characteristics such as severity of illness. If one of the groups is, in general, more severely ill than the other, the sicker group is less likely to do well regardless of the treatment. If the more severely ill group receives the active medication, the difference between the medication group and the placebo group after treatment does not appear as great because the placebo (control) group was less ill at the beginning. Thus, it may appear that the medication was not effective.

Double-blinding. "Blinding" of both the patients and the physicians is necessary because of their expectations and beliefs. Patients usually seek treatment in the expectation that the physician will prescribe or recommend something that will cure or improve their condition. Hence, patients who receive placebos often feel better. Therefore, if a placebo control is not used, one might conclude that a new treatment works when one is only observing the placebo kind of response. (Conversely, patients often report side effects when they take placebos. So not all side effects are necessarily due to an active medication.) Physicians often believe very strongly that the new drug will be the effective treatment they are searching for, and their objectivity is diminished by this bias. To remove this influence on their perception of the outcome of treatment, the physicians treating the patients are "blinded" as well as the patients, hence a double-blinding is effected.

Accurate Assessment. If the methods used to assess the response to treatment do not accurately measure the response to the medication, erroneous conclusions may be drawn (Fuller, Lee, & Gordis, 1988). Patients' self-reports about their response to treatment should not be used without corroborating data in controlled clinical trials unless no other means for obtaining information is available. Such reports may be inaccurate for a variety of reasons, including inaccurate memory and the tendency to give socially desirable answers.

It is also important to know whether the patients actually took the medications. Often, patients do not take their medications or take them erratically, particularly if they are being treated for an asymptomatic condition for a long period of time and/or if the medication has a high incidence of unacceptable side effects (Haynes, Taylor & Sackett, 1979).

Patients who drop out of treatment frequently are atypical of all patients in treatment. In alcoholism treatment, the dropouts are usually drinking and having problems because of their drinking. So, if a study bases its conclusions only on those who stay in treatment, the results of the therapy are likely to be exaggerated. Therefore, it is important to locate and assess treatment response in all or almost all who initially began treatment. For an excellent description of clinical trials, their methods and issues, see Byar et al. (1976).

If control groups were used, methods other than randomization were used to assign patients to the disulfiram group or the control group. Hence, the groups were not comparable, and placebo groups were rarely used. "Blinding" was not done. No attempts were made in most of the studies to determine whether patients took the medication. The alcoholic's report on abstention from alcohol was the only information obtained to judge whether disulfiram was effective. In some studies, only about half the patients were available for follow-up.

Multi-Site Trial. During the past decade, more rigorously designed clinical trials of disulfiram have been done, and these give more precise information about the efficacy of disulfiram. The largest of these was a multi-site clinical trial done in nine Veterans Administration clinics (Fuller et al., 1986). In this study, 605 men were randomly assigned to three groups: 1) a 250-milligram disulfiram group (the usual dose); 2) a 1-milligram disulfiram group; and 3) a no-disulfiram group. The 1-milligram group was equivalent to a placebo, because this dose is not sufficient to cause a disulfiram–ethanol reaction (DER) but controls for the expectation that one will get sick if one drinks alcohol while taking disulfiram. The no-disulfiram group was told they were not receiving Antabuse; it was a control for the standard counseling that alcoholics receive in treatment. The patients in the two disulfiram conditions were "blinded" as to whether they were receiving the 250-milligram or the 1-milligram dose. The data to judge the effect of treatment were collected by research personnel who had no involvement in the treatment of the patients and were "blinded" to group assignment. The research staff members interviewed the patients, cohabiting relatives, and friends (*collaterals*) every two months during the year of follow-up. Urine specimens were collected every time the patients returned to the clinic and were analyzed for the presence of alcohol. A vitamin, riboflavin, was incorporated into the 250-milligram and 1-milligram tablets. The no-disulfiram patients received a tablet identical in appearance to the disulfiram tablets but containing only riboflavin. The urine specimens were also analyzed for riboflavin. This allowed the investigators to tell whether the patients were taking their medications regularly.

In contrast to most of the previous studies, this tightly designed study did not find that more of the patients who received disulfiram stayed sober for the year than those who received the placebo or counseling only. Nor was disulfiram associated with better employment or social stability; however, in about 50 percent of the men who relapsed, drinking frequency was significantly less for those who received disulfiram than for those who received either the placebo or no disulfiram. This subset of men who relapsed by drinking less frequently if assigned to disulfiram were slightly older and had more social stability (as indicated by longer residence at their current address) than the other men who relapsed. These results indicate that disulfiram is not more effective than routine treatment for most male alcoholics—female alcoholics were not included in the study—but may have some benefit for socially stable male alcoholics.

In the multi-site study, only 20 percent of the patients took the medication regularly; however, abstinence for the year was highly associated with compliance with the disulfiram regimen. This suggests that if ways were found to get patients to take disulfiram regularly, the effectiveness of the drug would be greatly improved. This concept has to be tempered by the finding that those who regularly took the 1-milligram placebo or the vitamin without disulfiram, as well as those who took disulfiram, were much more likely to remain sober than those who were less adherent to their regimens. Nevertheless, alcoholism treatment researchers have studied various methods for improving compliance with disulfiram, and preliminary results suggest that these may be beneficial. These treatment strategies have included having the spouse or a treatment facility staff member observe the patient ingesting the medication, establishing a contract with the patient about taking it, and/or building in positive (rewards) or negative (loss of privileges) incentives to take it. A recent controlled study of disulfiram taken in the presence of a relative, friend, or member of the clinic staff found that this method of administration resulted in significantly less alcohol being consumed during a six-month period (Chick et al., 1992). More well-designed studies of these measures to improve compliance with the disulfiram regimen are needed before we know if they will improve the effectiveness of disulfiram as a treatment for alcohol dependence.

On the basis of the large well-designed studies done to date, it seems prudent to recommend that disulfiram should not be used initially in the treatment for alcoholism. However, if the patient relapses and has indicators of social stability, a discussion with the patient about the possible benefits and the possible risks of disulfiram is warranted, and if the patient is willing to take disulfiram, a trial course is warranted. During the first six months of treatment, it is important that liver tests be monitored closely. The effectiveness of the drug may be enhanced if the patient agrees to take it under supervision.

ALCOHOL-SENSITIZING AGENTS

The most commonly used alcohol-sensitizing agent is disulfiram, which has been used in clinical practice since the 1950s to deter alcoholics from drinking. It is not an aversive drug in the strict sense of the word, since it is not used to condition individuals to have an aversive response at the sight or smell of alcohol. Rather, its objective is to deter drinking by the threat of having a very unpleasant reaction if one does drink alcoholic beverages. Its severity depends on the amount of alcohol and disulfiram in the blood. The symptoms of the reaction include facial flushing, tachycardia (rapid heart beat), palpitations, dyspnea (indigestion), hypotension (lowered blood pressure), headaches, nausea, and vomiting. Deaths have occurred with severe disulfiram–ethanol reactions (DERs).

A DER results when alcohol is ingested because disulfiram inhibits the functioning of an enzyme, aldehyde dehydrogenase. This enzyme is needed to convert the acetaldehyde—the first metabolic product in the catabolism of ethanol—to acetic acid. If aldehyde dehydrogenase is inhibited, an elevation in blood acetaldehyde results. The increased circulating acetaldehyde is believed to cause most of the symptoms and signs of the DER.

Disulfiram is given orally. The usual dose is 250 milligrams, although larger doses have been used. Doses of less than 250 milligrams may fail to cause a DER, while doses of more than 250 milligrams have a greater risk of producing serious side effects. Adverse effects of disulfiram range from mild symptoms such as sedation, lethargy, and a garliclike or metallic taste in the mouth to more serious side effects such as major depression, psychotic reaction, or idiosyncratic toxic hepatitis—which may be fatal. A dose between 250 milligrams and 500 milligrams is usually adequate to cause a DER if alcohol is ingested but not so high as to cause major side effects. The dose should be individualized for each patient.

Although disulfiram has been used for more than four decades, only a few controlled studies have been done to determine its efficacy.

Alcohol-sensitizing agents other than disulfiram also exist. CALCIUM CARBIMIDE has been used clinically, although it is currently not available in the United States. Calcium carbimide produces physiological reactions with alcohol similar to those produced by disulfiram, but the onset of action is quick—within one hour after administration—compared to twelve with disulfiram. Also, the duration of action is short—approximately twenty-four hours—versus up to six days with disulfiram. Calcium carbimide, with its faster onset of action, might be especially helpful with impulsive drinkers. A possible side effect of calcium carbamide is reduced thyroid function, however, thus making its use problematic in patients with thyroid problems. There is a paucity of randomized clinical trials comparing calcium carbamide to placebo—so, its efficacy is uncertain.

AGENTS THAT ATTENUATE DRINKING BEHAVIOR

The development of medications to curb drinking behavior is one of the important and exciting areas of alcohol research. In developing such medications, researchers have relied on new information about the biological bases of drinking behavior and alcohol craving. This process is complex and involves the interactions among several neurochemical mechanisms, including NEUROTRANSMITTERS, hormones, neuropeptides, RECEPTORS, second messenger systems, and various ion channels in multiple regions of the brain.

Recent research has focused on medications that alter the functional activity of several neurotransmitter systems. In this section, we discuss medications that directly attenuate drinking by acting on the following neurotransmitter systems: SEROTONIN, OPIOIDS, DOPAMINE, and GAMMA-AMINOBUTYRIC ACID (GABA).

Agents That Affect the Serotonin System. Several lines on animal and human research suggest that brain serotonin is associated with alcoholism. Serotonin levels are lower in several regions of the brain in rats selectively bred to drink alcohol than in rats that do not prefer alcohol. In humans, measurements of cerebral spinal fluid levels of 5-hydroxyin-doleacetic acid (5-HIAA), a metabolite of serotonin, revealed lower levels of 5-HIAA in alcoholics who were abstinent for four weeks than in nonalcoholics. Also, the availability of the serotonin precursor, tryptophan, appears to be lower in alcoholics, particularly those in early onset of alcoholism (drinking before twenty years of age).

SEROTONIN-UPTAKE INHIBITORS, commonly used to treat depression, seem effective in reducing alcohol consumption in both animal models and humans. Serotonin-uptake inhibitors act by preventing the uptake of serotonin during synaptic transmission, resulting in a prolonged action. They are easily administered (orally) and require only a single daily dose.

The serotonin-uptake inhibitors available for clinical testing include fluoxetine (Prozac), fluvoxamine, citalopram, and viqualine. Several double-blind, placebo-controlled studies of these agents in various types of subjects—ranging from social drinkers to chronic alcoholics—have shown an increase in the number of abstinent days and a decrease in the number of drinks on drinking days (Gorelick, 1989). The effect of the serotonin-uptake inhibitors studied has been modest (a 25 percent decrease in alcohol intake), but newer compounds in this group may be more effective.

The precise mechanism of action of the serotonin-uptake inhibitors on drinking behavior is unknown. One of the most plausible explanations offered is their ability to suppress appetitive behaviors in general. However, consummatory behaviors are quite complex, and even this hypothesis may be an oversimplification.

In addition to the serotonin-uptake inhibitors, agents that selectively block (antagonists) or activate (agonists) the subtypes of serotonin receptors also appear promising. At least four major types of serotonin (5-hydroxytryptamine, or 5-HT) receptors exist: 5-HT_1, 5-HT_2, 5-HT_3, and 5-HT_4. In turn, 5-HT_1 has several subdivisions, including 5-HT_{1A} receptor. Recent research has shown that a 5-HT_3 antagonist, ondansetron, reduces alcohol consumption in alcohol abusers (Toneatto et al., 1991). Also, 5-HT_{1A} and 5-HT_2 receptors are believed to influence alcohol intake. For example, buspirone, a 5-HT_{1A} agonist and an antianxiety agent, has been shown in some studies to reduce alcohol consumption in humans.

Finally serotonergic agents (e.g., fenfluramine) that cause a release of serotonin from presynaptic neurons may also have clinical efficacy in reducing alcohol intake. In addition, the administration of serotonin precursors appears to alter drinking behavior. Several animal studies have shown that tryptophan (precursor to serotonin) and 5-hydroxtryptophan (hydroxylated form of tryptophan) reduce the amount of alcohol consumed.

In summary, medications that alter the serotonin system seem to offer promise as pharmacological treatments of alcoholism. Future research will continue to examine other functional serotonergic agents as well as to assess long-term efficacy, determine long-term risks/side effects, and identify the individuals who will respond most favorably to the medication. For example, the antidepressant actions of the serotonin-uptake inhibitors may be particularly effective with alcoholics suffering depression.

Agents That Affect the Dopamine System. DOPAMINE is another neurotransmitter identified as influencing drinking behavior. Dopamine is thought to play a major role in the stimulant and reinforcing properties of alcohol as well as other drugs. Decreased levels of dopamine are observed in the NUCLEUS ACCUMBENS of alcohol-seeking rats (as compared with nonalcohol-seeking rats). The nucleus accumbens is the region of the brain believed to be involved with alcohol craving. Studies have demonstrated that the application of alcohol to the nucleus accumbens and striatum of a rat brain causes a release of dopamine (Wozniak et al., 1991; Yoshimoto et al., 1991).

The administration of medications that increase brain dopamine levels (bromocriptine, GBR 12909, and amphetamine) results in a reduction of alcohol intake in alcohol-preferring rats. Several studies have been conducted in humans using the dopamine type 2 agonist (D_2) bromocriptine. One study (Borg, 1983) indicated that bromocriptine reduced alcohol craving and consumption in severe alcoholics, while another (Dongier et al., 1991) found a reduction in alcohol consumption and an improvement in psychological problems in both bromocriptine-treated and placebo alcoholics, although no significant differences were observed between the two groups.

The efficacy of the dopaminergic medications in the long-term management of alcoholism is currently unclear. Further research needs to be conducted on the two major subtypes of dopamine receptors, D_1 and D_2. In addition, their interaction with other neurotransmitter systems needs to be investigated. An illustration that neurotransmitter systems do not work in isolation and that a medication affecting one may also alter another is present in several studies, which have shown that blocking the serotonin $5-HT_3$ receptor with the antagonist ICS 205-930 results in an attenuation of alcohol-induced release of dopamine in the nucleus accumbens and corpus striatum of the rat brain (Wozniak et al., 1990; Yoshimoto et al., 1991).

Agents That Affect the Opioid System. Studies have shown that the opioid system also plays a role in modifying drinking behavior. Many researchers believe that alcohol craving and increased drinking behavior are related to low brain levels of endogenous opioids (compounds with opium or morphine-like properties, e.g., ENDORPHINS and ENKEPHALINS). Subsequently, increasing the opioid levels causes a decrease in drinking. This is supported by several studies. For example, administration of the opioid agonist [D-Ala^2, MePhe^4, Met(O)^5-ol]-enkephalin decreases alcohol consumption in alcohol-preferring mice. Large doses of morphine (a classic opioid agonist) also result in a significant reduction in alcohol intake. In addition, increasing the availability of endogenous enkephalins by injecting mice with the enkephalinase inhibitor kelatorphan (which prevents breakdown of endogenous enkephalins) results in decreased alcohol consumption. Finally, a recent study demonstrated that high-risk individuals (those who have a family history of alcoholism) have lower plasma levels of beta-endorphin than do low-risk individuals (no family history of alcoholism for at least the three preceding generations).

Some researchers have challenged the hypothesis that excessive drinking is related to decreases in endogenous opioid levels. Experimental evidence includes the observation that low doses of morphine cause an increase in alcohol intake in rats.

Regardless of the mechanism of action, the opioid ANTAGONISTS NALTREXONE and NALOXONE—currently used to treat opiate abuse—have been shown to influence alcohol consumption. Both agents reduce voluntary alcohol intake in rats and monkeys. In humans, studies have shown that alcoholics treated with naltrexone have fewer drinking days, fewer relapses, and less subjective craving for alcohol (Volpicelli et al., 1992; O'Malley et al., 1992). In addition, naltrexone appears to cause few side effects. Interestingly, naltrexone-treated alcoholics who did have one or two drinks were less likely to continue drinking. This is important, since some alcoholics appear to lose control of drinking after one or two drinks.

Further research on identifying the specific mechanisms of the opioid system responsible for altered drinking behavior will undoubtedly be continued.

Agents That Affect the GABA System. Several studies have now investigated the GABA system as a modulator of drinking behavior. The number of GABAergic receptors appears to be greater in the nucleus accumbens region of the brain of alcohol-preferring rats than in those of the alcohol-nonpreferring rats. Administration of RO 15-4513, a benzodiazepinelike agent that binds to the GABA receptor, leads to a dose-related reduction in alcohol intake by rats. This compound also attenuates some of the acute intoxicating effects of alcohol. This experimental agent will probably not be used in humans, however, because it tends to cause convulsions. In a preliminary report, a GABA agonist appears to improve the rates of abstinence in severe alcoholics.

AGENTS TO IMPROVE COGNITIVE FUNCTION

Chronic heavy drinking can lead to impairment of most cognitive functions, including abstract thinking, problem solving, concept shifting, psychomotor performance, and memory. The two most common diseases of cognitive impairment in alcoholism are alcoholic amnestic disorder (WERNICKE-Korsakoff syndrome) and alcoholic dementia. Alcoholic amnestic disorder is associated with prolonged and heavy use of alcohol and is characterized by severe memory problems. Though the exact cause is unknown, this disease is thought to be preventable by proper diet, including vitamins, particularly the B vitamin thiamine. The other impairment, alcoholic dementia, has a gradual onset and thus displays various degrees of cognitive impairment, including difficulties in short-term and long-term memory, abstract thinking, intellectual abilities, judgment, and other higher cortical functions.

Most studies indicate that alcoholics with impaired cognitive function will have poorer treatment outcome. This, of course, depends on the severity of impairment. Little research has been conducted with medications to improve cognitive function. Serotonin-uptake inhibitors have shown some promise in improving learning and memory. A recent study with the serotonin-uptake inhibitor fluvoxamine has demonstrated improvement in memory in patients suffering from alcohol amnestic disorder, but not in patients with alcoholic dementia.

AGENTS TO TREAT PSYCHIATRIC DISORDERS CONCOMITANT TO ALCOHOLISM

Alcoholism may be accompanied with various psychiatric problems including anxiety, depression, antisocial behavior, panic disorders, and phobias. Part of the problem in treatment is to determine if the psychiatric disorder developed before alcoholism (primary), or after (as a result of) alcoholism (secondary). Nevertheless, several studies have been conducted predominately with medications used to treat depression and anxiety.

Agents to Treat Alcoholics with Depression. Depression has been associated with alcoholism, especially with relapse to drinking. A frequent pharmacologic treatment of depression is with a group of medications called tricyclic ANTIDEPRESSANTS (desipramine, imipramine, amitriptyline, and doxepin). Their efficacy in treating alcoholics with depression is, however, largely unknown. This is in part because of poor methodological studies. A recent study of desipramine was conducted on alcoholics with and without secondary depression (Mason & Kocsis, 1991). Preliminary findings showed that desipramine is effective in reducing depression in the depressed group and may also prolong the period of abstinence from alcohol in both depressed and nondepressed patients. This implies that both nondepressed and secondary depressed alcoholics may be responsive to desipramine treatment. Preliminary results of another study suggest that imipramine both improves mood and reduces drinking in alcoholics suffering from major (primary) depression.

In addition to the tricyclic antidepressants, the serotonin-uptake inhibitors are used to treat depression. One of these inhibitors, fluoxetine, is widely used as an antidepressant. As discussed earlier, fluoxetine has been studied to see whether it attenuates drinking behavior in nondepressed alcoholics. To date, its effectiveness in treating depressed alcoholics has not been determined, though several studies are ongoing.

Lithium, an effective medication for the treatment of manic-depressive disease, has also been studied as a pharmacologic agent in the treatment of alcoholic patients. In a recent multi-site clinical study of lithium in depressed and nondepressed alcoholics, lithium therapy was not effective in reducing the number of drinking days, improving abstinence, decreasing the number of alcohol-related hospitalizations, or reducing alcoholism dependence (Dorus et al., 1989). This investigation as well as other studies did not address the effectiveness of lithium in other types of psychiatric disorders that may respond—including hypomania (a mild degree of mania), bipolar manic-depressive illness, and other mood disorders. If lithium is found effective in treating some types of alcoholics, it will be necessary to monitor the associated toxic reactions and side effects (e.g., vomiting, nausea, diarrhea, and trouble walking).

Agents to Treat Alcoholics with Anxiety Disorders. Recent studies have indicated that a sizeable proportion of individuals who abuse alcohol also suffer from anxiety disorders. Buspirone, an agent commonly used to treat anxiety, has shown potential in reducing alcohol consumption. As discussed earlier, buspirone acts as an agonist on the serotonin 5-HT_{1A} receptors and also alters the dopamine and norepinephrine systems.

An attractive feature of buspirone is that its use does not lead to physical dependence on the drug, as with antianxiety drugs, particularly with BENZODIAZEPINES. Furthermore, buspirone lacks side effects often found with anxiolytic medications. For example, buspirone lacks sedative, anticonvulsant, and muscle-relaxant properties, does not impair psychomotor, cognitive, or driving skills, and does not potentiate the depressant effects of alcohol.

Administration of buspirone to rats and monkeys has resulted in a decrease in alcohol intake (Litten & Allen, 1991). In humans, one study reported that buspirone diminished alcohol craving and reduced anxiety. Another study found buspirone to be more effective with alcoholics suffering from high anxiety than those with low levels of anxiety. A third study on more severe alcoholic patients found no effect. Thus, further research is needed before this drug's efficacy can be accurately evaluated.

In summary, early evidence indicates that effective treatment of a psychiatric disease may also be beneficial to the treatment of alcoholism, particularly in alcoholics with coexisting psychiatric disorders. In addition, it may be important to make the distinction between primary and secondary psychiatric disorders (disorders occurring before or after alcoholism) before efficacious treatment can be applied.

CONCLUSIONS

Development of new medications to decrease drinking, prevent relapse, and restore cognition will likely have a role in alcoholism treatment in the future—but as a part of treatment regimens—given with other nonpharmacological therapies. Advances in understanding the mechanisms responsible for alcohol craving, drinking behavior, cognition, and even some of the psychiatric disorders such as depression and anxiety disorders will, undoubtedly, produce new pharmacotherapeutic possibilities for treating alcoholism. Currently, the most promising of these agents are the serotonergic agents and the opioid antagonists.

Once research has identified medications as promising, it is important to determine their long-term efficacy, specify subtypes of alcoholics who may respond favorably to the medication, and assess risks. Afterward, proper treatment strategies need to be developed to integrate the pharmacologic treatment with nonpharmacological interventions to optimize treatment outcome. Finally, methods to enhance compliance also need to be developed if the full benefits of pharmacotherapy are to be obtained. These methods include advancements in behavioral and psychosocial techniques as well as the development of new physical modes of drug delivery, such as the transdermal patch.

(SEE ALSO: *Comorbidity and Vulnerability; Complications; Contingency Contracts; Disease Concept of Alcoholism and Drug Abuse; Drug Interactions and Alcohol; Drug Metabolism; Treatment, History of*)

BIBLIOGRAPHY

BORG, V. (1983). Bromocriptine in the prevention of alcohol abuse. *Acta Psychiatrica Scandinavica, 68*, 100–110.

BYAR, D. P., ET AL. (1976). Randomized clinical trials. *New England Journal of Medicine, 295*, 74–80.

CHICK, J., ET AL. (1992). Disulfiram treatment of alcoholism. *British Journal of Psychiatry, 161*, 84–89.

DONGIER, M., VACHON, L., & SCHWARTZ, G. (1991). Bromocriptine in the treatment of alcohol dependence. *Alcoholism: Clinical and Experimental Research, 15*, 970–977.

DORUS, W., ET AL. (1989). Lithium treatment of depressed and nondepressed alcoholics. *Journal of the American Medical Association, 262,* 1646–1652.

FULLER, R. K., LEE, K. K., & GORDIS, E. (1988). Validity of self-report in alcoholism research: Results of a Veterans Administration Coooperative Study. *Alcoholism: Clinical and Experimental Research, 12,* 201–205.

FULLER, R. K., ET AL. (1986). Disulfiram treatment of alcoholism: A Veterans Administration cooperative study. *Journal of the American Medical Association, 256,* 1449–1455.

GORELICK, D. A. (1989). Serotonin uptake blockers and the treatment of alcoholism. In *Recent Developments in Alcoholism: Treatment Research,* Vol. 7. New York: Plenum Press.

HAYNES, R. B., TAYLOR, D. W., & SACKETT, D. L. (1979). *Compliance in health care.* Baltimore: Johns Hopkins University Press.

LISKOW, B. I., & GOODWIN, D. W. (1987). Pharmacological treatment of alcohol intoxication, withdrawal and dependence: A critical review. *Journal of Studies on Alcohol, 48,* 356–370.

LITTEN, R. Z., & ALLEN, J. P. (1991). Pharmacotherapies for alcoholism: Promising agents and clinical issues. *Alcoholism: Clinical and Experimental Research, 15,* 620–633.

MASON, B. J., & KOCSIS, J. H. (1991). Desipramine treatment of alcoholism. *Psychopharmacology Bulletin, 27,* 155–161.

O'MALLEY, S. S., ET AL. (1992). Naltrexone and coping skills therapy for alcohol dependence. *Archives of General Psychiatry, 49,* 881–887.

TONEATTO, T., ET AL. (1991). Ondansetron, a 5-HT$_3$ antagonist, reduces alcohol consumption in alcohol abusers. *Alcoholism: Clinical and Experimental Research, 15,* 382.

VOLPICELLI, J. R., ET AL. (1992). Naltrexone in the treatment of alcohol dependence. *Archives of General Psychiatry, 49,* 876–880.

WOZNIAK, K. M., PERT, A., & LINNOILA, M. (1990). Antagonism of 5-HT$_3$ receptors attenuates the effects of ethanol on extracellular dopamine. *European Journal of Pharmacology, 187,* 287–289.

WOZNIAK, K. M., ET AL. (1991). Focal application of alcohols elevates extracellular dopamine in rat brain: A microdialysis study. *Brain Research, 540,* 31–40.

YOSHIMOTO, K., MCBRIDE, W. J., LUMENG, L., & LI, T.-K. (1991). Alcohol stimulates the release of dopamine and serotonin in the nucleus accumbens. *Alcohol, 9,* 17–22.

RICHARD K. FULLER
RAYE Z. LITTEN

Alcohol, Psychological Approaches   The use of behavioral and other psychological treatments for alcohol abuse has a long history. In the nineteenth century, Benjamin Rush, often regarded as the founder of American psychiatry, described a variety of social and psychological cures for chronic drunkenness. Treatment procedures derived from principles of learning and conditioning were being tested in the 1920s, prior to the development of modern pharmacologic approaches. Currently, there is a large scientific literature documenting the effectiveness of various psychological treatments for alcohol problems.

WHY USE PSYCHOLOGICAL APPROACHES?

The most obvious argument for the use of behavioral and other psychological approaches in treating alcohol abuse is that the drinking of alcohol or ethyl alcohol is a *behavior.* Regardless of the therapeutic approach used, the criterion for success or failure in treatment studies is typically behavioral—whether and how much a person continues to drink. Research amply demonstrates that drinking behavior is substantially influenced by a wide variety of psychological processes, including beliefs and EXPECTANCIES, the examples of friends and family, the customs and norms for drinking within one's society or subgroup, emotional states, family processes, and the positive and negative consequences of drinking. Treatments that address these factors directly, then, might be expected to be helpful in overcoming alcohol problems.

In fact, dozens of well-controlled studies since the 1960s do support the effectiveness of behavioral and other psychological treatments. The benefits of such treatment have typically been larger than those reported for pharmacologic approaches and have been shown in some studies to endure over follow-up periods of several years. This research in itself provides a convincing reason to use psychological methods in treating alcohol abuse.

Still another reason is the finding that psychosocial processes strongly influence whether or not a person will relapse after treatment. The likelihood of relapse is decreased by factors such as marital stability, social support, personal coping skills, employment, and confidence in one's abilities to deal with problems. Factors like these in a person's life *after* treatment are important determinants of outcome. Treatment methods that anticipate and address these posttreatment adjustment challenges are thus important.

There is, however, little reason to argue for psychological versus pharmacological treatment approaches, since these two approaches can be used together with good result. Psychological methods play a key role in addressing psychosocial aspects of drinking problems and are compatible with the use of medications, where they are appropriate.

ALTERNATIVE PSYCHOLOGICAL METHODS

A psychological approach to treating alcohol abuse does not involve just one method. Rather, a variety of strategies can be used to accomplish the central goal—to change drinking behavior—and several methods are typically employed in a treatment program.

Treatment methods should not be confused with treatment goals. The general psychological methods described below can be applied in pursuit of different goals. Sometimes the goal of treatment is the complete elimination of alcohol drinking for the rest of a person's lifetime (total and permanent abstinence). For others, the goal may be to reduce alcohol use to a level that will no longer threaten a person's physical or psychological health. The goals of treatment may also include other important dimensions besides drinking—to get and hold a job, to have a happier marriage and family life, to learn how to deal with anger, and to find new ways of having fun that do not involve drinking. Finally, it is worth noting that clients may have treatment goals that differ from those of the therapist. Psychological treatment methods do not inherently dictate outcome goals, but they can be used to achieve goals once chosen.

Teaching New Skills. Alcohol is often used in an attempt to cope with life problems. People may drink to relax or loosen up, to get to sleep, to feel better, to enhance sexuality, to build courage, or to forget. In truth, alcohol rarely works as an effective coping strategy for dealing with emotional and relationship problems. In the long run, it often makes such problems worse. Yet the seeming immediate relief can make alcohol appealing when a person is faced with bad feelings or social problems. To the extent that a person comes to rely upon drinking to cope, that person is termed *psychologically dependent* on alcohol.

One psychological approach, sometimes called *broad-spectrum treatment*, directly addresses this problem by teaching the person new coping skills. Ten controlled studies, for example, have found that the addition of *social-skills training* increases the effectiveness of treatment for alcohol abuse. People are taught skills for expressing their feelings appropriately, making requests, refusing drinks, and carrying on rewarding conversations. *Stress-management training* has also been shown to help prevent relapse to drinking. People learn how to relax and deal with stressful life situations without using drugs.

Self-Control Training. Another well-documented psychological approach is *self-control training*, which teaches methods for managing one's own behavior. Some common elements in self-control training include: (1) setting clear goals for behavior change; (2) keeping records of drinking behavior and urges to drink; (3) rewarding oneself for progress toward goals; (4) making changes in the way one drinks, or in the environment, to support new patterns; (5) discovering high-risk situations where extra caution is required; and (6) learning strategies for coping with high-risk situations. Although often used to help people reduce their drinking to a moderate and nonproblematic level, self-control training can also be used when total abstinence is the goal. This method has been found to be particularly helpful for less severe problem drinkers. It has also been found to be more effective than educational lectures for drunk-driving offenders.

Marital Therapy. There are several reasons to consider treating not only the excessive drinker, but also the spouse. First, problem drinking commonly affects the drinker's partner in adverse ways. Secondly, the spouse may be quite helpful during treatment in clarifying the problem and in developing effective strategies for change. Thirdly, the spouse can provide continuing support for change after treatment. Finally, marital distress may be a significant factor in problem drinking, and direct treatment of marital problems can help to prevent relapse.

Research indicates that problem drinkers treated together with a spouse fare better than those treated individually. Behavioral marital therapy in particular is well supported by current outcome research.

Aversion Therapies. Another set of treatment strategies applies the learning principle of aversive counterconditioning (called AVERSION THERAPY). The idea here is that if drinking is paired with unpleasant images and experiences, the desire for alcohol is diminished, and drinking decreases. There is sound evidence that it is possible to produce a conditioned aversion to alcohol in both animals and humans. The taste and even thought of alcohol become unpleasant. There is also evidence that aversion therapy is successful to the extent that this kind of conditioned aversion is established during treatment. Some forms of aversion therapy pair the taste of alcohol with unpleasant sensations such as nausea, foul odors, or electric shock. A newer form, termed *covert sensitization*, uses no physical aversion of this kind but instead pairs alcohol with unpleasant experiences in imagination. These approaches may be particularly useful for those who continue to experience craving or a strong positive attachment to alcohol.

Psychotherapy. Many kinds of psychotherapy have been tried with alcohol abusers. In general, studies suggest that individual psychotherapies with a goal of insight into unconscious causes of drinking have been largely unsuccessful. Likewise, group psychodynamic psychotherapies have had a poor track record in treatment-outcome studies. As a distinct element, confrontational group therapy, a common element of U.S. treatment programs, is also unsupported by current research. More recently, *cognitive therapies* have gained popularity, and some controlled trials supporting their efficacy.

Changing the Environment. Yet another psychological approach is behavior modification by changing the consequences of drinking. The goal here is to eliminate positive reinforcement for drinking, and to make alternatives to drinking more rewarding. Studies have reported success in working unilaterally with a drinker's spouse to make changes that discourage drinking and reinforce alternatives. A complex treatment known as the *community-reinforcement approach* (CRA) has fared well in comparisons with traditional methods. The CRA systematically encourages rewarding alternatives to drinking, teaching skills needed for living without alcohol. The CRA incorporates a number of treatment elements, including marital therapy, social-skills training, the taking of disulfiram (Antabuse—a medication that causes aversive effects when alcohol is ingested), and job-finding training. The use of *behavioral contracting*—drawing up a specific

agreement about future drinking and its consequences—has been found to be an effective component of treatment in several studies.

Brief Motivational Counseling. An interesting and unexpected finding in more than a dozen well-controlled studies is the effectiveness of relatively brief motivational counseling. Certain treatments, consisting of one to three sessions, have been found to be significantly more effective than no treatment and often as effective as more extensive treatment regimens. These motivational approaches, now studied in several nations, typically include a thorough assessment, feedback of findings, clear advice to change, and an emphasis on personal responsibility and optimism. The key seems to be to trigger a decision and commitment to change. Once this motivational hurdle has been crossed, people frequently proceed to change their drinking on their own without further professional assistance. In fact, treatment approaches that proceed directly into strategies for changing drinking may fail because they do not address this motivational prerequisite for change.

Therapist Style. Other recent research indicates that the skills and style of the therapist have important effects on treatment outcome. With impressive consistency, therapist success has been linked to an empathic and supportive style, rather than an aggressive and confrontational approach. Directive and confrontational tactics tend to elicit resistance and defensiveness from clients, which in turn are predictive of a lack of therapeutic change. It is clear that the same treatment approach can have dramatically different outcomes when administered by different therapists.

HOW IS SUCCESS JUDGED?

In one sense, judging the outcome of treatment would seem simple: Either the person is or is not still drinking in a problematic manner. A closer examination of treatment-outcome research quickly reveals a number of complexities.

First is the question of the standard against which a treatment is to be judged. Is a "success" rate of 60 percent spectacularly good or shameful? This is decided relative to the expected outcome without the same treatment. This is why the usual standard for judging effectiveness in medical research is the *controlled trial* in which clients are randomly assigned to different treatment methods. In the absence of proper controls, one cannot judge adequately whether the outcome of a treatment is better or worse than it would have been without the special treatment. Evidence from properly controlled trials is more consistent than the results of uncontrolled trials, presenting a clearer picture of effectiveness.

A second complexity is: What constitutes success? When success is defined very conservatively, as total abstinence from alcohol (not even one drink) since the end of treatment, low success rates can be expected. Yet if some drinking is permitted among "successes," it is necessary to define the acceptable limits for how much, how often, and with what consequences. Some studies have reported only a category of "improved" cases without adequate definition.

Once successful outcome is clearly defined, there is the problem of how to measure it. Should a researcher accept the client's self-report? Should friends and family members be interviewed? Should blood, breath, or urine samples be required? If multiple outcome measures are used, how does one decide which is the truth?

Still another example is the issue of length of follow-up. Success rates are typically highest within a few weeks or months from the time of treatment. A large percentage of relapses occur between three and twelve months after treatment. Short follow-up periods, then, overestimate success rates. Longer follow-ups raise the additional problem of how to deal with lost cases. If one studies only those who can be easily found two years later, success rates may be inflated.

For these reasons, the effectiveness of treatment approaches is best judged by accumulating evidence from several properly controlled studies. Conclusions presented above, regarding the efficacy of different psychological treatment approaches, were drawn on this basis.

MATCHING PEOPLE TO TREATMENTS

It is unlikely that research will ever identify a single superior treatment for alcohol abuse. Drinking and alcohol-related problems are far too complex. The cause for real optimism is found in the number of different approaches with reasonable evidence of effectiveness. For a given person, then, the chances of eventually finding an effective approach are good.

Recent research indicates that these various treatment approaches work best for different kinds of people. As such evidence accumulates, it will be increasingly possible to choose optimal treatment strategies for people based on their individual characteristics. Treatment systems, therefore, should work toward providing a range of different approaches, rather than offering the same basic treatment to everyone with alcohol problems.

(SEE ALSO: *Causes of Substance Abuse; Disease Concept of Alcoholism and Drug Abuse; Treatment Types*)

BIBLIOGRAPHY

COX, W. M. (ED.). (1987). *Treatment and prevention of alcohol problems: A resource manual.* Orlando, FL: Academic Press.
HESTER, R. K., & MILLER, W. R. (EDS.). (1989). *Handbook of alcoholism treatment approaches: Effective alternatives.* New York: Pergamon Press.
MILKMAN, H. B., & SEDERER, L. I. (EDS.). (1990). *Treatment choices for alcoholism and substance abuse.* Lexington, MA: Lexington Books.
MILLER, W. R., & ROLLNICK, S. (1991). *Motivational interviewing: Preparing people for change in addictive behaviors.* New York: Guilford Press.
MONTI, P. M., ET AL. (1989). *Treating alcohol dependence.* New York: Guilford Press.

WILLIAM R. MILLER

Cocaine, An Overview COCAINE abuse and dependence should be approached as chronic disorders that require long-term treatment. In evaluating a patient for treatment, many factors must be taken into consideration. First, at the time patients present for treatment there are often complicating factors, such as coexisting psychiatric disorders, family problems, employment problems, and medical problems.

Coexisting psychiatric disorders are extremely common. To a great extent the presence of these disorders depends on the length of time the individual has been using cocaine, the dose of cocaine taken, and the route of administration.

COCAINE USE PATTERNS

Cocaine may be taken in various ways that differ in speed of onset, in blood levels, and, consequently, in brain levels. These ways of administration, in ascending order of efficiency, are chewing coca leaves (absorption through the mucous membranes of the mouth), oral ingestion of cocaine hydrochloride, intranasal absorption of cocaine hydrochloride, smoking of alkaloidal (freebase) cocaine (CRACK), and intravenous injection of cocaine hydrochloride. There are also different use patterns. Some patients rarely use cocaine except at parties and in relatively low doses. Some ethnic and social groups are particularly likely to use cocaine by the intranasal route, a method that does not achieve as high brain levels as administration via crack (freebase inhalation) or the intravenous route. Women and adolescent users are more likely to use crack, which is inexpensive per unit dose. A vial of crack sufficient to produce a brief, intense period of euphoria averages 2 to 3 dollars in some large East Coast cities.

Other users tend to administer cocaine in intense bursts several times per week. A binge may last several hours or even several days. In these individuals the binge is usually terminated by exhaustion of supplies or by behavioral, cardiovascular, or neurological side effects. Regular low- or moderate-dose daily use is not a typical pattern.

The higher the dose of cocaine reaching the nervous system and the longer the period of use, the more likely that there will be some form of behavioral toxicity. Personality change consisting of irritability, suspiciousness, and paranoia may occur. Psychosis consisting of HALLUCINATIONS and persecutory delusions with increasing likelihood of violence is also seen in heavy cocaine users. Auditory hallucinations are the most common, but tactile an gustatory hallucinations are occasionally reported. During the crash period after termination of a binge of cocaine use, there is often DEPRESSION. The period of depression is usually brief, but in some patients it can trigger a major affective disorder. This is a psychiatric syndrome consisting of a period of continuing depression, often with accompanying somatic symptoms and requiring ANTIDEPRESSANT medication. For most patients, cocaine withdrawal consists of several days of gradually decreasing depression and fatigue with episodes of craving for cocaine.

PHASES OF TREATMENT

Treatment can be divided into three phases: (1) achievement of initial abstinence or detoxification; (2) rehabilitation; (3) aftercare. The treatment of cocaine abuse or dependence should always be thought of in terms of these phases, and the patient and the patient's family should be told to anticipate a period of treatment lasting at least 18 months and often 3 years or longer.

Achievement of Initial Abstinence. For most patients no pharmacological assistance is required to achieve initial abstinence. This is because most patients use irregularly; thus, they stop and restart cocaine use frequently. Although there is a definite cocaine withdrawal syndrome, it has an irregular pattern and does not fit neatly into distinct phases. Careful studies of patients going through cocaine withdrawal reveal an early severe period of dysphoria, depression, fatigue, and sleepiness. Over the ensuing hours and days there is a gradual improvement. There may also be physical signs, such as a bradycardia (slow heart rate) that gradually returns to normal. These withdrawal symptoms may be accompanied by periodic severe craving for cocaine. If a person is being treated on an outpatient basis, achieving abstinence can be very difficult.

Some clinical investigators have recommended medication to aid in the achievement of initial abstinence. Bromocriptine has been recommended because it stimulates DOPAMINE receptors. There is some evidence that the withdrawal period is characterized by a downregulation of RECEPTOR sensitivity and a reduction in the production of the NEUROTRANSMITTER dopamine. Thus bromocriptine may reverse some of these effects; however, there is also a potential for side effects from this medication. Most clinicians do not recommend the routine use of bromocriptine to treat acute cocaine withdrawal.

Another dopaminergic medication, AMANTADINE, has been used in an outpatient study to help patients achieve initial abstinence. It is very important to evaluate potential medications for any disorder by using a comparison or control group. Typically a group of patients is randomly assigned to receive either the drug to be tested or a placebo. Patients are given identical-appearing capsules so that neither the patients nor the physicians know who is receiving the test drug and who is getting the placebo. Such a double-blind trial found a significant advantage for patients randomly assigned to amantadine as compared with the group receiving a placebo. This was found only during the initial two-week phase of treatment, when the goal is achievement of abstinence. Another outpatient study found desipramine to be helpful in achieving early abstinence and maintaining it for six weeks. This was relatively early in the cocaine epidemic, and the patients were all intranasal users. More severely cocaine-dependent patients have generally failed to show these advantages for desipramine.

Rehabilitation Phase. The major emphasis of treatment should be prevention of relapse to compulsive cocaine use. Some clinicians recommend inpatient treatment to obtain initial abstinence and to begin the process of rehabilitation. In no case should inpatient treatment be considered adequate by itself. It is an expensive modality, and sooner or later the patient has to return to his or her usual environment. Thus the treatment must be continued on an outpatient basis. Other clinicians recommend giving all patients an initial trial of outpatient treatment. They would reserve inpatient treatment only for those who repeatedly fail in less expensive outpatient programs.

One study made a direct comparison between outpatient and inpatient rehabilitation. Patients at the Philadelphia Veterans Administration Medical Center were randomly assigned to either inpatient treatment for achievement of initial abstinence and for a 28-day rehabilitation program or outpatient treatment for achievement of initial abstinence and a hospital day program for rehabilitation. The hospital day program was similar to the inpatient rehab program in that it was based on the TWELVE STEPS with emphasis on group therapy, although some individual counseling was available. Patients came to the day hospital 5 days per week for a total of 27 hours of therapy per week. Those in the inpatient program remained in treatment 7 days per week, 24 hours per day. Those in the day program returned home in the evening.

At the end of the 28-day program, both groups were encouraged to continue treatment in an aftercare program consisting of weekly visits to the outpatient clinic. At the end of four months and at the end of seven months, evaluations were conducted on all patients initiating the study, even if they had dropped out immediately after beginning. The results showed that there were fewer dropouts in the inpatient program, but even including all dropouts,

there was no significant difference between the two groups. Both had a 50–60 percent success rate at the two follow-up periods. Success was defined as no cocaine use for the prior 30 days, supported by a negative urine test at the time of the interview.

A few drug-dependent patients stop their drug use and remain permanently abstinent. In most cases, there are slips during which the patient again uses cocaine or other drugs. A slip does not denote a relapse or a failure of treatment. It does require additional counseling to determine why it occurred and how it might have been prevented.

Medications have been used during the rehabilitation period in an attempt to improve the results of counseling and psychotherapy. Desipramine has been reported to be of some benefit in this phase of treatment, although more recent studies with severe cocaine dependence have failed to replicate early reports of success. Carbamazepine has been reported to have benefits because it has been shown in animals to block the development of subcortical seizure activity produced by cocaine. Controlled studies, however, have failed to show any benefit for this anticonvulsant medication in prevention of relapse. There have also been claims of benefit for ACUPUNCTURE, but there is no scientific evidence to support them. There have also been reports in the lay literature that the hallucinogenic drug IBOGAINE produces a long-term loss of craving for cocaine. There are no follow-up studies to support this claim, and the lay press has reported three deaths from the use of this drug. Also, there have been animal studies showing toxicity to cerebellar neurons after ibogaine administration. Moreover, animal models have failed to consistently demonstrate a reduction in cocaine self-administration in rodents given ibogaine.

Based on knowledge of the pharmacological effects of cocaine, there has been an intensive search for medications that might ease cocaine withdrawal or reduce craving for the drug in recovering addicts. Cocaine is known to block the dopamine transporter, a specialized membrane protein that serves to clear cocaine from the synaptic space after it has been released, thus helping to terminate neurotransmission. Cocaine use produces excessive dopaminergic stimulation. Cocaine also increases the availability of other neurotransmitters, such as serotonin. The search for a medication to improve the results of treatment for cocaine dependence has focused on substances that influence dopamine mechanisms, either presynaptic or at the receptor level. There have also been studies of medications that influence the serotonin system. As of the mid-1990s, no medication has been consistently helpful in the treatment of cocaine dependence.

Psychotherapy During Rehabilitation. In addition to standard treatments provided in most rehabilitation programs, such as the twelve-step program, group therapy, vocational counseling, and family therapy, there have been studies using specific psychotherapy and behavioral therapy. A large-scale multiple-clinic trial using supportive expressive psychotherapy in the treatment of cocaine dependence is under way, as of the mid-1990s. The pilot trial indicated significant benefits from individual psychotherapy in selected cocaine patients.

Reinforcement of Clean Urine. Another treatment approach that has resulted in significant success is using systematic reinforcement of cocaine abstinence. Researchers arranged for patients to be rewarded with vouchers that could be exchanged for desirable goods, restaurant meals, or other constructive purchases when they presented drug-free urine. This treatment approach was accepted well by patients, and the results were significantly better than a control group receiving counseling alone. A one-year follow-up of patients previously treated for six months in this manner showed that 74 percent were abstinent during the 30 days prior to the follow-up interview.

A similar study has reported on opiate-dependent patients on a methadone program who were also abusing cocaine. A program of reinforcement of clean urine using vouchers that could be exchanged for desirable objects produced a significant reduction in their cocaine use.

Extinction of Cocaine-Related Cues. Even highly motivated former cocaine-dependent patients experience craving after stopping the use of cocaine. While they are in a protected hospital environment, they may feel confident that they can remain abstinent, but upon returning to their neighborhood, they encounter environmental cues that may produce excitement and cocaine craving. These cues usually are people, places, and things that had previously been linked to cocaine use. Many patients say they become so conditioned to the effects of cocaine that simply seeing their drug dealer or a vial of cocaine produces a rush long before the drug gets into their body. Treatments have been designed to reduce or extinguish these conditioned responses. They consist of repeatedly reviewing drug-related stimuli and

learning various coping skills, such as the relaxation response, visual imagery, and mastery techniques. These techniques are used by behavioral therapists to reduce the symptoms of other disorders, such as phobias or obsessive–compulsive disorder. For cocaine dependence, the patient can be taught the techniques by the therapist. Later he or she can practice the techniques in the clinic by viewing videos of cocaine use. There is now evidence that patients randomly assigned to these behavioral treatments do significantly better in outpatient treatment than do control subjects assigned to standard treatment with the same amount of attention.

Aftercare. After about a month of intense rehabilitation treatment, a patient can graduate to an aftercare program of variable intensity. Sessions may initially be once or twice a week, decreasing gradually to once or twice per month. There should continue to be urine testing to monitor drug use. The cocaine metabolite, benzoylecgonine, remains in the urine for several days and can effectively signal a resumption of cocaine use. Patients who admit to a slip or show evidence from urine tests that they have used cocaine should resume intensive counseling. Every attempt should be made to determine why the slip occurred, so that it can be avoided in the future. A slip of this nature should not be considered to be a treatment failure, even if it results in a significant binge. It is simply a signal that the patient requires a resumption of intensive treatment. Thus far there is no evidence that any medication is helpful in this phase of treatment. Of course, if the patient remains depressed or anxious, or has symptoms of another psychiatric disorder, specific treatment such as antidepressants should be employed.

SUMMARY

Cocaine abuse and dependence represent chronic disorders that require long-term treatment. A brief initial inpatient phase may be necessary, but the major part of treatment consists of long-term outpatient care. Various techniques can be used. Most patients receive group therapy and counseling based on the Twelve Steps developed by Alcoholics Anonymous. Professional psychotherapy may be helpful in selected cases, but data are still preliminary. There are also data showing efficacy for behavioral treatments, such as contingent reinforcement of clean urine and for extinction of craving and arousal responses produced by cocaine-related stimuli. Only a few studies so far show long-term results of treatment. After seven to twelve months of treatment, 50 to 60 percent of patients do well; one study reported a 74 percent success rate.

(SEE ALSO: *Abstinence Violation Effect; Addiction: Concepts and Definitions; Comorbidity and Vulnerability; Complications; Drug Metabolism; Pharmacokinetics; Relapse Prevention; Treatment Types; Wikler's Pharmacologic Theory of Drug Addiction*)

BIBLIOGRAPHY

ALTERMAN, A. I. ET AL. (1992). Amantadine may facilitate detoxification of cocaine addicts. *Drug and Alcohol Dependence, 31,* 19–29.

ALTERMAN, A. I., & McLELLAN, A. T. (1993). Inpatient vs. day hospital treatment services for cocaine and alcohol dependence. *Journal of Substance Abuse Treatment, 10,* 269–275.

CHILDRESS, A. R. ET AL. (1994). Using active strategies to cope with cocaine cue reactivity: Preliminary treatment outcomes. NIDA Research Monograph Series (in press, 1995).

HIGGINS, S. T., ET AL. (1991). A behavioral approach to achieving initial cocaine abstinence. *American Journal of Psychiatry, 148*(9), 1218–1224.

SILVERMAN, K, ET AL. (1994). Contingency management of cocaine use in a methadone maintenance program. Paper presented at the annual meeting of the College on Problems of Drug Dependence, Palm Beach, FL, June 1994. NIDA Research Monograph Series (in press, 1995).

WEDDINGTON, W. W. (1990). Changes in mood, craving, and sleep during short-term abstinence reported by male cocaine addicts. *Archives of General Psychiatry, 47,* 861–868.

CHARLES P. O'BRIEN

Cocaine, Pharmacotherapy The pharmacological treatment of COCAINE abuse is defined as the use of medication to facilitate initial abstinence from cocaine abuse and to reduce subsequent relapse. The initiation of abstinence from cocaine abuse involves reduction in the withdrawal symptoms associated with cessation of cocaine. This WITHDRAWAL syndrome resembles depression but includes a great deal of anxiety and craving for co-

caine. CRAVING for cocaine often persists for several weeks after abstinence has been attained, and places or things associated with cocaine use in the past, called *cues,* can continue to stimulate cocaine craving for many months. Because of this persistence of what is known as *conditioned craving,* relapse to cocaine abuse can occur after the patient has become abstinent. Preventing relapse is a second important function of medication treatment.

An objective of the use of medications in cocaine dependence is to reverse changes that are caused in the brain after chronic cocaine use. These brain changes, called *neuroadaptation,* are easily demonstrated in animal models of cocaine dependence. Chemical analyses of animal brains exposed to cocaine chronically show abnormalities in the NEUROTRANSMITTER receptors on brain cells. The brain cell receptors that are affected by cocaine include DOPAMINE receptors and SEROTONIN receptors.

Direct and indirect evidence that there are changes in these brain receptors can also be found in human studies. Prolactin is a hormone that is controlled by the neurotransmitters dopamine and serotonin. In some heavy cocaine abusers, prolactin levels are abnormally high after abuse has stopped and remain elevated for a month or more. This evidence suggests that both dopamine and serotonin brain systems are perturbed by cocaine and that the abnormality persists for some time. Other evidence of persistent abnormalities in the dopamine systems comes from brain imaging studies directly examining dopamine receptors. Positron-emission tomography (PET) studies have shown a marked reduction in dopamine receptors on brain cells that are ordinarily very rich in such receptors. This abnormally low amount of dopamine receptors persists for at least two weeks after a patient stops using cocaine. That several medications may reverse these neurochemical receptor changes has been an important rationale for their use.

These brain changes are thought to occur after chronic heavy cocaine use and not after single, small amounts of cocaine have been ingested. These brain changes are also most likely to occur when cocaine enters the body by intensive routes of administration such as by smoking or injecting. Much larger dosages of cocaine get to the brain through these routes than through others such as orally or intranasally ingested (snorted) cocaine. In addition to these direct biological indicators of neuroadaptation, neuropsychological tests have documented sustained deficits in thinking, concentration, and learning among chronic cocaine abusers. These deficits may persist for months after cocaine use has stopped.

The biological abnormalities in the brains of abusers clinically may be manifest by a characteristic withdrawal syndrome. The very early phases of this syndrome, commonly called the "crash," may involve serious psychiatric complications, such as paranoia with agitation and depression with suicide. These complications require medications for symptomatic management, including ANTIPSYCHOTIC agents, such as chlorpromazine and haloperidol, or large dosages of BENZODIAZEPINES to calm highly agitated patients. Many patients self-medicate these crashes using such sedating substances as benzodiazepines or alcohol. Because this crash phase is usually relatively brief, rarely lasting more than several days, there is generally no role for sustained medication. The more important role for medications occurs during the later phase of withdrawal from cocaine, which may persist for several weeks. This later phase resembles a depressive syndrome, with substantial anxiety and craving to use cocaine. The neurobiological changes noted in both human and animal studies after chronic cocaine use correspond in time to the occurrence of this syndrome. This temporal correspondence has provided a further rationale for the use of ANTIDEPRESSANT medications in the treatment of cocaine dependence and withdrawal.

A wide range of pharmacological agents besides antidepressants have been tried as treatments for cocaine withdrawal and as agents to reduce relapse back to cocaine dependence. Other agents besides antidepressants include dopamine agonists, dopamine blockers, lithium, monoamine-oxidase inhibitors, amino acids, and opioid-derived medications. The use of opioid-derived medications to treat cocaine dependence has an ironic twist, because Sigmund Freud had suggested that cocaine might be an appropriate treatment for morphine (an opioid) addiction. Clearly substituting one drug of abuse for another drug of abuse is a risky treatment approach, but new ideas are emerging on the use of opioids with lower abuse potential than morphine, such as BUPRENORPHINE for patients dependent on both opioids and cocaine.

Evaluation of medications in controlled studies using double blinding and random assignment is very important, because a substantial placebo response may occur in cocaine abusers when they enter treatment, even if they are given a simple sugar pill. In double-blind, placebo-controlled studies, neither the patient nor the physician knows whether the patient is receiving active medication or placebo. Controlled studies provide the clearest indication of an efficacious medication when it is found to be significantly better than a placebo given to similar patients in a randomized and blinded manner. Randomization simply means that patients who are potential subjects for a study are randomly assigned to get either the active medication or the placebo. Choices about who will get active medication and who will get the placebo are made by chance alone and not decided by the physician based on drug-abuse severity or any other criteria. Relatively few medications have had this rigorous testing of their efficacy for cocaine abuse treatment, and most medications have undergone only uncontrolled testing. In uncontrolled tests, patients are given the medication and their response is compared with their behavior before starting treatment.

ANTIDEPRESSANTS

In a controlled study, desipramine, an antidepressant, has been found clearly superior to the placebo or to the active medication lithium. In a study of 72 cocaine-dependent patients, 24 of whom received desipramine, desipramine was found to reduce significantly cocaine craving and use. By the second week of this 6-week study, patients given desipramine reported a significant reduction in cocaine use compared with those given the placebo. Self-reported cocaine use was reduced from an average of 3.5 grams a week at intake into the study to less than 0.25 gram during week two, and many patients reported complete cocaine abstinence. The placebo and lithium groups never dropped below 1.5 grams per week. Moreover, 65 percent of the patients on desipramine attained at least 3 weeks of abstinence compared with only 20 percent of the placebo group and 30 percent of the lithium group. Statistical differences were highly significant, thus supporting the efficacy of desipramine as an agent to facilitate initial abstinence from cocaine. However, this study has been criticized because it relied primarily on self-report rather than on monitored urine tests.

Desipramine was felt to increase cocaine abstinence by reducing craving for cocaine, and reported cocaine craving was significantly less in the desipramine group than in the placebo and lithium groups.

Craving decreased by more than 80 percent of baseline by the fifth week of treatment, but, interestingly, cocaine use declined several weeks before cocaine craving was reduced. This delay suggested that desipramine reduced the recurrence of craving after cocaine abstinence had been attained, and thus its anticraving action might be more important for the prevention of relapse than for the initiation of abstinence. Neurobiological actions of desipramine, such as the decrease in dopamine receptor sensitivities, might be important for the reduction in cocaine use and the initial facilitation of abstinence, independent of this decrease in craving. A 6-month follow-up of these patients after the medication was discontinued found that many patients relapsed to cocaine abuse in all three of the treatments: placebo, desipramine, and lithium. Although the rate of abstinence dropped from 75 percent to 55 percent at the follow-up for the desipramine group, a similar or greater rate of relapse was found in the other two groups. For other drugs of abuse, such as benzodiazepines, alcohol, and opiates, there was no increase from the patients' use before entering treatment; however, some patients who had relapsed to cocaine dependence also began abusing alcohol. No evidence of switching to other drugs of abuse was shown.

Two other studies have examined desipramine in the treatment of cocaine-abusing heroin addicts. In a study with male military veterans at the University of Pennsylvania, cocaine abusers on METHADONE MAINTENANCE were found to have no significant benefit from desipramine compared with placebo. At a 6-month follow-up after discontinuing desipramine, however, the desipramine group appeared to have a worse outcome than patients who had received the placebo.

In a similar study done at Yale University involving both male and female cocaine abusers on methadone maintenance, desipramine also showed little efficacy relative to a placebo. In this study, 94 patients were treated with desipramine, amantadine, or a placebo. Both medications showed some superiority to the placebo in self-reported cocaine use during the first 5 weeks of treatment in this 8-week trial, but urine toxicologies to confirm self-reports showed no significant effect of the medication. During the first 5 weeks of treatment, the patients who received the placebo reported an increase in cocaine use to 150 percent above the baseline. In contrast, the pa-

tients who received active medication showed a steady decline in their cocaine use to about 30 percent of baseline use by week four. This reduction in cocaine use on medications was not as robust as the reduction found in pure cocaine abusers and could not be supported by urine tests. This lesser efficacy in methadone-maintained patients may be partially accounted for by the interference of methadone with desipramine metabolism. This metabolic interference reduces the production of an active desipramine metabolite (2-hydroxy-desipramine) that may be critical for desipramine's anticraving effect.

Other treatment trials with desipramine using an adequate dose and sufficient duration also failed to show its superiority to a placebo. Thus studies with this cocaine pharmacotherapy have not been uniformly supportive.

Pilot studies with other antidepressants have suggested their efficacy, but controlled studies are usually lacking. Suggestive agents include imipramine, maprotiline, phenelzine, and trazadone. Recent controlled trials of SEROTONIN-UPTAKE INHIBITORS such as fluoxetine (Prozac) and of a different type of serotonergic antidepressant, gepirone, have shown no efficacy compared with placebo.

DOPAMINERGIC AGENTS

In theory, dopaminergic agents may be useful in ameliorating early withdrawal symptoms after cocaine binges, because these agents appear to have their onset of action within a day of starting. These agents include AMANTADINE, bromocriptine, L-dopa, mazindol, METHYLPHENIDATE, and pergolide. Few adequate, placebo-controlled, double-blind studies have been done with any of these agents, but their acute efficacy has been suggested in several single-dose, placebo crossover trials. Bromocriptine has been studied by several groups of investigators and has shown efficacy for some and not for others. Several trials have examined amantadine at 200 and 300 milligrams (mg) daily and found that it reduces craving and use for several days to a month. A controlled trial by Weddington et al. (1991) did not show amantadine to be more efficacious than a placebo. Four trials have examined bromocriptine at doses varying from 0.125 to 0.6 mg three times daily but have had contradictory results on efficacy, mostly because of dropout from side effects. Two open studies of

L-dopa and methylphenidate have shown short-term (2-week) reductions in cocaine craving, although a second trial using methylphenidate found that it increased cocaine use over several weeks of treatment. Mazindol has shown promise in an open trial, but a relatively brief crossover study showed no significant improvement in cocaine craving or use. Pergolide, which has more specific dopamine D2 receptor agonist properties than bromocriptine, has been shown to decrease significantly cocaine craving, without the excessive side effects that often lead to early dropout in bromocriptine studies. This may be the result of its lack of D1 receptor antagonism, which characterizes bromocriptine.

MISCELLANEOUS AGENTS

A number of other agents recently have been utilized to treat different aspects of cocaine abuse and dependence. Several authors report a decrease in euphoria and/or paranoia with neuroleptics (ANTIPSYCHOTIC medications). Of these, only flupenthixol (not available in the United States at this time) was acceptable to the patients. In this study, flupenthixol decanoate was injected in a population ($n = 10$) of treatment-resistant patients, resulting in a dramatic decrease in cocaine craving and use, and an increase in treatment retention. Buprenorphine, a mixed opiate agonist-antagonist, showed initial promise for cocaine-abusing methadone-maintained patients in an open trial, but subsequent double-blind, controlled studies were not able to replicate these positive findings. Another open study demonstrated initial promise with the anticonvulsant carbamazepine in the treatment of cocaine-dependent patients. Its use was based on animal models and suggested a "kindling" mechanism in cocaine abuse, especially in high-dose cocaine binging. But as was the case with buprenorphine these initial findings were not confirmed in double-blind studies.

(SEE ALSO: *Causes of Substance Abuse; Drug Metabolism; Research, Animal Model*)

BIBLIOGRAPHY

GAWIN, F. H., ET AL. (1989). Desipramine facilitation of initial cocaine abstinence. *Archives of General Psychiatry, 46,* 117–121.

JAFFE, J. H. (1985). Drug addiction and drug abuse. In A. G. Gilman et al. (Eds.), *Goodman and Gilman's the pharmacological basis of therapeutics,* 7th ed. New York: Macmillan.

KOSTEN, T. R. (1989). Pharmacotherapeutic interventions for cocaine abuse: Matching patients to treatment. *Journal of Nervous and Mental Disease, 177*(7), 379–389.

KOSTEN, T. R., & KLEBER, H. D. (EDS.). (1992) *Clinician's guide to cocaine addiction.* New York: Guilford Press.

LOWINSON, J. H., RUIZ, P., & MILLMAN, R. B. (EDS.). (1992). *Substance abuse: A comprehensive textbook.* Baltimore: Williams & Wilkins.

MILLER, N. S. (ED.). (1991). *Comprehensive handbook of drug and alcohol addiction.* New York: Marcel Dekker.

WEDDINGTON, W. W., ET AL. (1991). Comparison of amantadine and desipramine combined with psychotherapy for treatment of cocaine dependence. *American Journal of Drug and Alcohol Abuse, 17,* 137–152.

THOMAS R. KOSTEN

Cocaine, Psychological Approaches No consensus exists about how to treat COCAINE dependence. This statement is particularly alarming given that in 1990 it was estimated that somewhere between 1.5 and 2.2 million persons in the United States were dependent on cocaine. The abuse of cocaine was first recognized in the medical literature in the late 1800s. Early proposed treatments included various herbal and medical potions, nutritional supplements, hot baths, substitution of MORPHINE, long stays in sanitoriums, education, and psychotherapy. Systematic evaluation of the effectiveness of these early treatments did not occur.

The goals and focus of psychological treatments for cocaine dependence vary greatly depending on the beliefs held by the treatment provider regarding the causes of cocaine dependence. The efficacy of the various treatments is only beginning to be evaluated. This article describes the primary psychological approaches used to treat cocaine dependence, as of the mid-1990s, and discusses the efficacy of those interventions. Although numerous psychologically based interventions are being used as treatments for cocaine dependence, this article is limited to providing an overview and discussion of approaches that have received attention in the scientific literature.

OUTPATIENT VERSUS INPATIENT TREATMENT

Studies suggest that inpatient rehabilitation is neither necessary nor cost-effective in most cases of drug or alcohol dependence. With cocaine dependence, inpatient treatment may be indicated if the patient (1) fails to make progress or deteriorates during outpatient treatment; (2) has severe medical or psychiatric problems; or (3) is physically dependent on other drugs. Some professionals suggest that individuals who regularly smoke cocaine (CRACK or freebase) or use cocaine intravenously need an initial period of hospitalization to get impulsive use under control; recent data, however, suggest that this may not be necessary (Budney et al., 1993).

No well-controlled treatment studies have compared the effectiveness of inpatient versus outpatient treatment for cocaine dependence. Results from a quasi-experimental study and preliminary results from a controlled study are available. Rawson et al. (1986) followed fifty-three cocaine users who self-selected an outpatient or inpatient treatment program. Outpatient treatment resulted in a lower relapse rate to cocaine after treatment than did the inpatient program. The primary problem with the inpatient program was that most patients did not follow through with scheduled treatment after their release from the hospital. O'Brien et al. (1990) examined forty-two male cocaine abusers randomly assigned to a twenty-eight-day intensive outpatient program or a twenty-eight-day inpatient program. Although the inpatient program had a higher rate of persons remaining in treatment the full twenty-eight days, no difference between groups in cocaine use at a four-month follow-up point was reported. No urinalysis results, however, were reported to confirm these findings.

The methodological limitations (self-selection of treatment and the lack of urinalysis data) prevent drawing strong conclusions from these studies. Instead, studies using (1) random assignment of patients to inpatient or outpatient treatment; (2) similar approaches to treatment in both settings; and (3) frequent urine testing are needed to help determine the difference in outcome between outpatient and inpatient treatments.

In general, inpatient treatment should be considered an initial stage of treatment. Learning to cope with the multitude of environmental circumstances that have contributed to the initiation and mainte-nance of cocaine abuse is the most important task of the abuser. This task can be accomplished effectively only outside the hospital.

COCAINE ANONYMOUS

COCAINE ANONYMOUS (CA) is a community-based self-help group organization modeled after ALCOHOLICS ANONYMOUS (AA). CA groups, however, are not as plentiful as AA groups or NARCOTICS ANONYMOUS (NA) groups, so many cocaine dependent persons who seek out a SELF-HELP group use NA or AA. The basic principles are the same as AA's. The program is based on the "disease" model of substance dependence. Achievement and maintenance of abstinence from cocaine is presumed to be facilitated by following the Twelve Steps of CA (which are based on the original TWELVE STEPS of AA).

CA is available to anyone who expresses a desire to stop using cocaine. All that is necessary to become a group member is that one attend meetings. Meetings vary from large open ones that anyone can attend to small, closed discussions reserved for specific groups. For example, a group of young people, professionals, or women is organized to address specific concerns. At most meetings, experiences are shared and advice and support are given. Two other components of the CA program are sponsorship and education. A sponsor is a person who has been in recovery for a substantial period of time and who is available at any time to provide support and guidance to the person attempting to recover. Education about the "disease" is provided through pamphlets, books, films, and other literature.

It is difficult to evaluate the efficacy of CA. Research on the topic has not been conducted due to the anonymous nature of CA membership. No objective data are available on its effectiveness. The limited research on AA suggests that a high dropout rate exists; this is likely true of CA as well. Possible reasons for dropout are the religious overtones of the program, social anxiety concerning group participation, and the potential loss of confidentiality associated with group situations. In general, one can reasonably assume that a number of cocaine-dependent persons can use CA to help facilitate recovery, while others either do not benefit from or refuse to attend such a program. Despite the absence of empirical evidence of its efficacy, CA is recommended by many treatment professionals as the treatment for or as an important adjunct of treatment for persons with cocaine problems.

GROUP THERAPY

Many professionals suggest that group therapy is an invaluable component of cocaine abuse treatment (e.g., Carroll et al., 1987; Washton, 1987). Most groups are structured to include persons of different backgrounds and at different stages of recovery (1) to help deal with feelings of uniqueness, (2) to expose those in the early stage of treatment to positive role models, and (3) to help instill hope for success. Those who promote group therapy view peer pressure and support as necessary to overcome ambivalence about abstaining from cocaine. Providing support for others and the development of intimate social interaction (e.g., sharing of feelings) is facilitated and presumed to be therapeutic.

Topics of discussion in group therapy vary depending on the group members and the orientation of the therapist. Topics may include early abstinence issues, guilt resolution, marital conflict, or lifestyle changes. Education about adverse effects of cocaine is often included. Group therapy occurs in outpatient or inpatient settings. It is sometimes used as the sole source of treatment or combined with individual counseling and other treatment components.

Although group therapy is highly recommended by many professionals, the only study that has evaluated its efficacy demonstrated poor outcome (Kang et al., 1991). Limitations of group approaches, such as loss of confidentiality, avoidance because of social anxiety, and negative peer influences are rarely discussed in the literature. Treatment research that focuses on assessing the efficacy of different types of group therapy and group versus individual therapy is needed.

SUPPORTIVE-EXPRESSIVE AND INTERPERSONAL PSYCHOTHERAPY

Psychotherapy is usually suggested as a component of cocaine-treatment programs, both inpatient and outpatient. Typically, the therapy is based on psychodynamic theories of substance abuse. This means that intrapersonal factors and underlying personality disturbances are considered causes of cocaine abuse. It is presumed that cocaine is used to cope with painful emotional states, and that issues

such as separation-individuation, depression, and dependency must be resolved to maintain abstinence. The therapist tends to adopt an exploratory role that promotes insight into interpersonal and intrapersonal conflict underlying the cocaine dependence. Increased insight is presumed to result in a reduction in the underlying problems, which, in turn, should result in cocaine abstinence.

Probably the most common therapy offered for cocaine dependence is supportive or supportive-expressive psychotherapy. This therapy in combination with pharmacotherapy has demonstrated some efficacy in previous research with HEROIN-dependent persons (Woody et al., 1983). Initially, supportive psychotherapy focuses on acknowledging the negative consequences of cocaine use, accepting the need to stop using, and helping manage impulsive behavior. The therapist and user explore ways to stay away from other users and high-risk environments. The focus of treatment then shifts to insight-oriented psychotherapy in which the therapist's role is to facilitate the exploration of underlying reasons for the cocaine abuse. Long-term abstinence depends on the degree to which the underlying psychic disturbances are resolved.

Seigel (1982) treated thirty-two heavy cocaine smokers with supportive psychotherapy, self-control strategies, exercise therapy, and hospitalization during the initial detoxification phase. Sixteen of the thirty-two abusers dropped out of treatment. Thirteen of the sixteen who remained in the treatment were reported to be cocaine-free at a nine-month follow-up, but no urinalysis results were provided to confirm this.

Kang et al. (1991) reported data from a study examining the effect of once-a-week supportive-expressive therapy, family therapy, and group therapy with 148 cocaine-dependent persons. Treatment was twenty-four weeks in duration. No significant group differences in outcome resulted. Only 40 percent of the subjects attended 3 or more sessions; only 25 percent of those subjects (9% of the total sample) did not use cocaine during treatment. The authors concluded that low-intensity psychotherapy was ineffective with the majority of their subjects.

Interpersonal psychotherapy (IPT) was originally developed for and found to be effective with DEPRESSION (Klerman et al., 1984). The rationale for using this approach for cocaine dependence is based on theory relating disturbances in interpersonal functioning to the onset and/or maintenance of cocaine abuse (Rounsaville et al., 1985). The focus of IPT is the improvement of interpersonal functioning. The therapist helps the patient to recognize the function or role cocaine has in meeting the needs of his or her life. Other means of meeting these needs are explored. Four types of interpersonal problems are addressed: interpersonal disputes, role transitions, grief, and interpersonal deficits.

Carroll et al. (1991) compared IPT to a relapse-prevention therapy (RPT) with forty-two cocaine dependent subjects. Both therapies were twelve weeks in duration and involved once-a-week individual therapy sessions. The RPT treatment is based on behavioral principles. It focuses on the development of skills to identify and cope with potential relapse situations. Cocaine use was determined primarily by self-report, although some random urine testing was used in an attempt to confirm self-report. The use of RPT resulted in a treatment retention rate of 67 percent compared to only 38 percent for IPT. Also, RPT resulted in a greater number of subjects achieving a continuous three-week period of abstinence at any point during treatment (57%) and during the last three weeks of treatment (43%) than IPT (33% and 19%, respectively). These differences, however, did not reach statistical significance. In addition, RPT was more effective with the more severe cocaine users as well as with those with higher psychiatric severity.

As of the mid-1990s, the limited treatment-outcome research on supportive and interpersonal psychotherapies has not produced promising results for the treatment of cocaine dependence.

BEHAVIORAL TREATMENT

Behavioral perspectives of cocaine dependence view drug taking as a learned behavior that begins and continues because of the reinforcing effects of the drug. These reinforcing effects are determined, in part, by basic biological events in the brain. This means that, to some extent, most persons are susceptible to becoming dependent because cocaine produces a reaction in the brain that increases the likelihood that drug taking will recur. The other factors that determine whether a person will become dependent on cocaine are environmental factors (e.g., peers, acceptance by others, and no apparent negative consequences). In contrast to other theories of the development of cocaine or other drug abuse, behavioral theory is based on experimental evidence from research on drug-taking behavior in humans and laboratory animals. This research clearly demonstrates that cocaine seeking and use are learned responses that occur regularly under specific conditions (e.g., certain times of day, events, internal states). This theory translates into treatment that focuses on changing these "using" conditions and creating new conditions that encourage abstinence from cocaine.

An initial study (Anker & Crowley, 1982) evaluated the efficacy of including a CONTINGENCY CONTRACT with supportive therapy for primarily white, employed, highly educated, male cocaine abusers. The contract first required participation in a urinalysis program designed to detect cocaine use. Second, the patient had to agree to a predetermined negative, severe consequence if cocaine was used. An example of a negative consequence was sending a letter to a professional licensing board informing it of a physician's or nurse's cocaine use. Thirty-two of the sixty-seven patients presenting themselves for treatment entered the contract for three to six months. Thirty-one remained abstinent from cocaine throughout that period. Of the thirty-five patients who did not sign a contract and who received only the standard treatment, none were abstinent for more than four weeks. Follow-up revealed, however, that half of the contract patients relapsed after the contract expired.

This study is important because it demonstrates that cocaine use can be influenced by manipulating the consequences of using. There are, however, limitations of this study that must be noted. First, patients self-selected whether to enter into a contract. This makes it impossible to determine whether they would have done well without the contract. Second, the contract was unacceptable to more than 50 percent of those seeking treatment. Third, ethical issues arise from using severe consequences without first trying a less severe consequence.

Higgins et al. (1991) evaluated a behavioral treatment that provided positive consequences for abstinence, rather than negative consequences for using cocaine. Treatment included individual counseling sessions that addressed the following areas: employment, finances, relationship problems, problem-solving, assertiveness, controlling urges to use, time management, and social and recreational skills. In addition, cocaine abstinence was reinforced with points that could be exchanged for social and recreational activities or services. These interventions were designed for the purpose of assisting the patient develop a rewarding lifestyle that could compete with the reinforcing effects the person was deriving from cocaine. A control group was assigned to a treatment that used more traditional drug counseling including AA/NA participation. Both treatments lasted three months. Behavioral treatment was superior to the drug-counseling treatment in retaining individuals in treatment and in fostering cocaine abstinence.

It should be noted that the subjects were not randomly assigned to the two treatments; therefore, other factors such as unmeasured differences in subject characteristics or time of treatment may have affected the results. Higgins et al. (1993) have however replicated those results in a randomized controlled trial that compared those same two treatments. Moreover, in a third study, Higgins & Budney (1993) demonstrated that the procedure, which involved reinforcing cocaine abstinence with social and recreational items, was an effective component of the treatment package. Thus, the results of that behavioral treatment appear promising.

Another behavioral approach focuses on the conditioned stimuli (environmental events) associated with cocaine use and the way those events affect relapse and deter abstinence attempts. This approach focuses intensely on the persons, places, and things that have frequently been paired with cocaine use. Theoretically, things like drug-using friends, paraphernalia, white powder, and places where cocaine is used can produce cravings for cocaine and ultimately result in cocaine use. Therefore, with repeated exposure to those events under conditions where cocaine is not available (i.e., an extinction procedure), the events gradually lose their ability to elicit the cocaine craving and presumably reduce the probability of cocaine use.

Preliminary reports are available addressing the efficacy of this extinction procedure as an adjunct to psychotherapy for cocaine dependence. Childress et al. (1993) reported that those patients who received extinction training in conjunction with psychotherapy continued involvement with treatment longer after release from the hospital and had more weeks with cocaine-free urine results than those who received psychotherapy without extinction training.

One other behavioral approach that has received increasing attention is Relapse Prevention Treatment (RPT), originally formulated for treating alcohol dependence (Marlatt & Gordon, 1985). The first test of its efficacy with cocaine dependence was the

Carroll et al. (1991) described earlier. RELAPSE PREVENTION was superior to IPT in retaining individuals in treatment and in facilitating greater rates of cocaine abstinence. A second trial of RPT provided additional support for its efficacy. One-year follow-up data showed RPT to be superior to case management in facilitating higher levels of cocaine abstinence (Carroll, 1993).

Recently, two additional behavioral approaches, coping-skills training and neurobehavioral treatment, have received preliminary support as potentially effective treatments. Rohsenow (1993) modified and tested a coping-skills training approach that was originally developed for alcoholism treatment. This treatment is similar to RPT in that it involves teaching specific drug refusal and coping skills important for accessing alternatives to drug use and for coping with events that place the abuser at high risk. Neurobehavioral treatment, developed by Rawson et al. (1993), emphasizes many of the elements of RPT and coping-skills training to assist the abuser to abstain from cocaine and avoid relapse. The "neuro" prefix denotes specific treatment focus on difficulties that may arise due to the neurobiological changes that accompany abstinence from cocaine.

ECLECTIC TREATMENT

Many treatment providers use an eclectic approach to treat cocaine dependence—that is, a combination of approaches. For example, many programs based on a disease or a psychodynamic model may use certain behavioral procedures such as contingency contracting or relapse prevention strategies.

Washton (1987) treated sixty-three well-educated, high-functioning cocaine abusers. Forty-seven were treated in a six-month outpatient program, while sixteen received four to ten weeks of hospitalization prior to entering the outpatient program. Treatment included supportive, educational, and behavioral components provided in individual and group-therapy settings. Urine testing was conducted two to three times a week. Sixty-seven percent of the patients completed six months of treatment. Fifty-one of the patients did not report a "major relapse"; however, approximately half of the fifty-one reported occasional cocaine use. Limitations of this study were that patients self-selected treatment, they were middle-to-high socioeconomic status, and precise data on cocaine use were not reported. Therefore, the efficacy of this treatment for the general cocaine-abusing population cannot be determined until a more rigorous, controlled test of its efficacy has been conducted.

In general, a limitation of eclectic approaches is that mixed messages may be given to the patient. Moreover, the intensity and quality of each component may not be as high as approaches that are more unilateral in focus. For example, behavioral approaches spend a great deal of time counseling and assisting the abuser to make the behavioral changes needed to achieve and maintain abstinence. Eclectic approaches may spend only a small portion of time on those changes. The small time spent focused on those changes may not be sufficient to facilitate change—and it may give the abuser the message that those changes are relatively unimportant.

CONCLUSIONS AND FUTURE DIRECTIONS

This overview of the cocaine-treatment literature indicates that there has not been enough well-conducted research to draw strong conclusions regarding the efficacy of the different approaches. It is equally clear that more has been written about what professionals think works for cocaine dependence than has been demonstrated using objective data. That said, a number of promising findings have been produced by the behavioral approaches. This is of particular importance because behavioral treatments are the only approaches to cocaine dependence that are based on a consistent body of scientific principles.

In addition to the need for basic treatment-outcome research on cocaine dependence, some specific issues that may influence treatment outcome need to be addressed in future studies. These issues include (1) the use of other drugs including ALCOHOL, (2) the presence of other psychiatric problems, (3) the route of administration (intranasal, smoked, or intravenous), and (4) the severity and duration of the abuse. It is unclear at this time how many of these variables affect treatment outcome.

Clearly, more research and evaluation of cocaine-dependence treatment is necessary. Future investigations, at a minimum, should include urinalysis results as an objective measure of cocaine use. Without those data, it is impossible to evaluate any treatment program adequately.

(SEE ALSO: *Adjunctive Drug Taking; Causes of Substance Abuse; Comorbidity and Vulnerability; Disease Concept of Alcoholism and Drug Abuse; Treatment Types*)

BIBLIOGRAPHY

ANKER, A. L., & CROWLEY, T. J. (1982). Use of contingency contracts in specialty clinics for cocaine abuse. In L. S. Harris (Ed.), *Problems of drug dependence 1981.* NIDA Research Monograph no. 41. Washington, D.C.: U.S. Government Printing Office.

BUDNEY, A. J., HIGGINS, S. T., BICKEL, W. K., & KENT. L. (1993). Relationship between intravenous use and achieving initial cocaine abstinence. *Drug and Alcohol Dependence, 32,* 133–142.

CARROLL, K. M. (1993). Psychotherapy and pharmacotherapy for ambulatory cocaine abusers. Paper presented at the NIDA Technical Review Meeting on Outcomes for Treatment of Cocaine Dependence, September, Bethesda, MD.

CARROLL, K. M., ROUNSAVILLE, B. J., & GAWIN, F. H. (1991). A comparative trial of psychotherapies for ambulatory cocaine abusers: Relapse prevention and interpersonal psychotherapy. *American Journal of Drug and Alcohol Abuse, 17,* 229–247.

CARROLL, K. M., ET AL. (1987). Psychotherapy for cocaine abusers. In D. Allen (Ed.), *The cocaine crisis.* (pp. 75–105). New York: Plenum.

CHILDRESS, A. R. ET AL. (1993). Cue reactivity and cue reactivity interventions in drug dependence treatment. In L. S. Onken, J. D. Blaine, & J. J. Boren (Eds.), *Behavioral treatments for drug abuse and dependence.* NIDA Research Monograph no. 137. Washington DC: U.S. Government Printing Office.

HIGGINS, S. T. & BUDNEY, A. J. (1993). Treatment of cocaine dependence through the principles of behavior analysis and behavior pharmacology. In L. S. Onken, J. D. Blaine, & J. J. Boren (Eds.), *Behavioral treatments for drug abuse and dependence.* NIDA Research Monograph no. 137. Washington DC: U.S. Government Printing Office.

HIGGINS, S. T., ET AL. (1993). Achieving cocaine abstinence with a behavioral approach. *American Journal of Psychiatry, 150,* 763–769.

HIGGINS, S. T., ET AL. (1991). A behavioral approach to achieving initial cocaine abstinence. *American Journal of Psychiatry, 148* 1218–1224.

KANG, S.-Y., ET AL. (1991). Outcomes for cocaine abusers after once-a-week psychosocial therapy. *American Journal of Psychiatry, 148,* 630–635.

KLERMAN, G. L., ET AL. (1984). *The theory and practice of interpersonal psychotherapy for depression.* New York: Basic Books.

MARLATT, G. A., & GORDON, J. R. (1985). *Relapse prevention: Maintenance strategies in the treatment of addictive disorders.* New York: Guilford Press.

O'BRIEN, C. P., ET AL. (1990). Evaluation of treatment for cocaine dependence. In L. S. Harris (Ed.), *Problems of drug dependence 1989*, NIDA Research Monograph no. 95. Washington, D.C.: U.S. Government Printing Office.

RAWSON, R. A., ET AL. (1993). Neurobehavioral treatment for cocaine dependency: A preliminary evaluation. In F. M. Tims & C. G. Leukefeld (Eds.), *Cocaine treatment: Research and clinical perspectives.* NIDA Research Monograph no. 135. Washington DC: U. S. Government Printing Office.

RAWSON, R. A., ET AL. (1986). Cocaine treatment outcome: Cocaine use following inpatient, outpatient, and no treatment. In L. S. Harris (Ed.), *Problems of drug dependence 1985.* NIDA Research Monograph no. 67. Washington, D.C.: U.S. Government Printing Office.

ROHSENOW, D. J. (1993). Coping skills training for cocaine dependent individuals. Paper presented at the NIDA Technical Review Meeting on Outcomes for Treatment of Cocaine Dependence. September, Bethesda, MD.

ROUNSAVILLE, B. J., GAWIN, F., & KLEBER, H. (1985). Intrapersonal psychotherapy adapted for ambulatory cocaine abusers. *American Journal of Drug and Alcohol Dependence, 11,* 171–191.

SEIGEL, R. K. (1982). Cocaine smoking. *Journal of Psychoactive Drugs, 14,* 271–359.

WASHTON, A. M. (1987). Outpatient treatment techniques. In A. M. Washton & M. S. Gold (Eds.), *Cocaine: A clinician's handbook.* New York: Guilford Press.

WASHTON, A. M., GOLD, M. S., & POTTASH, A. C. (1987). Treatment outcome in cocaine abusers. In L. S. Harris (Ed.), *Problems of Drug Dependence.* NIDA Research Monograph no. 76. Washington, D.C.: U.S. Government Printing Office.

WOODY, G. E., ET AL. (1983). Psychotherapy for opiate addicts: Does it help? *Archives of General Psychiatry, 42,* 1081–1086.

ALAN J. BUDNEY
STEPHEN T. HIGGINS

Heroin, Pharmacotherapy HEROIN abuse has been a social problem for many years. Heroin was trademarked after its first synthesis and use by the Bayer pharmaceutical company in Germany in 1898. It is derived from MORPHINE, the natural alkaloid complex that is found in opium. Although heroin is taken into the body by a number of routes, the most common is injection. The rapid absorption of injected heroin into the bloodstream causes a large "high" and a "rush," at first (before tolerance occurs), and all the heroin is absorbed by this route.

Another method, smoking heroin, has been called "chasing the dragon," perhaps as an allusion to Chinese opium smoking; in this method, heroin is placed on a metallic foil and a match lit under it. When the heroin vaporizes, the vapor is inhaled through a straw; liquid heroin rolls around on the foil—hence the chase. A third method of heroin use, which waxes and wanes in popularity as the purity of illicit street heroin changes, is insufflation (snorting). This method minimizes the risks of intravenous drug use, including blood-borne infectious diseases such as hepatitis and HIV/AIDS, but it does not produce a rush because absorption into the bloodstream is slow. Heroin can also be injected into a muscle or under the skin (known as skin popping).

At first, heroin users have few lingering effects after a dose. The drug effects wear off after about six hours. Over time, however, addicts develop tolerance to the dose and dependence on the drug. Addicts will begin using heroin because they see people (friends, family, peers, role models) using it or because they feel a need to try it. As the frequency of use increases, they begin to experience withdrawal symptoms when they are not using the drug. At this point, they are physically dependent on heroin and will require larger and larger doses of heroin to achieve the same high or any high at all. Many addicts report that tolerance develops to such an extent that they cannot use enough for a high but must continue to use it to just feel normal (i.e., not be in withdrawal). It takes several weeks for a naive user to become dependent with this type of regular use.

HISTORICAL OVERVIEW OF TREATMENTS

When heroin was first commercially marketed by the Bayer Company as a morphine-like cough suppressant, it was thought to have fewer side effects than morphine. It was also used in the "treatment" of morphine addiction since it enters the brain more rapidly than does morphine. Instead, heroin introduced a new, more potent addiction. An over-the-counter industry in the legal sale of morphine and codeine elixirs also existed until opiates were outlawed by the HARRISON NARCOTICS ACT of 1914 and subsequent laws were passed during World War I (1914–1918).

Treatment of heroin abuse in the United States was initially targeted at removing the drug user from the environment of use. The federal prison in Lexington, Kentucky, became the site where incarcerated heroin addicts in federal custody were sent. Much of the current knowledge about opiate abuse was gained from the careful observations and carefully controlled studies of the researchers there. After incarceration, the addicts often returned to their towns of origin, and most of them turned back to drug abuse. The resulting clinical observation has been that imprisonment alone (with no drugs available) is an ineffective treatment of heroin abuse.

Historically, many of the medications used to treat heroin withdrawal in the general public have been largely ineffective; in some cases, the cure has been worse than the disease. Among the numerous ineffective treatments have been Thorazine, BARBITURATES, and electroshock therapy. In one method, belladonna and laxatives were used, because of the incorrect supposition that narcotics needed to be "rinsed" from the bodily tissues in which they were stored. At one institution that used this treatment, six of 130 addicts died during such opiate detoxification. Commenting on these methods, two of the researchers at Lexington noted: "The knockout feature of these treatments . . . doubtless had the effect of holding until cured many patients who would have discontinued a withdrawal treatment before being cured, and the psychological effect of doing something for patients practically all the time has a tendency, by allaying apprehension, to hold them even though what is done is harmful" (Kolb & Himmelsbach, 1938). Since the research conductetd at Lexington from the 1930s to the 1950s, which showed that opiate withdrawal was not fatal (unless complicated by other disorders or treatments), more standardized methods of detoxification have been developed.

A true advance was the development of methadone as a long-acting, orally effective opioid. Methadone was developed in Nazi Germany and was given the trade name Dolophine by the Eli Lilly company (from *dolor*, pain). The advantages of methadone over heroin include methadone's effec-

tiveness when taken by mouth; its long action, which allows single daily doses; and its gradual onset and offset, which prevents the rapid highs and withdrawal seen with heroin. Methadone-maintenance treatment was developed in the 1960s in New York City and has become an accepted treatment for opioid dependence. With the discovery that HIV infection can be transmitted by intravenous drug users, the benefits of methadone in decreasing intravenous heroin use have become even more evident.

PHARMACOLOGICAL TREATMENT APPROACHES

The most common and first-line treatment approach is to try to get the addict to stop using heroin by detoxification. *Detoxification* refers to using medications to treat withdrawal symptoms. The heroin withdrawal symptoms are similar to the symptoms of a severe flu. Although these withdrawal symptoms are rarely medically dangerous for those in good health, they are extremely uncomfortable, and, in many addicts, they make the alternative, using heroin, more attractive than detoxification. Severe withdrawal is associated with signs of sympathetic nervous system arousal as well as increased pulse, blood pressure, and body temperature. Addicts experience sweating, hair standing on their arms (i.e., gooseflesh—hence the expression "cold turkey"), muscle twitches (from which the expression "kicking the habit" comes), diarrhea, vomiting, insomnia, runny nose, hot and cold flashes, and muscle aches. A host of psychological symptoms accompany the withdrawal distress. After addicts have been detoxified, they may be treated with medications that make it less likely they will use heroin again; these medications that prevent relapse may work by blocking heroin's effects. Medications can also be used to treat underlying psychiatric problems that contributed to the addict's use of drugs.

An alternative approach is METHADONE MAINTENANCE, which does not initially aim to stop the addict from using opioids but instead to substitute oral methadone use for heroin abuse. Methadone is a clear liquid, usually dissolved in a flavored drink, that is given once a day and is prescribed by a physician. Used as a way to treat addicts' withdrawal symptoms and drug craving, the prescription of methadone is closely controlled by state and federal regulations.

Opiate Detoxification. The simplest approach to detoxification is to substitute a prescribed opioid for the heroin that the addict is dependent on and then gradually lower the dose of the prescribed opioid. This causes the withdrawal to be less severe, although the withdrawal symptoms may last longer. A typical procedure entails first verifying that addicts are dependent on opioids (by some combination of observed withdrawal, a withdrawal response to naloxone, or evidence of heavy opioid use). The addicts are then given an appropriate dose of methadone, which treats the withdrawal symptoms. They are monitored for oversedation due to methadone or undermedication of withdrawal symptoms. Intravenous users of street heroin admitted to the hospital usually tolerate well a starting methadone dose of 25 milligrams. The methadone dose is then gradually lowered over the next several days. It is typical to taper a starting methadone dose of 25 milligrams over a period of seven days.

Another approach avoids the difficulties of prescribing an opioid to an addict. It involves using the antihypertensive CLONIDINE to treat withdrawal symptoms after the addict has stopped using the opiates. Clonidine suppresses many of the physical signs of opiate withdrawal, but it is less effective against many of the more subjective complaints during withdrawal such as lethargy, restlessness, and dysphoria. Clonidine's side effects of low blood pressure, sedation, and blurry vision make it unpleasant to take and unlikely to be abused by addicts. Although clonidine has not been approved by the Food and Drug Administration for opiate detoxification, it is widely used for this purpose and has demonstrated efficacy. It is most effective when used in addicts who are not addicted to large doses of opioids.

Opiate Antagonists. The opiate antagonist NALTREXONE is used clinically to accomplish rapid detoxifications and to help detoxified addicts stay off opioids. Naltrexone binds more strongly than heroin to the specific brain receptors to which heroin binds. If, therefore, addicts who are dependent on heroin take a dose of naltrexone, the naltrexone will replace the heroin at the brain receptor and the addicts will feel as if all the heroin has been suddenly taken out of their body. The effect of this rapid reduction in effective heroin (at the receptor) is withdrawal. The withdrawal is usually more severe than that which comes from simply stopping the heroin, but it also has the effect of accomplishing a detoxification more

quickly. Thus, a combination treatment of clonidine to suppress the intensity of withdrawal symptoms and naltrexone to accelerate the pace of withdrawal has been used for rapid detoxification.

Naltrexone is primarily used after detoxification to prevent addicts from returning to opioid use. Because naltrexone binds to opioid receptors more tightly than does heroin, opioid addicts on naltrexone who use heroin will find the heroin effect blocked by naltrexone. Addicts maintained on naltrexone who use heroin will only be wasting their money. One effect of naltrexone is thus to extinguish the conditioned response to heroin injection. Naltrexone is prescribed in the form of a pill that can be given as infrequently as three times a week. It has few side effects in the majority of patients who take it, and, contrary to some rumors, it does not suppress other "natural highs."

Opioid Maintenance. Methadone is the most common opioid used for the maintenance treatment of opioid addicts. Methadone satiates the heroin user's craving for heroin in order to prevent heroin withdrawal. The more important therapeutic effect of methadone, however, is tolerance to it. Addicts maintained on a stable dose of methadone do not get high from each dose because they are tolerant to it. This tolerance extends to heroin, and methadone-maintained addicts who use heroin experience a lesser effect because of the tolerance. Tolerance accounts for the fact that methadone-maintained addicts can take methadone doses that would cause a naive (i.e., first-time) drug user to die of an overdose. Generally, methadone-maintained addicts do not appear to be either intoxicated or in withdrawal. Tolerance is admittedly incomplete, and methadone-maintained addicts have some opioid side effects that they do not become tolerant to—for example, constipation, excessive sweating, and decreased libido. There is no known medical danger associated with methadone maintenance, however.

Methadone is dispensed as part of licensed programs, usually on a daily basis. It is generally well received by addicts, and the risk of incurring withdrawal symptoms if methadone treatment is interrupted provides a strong incentive for addicts to keep appointments. The ritual of daily clinic attendance has the additional therapeutic benefit of beginning to impose structure on the chaotic lives of most opiate addicts. Methadone treatment is often augmented with medical, financial, and psychological support services to address the many needs of opioid addicts.

Despite the philosophical debates about the appropriateness of using methadone, there is a large body of evidence indicating that methadone-maintained addicts show decreases in heroin use, crimes committed, and psychological symptoms. The major drawbacks to methadone maintenance include the great difficulty of achieving detoxification from methadone, the methadone side effects, and the possibility of increased use of other illicit drugs such as cocaine.

An opiate addict initially coming in for treatment will usually be put through detoxification and possibly put on naltrexone maintenance. Addicts with intact family supports, good jobs, or strong motivation are more likely to benefit from naltrexone maintenance than those who are more impaired. Younger addicts and adolescents are urged to try nonmethadone approaches, so as to avoid developing a methadone addiction. Methadone maintenance is usually reserved for patients who have failed at previous detoxifications. An exception is made for pregnant women, in whom methadone maintenance is the treatment of choice, with detoxification of the infant from methadone accomplished after birth. Opiate detoxification is risky in pregnant women because of the adverse effects on fetal development in the first and second trimesters, and the risk of miscarriage.

Other nonmethadone medications for maintenance treatment of opioid dependence have not yet been widely used. BUPRENORPHINE is a partial opioid agonist medication that has the advantages of being safe, even at higher doses, and being associated with less severe withdrawal symptoms than methadone after discontinuation. Another medication recently approved for treating opioid dependence is LAAM (levo-alpha-acetylmethadol). LAAM is broken down in the body to very long-acting active metabolites, and therefore it can be prescribed as infrequently as three times a week.

THE INTEGRATION OF PHARMACOLOGICAL AND PSYCHOSOCIAL TREATMENTS

No medication will prevent an addict who wants to use heroin from doing so. Naltrexone maintenance can be discontinued, and addicts who discontinue it are able within one to three days to use heroin without the naltrexone blockade. Similarly, methadone maintenance is ineffective in addicts who are unable or unwilling to meet the requirements of clinic attendance (which sometimes requires payment of fees) and staying out of prison. Addicts whose lives are in disarray require medications as part of a comprehensive treatment program that also addresses their other needs. In a street addict who chronically uses drugs, these may include needs for counseling, medical attention, vocational rehabilitation, and a host of other services. There is evidence that methadone treatment is more effective if a higher "dose" of psychosocial treatment is provided along with it.

Detoxification is a first step toward recovery because it makes the addict available to further psychosocial and medical treatments. There is evidence that mild physiological abnormalities due to withdrawal of opiates linger for as long as three months after detoxification. This "long-term abstinence syndrome" is thought to contribute to the craving for opiates that occurs after detoxification. Naltrexone maintenance is most effective in addicts who have jobs and stable social supports—for example, in anesthesiologists who have become addicted to hospital medications. Because naltrexone itself is not reinforcing and many heroin addicts have a host of psychosocial problems, many clinics have reported that naltrexone maintenance alone was minimally effective in the treatment of long-term addicts.

SUMMARY

Opioid addiction is, in many ways, a physical problem as well as a psychological and behavioral problem. Addicts become physically addicted to opiates and, in the later stages of addiction, become preoccupied with relieving the physical symptoms of withdrawal. They become highly attuned to the bodily signals that withdrawal is coming. Heroin addicts spend most of their waking life procuring, using, and withdrawing from heroin—three times a day, seven days a week, fifty-two weeks a year—for years.

The medications used to treat opioid abuse are powerful agents that interrupt this cycle. Although medications alone rarely cure an addiction, they are critically important to breaking the cycle of preoccupation with opioid use and enabling addicts to benefit from comprehensive drug-abuse treatment.

(SEE ALSO: *Coerced Treatment for Substance Offenders; Ibogaine; Opioid Dependence; Opioid Complications and Withdrawal; Pregnancy and Drug Dependence; Substance Abuse and AIDS; Treatment Types*)

BIBLIOGRAPHY

KOLB, L., & HIMMELSBACH, C. K. (1938). Clinical studies of drug addiction. III. A critical review of the withdrawal treatments with methods of evaluating abstinence syndromes. *Public Health Reports, 128*(1). Cited in H. D. Kleber (1981), Detoxification from narcotics. In J. H. Lowinson & P. Reiz (Eds.), *Substance abuse: Clinical problems and perspectives.* Baltimore: Williams & Wilkins.

MARC ROSEN

Heroin, Psychological Approaches Psychological treatments are an important component of comprehensive drug-abuse treatment. Medications such as METHADONE can be used to address physical dependence and other biological aspects of addiction, but HEROIN abuse is also a disorder involving maladaptive learned behavior that must be stopped and replaced by healthier behaviors. Psychological therapies help drug abusers to understand their feelings and behaviors and to make changes in their lives that will lead to ending drug use and maintaining abstinence. Drug abusers also may have psychiatric problems, such as DEPRESSION and ANXIETY, and they may have problems interacting with other people or dealing with anger and frustration. These problems can also be addressed by psychological therapies. In addition, heroin abuse is a chronic relapsing disorder (i.e., many people who try to stop end up returning to drug use). Relapse to drug use following treatment is commonly attributed to environmental (e.g., associating with drug-using friends), psychological (e.g., feeling depressed or angry), and/or behavioral (e.g., having poor social skills) factors that are typically the focus of psychological interventions.

A variety of psychological treatments, often in combination with pharmacological approaches, have demonstrated effectiveness in the treatment of heroin abuse. The purpose of this article is to survey the most prominent psychological interventions currently used in the treatment of heroin abusers. Following a brief discussion of the development of heroin abuse, we describe the factors that lead people to seek treatment, the range of problems that may be characteristic of heroin abusers, and the psychological treatments—including THERAPEUTIC COMMUNITIES, counseling, psychodynamic and cognitive-behavioral psychotherapies, family therapy, and SELF-HELP approaches. The chapter concludes with a discussion of the effectiveness of these interventions.

DEVELOPMENT OF HEROIN ABUSE

Initial heroin use is motivated by curiosity and the desire to use it without becoming addicted. Heroin is injected into a vein (although it is sometimes inhaled), and the user experiences an immediate rush, characterized by feelings of relaxation and well-being. As use escalates, withdrawal symptoms (e.g., cramps, irritability) may appear as the drug is eliminated from the body. At this point, individuals may start using the drug both for its positive effects and for alleviating uncomfortable withdrawal symptoms. Drug use may also be motivated by an attempt to cope with feelings of STRESS, hopelessness, or depression. Whatever the causes of initial use, the frequent and repeated acquisition of heroin soon becomes a priority; some addicted individuals may resort to illegal activity (e.g., stealing; prostitution) to buy illicit drugs. In addition heroin abusers are often concurrently addicted to ALCOHOL and/or other drugs, including COCAINE and BENZODIAZEPINES (e.g., Valium) that they may have started taking before or after they began using heroin. It is in the context of this addictive lifestyle that heroin abusers come to the attention of treatment providers. Heroin abusers are usually ambivalent about seeking treatment; they like taking drugs and have difficulty seeing any reason to stop. They are most likely to begin treatment following a crisis of some sort—a legal, physical, family, financial or job-related problem caused by their drug use. They are typically referred to specific treatment sites by friends, family, or the legal system, which may mandate treatment as a part of probationary sentences. The cost, location, and availability of treatment slots are all factors that affect selection of treatment setting.

TREATMENT SETTINGS

Treatment for heroin dependence is offered in publicly funded clinics that accept patients with limited resources, including those who receive public assistance. It is also treated in private programs that take patients with higher incomes and/or medical insurance. Treatment for heroin abuse is often defined by the setting in which it is delivered, not by the actual content of treatment, which may or may not differ across treatment settings. For example, outpatient and inpatient clinics may offer remarkably similar services for drug abusers. One exception is the THERAPEUTIC COMMUNITY, where the treatment philosophy and approach is uniquely associated with long-term recuperation in a residential setting. Treatments are also labeled with regard to the relative role of psychological versus pharmacological interventions used. With METHADONE MAINTENANCE, for example, counseling and psychotherapy are viewed as secondary, although complementary, to the daily oral administration of methadone—a drug that replaces heroin within the dependence mode. At the opposite end of the spectrum are residential therapeutic communities and TWELVE-STEP self-help programs, where the entire intervention consists of social and behavioral modeling, with no use of medications. Drug-abuse treatment may also be distinguished by whether it is offered in a hospital versus a community clinic outpatient setting. Outpatient clinics usually emphasize psychological techniques, by providing counseling and psychotherapy services. Hospital chemical dependency units usually offer medical detoxification that involves prescribed medications along with some combination of psychological approaches. In this chapter, we will describe the content of psychological interventions for heroin abuse independent of the settings in which they are typically administered.

ASSESSMENT

By the time drug abusers seek treatment, they often have a number of problems that need to be solved, only the first of which is stopping drug use. Within any treatment setting, comprehensive assessment is essential to focus treatment on the areas where change is needed. It is first important to understand the types and amounts of drugs that are typically taken in order to assess severity of the drug abuse problem. Drug use information is assessed through the patient's self-report and urinalysis testing. Urinalysis testing provides objective information about whether the individual has or has not used drugs recently and can also be used to verify the truthfulness of self-reports. An understanding of psychological and environmental factors that precede and follow drug use (e.g., when, where and why drugs are taken; where and how the drugs are acquired), also known as a functional analysis, is also necessary for the development of strategies to initiate abstinence and prevent relapse. Evaluation of psychiatric disorders is essential for determining appropriate treatment intervention. Depression and ANTISOCIAL PERSONALITY, for example, are quite common among heroin abusers (Rounsaville et al., 1982). Some problems, however, such as depression, may go away when drug use stops. Finally, social functioning, employment history, and illegal activity all have implications for psychological interventions and treatment prognosis and need to be thoroughly assessed. Indeed, being employed and having good social support (e.g., from a spouse who does not abuse drugs) are excellent predictors of treatment success if they are already present, and areas that need attention in treatment if they are not. The recently developed ADDICTION SEVERITY INDEX (ASI; McLellan, Luborsky, & O'Brien, 1980), a structured interview that assesses drug use, physical and emotional health, employment, social support, and legal status, is often used by clinicians and researchers to evaluate the broad range of factors that are related to drug abuse and may improve with treatment.

PSYCHOLOGICAL TREATMENTS OF HEROIN ABUSE

This section will survey common psychological approaches to the treatment of heroin abuse. Although each differs in regard to its philosophy and goals, all share an interest in eliminating the drug use of the heroin abuser and the substitution of healthier behaviors.

Therapeutic Communities. Therapeutic communities (TCs) are long-term (6–24 month) residential programs developed specifically for helping drug abusers change their values and behaviors in order to sustain a drug-free lifestyle. The assumption behind these communities is that drug abusers, who have typically been involved in a special illicit subculture for most of their lives, need to learn how nondrug-abusing individuals function in society. The goal is to rehabilitate the drug abuser into a person who can conform to society's values and goals, assume social and job responsibilities, and make contributions to the community. During treatment, the drug abuser lives in a special residential community with other drug abusers and with therapists who may be ex-addicts in recovery. A behavioral shaping/incentive system is set up so that desirable behaviors are rewarded through community privileges and increased responsibilities. In addition, patients learn through observing peers and staff, who serve as role models for appropriate behavior, sometimes called "right living."

Patients progress through three stages. In the first stage, Orientation (0–2 months), the patient assimilates within the therapeutic community by attending seminars concerning the philosophy and rules of the program. The second stage is called Primary Treatment (2–12 months) and is characterized by increasing work responsibilities and group leadership roles. This stage includes three phases. In the first phase (2–4 months), patients conform to the TC policies by following the rules, engaging in low-level work assignments, and attending group meetings. By the second phase (4–8 months), patients work at more responsible jobs, actively participate in group meetings, and begin to assume the responsibility of a role-model for other patients. In the third phase (8–12 months), patients engage in top-level jobs (e.g., co-ordinating services in the program), co-lead support and treatment groups, and become social leaders in the community. The final stage, Reentry (12–24 months), focuses on preparing the patient to separate from the TC and rejoin the outside community. It is expected that after leaving patients will establish their own households and obtain regular employment or continue their education. In summary, TCs attempt to rehabilitate the drug abuser by instilling a whole new set of attitudes and behaviors that conform to those expected by a nondrug-abusing society.

Drug-Abuse Counseling. This intervention approach is practiced in methadone maintenance programs, where patients are required to see a counselor throughout the course of treatment—and may also be provided in outpatient community-clinic programs. Counselors are usually professionals with a college degree in counseling, although ex-addicts who have personal experience with recovery from

drug abuse may also provide counseling. Counselors have several roles. First, they monitor treatment compliance (that the patient is attending regularly and providing urine specimens for drug testing as requested), confront any violations of program rules, and enforce penalties and privileges. Second, based on problems and deficits identified during the assessment phase, counselors formulate a treatment plan that specifies goals for the patient. For example, a treatment plan may contain recommendations to abstain from drug use, obtain employment, and participate in self-help groups. Counselors work with their patients using several strategies to implement such a treatment plan. Goal setting helps patients learn to set reasonable goals (e.g., starting a bank account; obtaining a driver's license) and to outline specific steps required to attain chosen goals. In problem-solving training, counselors and patients work together to address both immediate and long-standing problems in the patient's life. The primary goal is for patients to learn the strategies for solving everyday problems and for making decisions. Recreational planning may be used to encourage patients to engage in new social and recreational activities that might substitute for their typical lifestyle of searching for drugs or hanging out with drug-using friends. Finally, counselors are expected to refer patients to other community-helping agencies for services that they cannot provide themselves. For example, patients who are unemployed may be referred to an employment-counseling service. In summary, counseling attempts to comprehensively address the problems of drug abusers using practical, goal setting, and problem-solving techniques.

Psychotherapy. This type of psychological treatment, usually practiced by trained clinical psychologists, psychiatrists, or psychiatric social workers during a one-on-one interaction with the patient, uses interpersonal skills to promote insight and behavior change. Psychotherapy was developed for use with neurotic and emotional disorders, but has been adapted for use with drug abusers. Several specific types of psychotherapy are practiced by various therapists, depending on their training, with psychodynamic and cognitive-behavioral being two prominent types. In each of these therapies, comprehensive assessment, empathic listening, nonjudgmental understanding, and patience are necessary tools to help the patient become involved in a therapeutic relationship and provide a context for behavior change.

Psychotherapy can also be practiced in groups, and group treatment is frequently defined as a separate type of treatment. Groups are a popular way to conduct treatment and may be found in virtually any treatment setting, including hospital and outpatient chemical-dependency programs, methadone programs, and therapeutic communities. The content of therapy, however, can vary widely from one group to another in the same way that differing approaches are used for individual psychotherapy. Regardless of therapeutic approach, group therapies do differ from individual therapies in some specific ways. Groups provide a context for mutual empathy, encouragement, and support among people who share similar problems. Patients in groups may benefit from the experience of others in solving these problems and by entering reciprocal helping relationships. The interactions among group members also provide a context in which the therapist can facilitate improved social skills for those who may need them.

Psychodynamic Therapy. Psychodynamic therapy with heroin abusers employs supportive, analytical techniques to explore heroin use and the addictive experience from the patient's point of view. Drug use is viewed as a symptom of underlying emotional problems and/or relationship difficulties. Thus, psychodynamic therapy rarely confronts or attempts to modify drug use directly, and for this reason, it is usually implemented after stable abstinence from drugs has been achieved. Therapy focuses instead on the patient's thoughts, feelings and relationships (past and present) with parents, spouse, friends, and other significant individuals—from which the therapist tries to identify common patterns or themes. As therapy progresses, the therapist–patient relationship becomes the focal point, as this relationship often replicates themes from interactions with others, which the therapist points out. The primary means of behavior change results from the patient recognizing these common, often maladaptive, interaction themes and determining to change them. Thus, the goal of treatment is for the patient to understand the origin and function of their feelings and behavioral patterns, and to use this awareness to change the manner in which they cognitively interpret, emotionally respond, and behaviorally interact with individuals in their environments. For example, a psychodynamic therapist might observe that anger is a continuing theme in a patient's life and be sensitive to situations when the patient shows anger to-

ward the therapist. When this happens, the therapist will help the patient understand the circumstances leading to the anger and relate these circumstances to other situations when the patient had been angry. Eventually, the patient and therapist might explore the origins of the patient's anger (perhaps toward his or her parents) and the relationship between the patient's anger and engaging in self-destructive behavior (e.g., drug use). As the patient develops more adaptive ways of coping with thoughts, emotions, and relations to others, heroin and other substance use becomes less necessary and desirable. In summary, psychodynamic therapy views long-term abstinence from drug use as an indirect result of resolving the causes of drug use. In this way, it is believed that a more permanent cure will result.

Cognitive-Behavioral Therapy: Relapse Prevention. Cognitive-behavioral therapists are concerned with direct interventions that will change behavior and thinking without necessarily requiring or expecting insights into the causes of behavior. Recognizing that relapse is a serious problem in drug abuse, these therapy approaches have been specifically adapted for use with heroin and other drug abusers in a therapy called RELAPSE PREVENTION, to teach them the skills necessary to initiate and sustain abstinence (Marlatt & Gordon, 1985). A functional analysis derived in the assessment phase allows the therapist to understand the thoughts, behaviors, and environmental conditions that precede and follow heroin and other drug use and to help the patient recognize the environmental (e.g., drug-using friends), cognitive (e.g., irrational thinking), emotional (e.g., anger), and behavioral (e.g., starting arguments) factors that may either reduce the likelihood of stopping or increase the likelihood of returning to drug use. Based on this functional analysis, the cognitive-behavioral therapist and the patient decide which factors (e.g., thoughts, places, people) are most likely to sustain ongoing drug use or act as triggers for relapse during abstinence; then specific treatments are based on this analysis (Stitzer, Bigelow, & Gross, 1989).

Patients and therapists may work together to devise strategies for avoiding drug-using friends and staying away from places in which the patient has bought and used drugs in the past. In some cases, patients may even want to change their phone numbers or move to new locations. In addition to environmental changes, heroin abusers may be taught new skills designed to help them cope with high-risk situations that could trigger relapse. For example, patients who use drugs when they feel stressed may be taught specific relaxation techniques that can counteract stressful feelings. Patients may also learn drug-refusal skills to handle situations where they actually encounter drugs (although it is better to avoid such situations altogether) and to use specific strategies for coping with situations in which the return to drug use is likely (e.g., calling a nonusing friend; leaving the situation; making an appointment with their therapist). In addition, cognitive-behavioral therapists may address the patient's thought patterns that precede heroin use and call attention to dysfunctional thinking. For example, patients may have unrealistic thoughts ("I must be loved and accepted by everybody or else I am a failure and might as well use drugs") or illogical thoughts ("I will never be able to stop using drugs because I am an addict"). The cognitive-behavioral therapist aims to change negative cognitions to adaptive, positive thinking ("I do not need everybody's approval"; "I can learn to gain control over my behavior").

Sometimes a pervasive maladaptive behavior pattern underlies drug abuse that can be addressed with a cognitive-behavioral approach. For example, with a patient who has trouble controlling anger and tends to use drugs after angry confrontations, the cognitive-behavioral therapist may place the patient on an anger-control skills-training program. The patient would be instructed to avoid situations likely to induce anger (e.g., confrontations with a supervisor) and would be taught specific strategies for dealing with potential anger-producing situations. For example, relaxation might be employed to gain control over anger. Further, the patient might be taught new self-statements to replace thoughts that have typically preceded feelings of anger (e.g., "It would be nice to get a raise, but it isn't the end of the world if I do not get it"). In summary, cognitive-behavioral therapy focuses directly on behavior change without expecting or requiring insight into the cause of the problem. To the extent that underlying emotional and interactional dysfunctions often exacerbate drug use, however, both the cognitive-behavioral and the psychodynamic therapist will end up dealing with the same issues—albeit in slightly different ways.

Family Therapy. Heroin abusers are often raised in dysfunctional families and may replicate the maladaptive behavior patterns learned from their families within their own personal and romantic relation-

ships. In addition, the patient's heroin abuse may have had a disruptive effect on that family. These observations suggest the importance of including the family in the treatment process, and this is particularly true for adolescents who become involved with drugs while still living with their families. For older drug abusers, it is often difficult to involve the family in treatment, and family resistance/avoidance is one of the first issues that the therapist must address. Family therapy is a specialized type of psychotherapy that has its own methods, in which practitioners must be trained. Thus, it is generally conducted by a psychologist or other health professional who has been trained in one of several specific familial treatment approaches. Although there are several theoretical perspectives to family therapy (e.g., psychodynamic, cognitive-behavioral, family systems, etc.), the goals of these types of interventions are to help the family recognize maladaptive patterns of behavior, to learn better ways of solving family problems, to better understand each other's needs and concerns, and to identify and modify family interactions that may be helping to maintain drug use in the targeted family member (or members).

Self-Help Groups. ALCOHOLICS ANONYMOUS (AA) was created in 1935 by recovering alcoholics so that alcoholics could help each other abstain. NARCOTICS ANONYMOUS (NA) and COCAINE ANONYMOUS (CA) were later based on the tenets of AA but geared toward drug addictions. The newest group is Methadone Anonymous (MA), which accommodates drug addicts who use methadone. The core beliefs espoused by self-help groups are commonly adopted by many treatment programs, and drug-abuse patients are often referred to self-help groups as an adjunct to other treatments. Active members of self-help groups attend frequent meetings, some as often as once per day. At these meetings, members speak to each other about their drug use and drug-related problems; they offer mutual advice and support without the help of any trained therapists.

The philosophy, treatment goals, and procedures of self-help groups are contained in a book called *The 12 Steps to Recovery*. This book, often referred to as "The Big Book," outlines a series of tasks designed to promote abstinence and long-term recovery among alcoholics and drug abusers. The first step in recovery is to admit that one has a problem with drugs and/or alcohol and that outside help is needed to solve the problem. The sources of help to be called upon are other group members and a higher spiritual power (e.g., God), who will supply the spiritual strength necessary to stop drug use. The twelve-step program also advocates specific practical changes in lifestyle; these revolve around regular and frequent attendance at group meetings and concentration on the goal of abstinence (e.g., remembering the motto "one day at a time"). Once stable abstinence is achieved, the drug user is encouraged to restore relationships with friends and family that have been damaged by former drug use. For some, however, the self-help community becomes the primary source of friendships and social support.

Sponsorship is another technique used to promote and sustain abstinence. Specifically, all group members are encouraged to work with a sponsor who is typically an older, long-standing, group member who models appropriate behavior, guides new members through the twelve-step process, and provides a source of support for the new member to turn to in times of crisis. Later, the new member may sponsor someone else. To the extent that self-help programs permit former drug abusers to receive support from peers, associate with new groups of non-drug-using friends, and engage in alternate recreational activities with newly developed social contacts, the goals and even processes are similar to therapy. However, these goals are accomplished through group support and modeling using a treatment plan laid out in the twelve-step code rather than through formal meetings with a professional therapist.

EFFECTIVENESS OF PSYCHOLOGICAL TREATMENTS FOR HEROIN ABUSE

An end to drug use is the primary outcome measure for evaluating the effectiveness of drug-abuse treatment. Urine testing is usually included as a routine part of any drug-abuse treatment, to provide objective information on whether the treatment is being successful at motivating the patient to stop drug use and maintain abstinence. Changes in criminal behavior, employment status, family problems, and physical and emotional health are also relevant to understanding the effectiveness of treatment. Many of these collateral difficulties improve once drug use is stopped, although more improvement would be expected in treatment programs that offer services to specifically address these collateral problems. Using this array of outcome measures, studies

have been conducted to evaluate the relative efficacy of treatments for heroin abusers. These studies have typically focused on the treatment setting rather than the content of treatment that is delivered within each setting. Further, some treatment settings have received much more evaluation than others. Methadone maintenance and TCs, for example, have received lots of attention, while hospital chemical-dependency programs have been infrequently evaluated and self-help programs have not been evaluated at all (Gerstein & Harwood, 1990).

Large scale follow-up studies such as the TREATMENT OUTCOME PROSPECTIVE STUDY (TOPS) and the DRUG ABUSE REPORTING PROGRAM (DARP), which have surveyed outcomes from methadone, therapeutic community, and outpatient modalities, have found that drug abusers who enter treatment display less drug use and better social adjustment during and following treatment than they did prior to treatment and also have better outcomes than groups of patients who applied for treatment but never followed through (Hubbard et al., 1989; Simpson & Sells, 1990). These studies also found that effectiveness does not seem to be related to type of treatment but rather to duration of stay in treatment. Several types of treatment can be effective, but only with those patients who remain for prolonged periods of time. Thus, methadone maintenance and therapeutic-community treatments produce similar degrees of success with those who stay—but more patients tend to stay in methadone than in TC treatment. Finally, the success of drug-abuse treatment in general is better for patients who exhibit the fewest psychiatric symptoms and the greatest social stability (McLellan, 1983).

When evaluation focuses on treatment setting rather than on treatment content, it becomes difficult to determine which components of treatment are responsible for outcome results. This is especially true since treatment programs for heroin abuse are typically comprehensive and multimodal, encompassing a variety of techniques that may include psychological and behavioral interventions, medications, and self-help. The few well-executed studies that have attempted to evaluate the impact of specific psychological interventions on heroin abusers have been conducted with methadone maintenance programs. These studies have shown that methadone-maintenance treatment outcome is enhanced by a variety of psychological interventions, including counseling (McLellan et al., 1988), individual psychotherapy (Woody et al., 1983), family therapy (Stanton & Todd, 1982), and cognitive-behavioral/relapse prevention aftercare (McAuliffe, 1990), as evidenced by reduced drug use and crime, plus improved social and psychological functioning.

SUMMARY

Research has shown that several different types of treatment for heroin abusers can be effective. Heroin abusers who enter treatment do better than those who apply but do not follow through with treatment. Heroin abusers who remain in treatment the longest achieve better treatment outcomes than those who drop-out early. In addition, heroin abusers who exhibit the fewest psychiatric symptoms and demonstrate the most social stability appear to benefit most from treatment. Finally, specific psychological interventions have enhanced the effectiveness of methadone-maintenance treatment. As previously noted, heroin abuse is a chronic, relapsing disorder: It appears that long-term treatment and perhaps repeated treatment may be necessary to eliminate drug use and to successfully address the broad range of psychosocial difficulties that usually accompany this disorder.

(SEE ALSO: *Addiction: Concepts and Definitions; Causes of Substance Abuse; Coerced Treatment for Substance Offenders; Comorbidity and Vulnerability; Drug Testing and Analysis; Opioid Dependence; Opioid Complications and Withdrawal; Tolerance and Physical Dependence; Treatment, History of; Treatment Types; Wikler's Pharmacologic Theory of Drug Addiction*)

BIBLIOGRAPHY

GERSTEIN, D. R., & HARWOOD, H. J. (EDS.) (1990). *Treating drug problems*, Vol. 1. Washington, DC: National Academy Press.

HUBBARD, R. L., ET AL. (1989). *Drug abuse treatment: A national study of effectiveness*. Chapel Hill: University of North Carolina Press.

MARLATT, G. A., & GORDON, J. (1985). *Relapse prevention*. New York: Guilford Press.

MCAULIFFE, W. E. (1990). A randomized controlled trial of recovery training and self-help for opiate addicts in New England and Hong Kong. *Journal of Psychoactive Drugs, 22*, 197–209.

MCLELLAN, A. T. (1983). Patient characteristics associated with outcome. In J. R. Cooper et al. (Eds.), *Research on the treatment of narcotic addiction: State of the art.* Rockville, MD: National Institute on Drug Abuse.

MCLELLAN, A. T., LUBORSKY, L., & O'BRIEN, C. P. (1980). An improved diagnostic evaluation instrument for substance abuse patients: The Addiction Severity Index. *Journal of Nervous and Mental Disease, 168*, 26–33.

MCLELLAN, A. T., ET AL. (1988). Is the counselor an "active ingredient" in substance abuse rehabilitation? *Journal of Nervous and Mental Disease, 176*, 423–430.

ROUNSAVILLE, B. J., ET AL. (1982). Heterogeneity of psychiatric diagnosis in treated opiate addicts. *Archives of General Psychiatry, 39*, 161–166.

SIMPSON, D. D., & SELLS, S. B. (1990). *Opioid addiction and treatment: A 12-year follow-up*. Malabar, FL: Robert E. Krieger.

STANTON, M. D., & TODD, T. C. (1982). *The family therapy of drug abuse and addiction*. New York: Guilford Press.

STITZER, M. L., BIGELOW, G. E., & GROSS, J. (1989). Behavioral treatment of drug abuse. In *Treatments of psychiatric disorders*, Vol. 2. Washington, DC: American Psychiatric Association.

WOODY, G. E., ET AL. (1983). Psychotherapy for opiate addicts: Does it help? *Archives of General Psychiatry, 40*, 639–645.

MAXINE L. STITZER
MICHAEL KIDORF

Polydrug Abuse, An Overview   Polydrug abuse (also called multiple-drug abuse) refers to the recurring use of three or more categories of PSYCHOACTIVE substances. It is a pattern of substance abuse that is most commonly associated with illegal drug use and youth. Most polydrug users also smoke TOBACCO, but NICOTINE has only recently begun to be recognized as a drug of abuse to be addressed with polydrug users.

While the term *polydrug user* is usually reserved for people with a rather varied and nonspecific pattern of drug use, many drug users who have a preferred (a primary) drug of abuse are also polydrug users. In fact, it is uncommon for users of any illicit drug to restrict their substance use to only the one drug. For example, an individual may be a regular

COCAINE user but also use ALCOHOL, TRANQUILIZERS, and MARIJUANA.

WITHDRAWAL

The intensity of withdrawal symptoms and their medical risk depends on the particular substances used and the degree to which dependence has developed. Withdrawal is most often clinically significant in those who have developed severe dependence on a primary drug of abuse; the medical risks of such withdrawal vary substantially with the type of drug. For example, much greater risks exist for BARBITURATE than for HEROIN withdrawal. The recent use of other drugs in addition to the primary drug of abuse complicates the withdrawal process. In such cases, careful medical assessment is important in the planning of withdrawal management for polydrug users.

Polydrug users who typically dabble among the available drugs without developing severe dependence on any of them usually have no clinically serious problems when they stop using drugs. They may experience some discomfort, agitation, or sleeplessness but they do not normally require medical treatment. Social stability and support would be important, however, as the risk of relapse could be high during this period of discomfort.

ASSESSMENT

There are two main purposes of assessment: (1) to determine what specific treatment would be most suited to the specific needs of the polydrug user; and (2) to determine baseline levels of functioning against which progress in treatment can be measured. Assessment must address many areas of functioning in addition to drug use. These include the following: medical and psychiatric problems; family and other social relationships; school or work problems; leisure activities and skills; criminal activities and legal problems; and financial status.

Drug use must be carefully assessed in the polydrug user, because of the variety of drugs used and the need to evaluate the risks associated with the particular pattern of use. The usual procedure is to divide drug use into categories based on pharmacological similarities. These categories typically in-

clude: alcohol; marijuana; HALLUCINOGENS (e.g., LSD); heroin; other OPIOIDS (e.g., CODEINE); cocaine; other STIMULANTS (e.g., AMPHETAMINES); TRANQUILIZERS (e.g., BENZODIAZEPINES such as Valium) and other sedative hypnotics (e.g., barbiturates); and solvents (including glue). Because accurate estimates of doses are very difficult to obtain from polydrug users, their drug use is usually assessed as the number of times each drug has been used within a specified time period. Other important factors to consider in assessing drug use are risks related to HUMAN IMMUNODEFICIENCY VIRUS (HIV) infection—especially injection drug use, and drugs used in combination.

A further consideration in assessment is the client's commitment to change. Polydrug users may be, at best, ambivalent about the need for change. The assessment process offers an excellent opportunity to enhance the polydrug user's motivation for change by providing feedback and support, as well as by helping the person to clarify goals and values.

TREATMENT APPROACHES

Many different treatment approaches are available, but they reflect differing conceptual or theoretical perspectives on the origins of drug-use problems as well as on the best ways to treat them. Most of these approaches were not developed for the polydrug user but, instead, were adapted from other substance-abuse treatments. The approaches described may be presumed to be quite widely available except where restrictions are noted. Research evidence concerning their comparative effectiveness for polydrug users is extremely limited.

Approaches Based on the Disease Concept. According to one variant of the disease concept, alcoholism and drug addiction are incurable diseases. Those affected are considered unable to control their use of the substance, because of an allergiclike, biological reaction. This approach has only one solution to the problem—to get the user to abstain from any use of the drug.

Twelve-Step Groups. The treatment approaches most commonly associated with the disease concept are those based on ALCOHOLICS ANONYMOUS (AA), which was started in 1935. The TWELVE-STEP approach developed by AA has been adapted for application to other primary drugs of abuse, e.g., NARCOTICS ANONYMOUS (NA) and COCAINE ANONYMOUS (CA). Like

AA, these approaches rely exclusively on self-help peer-group procedures. Members voluntarily embark on a lifetime journey of recovery, armed with a set of principles and the support of peers who share a common problem and a desire for change. The central features of these approaches are the following: an acceptance of being powerless over the drugs; a belief in a higher power; a commitment to make restitution to those who have been harmed; and personal responsibility to maintain abstinence. Polydrug users may affiliate with any of such groups, depending on the particular drugs most commonly used. They may, however, have some difficulty in identifying with the majority of group members as peers. Often a buddy or two with the same problems and concerns become a special subgroup.

Chemical-Dependency Programs. Some treatment programs, most notably residential programs, have adapted the twelve-step approach as the basis of their treatment. Chemical-dependency (CD) programs are the most prominent example. These programs are an extension of the four week MINNESOTA MODEL (for ALCOHOLISM) to a broader range of substances of abuse. Some have a particular focus for young polydrug users.

The CD approach usually involves a three- to six-week structured and intensive residential-treatment phase, which includes lectures and discussions about the harmful effects of drug use; group-therapy sessions that focus on breaking down denial and personal issues related to drug use; an orientation to the twelve-step approach; recreational and physical activity; and family counseling sessions. The residential phase is followed by an extended aftercare program, typically involving attendance at AA, NA, or CA meetings. Many CD programs specialize in the treatment of polydrug users who also have coexisting psychiatric problems.

The number of CD programs has grown rapidly in the past decade, particularly in private hospitals. Because of their residential phase, these CD programs are among the most expensive form of treatment available to polydrug users.

Systems Theory–Based Approaches. Systems theory holds that individuals function within a variety of social systems (e.g., the family and peer groups) and that these systems act to influence behavior and to resist changes that are not in the interest of the broader system. From this perspective, drug use may be seen as serving some useful purpose within the "identified client's" social systems. Attempts to change that drug-use behavior without ensuring that the system will support and maintain such a change may be doomed to failure.

Family Therapy. Family therapy is the most common application of systems theory to the treatment of polydrug users. This is because research has linked various forms of family dysfunction to the development of drug-use problems. Also, many polydrug users are children and young adolescents and their drug use is a major family issue.

In family therapy, the family rather than the polydrug user becomes the client. Treatment addresses family-system issues, which include family roles, patterns of communication, and structural factors such as the alliances that may exist within and among parts of the family system. The presenting problem of drug abuse may be dealt with directly within the framework of the family approach. It may otherwise be treated as a symptom of the family's dysfunction—where the expectation is that the drug use will disappear with resolution of the more fundamental family problems.

In family therapy, all or most of the family members typically attend the treatment sessions. One-person family therapy is a variation on this practice, in which the treatment focuses on changes to the family system via one member of that system. This practice is, however, very limited in comparison with the more common approach of involving most or all the other family members.

Peer-Network Therapy. Peer-network therapy focuses on the peer or friendship social system. Polydrug users are typically young and their drug use is often a social activity. Much research evidence links all drug use to peer associations. This may be caused by peer influence or because drug users seek out other drug users. Either way, it is widely believed that changes in peer associations are a necessary step for polydrug users who would attempt to discontinue drug use.

Peer-network therapy involves systematically examining the relationship of drug use to association with particular peers. Strategies involve avoiding certain peers; strengthening peer relationships in which drug use is not a factor; reestablishing old relationships that may have been ignored while drug use was occurring; using a buddy system to facilitate developing new peer relationships; and structuring leisure activities to help the client meet new friends who share similar attitudes and goals concerning drug use. Typically, changes in the peer system are introduced via the identified client, but peer-network therapy may also involve sessions that include other members of the peer network.

Peer-network therapy is still a relatively novel approach to the treatment of polydrug users, although many treatment programs are placing increased emphasis on changes to peer networks as part of their overall treatment strategy.

Peer Counseling. Polydrug use is the most common pattern of substance abuse for many novice drug users. For such individuals, early intervention programs based on peer counseling, and provided in school or neighborhood settings, may be appropriate. Peer counseling capitalizes on the tendency for adolescents to be most influenced by their peers. Peer counselors are selected on the basis of their ability to act as good role models. They are trained to emphasize practical strategies to assist polydrug users to change their lifestyles in ways that support becoming drug free. They also act as facilitators or group leaders in peer counseling groups, in which adolescents learn from each other.

Social Learning Theory–Based Approaches. Social learning theory suggests that drug use is a learned behavior and that it may be changed by the therapeutic application of principles of learning theory. Treatments based on social learning theory usually begin with a functional analysis of the drug use. This involves a detailed analysis of the circumstances in which drug use occurs and the apparent benefits to the user. The basic assumption is that drug use serves useful purposes (functions) in the life of the user and that understanding these functions of drug use is a critical step in planning treatment.

Coping Skills Training. One such treatment approach is based on substituting alternative methods of obtaining the same benefits that drug use provides. If the individual becomes more sociable and outgoing on drugs, social-skill training is provided; if drug use reduces tension, stress-management techniques are offered. This approach is sometimes referred to as coping-skills training, because improved coping in one or more life areas usually becomes the primary treatment goal. Coping-skills training can address a variety of skill deficits from improved problem solving, to coping with depression, to increased assertiveness. The objective is to

provide the polydrug user with alternative methods of coping with difficult life situations.

Since the 1970s, this type of approach has become the primary alternative to more traditional approaches based on the disease concept or psychotherapy.

Contingency Management. Contingency management involves structuring unpleasant consequences to occur when drugs are used. The assumption is that these adverse consequences will compete with the benefits the user gets from the drug use, thereby reducing the likelihood that drug use will continue. Contingency management procedures are most effective when the occurrence of the drug use behavior can be reliably determined and the prescribed consequences reliably administered. Urine screening is the most common means of monitoring whether any drug use has occurred. Clients are typically required to provide urine specimens according to a random schedule that minimizes the opportunity to plan drug use to escape detection. A variety of types of consequences can be used. For example, clients may avoid the loss of a job, regain custody of children, or avoid breach of probation by consistently providing "clean" urines. While many treatment programs emphasize the consequences of drug use, few do so in the very systematic way required by contingency management.

Cue Exposure. Cue-exposure techniques focus on the circumstances that precede or "cue" drug use. Frequent repetition of patterns of drug taking may result in certain cues becoming conditioned so that the user experiences cravings for the drug in the presence of these cues. For example, observing drug-use paraphernalia or being in a setting in which drugs have frequently been used in the past, may cause the polydrug user to experience cravings. These cues can be the cause of relapse. Treatment involves repeatedly exposing the individual to these cues in a controlled manner (e.g., with a supportive person present) until the cue no longer elicits the craving response. Conditioning is more apt to occur for a specific drug than across a variety of drugs. Hence cue exposure may be most relevant for polydrug users with a pronounced primary drug of abuse.

Approaches Aimed at Major Psychological Change. These approaches assume that the cause of drug use lies in the psychological makeup of the polydrug user. From this perspective, drug use is a self-destructive or deviant act brought about by serious underlying psychological problems or the adoption of anti-social values. Treatment is aimed at correcting the underlying problem for which drug use is thought to be merely a symptom.

Psychotherapy. Psychotherapy is an intensive and extended counseling approach in which the therapist explores the past events in the client's life with the aim of uncovering emotionally upsetting events or identifying themes or patterns of behavior that interfere with the effective social and psychological functioning of the individual. The drug use itself would seldom be the focus of the treatment sessions. Rather, the goal of psychotherapy would be psychological growth to change the personality of the polydrug user.

Psychotherapy can be provided on a one-to-one or group basis. It is typically provided on an outpatient basis but has also been provided within the framework of long-term residential programs for young drug users. Psychotherapy can be a comparatively expensive form of treatment, because it requires highly skilled therapists and typically takes longer to complete than other therapies. It may be most relevant when the polydrug user also has a psychiatric problem (e.g., depression).

Therapeutic Communities. THERAPEUTIC COMMUNITIES (TC's) are long-term residential programs of twelve to twenty-four months duration. There are several types of TC, all of which share a common belief that clients gain from living together in a therapeutic environment for an extended period of time. The most prominent TC model is based on the Synanon program developed for heroin addicts in the late 1950s. Since that time, many variations of this model have evolved and the target treatment population has been broadened to include polydrug users.

The treatment approach is typically targeted to hard-core drug users who are judged to have serious personality deficits or chronic antisocial values. The problem is presumed to be the person, not the drug or the individual's social environment. The treatment is extremely intensive, often involving harsh confrontation and emotionally charged encounters. The intent is to break through the protective shell that the polydrug user has developed—in response to past deprivations and abuse—and to resocialize the individual to adopt new values and patterns of behavior. Consistent with its self-help origins, treatment within the TC is usually provided by recovered addicts.

Psychobiological Approaches. Psychobiological approaches involve interventions which have a biological (often neurological) mechanism of action. Examples include treatments that involve the administration of a drug (pharmacotherapies) and ACUPUNCTURE, although the latter has had little application to the treatment of polydrug users. These approaches are based on the assumption that it is possible to change drug-use behavior by biological methods even though the drug-use problem may not have biological origins. For example, a drug may be used in treatment to eliminate the positive effects of an abused drug, thereby reducing the likelihood that its use will continue.

Pharmacotherapies. Drugs are used in the treatment of substance-abuse problems for a variety of purposes. These include substituting for the drug effect; blocking or changing the drug effect; or treating a condition that is believed to underlie, or at least contribute to, the substance-abuse problem. Most pharmacotherapy approaches are intended to address the misuse of specific substances, which limit their application to polydrug users; however, many polydrug users have preferred drugs of abuse for which a pharmacotherapy approach may be appropriate. In such instances, it will usually be necessary to combine the pharmacotherapy treatment with some other approach to ensure that treatment addresses all the individual's drugs of abuse.

Methadone treatment is the best-known of the drug-substitution approaches. Methadone substitutes for heroin (and other opioid drugs) prevent the onset of withdrawal symptoms in addicts. This serves to stabilize the user with regard to the desire or need to continue heroin use until the addict develops sufficient confidence and a strong enough support system to become drug free.

Other drugs used in treatment (e.g., NALTREXONE) act on the brain to block or reduce the pleasant sensations associated with the use of particular drugs. The assumption is that if the so-called beneficial effects of the drug are eliminated or reduced, it is less likely to be used. So-called anti-alcohol drugs (ANTABUSE and Temposil) take this notion one step further, by altering the metabolism of alcohol so that its effects become very unpleasant (the individual gets sick if alcohol is consumed while the drug is in effect). For all these approaches, strategies to ensure that the individual actually takes the prescribed drug are very important since the polydrug user can easily obtain the desired drug effects just by not taking the treatment drug.

Finally, some polydrug use reflects an attempt at self-medication to cope with symptoms of untreated psychiatric problems. The appropriate diagnosis and treatment (with medication) of such problems may reduce the client's need to self-medicate. Examples of this form of pharmacotherapy include medications for the treatment of anxiety, mood disorder, and psychotic disorders.

THE IMPORTANCE OF MATCHING TREATMENT TO CLIENT NEEDS

This chapter has described a broad range of treatment approaches available to the polydrug user. In practice, treatment programs often combine elements of the various approaches described. None of the approaches can claim general superiority over any other. Any one of them may be the most appropriate treatment choice for a particular individual under certain circumstances. It is important to assess the needs and wishes of the polydrug user carefully before selecting the treatment that seems most likely to be most helpful.

(SEE ALSO: *Addiction: Concepts and Definitions; Adolescents and Drug Use; Causes of Substance Abuse; Comorbidity and Vulnerability; Contingency Contracts; Disease Concept of Alcoholism and Drug Abuse; Methadone Maintenance Programs; Prevention; Treatment Types*)

BIBLIOGRAPHY

BESCHNER, G. M., & FRIEDMAN, A. S. (1985). Treatment of adolescent drug abusers. *International Journal of the Addictions, 20*(6&7), 971–993.

DELEON, G., & DEITCH, D. (1985). Treatment of the adolescent substance abuser in a therapeutic community. In A. S. Friedman & G. M. Beschner (Eds.), *Treatment services for adolescent substance abusers*. Rockville, MD: National Institute on Drug Abuse.

HUBBARD, R. L., ET AL. (1989). *Drug abuse treatment: A national study of effectiveness*. Chapel Hill, NC: University of North Carolina Press.

INSTITUTE OF MEDICINE. (1990). *Treating drug problems*, vol. 1. Washington, DC: National Academy Press.

KAUFMAN, E. (1985). Family systems and family therapy of substance abuse: An overview of two decades of research and clinical experience. *International Journal of the Addictions, 20*(6&7), 897–916.

ONKEN, L. S., & BLAINE, J. D. (1990). *Psychotherapy and counseling in the treatment of drug abuse*. Rockville, MD: National Institute on Drug Abuse.

WILKINSON, D. A., & MARTIN, G. W. (1991). Intervention methods for youth with problems of substance abuse. In H. Annis & C. S. Davis. *Youth and drugs: Drug use by adolescents: Identification, assessment and intervention*. Toronto: Addiction Research Foundation and Health & Welfare Canada.

GARTH MARTIN

Polydrug Abuse, Pharmacotherapy Although many individuals present with abuse or dependence upon a single PSYCHOACTIVE SUBSTANCE, increasing numbers of drug users are presenting with dependencies upon two or more such substances. The DIAGNOSTIC AND STATISTICAL MANUAL of the American Psychiatric Association (DSM-IV) and the INTERNATIONAL CLASSIFICATION OF DISEASES of the World Health Organization (ICD-10) define a condition called "polydrug dependence" or "multiple drug dependence," in which there is dependence on three or more psychoactive substances at one time. Polydrug dependence is particularly common among adolescents and young adults. However, if one includes NICOTINE and CAFFEINE dependence, over half of patients with psychoactive-substance dependence are polydrug-dependent.

The use of specific, preferred combinations of drugs is typically seen in polydrug users. OPIOIDS and COCAINE are often used together, as are ALCOHOL and cocaine or nicotine and alcohol. Alcohol, BENZODIAZEPINES, and cocaine are often used together by opiate users, especially METHADONE users. Illicit-drug users often show nicotine and caffeine dependence. Some individuals will use whatever psychoactive substances are available. One useful distinction is the difference between simultaneous and concurrent polydrug use. In simultaneous polydrug use, the drugs are used together at the same time for a combined effect, such as heroin and cocaine mixed and injected as a "speedball." In concurrent polydrug use, the various drugs are used regularly but not necessarily together. An example is a heroin user who uses benzodiazepines and alcohol to get another kind of high.

TREATMENT

The treatment of the polydrug user presents a particular challenge to the clinician. The simultaneous and concurrent use of multiple drugs may increase the level of dependence, increase drug toxicities, worsen medical and psychiatric comorbidities due to the drugs, and intensify withdrawal signs and symptoms upon cessation of drug use. The basic principles of treatment of polydrug use are similar to those for the treatment of any single psychoactive-substance dependence. Patients require a complete medical and psychiatric assessment, treatment of active problems, detoxification, then rehabilitation with attempts to reduce subsequent use of the drugs.

In providing treatment for the polysubstance user, there are two options: (1) sequential treatment for the dependencies, with initial treatment of the major dependency or the dependency with greater morbidity; or (2) simultaneous treatment of all dependencies. Unfortunately, few objective data exist as to which type of treatment is optimal for which patients. Most clinicians rely on their own experience, the capabilities of the treatment setting, and the wishes of the patient.

The treatment of polysubstance dependence often involves more than one type of treatment modality. A common example is an alcohol-dependent, opioid-dependent, cigarette smoker who is receiving METHADONE MAINTENANCE for opioid dependence, abstinence-oriented treatment for alcoholism, and no specific treatment for nicotine dependence. The different treatment philosophies—methadone substitution, abstinence, and no treatment—necessarily conflict. In such cases, good communication and flexibility among the various treatment providers and with the patient are important to ensure optimal, coordinated treatment.

DETOXIFICATION

During the initial treatment of polysubstance abuse and dependence, the primary goals include cessation of substance use and the establishment of a substance-free state. If necessary, detoxification occurs, as well as management of medical and psy-

chiatric problems. Detoxification is the removal of the drug in a fashion that minimizes signs and symptoms of withdrawal. It can be pharmacological or drug free. Pharmacological methods for detoxification include (1) a slow decrease in the dose of the drug or of a cross-tolerant agent (e.g., methadone for heroin withdrawal, diazepam for alcohol withdrawal, NICOTINE GUM for smoking cessation) and (2) stopping the drug and using an alternative agent to suppress signs and symptoms of withdrawal (e.g., CLONIDINE for opioid withdrawal, atenolol for alcohol withdrawal). For many drugs, pharmacologically assisted detoxification is not necessary. Simple alcohol withdrawal can be treated with supportive care. However, the presence of polysubstance dependence usually increases the need for pharmacological agents to assist in withdrawal.

There are few controlled studies on the clinical course and optimal therapies for detoxification from multiple psychoactive substances. Patients can be detoxified from all psychoactive substances together, or maintained on one or more drugs while being detoxified from others. When the drugs used are all part of the same class (e.g., alcohol and sedatives; methadone, CODEINE, and heroin), a complete detoxification is more common. When the drugs used are from different classes, partial or sequential detoxification usually occurs. An example of the latter situation is an opioid, cocaine, alcohol, and nicotine user who is detoxified from alcohol and cocaine, but maintained on methadone and allowed to continue tobacco use. Sometimes a partial detoxification is indicated because of the need for continued psychotropic medication for medical or psychiatric illnesses, such as continued opioids for chronic pain or benzodiazepines for anxiety.

Given the cross-tolerance of most SEDATIVE-HYPNOTICS with ethanol, methods that are effective for the detoxification from alcohol or sedatives alone are usually effective for the combinations of alcohol and sedatives. Loading techniques, with long-acting benzodiazepines, such as diazepam or CHLORDIAZEPOXIDE, or with BARBITURATES, such as PHENOBARBITAL, are well documented as effective. The advantages of these methods include matching the medication used for withdrawal to the individual patient's tolerance and the avoidance of overmedication. The anticonvulsant carbamazepine has been shown to be effective for the treatment of combined alcohol and sedative withdrawal.

Although the mechanisms of action of various drugs differ, there are common neurological substrates of certain behavioral effects and of withdrawal signs and symptoms. The autonomic hyperactivity and some of the CNS excitation common to several withdrawal syndromes are mediated by the locus ceruleus of the brain. Medications such as alpha-2 antagonists (clonidine) and benzodiazepines, which inhibit locus ceruleus activity, have been shown to attenuate the symptoms of nicotine withdrawal. However, clonidine will not block the seizures that result from alcohol or sedative withdrawal.

LONG-TERM TREATMENT

In the long-term phase of treatment, the patient undergoes rehabilitation and reestablishment of a lifestyle free of drug dependency. Pharmacological treatment is sometimes used to assist rehabilitation. Pharmacotherapies may reduce drug craving, decrease protracted withdrawal symptoms, or decrease positive reinforcing effects of the drugs. Types of pharmacological therapies used in long-term treatment and rehabilitation include (1) maintenance (e.g., methadone maintenance for the treatment of opiate dependence); (2) blockade (e.g., NALTREXONE treatment for opioid dependence); (3) aversive therapy (e.g., DISULFIRAM for alcoholism, possibly naltrexone for alcoholism); and (4) psychotropic drug treatment of coexisting psychiatric disorders, such as lithium for bipolar alcoholics, or methylphenidate for cocaine-dependent patients with ATTENTION DEFICIT DISORDER.

The use of pharmacological agents as adjuncts in the treatment of polysubstance dependence is an area of active investigation. One medication that may prove useful in the treatment of combined cocaine and opioid dependence is buprenorphine. This partial mu agonist, used as a surgical analgesic, has shown efficacy as a substitute in the long-term treatment of opioid dependence. Compared with methadone, buprenorphine may produce less dependence and fewer withdrawal symptoms upon cessation. Buprenorphine treatment also may reduce cocaine use in some individuals dependent on both opioids and cocaine. Desipramine has been reported as being effective in reducing cocaine use in methadone patients. Disulfiram, which is efficacious in the treatment of alcoholism, may also reduce cocaine use in individuals using both alcohol and cocaine.

(SEE ALSO: *Comorbidity and Vulnerability; Treatment-Treatment Types*)

BIBLIOGRAPHY

GLASSMAN, A. H., ET AL. (1988). Heavy smokers, smoking cessation and clonidine. Results of a double-blind, randomized trial. *Journal of the American Medical Association, 259,* 2863–2866.

GOLD, M. S., POTTASH, A. C., SWEENEY, D. R., & KLEBER, H. D. (1979). Opiate withdrawal using clonidine: A safe, effective and rapid non-opiate treatment. *Journal of the American Medical Association, 234,* 343–346.

KOSTEN, T. R., KLEBER, H. D., & MORGAN, C. (1989). Treatment of cocaine abuse with buprenorphine. *Biological Psychiatry, 26,* 637–639.

LISKOW, B. I., & GOODWIN, D. W. (1987). Pharmacological treatment of alcohol intoxification, withdrawal and dependence. *Journal of Studies on Alcohol, 48*(4), 356–370.

LITTEN, R. Z., & ALLEN, J. P. (1991). Pharmacotherapies for alcoholism: Promising agents and clinical issues. *Alcohol Clinical Experimentation and Research, 15*(4), 620–633.

MALCOLM, R., BALLENGER, J. C., STURGIS, E. T., & ANTON, R. (1989). Double blind controlled trial comparing carbamazepine to oxazepam treatment of alcohol withdrawal. *American Journal of Psychiatry, 146*(5), 617–621.

MARTIN, C. S., ARRIA, A. M., MEZZICH, A. C., & BUKSTEIN, O. G. (1993). Patterns of polydrug use in adolescent alcohol abusers. *American Journal of Drug and Alcohol Abuse, 19*(4), 511–521.

MEYER, R. E. (1992). New pharmacotherapies for cocaine dependence . . . revisited. *Archives of General Psychiatry, 49* (11), 900–904.

PATTERSON, J. F. (1990). Withdrawal from alprazolam using clonazepam: Clinical observations. *Journal of Clinical Psychiatry, 51* (5, supp.), 47–49.

RESNICK, R. B., SCHUYTEN-RESNICK, E., & WASHTON, A. M. (1980). Assessment of narcotic antagonists in the treatment of opioid dependence. *Annual Review of Pharmacology and Toxicology, 20,* 463–474.

SELLERS, E. M., ET AL. (1983). Diazepam loading: Simplified treatment for alcohol withdrawal. *Clinical Pharmacology and Therapy, 6,* 822.

SENAY, E. (1985). Methadone maintenance treatment. *International Journal of Addictions, 20*, 803–821.

WESSON, D. R., & SMITH, D. E. (1977). A new method for the treatment of barbiturate dependence. *Journal of the American Medical Association, 231*, 294–295.

ROBERT M. SWIFT

Tobacco, An Overview Ever since tobacco use became popular, some users have been trying to quit. Sometimes they sought treatment because the tobacco was too expensive, because companions complained about the tobacco use, because they did not like the smoke in the air, or, in the case of SMOKELESS TOBACCO (chewing tobacco or spitting snuff), because they did not like the tobacco juice on the floor. Sometimes treatment was sought out of concern for health problems. It never took a genius, for example, to connect lung or breathing problems with the effects of inhaled tobacco smoke.

Cigarette smoking is the most common form of tobacco use, and smoking is viewed by the U.S. surgeon general as the "single most important environmental factor contributing to premature mortality in the United States" (USDHEW, 1979). This article focuses on the treatment of cigarette smoking but will include a brief discussion of the treatment of smokeless tobacco use, for which many of the same principles apply.

TRENDS IN SMOKING CESSATION

Cigarette smoking and smoking cessation need to be viewed as on a continuum. At any given time, some smokers are trying to stop smoking, but some will fail; Table 1 shows data on the trends in smoking cessation from 1978 to 1987. The minority of smokers have never tried to quit smoking, and this minority has become smaller in recent years (see row 1), dropping from 26 percent in 1978 to 19 percent in 1987. In 1987, 34 percent of current smokers had tried to quit at some point in that year (see row 9). Only a small percentage of smokers (7.8%) who were smoking in the prior year and who quit smoking were still not smoking at the time of the survey (see row 10).

Sex Differences. Males are somewhat more likely to be former smokers than are females (48.7% vs. 40.1%). Types of tobacco use (e.g., pipes, cigars, chewing tobacco) other than cigarettes are more common in males than in females. When adjustments are made for persons who have stopped smoking cigarettes but are using other forms of tobacco,

TABLE 1

Cigarette Smoking Continuum by Year, Percentage of Ever Cigarette Smokers, United States, 1978–1987, Adults Age 20 and Older

Cigarette Smoking Continuum	1978	1979	1980	1987
1. Current smokers who had never tried to quit	25.9	26.1	25.4	18.9
2. Current smokers who had quit previously but not in past year	22.7	21.4	23.1	20.0
3. Current smokers who had quit for <7 days in past year	6.6	6.0	5.9	7.0
4. Current smokers who had quit for ≥7 days in past year	8.5	8.6	7.8	8.4
5. Former smokers who had quit within past 3 months	1.3	1.6	1.4	1.8
6. Former smokers who had been abstinent for 3–12 months	2.7	2.6	2.7	2.8
7. Former smokers who had been abstinent for 1–5 years	9.0	10.0	9.5	10.4
8. Former smokers who had quit ≥5 years earlier	23.3	23.6	24.1	30.7
9. Percentage of those smoking during the year prior to the survey who tried to quit during that year (Categories 3+4+5+6 divided by 1+2+3+4+5+6)	28.2	28.4	26.8	34.0
10. Percentage of those smoking during the year prior to the survey who quit during that year and were still abstinent at the time of the survey (Categories 5+6 divided by 1+2+3+4+5+6)	6.1	6.3	6.2	7.8

Adapted from U.S. Department of Health and Human Services, 1990.

the sex difference in smoking cessation becomes smaller (males = 42.1% vs. females = 39.9%). The fashion and pressure to start smoking appeared earlier in males than in females. The same is true of the pressure to stop smoking. It is expected that the sex differences in smoking cessation will become smaller in the future.

Age-related Differences. The older the group of "ever smokers" (i.e., those who have ever smoked), the higher the chances that the smoker has quit. In 1987, only 24 percent of persons age 20–24 had quit smoking, but in the oldest group (65 or older), 69 percent had stopped. Older persons have had more time to try to give up smoking, had the chance to try more different methods of quitting, and they are more motivated since they are closely touched by the death and disability caused by smoking.

OTHER PATTERNS OF SMOKING CESSATION

African Americans are generally less likely to have quit smoking than are Caucasians (31.5% for African Americans, 46.4% for whites). Better-educated smokers are more likely to quit. In 1987, for example, 39.7 percent of those who did not finish high school had quit smoking, compared with 61.4 percent of those who had at least completed college.

RESEARCH ON CESSATION OF TOBACCO USE

Although the scientific study of smoking treatments is only about fifty years old, the "nonscientific" and "scientific" treatments often overlap. Until the 1980s, there were still many observers who doubted that tobacco use was based on an addiction to or dependence on nicotine (see USDHHS, 1989). In the 1950s and 1960s, many experts believed that smoking was "just a bad habit." Experts at that time failed to appreciate that tobacco use was a form of drug use; instead, they saw smoking as the kind of habit that could be broken by taking certain behavioral steps. This attitude was the origin of the so-called behavioral techniques for stopping smoking.

In the early part of the twentieth century, self-help movements were very popular and were directed against alcohol and other drug problems. Smoking was thought of as just another bad habit. The focus was on changing one's behavior. Making

themselves behave in particular ways is an ancient challenge for human beings. Just as people tend not to seek professional help for dealing with minor behavioral problems, it should not be surprising that much of the "treatment" for cigarette smoking has been self-given. By far, most of the successful treatments for smoking are, if not entirely self-devised, entirely self-administered. Practically speaking, this means that most persons who seek formal help for stopping smoking have already tried to quit on their own—even if for only a few hours—and have not succeeded. Therefore, formal programs are likely to get the harder cases. Who would spend the time and money for a formal treatment program if they could stop smoking on their own? Most smokers quit without attending any formal treatment program, yet the "subject self-selection" bias makes scientists reluctant to conclude that "no treatment" is the most effective treatment.

Randomized, controlled trials are needed for the fair comparisons of treatment A versus treatment B. Randomly assigning individuals to different treatments ensures that biased samples do not make one treatment, unfairly, look better and more effective than another. Unfortunately, there are very few studies that compare different forms of treatment with each other. More often, the research has compared one treatment with a control group or, if drugs are involved, with a placebo condition. This makes the research literature on smoking cessation difficult to interpret. Many different kinds of treatment appear to offer some help for those who want to stop smoking, yet no one treatment stands out as being the single best way for all smokers.

Table 2 derives from work by Jerome L. Schwartz for the National Cancer Institute. This table does not provide the results of randomized comparisons among the different treatment methods; rather, it gives a kind of "box score" on how the different programs have been doing. It is easy enough to get a smoker to stop for a few days or even a few months. Following participants for one year after treatment has become a standard in the evaluation of smoking cessation. This table shows both the wide variety of methods and the wide variation in results. The technique with the highest percentages is not necessarily the one that will give the most people the best results. For the most part, participants in these studies have been allowed to try whichever treatment most appeals to them.

TABLE 2
Summary of Follow-up Quit Rates of Smoking Cessation Trials with at Least 1-Year Follow-up by Selected Method (Reported 1959–1985)

	Number of Studies	Range	Median
Self-help	7	12–33%	18
Educational	12	15–55	25
Group	31	5–71	28
Medication	12	6–50	18.5
Nicotine chewing gum	9	8–38	11
Nicotine chewing gum plus behavioral treatment or therapy	11	12–49	29
Hypnosis–individual	8	13–68	19.5
Hypnosis—group	2	14–88	—
Acupuncture	6	8–32	27
Physician advice or counseling	12	3–13	6
Physician intervention—more than counseling	10	13–38	22.5
Physician intervention:			
Pulmonary patients	6	25–76	31.5
Cardiac patients	16	11–73	43
Rapid smoking	6	6–40	21
Rapid smoking and other procedures	10	7–52	30.5
Satiation smoking[a]	12	18–63	34.5
Regular-paced aversive smoking[a]	3	20–39	26
Contingency contracting[a]	4	14–38	27
Multimodal	17	6–76	40

Note: The median is the score that divides the sample in two halves. These quit rates suggest overall trends only, since most were based on self-reports and some include patients who either did not complete treatment or failed to reply to follow-up inquiries. (Adapted from Schwartz, 1987).

[a]Other procedures may have been used, and some trials may be included in more than one method.

GROUP VERSUS INDIVIDUAL THERAPIES

Much of the instruction and support that is part of smoking treatment can be done individually—one-on-one—with clients or can be delivered to a group of clients. Group programs have been used to provide hypnotism, educational therapies, behavioral therapies, and multimodal, combined therapies. There is no clear scientific evidence indicating which delivery system is best, but it is clear that group programs can be less expensive than individual programs and that some clients have strong personal preferences for how they wish to receive treatment. Some enjoy the group support and like to share their experiences in a group; others find such involvement with groups unpleasant or embarrassing. Women are more likely than men to prefer group programs.

Individual therapies involving psychoanalysis and other long-term psychotherapies have rarely been studied as smoking treatments. They are generally very expensive, and their cost-effectiveness is in doubt.

PHYSICIAN-BASED TREATMENTS

Physicians interested in preventive medicine make special efforts to encourage and support smoking cessation in their patients. In 1964, only about 15 percent of current smokers reported that a physician had advised them to quit smoking. By 1987, about

50 percent of current smokers had received such advice. Nicotine replacement therapies (discussed below) are available for use by prescription. Sometimes just the advice of a physician to quit and the setting of a quit date can lead to successful smoking cessation. Physicians can also be helpful by referring patients to smoking treatment programs. Specialists who deal with patients already suffering from a smoking-related disease can be in a good position to help those who are well motivated to quit, but, as can be seen in Table 2, individuals who are patients for cardiac or lung problems often fail to stop smoking. Having a smoking-related disease is no guarantee that the patient will quit smoking.

The Importance of "Minimal" Interventions. In medical settings, there has been research on the value of interventions (e.g., brief advice, pamphlets) that take only a few minutes of the physician's time. Although the effects of these interventions is small, they are generally viewed as worthwhile because they can reach so many smokers.

SMOKING CESSATION EFFORTS BY AGENCIES

Think of a disease caused by smoking—cancer, heart disease, lung disease—and consider agencies concerned with these diseases. The Cancer Society, the Lung Association, and the Heart Foundation are voluntary, charitable organizations. Each has developed materials and programs to promote smoking cessation. The measured treatment effects of simple stop-smoking pamphlets are small, but since they can reach many smokers at very low cost, they should be viewed as beneficial elements of the public-health efforts to support smoking cessation. U.S. government agencies concerned with smoking and smoking-related disease (e.g., the Office on Smoking and Health, the National Cancer Institute, the National Heart, Lung and Blood Institute, and the National Institute on Drug Abuse) have also developed and promoted materials and procedures to foster smoking cessation.

The voluntary agencies have supported smoking cessation efforts in the workplace, by providing smoking-treatment services and by promoting smoking bans in the workplace. EMPLOYEE ASSISTANCE PROGRAMS (EAPs) increasingly offer help for smokers who are trying to quit. Just as social pressures got many smokers to start smoking, social pressures can get them to stop. Once it was fashionable to be a cigarette smoker; now it is fashionable to stop smoking.

NICOTINE-REPLACEMENT THERAPIES

Nicotine-replacement therapies are available by prescription and require medical supervision. Patients with some conditions (e.g., pregnancy) are advised against using these products. These pharmaceutical tools give the physician an important role in smoking treatment. Available products include gum that releases nicotine as it is chewed, and a patch that slowly releases nicotine into the body through the skin. These techniques can help reduce the nicotine withdrawal symptoms after smoking cessation. Replacement therapies help individuals deal with their smoking in a stepped fashion by separating the behavioral and pharmacological components of smoking. While nicotine withdrawal is reduced, the individual can focus on dealing with the behavioral challenges in stopping.

Both nicotine-containing gum and nicotine-containing patches have been very popular and seem effective, in particular when they are combined with behavioral therapies. When the replacement therapies are used without careful instructions and without supportive behavioral therapies, their success rates are much lower. Use of a nicotine patch appears to be an extremely easy way to stop smoking, but the appearance is deceptive. None of the currently available nicotine replacement products takes away all desire to smoke. These products need to be viewed as tools that aid smoking cessation, but that cannot do all the work. Just as a jack can be a crucial tool for changing a tire, it does not perform that hard task all by itself.

Those smokers who give the most evidence of being dependent on nicotine (those who smoke over 20 cigarettes a day and who smoke their first cigarette of the day within minutes of waking) also give the most evidence of being helped by nicotine-replacement therapies.

OTHER DRUG THERAPIES

Over-the-counter, nonprescription products (e.g., those containing lobeline) have not been shown to work in promoting long-term smoking cessation, though they have at times been very popular. For someone who has tried repeatedly and yet failed to stop smoking for good, a medicine that can take away the desire to smoke would be welcome.

There has been recent research on the use of CLO-NIDINE to treat smoking (see also separate article on clonidine). This powerful antihypertensive drug has had some success, especially in women who have had a history of DEPRESSION. Some of this research has pointed to the possible role of depression in the failure to stop smoking. This is an important topic for future research. Overall, smokers who have had more depressive symptoms have less success quitting.

HYPNOSIS

HYPNOSIS is worth special mention because it is a popular form of smoking therapy. Careful evaluations of hypnotherapies show small or no treatment effects. One of the problems in studying hypnotherapies is that the actual hypnotic procedures involved are not standardized. The kind of procedures and suggestions (e.g., "You will not want a cigarette" vs. "The thought of a cigarette will make you feel sick") differ from therapist to therapist. It is important to deal with reputable therapists who charge reasonable fees for their services.

MULTIMODAL THERAPIES

A wide range of behavioral therapies have been tested, and no one stands out as particularly effective. Multimodal approaches have become widely used, in hopes that something loaded into the shotgun will hit its mark. We lack a good way to judge beforehand which smoker will be most helped by a particular technique (the exception being that heavier, more dependent smokers are likely to benefit from nicotine replacement). The multimodal, something-for-everyone approach is reasonable. There is not room in this article to discuss in detail the variety of behavioral therapies that have been used. They have in common the use of basic principles of learning.

Contingency contracting involves, for example, the preparation of detailed contracts that spell out punishments that will follow from the return to smoking (e.g., the person will give $100 to someone he or she dislikes).

Aversive conditioning procedures (e.g., rapid smoking, satiation) cause cigarette smoking to be associated strongly with the acute unpleasant effects (such as dizziness and nausea) of smoking very heavily.

Relapse Prevention and the Maintenance of Abstinence. RELAPSE-PREVENTION programs have been developed to reduce the problem of relapse or return to smoking. Many of the same behavioral techniques used in multimodal programs are applied to the task of helping prevent relapse and helping prevent the occasional slip back to smoking from becoming a permanent return.

Smoker's Anonymous Programs. Smokers have sometimes organized this type of program to support smoking cessation. The program allow smokers to support each other and teach each other techniques that will help them to stop smoking and to keep from returning to smoking. These programs have not generally become popular. This is in contrast to the great popularity of ALCOHOLICS ANONYMOUS (AA) groups. Some special nonsmoking Alcoholic's Anonymous groups have developed.

RELATION TO TREATMENT OF OTHER DRUG PROBLEMS

Heavy smoking is strongly linked to heavy alcohol and other drug use. Smoking is often found in those with ALCOHOL and other drug problems. Those smokers who fail to stop smoking may have serious alcohol or other drug problems that require treatment before the smoking problem can be resolved.

ON SELECTING A WAY TO STOP SMOKING

Smokers should be advised to take a long view of their efforts to stop smoking, understanding that if one method does not help them, they should try another, and another, until they have stopped smoking. Any one attempt to stop smoking can meet with poor success. With repeated attempts, the smoker may encounter some success. Also, repeated attempts can give the smoker experience with assorted treatment techniques, so that the individual begins to learn for himself or herself what helps and what does not help. Finally, there may be a kind of "no more nice guy" effect, so that the smoker gets fed up with failing to quit smoking.

It is also important not to think that one must go to an organized program to stop smoking. If one does so, it is important to realize that no two programs are delivered in exactly the same way. The individual characteristics of a therapist and the client's rapport with that therapist can contribute to a therapy's success. The person who wants help to stop smoking should shop around to see what is available in the community (or at the library). If one attempt fails, additional attempts should be planned.

A NOTE ON SMOKELESS TOBACCO

To the extent that chewing tobacco and dipping snuff can cause nicotine to be delivered to the brain in sufficient doses, they present a similar risk of nicotine dependence in the regular user. These products may prove more difficult to treat than cigarette use, because they are sometimes viewed as less risky alternatives to cigarettes. Once the "negative publicity" on smokeless tobacco use reaches a level close to the bad press on smoking, there should be a growing demand for using the smoking therapies as treatments for the use of smokeless tobacco.

(SEE ALSO: *Addictions: Concepts and Definitions; Nicotine Delivery Systems for Smoking Cessation; Tobacco; Treatment Types*)

BIBLIOGRAPHY

SCHWARTZ, J. L. (1987). *Review and evaluation of smoking cessation methods: The United States and Canada, 1978–1985.* Washington, DC: Division of Cancer Prevention and Control, National Cancer Institute.

U.S. DEPARTMENT OF HEALTH AND HUMAN SERVICES. (1990). *The health benefits of smoking cessation: A report of the surgeon-general.* Washington, DC: U.S. Government Printing Office.

U.S. DEPARTMENT OF HEALTH AND HUMAN SERVICES. (1989). *Reducing the health consequences of smoking: A report of the surgeon-general.* Washington, DC: U.S. Government Printing Office.

U.S. DEPARTMENT OF HEALTH AND HUMAN SERVICES. (1988). *The health consequences of smoking: Nicotine addiction: A report of the surgeon-general.* Washington, DC: U.S. Government Printing Office.

U.S. DEPARTMENT OF HEALTH, EDUCATION AND WELFARE. (1979). *Smoking and health: A report of the surgeon-general.* Washington, DC: U.S. Government Printing Office.

LYNN T. KOZLOWSKI

Tobacco, Pharmacotherapy Although tobacco use causes a powerful addiction, people who want to stop using it can be helped, and at far less expense than treatment of tobacco-caused diseases—which will kill approximately one in two smokers who do not quit. The effort to find pharmacological agents that would help tobacco users quit is not a new development. In the late 1890s and early 1900s, a number of potent medicines were advertised as being useful for reducing tobacco craving and helping break the habit. Such advertising was possible because at the time there were no regulations requiring a seller to demonstrate that the product was effective. None of the products offered to the public between the early 1900s and the late 1970s were demonstrably better than placebos in helping smokers quit. Effective pharmacological approaches to treating nicotine addiction, including transdermal patches that deliver nicotine through the skin, and resin complexes (gum) that release nicotine when chewed, were among the important medical advances of the 1980s and 1990s. To understand how pharmacotherapy works, it is necessary to understand the role of NICOTINE in the addiction to tobacco.

Nicotine is a naturally occurring alkaloid present in the tobacco leaf. It is a small lipid and water-soluble molecule, rapidly absorbed through the skin and mucosal lining of the mouth and nose or by inhalation in the lungs. In the lungs, nicotine is rapidly extracted from tobacco smoke within a few seconds because of the massive area for gas exchange in the alveoli; it is passed into the pulmonary veins, and pumped through the left ventricle of the heart into the arterial circulation within another few seconds. Within 10 seconds, a highly concentrated bullet (bolus) of nicotine-rich blood reaches organs such as the brain as well as the fetus of a pregnant woman. Arterial blood levels may be ten times higher than venous levels within 15 to 20 seconds after smoking. Nicotine arterial boli from smoking a single cigarette may be three to five times more concentrated than the low, steady levels obtained from nicotine gum or patch systems. These spikes probably contribute to the pleasure sought by the cigarette smoker, but, fortunately, they are not necessary to relieve withdrawal symptoms. NICOTINE GUM and patches, which provide more steady nicotine levels without arterial spikes, may selectively relieve withdrawal without the highly addictive nicotine spikes produced by cigarettes. Although SMOKELESS TOBACCO users do not obtain the same rapid nicotine increase as smokers, they may, by repeatedly putting new "pinches" in their mouths, achieve stable nicotine levels higher than those typical of smokers.

Most cigarettes on the U.S. market contain 8 to 9 milligrams (mg) of nicotine, and the average smoker obtains 1 to 2 mg per cigarette. In general, the type of cigarette or nicotine delivery rating reported by the manufacturer bears almost no relation to the level of nicotine obtained by the typical smoker, because smokers may change their behavior to compensate for differences in cigarette brands. For example, they may take additional puffs on low-nicotine brands.

Cigarette smoking produces rapid and large physiological changes, but, to a lesser extent, smokeless tobacco produces similar effects. Nicotine gum and patch treatments have the advantages of much slower nicotine delivery, and they produce less severe physiological changes. This slower delivery rate may be less pleasurable to the tobacco user, but the user is less likely to have difficulty giving up the gum or the patch after treatment.

Tobacco-caused cancer may be considered a side effect of nicotine dependence in much the same way that ACQUIRED IMMUNODEFICIENCY SYNDROME (AIDS) may occur as a side effect of heroin dependence. In both cases, the exposure to the disease-causing toxins or to HIV occurs repeatedly and often frequently because individuals are dependent on a drug that has reduced (if not nearly eliminated) their ability to abstain from the highly contaminated drug delivery system they know may lead to disease and premature death.

The physiological basis of drug dependence became increasingly well understood in the past few decades and especially with regard to nicotine dependence in the 1970s and 1980s. Awareness of the physiology of nicotine dependence can help researchers understand the problems faced by people attempting to give up tobacco and can provide a more rational basis for the development of treatment programs that may prevent the occurrence of cancer and other diseases or contribute to remission in people who have been treated for cancer.

TOLERANCE as a result of repeated nicotine exposure is a crucial factor in the development of lung and other cancers. Essentially, smokers self-administer much greater amounts of tobacco-delivered toxins than would be the case if they had not developed tolerance. In turn, with development of nicotine dependence, smokers come to feel normal, comfortable, and most effective when taking the drug and to feel unhappy and ineffective when deprived of the drug. This process makes it more difficult to achieve and sustain even short-term abstinence.

PHARMACOLOGICAL TREATMENTS

Most smokers have quit on their own or, rather, tried to quit. Although 18 million try each year, less than 7 percent do so successfully. Most of the efforts were "cold turkey," good for a start, but the least effective of all techniques. Long-term abstinence rates are low for people using this method. Treatment programs are helpful in increasing rates of success, and the availability of pharmacological interventions gives clinicians additional useful tools to help the smoker. The major pharmacological approaches are nicotine replacement, symptomatic treatment, nicotine blockade, and deterrent therapy. Nicotine replacement and symptomatic treatment have become part of general medical practice. Until further information is collected, blockade and deterrent therapy must be considered experimental.

Nicotine Replacement. The rationale for nicotine replacement is to substitute a safer, more manageable, and, ideally, less addictive (more easily discontinued) form of an abused drug to alleviate symptoms of withdrawal. An example of a less-addictive substitute is METHADONE MAINTENANCE for opiate abusers. Various forms of nicotine replacement have been developed including polacrilex (gum), transdermal delivery systems (patches), nasal vapor inhaler, nasal nicotine spray (gel droplets), and smoke-free nicotine cigarettes. The forms provide different doses and speeds of dosing. These parameters may be important in offering the smoker levels of nicotine necessary to alleviate withdrawal and cravings for nicotine. Currently, only the nicotine gum and patch are approved for use in the United States.

Several advantages exist in replacing nicotine from tobacco with non-tobacco-based systems such as gum or patches. First, they do not contain all the toxins present in tobacco or produced by burning tobacco. Second, total daily nicotine administration is lower for most patients on nicotine-replacement systems, and the high initial nicotine bolus doses produced by inhaling are not delivered. Third, the clinician can control doses more effectively than with tobacco-based products. The patient cannot, for example, take a few extra puffs per cigarette and defeat the purpose of gradual nicotine-reduction plans.

Nicotine gum may not be absorbed well if the client does not follow directions carefully. From 1984 until 1991, about 1 million prescriptions for nicotine gum, the only form of nicotine replacement then available, were filled per year. At the end of 1991, nicotine patches were introduced, and approximately 7 million prescriptions were filled for all replacement systems, with the nicotine patch accounting for nearly 90 percent of new prescriptions for nicotine replacement. The popularity of the nicotine patch can be measured by the higher rate of compliance than for the only currently available alternative, nicotine gum. Nicotine gum compliance rates tend to be lower because patients may dislike the taste and experience slightly sore mouths, throats, and jaws and gastrointestinal upset. Nevertheless, a study at the Addiction Research Center of the NATIONAL INSTITUTE ON DRUG ABUSE (NIDA) found nicotine gum to be effective in treating the cognitive function and corresponding brain electrical function changes of tobacco withdrawal. The effect was stronger at higher dose levels (e.g., 4 mg; see Figure 1). Because of current prescribing practices, this section will concentrate on the nicotine patch.

Four brands of nicotine patch are currently available in the United States. All deliver a given dose of nicotine transdermally, through the skin, over either a 24-hour (Habitrol, Prostep, and Nicoderm) or a 16-hour (Nicotrol) period. No clinical study has directly compared the four brands, but there is no evidence that any one brand leads to consistently higher rates of abstinence than any other. Variations in nicotine-delivery rate and skin contact effects may mean that certain patches work better for some people than others, but there is as yet no way to tell which patch will work better for an individual patient.

The nicotine patch is highly effective, resulting in an overall doubling of smoking cessation rates. Dif-

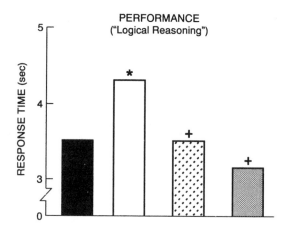

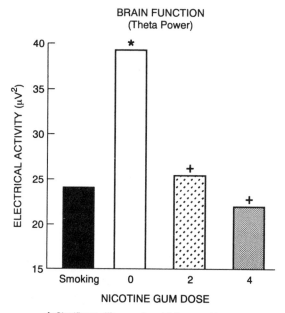

* Significant difference from Libitum smoking
\+ Significant difference from placebo gum

Figure 1

Cognitive Performance and an Electrophysiological Measure of Brain Function during Smoking and Abstinence with Nicotine- or Placebo-Delivering Gum Treatment.

ferent studies have reported cessation rates of between 22 percent and 42 percent after 6 months of use. The combination of intense counseling and patch use was associated with higher success rates.

Work is necessary to develop a list of characteristics of those patients most likely to benefit from nicotine patch use. The University of Wisconsin's

Center for Tobacco Research and Intervention suggests that patients may benefit if they are motivated to quit and fit into at least one of the following categories:

Smoke at least 20 cigarettes per day

Smoke first cigarette within 30 minutes of awakening

Have experienced a strong craving for cigarettes during the first week of previous attempts at quitting

The nicotine patch should be applied as soon as the patient awakens, and the user should stop all smoking during patch use. The patch should be applied to a hairless part of the body, with a different site every day. The same site should not be used again for one week. Side effects include a local skin reaction at the patch application site in 30 percent of patients and possibly sleep disruption. Because the tobacco-withdrawal syndrome also may include sleep disruption, it is sometimes difficult to determine whether the sleep disturbance is a result of tobacco withdrawal or nicotine patch therapy.

The four patches vary in their recommendations for length of treatment, from 6 to 16 weeks. Because no published studies have documented a benefit for longer treatment, some researchers recommend 6 to 8 weeks for most patients, but therapy should be individualized where appropriate. Other researchers have concluded that, in general, the chances of success appear better in longer-term use.

In patients with cardiovascular disease, the nicotine patch may be used cautiously, although there has been no documented association between patch use and acute heart attacks. It should be used in pregnant patients with caution—only after they have failed to quit using nondrug means. Nicotine replacement should not be given to people who continue to smoke, although the advisability of terminating therapy if only occasional cigarettes are smoked is subject to debate.

Nicotine delivered by tobacco products is one of the most highly addictive substances known. Even people highly motivated to quit may have profound difficulty doing so on their own. It is now known that people differ greatly in the severity of their addictions and their ability to cope. Our ability to treat nicotine addiction is continually improving. Even so, many people will require several repeated quitting attempts, regardless of treatment used. Therefore, long-term support by public health organizations and other facilities is essential if we are to prevent the serious diseases that will affect one in two untreated smokers.

Recent data from the 3 million people treated with the nicotine patch during its first seven months of availability in the United States increase optimism that the body can repair much of the damage caused by smoking. Epidemiological data indicate that 2,250 heart attacks would have occurred if these smokers had continued their habit. In fact, the Food and Drug Administration (FDA) received reports of only 33 severe cardiovascular problems. Even assuming underreporting, this decrease is so profound that it strongly supports the conclusion of the surgeon general in 1991 that risk of heart attacks rapidly declines after smoking cessation. These people were receiving nicotine via the patch, although probably at a lower level than if they continued smoking, and still their rate of heart attacks was significantly reduced.

Symptomatic Treatment. Nicotine administration and withdrawal produce a number of neurohormonal and other physiological effects. Symptomatic treatment methods are nonspecific pharmacotherapies to relieve the discomforts and mood changes associated with withdrawal. If the potential quitter relapses to escape the suffering of withdrawal, these methods should help to prevent such relapse. There is a long history of pharmacological treatment of smokers. To reduce withdrawal, sedatives, tranquilizers, anticholinergics, sympathomimetics, and anticonvulsants have all been tried at one time and were no more successful in helping smokers quit than was a placebo. CLONIDINE is one agent that has been tried in the treatment of nicotine withdrawal discomfort and is commonly used to treat opioid withdrawal. Glassman and his colleagues (1984; 1988) administered clonidine to heavy smokers on days they abstained from smoking and found that it reduced anxiety, irritability, restlessness, tension, and craving for cigarettes. When they gave clonidine to smokers trying to quit, 6 months later, 27 percent of those given clonidine and 5 percent of those given placebo reported abstinence. Surprisingly, clonidine seemed to be effective only for women. Among men, those given clonidine did no better than those given a placebo. Before recommending clonidine for smokers, practitioners should consider potential side effects. Clonidine has been used to treat hypertension, and abrupt termination has sometimes led to severe hypertension and in rare circumstances to hypertensive encephalopathy and death. More commonly, it may cause drowsiness, potentially dangerous to someone operating machinery or driving.

Among nicotine's effects is the regulation of mood. Smokers have been shown to smoke more than usual during stressful situations; therefore, those trying to quit often relapse (begin smoking again) during stressful situations. These observations suggest that treating the mood changes associated with abstinence with, for example, BENZODIAZEPINE tranquilizers, ANTIDEPRESSANTS, or psychomotor stimulants may improve abstinence rates. The benzodiazepine tranquilizer alprazolam was also examined by Glassman and his colleagues (1984; 1988) and found to reduce anxiety, irritability, tension, and restlessness, but it had no effect on cravings for cigarettes in heavy users abstaining from smoking for one day. More study is necessary on its effectiveness in maintaining tobacco abstinence.

Nicotine Blockade. Nicotine blockade therapy is based on the rationale that if one blocks the rewarding aspects of nicotine by administering an antagonist (or blocker), the smoker who seeks the pleasant effects nicotine produces will be more likely to stop. To be effective, the drug must be active in the central nervous system (brain and spinal cord). Thus mecamylamine, which acts at both central and peripheral nervous system sites, effectively increases rates of abstinence, whereas hexamethonium and pentolinium, which block peripheral nervous system receptors only, have no effect on abstinence. The problem is that there are no pure nicotine antagonists currently available. Drugs like mecamylamine produce side effects, such as sedation, low blood pressure, and fainting, that probably limit their role to that of an experimental tool, not appropriate for clinical treatment.

Deterrent Therapy. The rationale for deterrent therapy is that pretreatment with a drug may transform smoking from a rewarding experience to an aversive one if the unpleasant consequences are immediate and strong enough. DISULFIRAM treatment for alcoholism is an example of this type of treatment. After pretreatment, even a small quantity of alcohol can produce discomfort and acute illness. Silver acetate administration is a potential treatment for smokers. When silver acetate contacts the sulfides in tobacco smoke, the resulting sulfide salts are highly distasteful to most people. Although many over-the-counter deterrent products are available, their effectiveness has not been scientifically vali-

dated. Additionally, a severe limitation to this treatment is compliance. It may be difficult to ensure that patients continue to take the medication as needed.

BEHAVIORAL TREATMENTS

Characteristics of tobacco dependence and nicotine addiction suggest that combining nicotine replacement, to reduce the physiological disruptions of withdrawal, with behavioral treatments, to counter the conditioning cues, reinforcers, and social context cues associated with smoking, may be especially useful in helping people to quit. Adding behavioral treatments may increase both the rate of successful outcomes and the adherence to the pharmacological treatment. Behavioral interventions for smokers have been tried for many years. This section will focus on several of the current major approaches, but it is by no means comprehensive.

Social support has produced mixed results. Enlisting the help of the smoker's spouse and coworkers, or encouraging participation in a group, has yielded generally positive outcomes, but attempts to enhance social support further have been uniformly unsuccessful. Providing *skills training* in coping with stress and negative emotions has also been tried but generally as part of a multicomponent treatment plan. If the person smokes during times of stress and negative emotions, learning other means of dealing with these situations may lessen the need to smoke. Skills training appears beneficial in the short term, especially when combined with aversive smoking procedures (discussed below), but its long-term benefits are less clear. Mixed but generally negative results have been reported, but a problem in assessing skills training is that researchers have not controlled for the differences in treatments available. Some may be more effective than others. The techniques should be available for clients long after learning in order to be beneficial for long-term smoking cessation.

Contingency contracting uses operant conditioning techniques to reinforce quitting or punish smoking behaviors. Procedures include collecting monetary deposits from clients early in treatment and providing periodic repayment as nonsmoking goals are reached, having a client pledge to donate money to a disliked organization for every cigarette smoked, or similar procedures using nonmonetary rewards or punishers. Research indicates that contingency contracting aids quitting at least in the short term. *Stimulus control* procedures gradually eliminate situations in which the client smokes (e.g., only smoke outside) or the time the client smokes (e.g., only on the half hour) to reduce the number of cues for smoking.

Nicotine fading gradually changes brands or cigarette filters the smoker uses, in order to decrease tar and nicotine per cigarette before complete cessation. It is hoped this strategy will decrease later withdrawal symptoms when the client stops smoking. Problems are that the procedure may do nothing to reduce cravings (considered important for relapse prevention) and that the nicotine reduction is not as large as one would expect from ratings of the cigarettes' contents, because people change the way they smoke to receive more nicotine from each cigarette. Improved outcomes may occur with nicotine fading when it is part of multicomponent treatment approach.

Aversion treatments are designed to condition a distaste for cigarettes by pairing smoking with either unpleasant imagery (covert sensitization), electric shock, or unpleasant effects of smoking itself through directed smoking procedures. Directed smoking techniques include satiation, rapid smoking, and focused smoking. In satiation, clients smoke at least at twice their regular rate. Research indicates a low, 15 percent success rate when satiation is used by itself, versus 50 percent when it is part of a multicomponent program. In rapid smoking, clients inhale every 6 seconds until they will get sick, usually for six to eight sessions. As part of a multicomponent program, good outcomes are seen, but success is variable when rapid smoking is used alone, with high immediate abstinence rates, followed by low long-term rates. In focused smoking, clients either smoke for a sustained period at a slow or normal rate or do rapid puffing without inhaling. Long-term outcomes are similar to or slightly lower than for rapid smoking. The utility of aversion procedures is limited because the aversions are rarely permanent, and it is difficult to condition aversion to a substance that has had repeated past use.

CONCLUSIONS

Multicomponent interventions that combine pharmacological and behavioral components appear to be the best treatment strategies, often producing very high short-term (nearly 100% for the best programs) and impressive long-term success rates (at or above 50%). Ideally, the components should complement one another; however, it is not known how the separate components work in combination. It is possible that, because people smoke for different reasons (to prevent withdrawal, to ease anxiety, to relax, to achieve pleasant effects), a program that includes components that target enough different reasons for smoking will be successful in most cases. Second, it is not known which components work best together or how to target interventions for particular types of people. Third, a concern in designing a multicomponent treatment plan is that too many interventions may decrease patient compliance. Despite these gaps in our knowledge, smoking-cessation programs are improving constantly, and smokers do not have to go it alone in their attempts to quit.

(SEE ALSO: *Addiction: Concepts and Definitions; Nicotine Delivery Systems for Smoking Cessation; Relapse Prevention; Tobacco; Treatment Types*)

BIBLIOGRAPHY

BENOWITZ, N. L. (1992). Cigarette smoking and nicotine addiction. *Medical Clinics of North America, 76,* 415–437.

FIORE, M. C., ET AL. (1992). Tobacco dependence and the nicotine patch: Clinical guidelines for effective use. *Journal of the American Medical Association, 268,* 2687–2694.

GLASGOW, R. E., & LICHTENSTEIN, E. (1987). Long-term effects of behavioral smoking cessation interventions. *Behavior Therapy, 18,* 297–324.

GLASSMAN, A. H., ET AL. (1988). Heavy smokers, smoking cessation, and clonidine: Results of a double blind, randomized trial. *Journal of the American Medical Association, 259,* 2863–2866.

GLASSMAN, A. H., ET AL. (1984). Cigarette craving, smoking withdrawal, and clonidine. *Science, 226,* 864–866.

GRITZ, E. R., & JARVIK, M. E. (1977). Pharmacological aids for the cessation of smoking. In J. Steinfeld et al. (Eds.), *Health consequences, education, cessation activities, and governmental action.* Volume II, Proceedings of the Third World Conference on Smoking and Health. DHEW Pub. no. (NIH) 77-1413, 575–591.

HENNINGFIELD, J. E., LONDON, E. D., & BENOWITZ, N. L. (1990). Arterial-venous differences in plasma concentrations of nicotine from nicotine polacrilex gum. *Journal of the American Medical Association, 263,* 2049–2050.

HENNINGFIELD, J. E., ET AL. (1990). Drinking coffee and carbonated beverages blocks absorption of nicotine from nicotine polacrilex gum. *Journal of the American Medical Association, 264,* 1560–1564.

JARVIK, M. E., & HENNINGFIELD, J. E. (1993). Pharmacological adjuncts for the treatment of nicotine dependence. In J. D. Slade and C. T. Orleans (Eds.), *Nicotine addiction: Principles and management.* London: Oxford University Press.

PALMER, K. J., BUCKLET, M. M., & FAULDS, D. (1992). Transdermal nicotine: A review of its pharmacodynamic and pharmacokinetic properties, and therapeutic efficacy as an aid to smoking cessation. *Drugs, 44,* 498–529.

POMERLEAU, O. F., & POMERLEAU, C. S. (1984). Neuroregulators and the reinforcement of smoking: Toward a biobehavioral explanation. *Neuroscience Biobehavior Review, 8,* 503–513.

U.S. DEPARTMENT OF HEALTH AND HUMAN SERVICES. (1991). Strategies to control tobacco use in the United States: A blueprint for public health action in the 1990's. In D. R. Shopland et al. (Eds.), *Smoking and tobacco control monographs no. 1.* U.S. Public Health Service, NIH Pub. No. 92-3316. Washington, DC: U.S. Government Printing Office.

U.S. DEPARTMENT OF HEALTH AND HUMAN SERVICES. (1988). The health consequences of smoking: Nicotine addiction. A report of the surgeon general. U.S. Public Health Service, DHHS Pub. No. (CDC) 88-8406. Washington, DC: U.S. Government Printing Office.

LESLIE M. SCHUH
JACK E. HENNINGFIELD

Tobacco, Psychological Approaches

Treatment for tobacco dependence involves not only dealing with the pharmacological addiction to the drug but also the behavioral and psychological components associated with its use. Over the course of the addiction to tobacco, many behaviors become linked to either cigarette smoking or smokeless tobacco use. Furthermore, individuals become dependent on nicotine's psychological effects (e.g., reduction of anxiety or tension). Treatment of tobacco cessation involves intervention at various stages. These stages include preparation for cessation (reinforcing the motivation to quit), becoming abstinent from tobacco use, maintaining short-term abstinence, and relapse prevention.

BEHAVIORAL THERAPY

The principles behind behavioral treatment involve examining the antecedents as well as the consequences of the tobacco-use behavior. Consequences are events that occur after the use of tobacco. If the consequences increase behavior, then the process is termed reinforcement. There are two major types of reinforcement, positive and negative. Positive reinforcement involves the presentation of an event that then increases behavior. Negative reinforcement involves the removal of an event that also results in increased behavior.

Both positive and negative reinforcements initiate and maintain tobacco use. Positive reinforcement from smoking cigarettes, for example, may include improving concentration and providing stimulation. Negative reinforcement from smoking cigarettes may include reduction of tension or depressed mood, prevention of weight gain, or prevention of withdrawal symptoms.

If the consequence decreases behavior, then the process is termed punishment. Punishment can involve the presentation of an event or removal of an event. For example, the occurrence of social disapproval, negative physical consequences, and increased insurance premiums associated with cigarette smoking may reduce smoking. Similarly, the removal of privileges, such as being unable to participate in sports can decrease smoking behavior and serve as punishment. Tobacco use, however, often continues among a number of individuals in spite of many negative consequences associated with its use. The reason is that the strength of any reinforcer or punishment diminishes the further removed from the actual behavior. Thus the immediate reinforcing consequences of smoking (e.g., mood regulation, control of hunger) outweigh the later negative consequences such as lung cancer or heart disease.

There are many antecedents, or events that precede smoking or smokeless tobacco use, that begin to control or maximize the occurrence of tobacco use. In other words, these events begin to exert stimulus control. An individual learns that in certain situations, behavior is reinforced; in other situations, behavior is either not reinforced or punished. For example, a smoker may learn smoking in bars is reinforced socially, in addition to enjoying the rewarding effect of nicotine. On the other hand, smoking in church is not reinforced. Upon repeated experiences, going to bars begins to automatically elicit the behavior of or desire for smoking, whereas attending church does not.

Behavioral therapy involves manipulating these antecedents and consequences to reduce the probability of tobacco use. Further, skills that foster non-smoking behaviors such as stress management skills, relaxation techniques, physical activities, and assertiveness are also taught or encouraged.

PREPARATION

The first stage for tobacco use cessation involves encouraging the tobacco user to sufficiently be motivated to quit smoking. Motivation is often fostered by educating the tobacco users about the benefits of cessation and/or negative consequences from continuing to smoke. Having the tobacco user list personal reasons for not smoking is a technique that is frequently used. Not only is motivation important, but also the belief that one has the ability and tools to achieve abstinence from smoking. Part of this feeling of self-efficacy is knowing why one uses tobacco and what lies ahead. Therefore, this stage may also involve an overview to understand one's own dependence on nicotine, reasons for smoking, obstacles that may arise from smoking cessation, and skills and techniques to overcome these problems.

QUITTING

Monitoring the Smoking Habit. Before an individual quits using tobacco products, he or she is often required to monitor situations and feelings that are associated with tobacco use. After such a task is undertaken, the individual begins to recognize specific antecedent conditions that are associated with the use of tobacco. These antecedent conditions for a cigarette smoker often include smoking the first thing in the morning, with a cup of coffee, after a meal, while on the telephone, while in the car, and so on. Once the tobacco user understands these situations associated with his or her tobacco use, the smoker can better prepare for a quit date by learning how to handle these potentially difficult situations.

Choosing a Method of Quitting: Cold Turkey or Gradual Reduction. Several techniques have been developed to help the individual quit using tobacco products. These techniques include quitting abruptly ("cold turkey") or the gradual reduction of cigarettes or nicotine. The gradual reduction in the

number of cigarettes smoked per day can be accomplished by lengthening the time between cigarettes or delaying the onset of smoking. Another cigarette reduction technique involves systematically reducing the number of situations in which one smokes. The latter technique typically has the smoker avoiding cigarettes in situations which are least frequently associated with smoking and then proceeds to not smoking in the most frequent situations. These methods of gradual reduction work on the principle of systematically disassociating situations linked with tobacco use. Another principle behind these cigarette reduction methods is the reduction of withdrawal signs and symptoms from tobacco (e.g., irritability, frustration or anger, anxiety, difficulty concentrating, restlessness, increased appetite, and CRAVING for NICOTINE). In fact, gradual reduction of nicotine has been found to reduce WITHDRAWAL signs and symptoms; however, gradual reduction may prolong withdrawal symptoms for a longer period than abrupt cessation. Further, gradual reduction in the number of cigarettes has not generally been a successful technique. A significant number of smokers experience difficulty in reducing the number of cigarettes beyond a certain point.

Other techniques of gradual reduction based on the principle of minimizing withdrawal symptoms include systematically using cigarette filters that gradually reduce the nicotine/tar content of the cigarette or switching to a lower tar and nicotine cigarette. This technique is called *nicotine fading*. The nicotine-fading procedure has been found to be promising with some population of smokers (those in community settings or minimal-intervention programs).

Nicotine-replacement agents such as NICOTINE GUM or skin patch, have also been found to reduce significantly withdrawal signs and symptoms. Treatment with nicotine-replacement agents have also been found to enhance the success of behavioral treatments more consistently than have the gradual reduction or nicotine fading approaches. The technique chosen, however, may depend on the comfort and needs of the smoker. For example, the less physically dependent smoker may choose cold turkey without pharmacological aids or the gradual cigarette reduction/nicotine fading technique, and the more dependent smoker may choose to use nicotine replacements.

Aversions Techniques. Another technique during this phase of treatment that has been found to improve the success of smoking cessation is rapid smoking. Smokers are asked to smoke several consecutive cigarettes rapidly so that they will experience immediate adverse, punishing effects such as nausea from smoking, thereby reducing the desire to smoke. Similarly, reduced-aversion techniques may also facilitate smoking cessation by their unpleasant effects and improve the effectiveness of behavioral treatment. These techniques involve focusing on smoking while the person smokes for a sustained period of time, or on rapid puffing with no inhalation of the smoke.

Social Support. In helping the subject quit, the support from others may be solicited to provide some reinforcement. Studies have found that there is a relationship between positive support and successful treatment outcomes.

MAINTAINING SHORT-TERM ABSTINENCE

Maintaining abstinence involves developing both behavioral and cognitive skills. Smokers who tend to use both these skills have been found to experience more successes in smoking cessation. Behavioral principles are involved in the use of these skills. These principles are based primarily on changing the stimuli associated with smoking such that they are no longer associated with smoking and associated with behaviors other than tobacco use. Furthermore, consequences are manipulated such that not smoking is associated with reinforcement and smoking is associated with punishment.

Stimulus Control. There are many behavioral techniques used to affect stimulus control of smoking. These techniques include avoidance of the stimuli. For example, the tobacco user may avoid drinking ALCOHOL initially if alcohol has been associated with smoking. Other techniques include engaging in alternative behaviors (e.g., using toothpicks, eating celery) and other distractions (e.g., exercising, knitting). With the use of these skills, stimuli that have been associated with tobacco use become associated with nonsmoking behaviors.

Coping Skills. Smokers tend to smoke when experiencing negative affect (depression, tension, stress, boredom) and in social situations with smokers. Consequently, smokers who smoke under these conditions may need to be taught adaptive coping skills, that is, skills on how to deal with these situations more effectively. For example, tobacco users may need to be taught how to deal with stress without smoking. Techniques that have been suggested include exercise, relaxation exercises such as deep breathing or meditation, and problem solving. Problem solving includes learning how to assess the situation adequately, developing a number of solutions, and trying out these solutions. In other words, tobacco users are taught to use adaptive coping skills rather than to smoke cigarettes. Social situations involving smoking also can be problematic in attempts to maintain abstinence. Smokers trying to quit can avoid these situations or practice refusing cigarettes or asking people not to smoke.

Rewards and Punishments. Reinforcement and punishment are also used to increase the probability of not smoking or to decrease the probability of smoking, respectively. Typically reinforcements are rewards that can be material, leisure time, or mental statements. For example, successes in remaining smoke free can be increased by having individuals earn back money they have deposited for abstaining from cigarettes, save money that is typically spent on cigarettes, or buy magazines, record albums, or clothes. Rewards can also be leisure activities such as reading a book, going to a movie, and taking walks with friends. Finally, rewards can be mental statements such as, "I did really well today," or "It took some doing to refuse that cigarette, but I did it!" or thinking about the positive benefits derived from smoking cessation. Rewards are initially given for small successes, based on the occurrence of the goal behavior, and occur as soon as possible upon completion of this behavior.

Covert sensitization is a cognitive punishment technique that has been used in the treatment of tobacco users. This technique involves the use of negative imagery pertaining to the tobacco use behavior. They are asked to conjure up an image replete with aversive details relative to the smoking behavior. They may be asked to visualize about being notified that they have cancer and are about to die, or social ostracism due to their tobacco use. These images are conjured up whenever they have an urge to use tobacco. Similarly, the smoker can think of all the negative consequences associated with smoking. A behavioral punishment technique used in smoking cessation is the forfeiture of money if smoking occurs. This money can be given to smoking-related research or organizations, another individual, or organizations that are disliked.

Weight. The issue of weight gain is a frequent topic in treatment. Cigarette smokers who quit smoking are found to gain, on the average, 7 pounds. This weight gain is attributed to increased food intake and metabolic factors (how much energy the body uses). Although weight gain may not specifically play a large role in relapse, it may play an important factor in the maintenance of smoking. Treatment for weight gain has focused on monitoring the amount of food intake, specifically foods high in fats and sweets (sucrose), and an increase in physical activity. A significant reduction in calories and weight loss is not recommended, but rather an acceptance of some weight gain is encouraged, at least temporarily. Studies show that food reduction may in fact facilitate relapse to drugs in general. In fact, cigarette smokers actually are lower in weight compared to nonsmokers; therefore, when they quit smoking their average weight become the same as that of the general population.

RELAPSE PREVENTION

RELAPSE PREVENTION is also an integral part of treatment. Relapse is defined as return to regular tobacco use. Relapse is distinguished from a slip, which is smoking one or a few cigarettes after a period of abstinence. Various approaches have been developed to minimize the occurrence of relapse, such as planning for high-risk situations as well as learning how to deal with slips. Several high-risk situations for relapse have been described previously. These include negative emotions, interpersonal conflicts, and social pressure to smoke or specific smoking-related stimuli such as being with other smokers. The individuals in treatment are encouraged to anticipate these high-risk situations that may lead to a slip or relapse and plan for them using both cognitive and behavioral skills. Treatment also focuses on the issue of what to do if an individual has resorted to having a cigarette. One theory of relapse discusses the specific factors that may turn a slip into a relapse to smoking. These factors include the failure to cope with the situation, the focus on positive effects from smoking, and the general sense that one cannot succeed. Treatment involves having the individual become aware of these thoughts that facilitate relapse and teaching more effective ways to deal with them. For example, the individual may be asked to evaluate the impact of smoking one or a few cigarettes versus a returning to regular smoking. Further, dis-

cussions in treatment may focus on the mental attitude that the individual has toward the slip. An attitude that recognizes the previous success of the individual and the benefits from quitting rather than the failures may circumvent relapse. Another way to optimize long-term abstinence involves lifestyle changes that include the development of healthier habits such as increased levels of physical activities, proper eating, obtaining enough sleep and rest, and managing or changing levels of stress in adaptive ways.

OTHER TREATMENTS

Many commercially available methods of smoking cessation exist. However, very little is known about the treatment success associated with these products. Two noncommercial treatments that continue to be popular are HYPNOSIS and ACUPUNCTURE. Hypnosis is used to increase the smoker's motivation or ability to quit by offering them suggestions while under a state of deep relaxation. These suggestions can focus on the aversive and harmful aspects of smoking, the benefits of not smoking, or the ability and skills for tobacco cessation. Acupuncture involves the use of needles or staplelike attachments, usually to the ear. Acupuncture is thought to relieve withdrawal symptoms, to minimize the feeling of STRESS, or to induce a state of calmness. All these aspects of acupuncture are believed to help in smoking cessation. Studies do not clearly show that these techniques alone are useful in smoking cessation.

CONCLUSION

Many of the techniques used in behavioral therapy for smoking cessation have been described in this article. Studies show that smoking interventions are most effective when multiple techniques are used. On the other hand, too many behavioral procedures within a treatment program can lead to poorer treatment outcome. Unfortunately, nicotine is a highly addictive drug. Thus, the rate of relapse to smoking cigarettes or other tobacco use remains in spite of behavioral treatment interventions. The success rate for staying abstinent from tobacco products has ranged from 20 to 60 percent at six months after behavioral treatment, with rates decreasing further at one year. Consideration of differences in the individual's smoking habit may improve treatment outcome. For example, behavioral therapy is generally most effective in conjunction with a nicotine replacement agent. The nicotine replacement agent allows the tobacco user to focus on the behavioral or psychological aspects of smoking while some of the physical dependence aspects of smoking are reduced. However, it is important to keep in mind that some smokers may be more physically dependent on nicotine than others. Therefore, these individuals may require a nicotine replacement or other drug treatments; those less dependent may not require this approach. Likewise, smokers smoke for different reasons, and the same psychological treatment approach may not be effective for all smokers. Some individuals may require more focus on how to handle stress effectively, while other individuals may need to focus on the habit aspect of smoking. More recently, investigators have perceived smokers going through various distinct stages of change before actually attaining success at quitting. These stages of change include (a) precontemplation (no intention of quitting in the near future); (b) contemplation (intending to quit but not ready to take action); (c) preparation (planning to quit in the immediate future); (d) action (having quit recently and trying not to relapse); and (e) maintenance (having quit for a period of time). Individuals may go through some or all of these stages or even cycle through them again. Therefore, the use of behavioral techniques must be adapted not only to the different physical and psychological needs of individual smokers but also to each individual's stage of change.

Of final note, most cigarette smokers do not seek treatment but quit on their own. If smokers do seek treatment, they tend to obtain help from their physician or from health-care providers who often have little time to devote to intensive behavioral treatment. Therefore, availability and use of the behavioral techniques for smoking cessation must be adapted to these various methods or settings for tobacco cessation. In the long run, however, societal pressures (e.g., banning smoking in public places) and economic pressures (e.g., increasing taxes on tobacco products) will likely be among the more significant contributors to reducing smoking rates by discouraging individuals from beginning to use tobacco products.

(SEE ALSO: *Causes of Substance Abuse; Nicotine Delivery Systems for Smoking Cessation; Reinforcement; Tobacco; Treatment Types*)

BIBLIOGRAPHY

FIORE, M. C., ET AL. (1990). Methods used to quit smoking in the United States: Do cessation programs help? *Journal of the American Medical Association, 263,* 2760–2765.

HALL, S. M., ET AL. (1984). Preventing relapse to cigarette smoking by behavioral skill training. *Journal of Consulting and Clinical Psychology, 52,* 372–382.

HATSUKAMI, D. K. & LANDO, H. L. (1993). Behavioral treatment for smoking cessation. *Health Values, 17,* 32–40.

HUGHES, J. R. (1987). Combined psychological and nicotine gum treatment for smoking: A critical review. *Journal of Substance Abuse, 3,* 337–350.

KLEGES, R. C. & SHUMAKER, S. A. (1992). Understanding the relations between smoking and body weight and their importance to smoking cessation and relapse. *Health Psychology, 11* (suppl.), 1–3.

LICHTENSTEIN, E., & GLASGOW, R. E. (1992). Smoking cessation: What we have learned over the past decade? *Journal of Consulting and Clinical Psychology, 60,* 518–527.

LICHTENSTEIN, E. L., GLASGOW, R. E., & ABRAMS, D. B. (1986). Social support in smoking cessation: In search of an effective intervention. *Behavior Therapy, 17,* 607–619.

POMERLEAU, O. F., & POMERLEAU, C. S. (1987). *Break the smoking habit. A behavioral program for giving up cigarettes.* Ann Arbor: Behavioral Medicine Press.

PROCHASKA, J. O., DiCLEMENTE, C. L., & NARCROSS, J. L. (1992). In search of how people change: Applications to addictive behaviors. *American Psychologist, 47,* 1102–1114.

SHIFFMAN, S. (1984). Coping with temptations to smoke. *Journal of Consulting and Clinical Psychology, 52*(2), 261–267.

SHIFFMAN, S., ET AL. (1985). Preventing relapse in ex-smokers. In G. A. Marlatt and J. R. Gordon (Eds.), *Relapse prevention: Maintenance strategies in the treatment of addictive behaviors.* New York: Guilford Press.

U.S. DEPARTMENT OF HEALTH AND HUMAN SERVICES. (1994). *Tobacco and the clinician: Interventions for medical and dental practice.* NIH Publication no. 94-3693. Washington, DC: Author.

U.S. DEPARTMENT OF HEALTH AND HUMAN SERVICES. (1991). *Strategies to control tobacco use in the United States: A blueprint for public health action in the 1990s.* NIH Publication no. 92-3316. Washington, DC: Author.

DOROTHY HATSUKAMI
JONI JENSEN

TREATMENT TYPES This section provides the reader with brief descriptions of some of the diverse ways that people with substance-related problems can be helped. *Treatment Types* presents descriptions of distinct interventions that are applicable to dependence on each of a variety of drugs. In practice, though, treatment programs are hybrids, incorporating features from several distinct treatment modalities and adapting them to specific needs having to do with age, gender, ethnic, racial, and socioeconomic factors, provider preference, and the economic realities that govern delivery of treatment.

Neither this section nor the one above on *Treatment* is exhaustive. A number of substance dependence interventions employed in other countries and by certain U.S. ethnic groups (such as sweat lodges among some Native American tribes) are not covered. Nevertheless, the entries included here should allow the reader to become reasonably familiar with what is considered mainstream treatment in the United States in the 1990s.

This section contains the following articles:

> *An Overview;*
> *Art of Acupuncture;*
> *Assertiveness Training;*
> *Aversion Therapy;*
> *Behavior Modification;*
> *Cognitive Therapy;*
> *Group and Family Therapy;*
> *Hypnosis;*
> *Long-Term Approaches;*
> *Minnesota Model;*
> *Nonmedical Detoxification;*
> *Outpatient versus Inpatient;*
> *Pharmacotherapy, An Overview;*
> *Psychological Approaches;*
> *Self-Help and Anonymous Groups;*
> *Therapeutic Communities;*
> *Traditional Dynamic Psychotherapy;*
> *Treatment Strategies;*
> and *Twelve Steps, The.*

An Overview In the mid-1990s, MARIJUANA use is at its lowest level since the early 1970s, but COCAINE and HEROIN present a more mixed picture. The number of heroin addicts may be between 600,000 and 1 million individuals, and heroin's pu-

rity is rising and its price dropping. Nonaddictive cocaine use has significantly declined during the past seven years, but the casualties of frequent use continue rising. The number of cocaine addicts is estimated at 2 million individuals. Drug use cuts across ethnic and socioeconomic lines. Estimates of the total number of persons needing treatment for drug abuse are 5.5 to 6 million, but current treatment capacity, both public and private, is about 1.7 million treatment episodes a year—clearly short of the need.

Prior to referring an addicted patient to treatment, it is important to address certain questions: (1) What are the possible treatment alternatives? (2) What treatment modalities are best suited for a particular patient? (3) What is the efficacy of the preferred treatment? and (4) Is the chosen treatment available to the patient? As will be noted, the information base needed to answer these is often not available.

TREATMENT ALTERNATIVES

Treatment Setting. Excellent treatment can be delivered within both outpatient and inpatient settings. A more expensive inpatient program does not offer the best treatment for all individuals. The appropriate placement of a drug-dependent individual in a treatment program requires the consideration of several factors, including drugs that are being used, level of psychiatric distress, potential medical complications, family or other support, and availability of child care. Intensity of treatment is not necessarily a function of setting since some outpatient treatment programs provide more intense treatment than do inpatient ones.

Inpatient Programs. Usually, inpatient settings are of three types: (1) detoxification units within medical hospitals, (2) dual-diagnosis programs within psychiatric hospitals, and (3) rehabilitation programs. The first two settings are best utilized when there is a risk of serious medical problems (e.g., seizures) or psychiatric difficulties (e.g., suicidal ideation). Medical units generally employ pharmacologic detoxification protocols that are based on the type of drugs abused and the patient's concomitant medical condition. The length of stay is usually less than two weeks. Although many patients mistakenly believe that after detoxification no further intervention is necessary, detoxification is only the beginning of treatment. The next treatment placement should be based on the needs of the patient, but, unfortunately, it often depends on other factors (e.g., community resources or the patient's insurance coverage or ability to pay).

Dual-diagnosis programs are usually based in psychiatric hospitals and are designed to treat patients with both serious psychiatric illnesses and substance-use disorders. Treatment may include individual, group, and family therapy, pharmacotherapy, relaxation techniques, and education. ALCOHOLICS ANONYMOUS (AA) or NARCOTICS ANONYMOUS (NA) groups may also be offered. Individuals may reside in these hospital units from several weeks to several months.

Rehabilitation units are usually free-standing facilities that are often based on the AA TWELVE-STEP model of treatment. Some carry out uncomplicated pharmacologic detoxifications, but many patients are already detoxified at entry. Some rehabilitation programs are staffed to offer psychiatric evaluation or treatment (or both). Therapy usually consists of education, group therapy, individual meetings, and at times, specialized groups (e.g., a women's group), usually provided by drug or alcohol counselors. Social workers may provide family therapy. Traditionally, the standard length of stay was twenty-eight days, but lack of data to support the advantages of this length and reimbursement issues have often compelled programs to reduce treatment to less than fourteen days.

Outpatient Programs. Outpatient treatment generally consists of drug-free treatment or, in cases of opiate addiction, methadone treatment. The time for outpatient drug-free treatment can range from once a week to daily daylong activities. In comprehensive treatment programs, individuals may be initially enrolled in an intensive outpatient program consisting of many structured daily activities (e.g., group therapy, individual therapy, self-help groups, educational groups, stress-management groups) and "graduate" over a certain period (ranging from one to six months) to weekly or biweekly clinic visits. Random urine testing is usually an integral part of these programs. Completion of the intensive portion of the program is usually determined by documented behaviors such as length of abstinence, attendance in groups, and keeping scheduled appointments. Initiation of change—for example, the avoidance of drug-using friends or the desire to return to work or school—may suggest readiness for a less-intensive program.

Some outpatient programs have the necessary staff and expertise to provide medically supervised detoxification. Appropriate patient selection is crucial, however. There has been a growing recognition that many patients seeking drug treatment have additional psychiatric disorders (Rounsaville, Weissman, & Kleber, 1983; Weiss et al., 1986; Rounsaville et al., 1991), and, consequently, psychiatrists have been increasingly employed in drug-free outpatient settings to both assess patients and, when necessary, provide additional psychiatric treatment.

Methadone maintenance programs are designed for patients who have been addicted to opiates for at least one year. These patients often have lengthy drug-use histories and have been unable to maintain abstinence after repeated detoxifications. Verification of opiate addiction may be determined by using a naloxone challenge test or by observing withdrawal symptoms. Because of the risk of transmitting the human immunodeficiency virus (HIV), pregnant and HIV-positive opiate-dependent individuals may be given admission priority in some programs. As is the case with drug-free treatment programs, methadone programs vary in the comprehensiveness of their services. Some additional psychosocial services provided by a methadone program may include the teaching of job-hunting skills, family therapy, and parenting groups.

Residential Programs. Residential programs can be used as a bridge between inpatient and outpatient programs or as an alternative to them. Intermediate-care facilities, similar to those developed at HAZELDEN, allow individuals to live within a residential setting, be employed during the day, and receive comprehensive treatment, including group therapy, individual counseling and monitoring, and education. Both behavioral models and the principles of Alcoholics Anonymous are applied. The average stay is approximately four months.

THERAPEUTIC COMMUNITIES provide treatment within highly structured, hierarchical residential settings that stress the importance of community and recovering staff in treatment. More recently, professionals with or without prior drug histories are providing managerial expertise and treatment. Within therapeutic communities, behavior is shaped by using rewards and penalties (Kleber, 1989). Drug abusers are constantly confronted by their peers in a variety of situations regarding their functioning within the program. Jobs range from low to high status and are allocated to individuals on the basis of the length of their stay in the community, their competence, and their ability to behave responsibly. Traditional therapeutic communities recommend stays of twelve to twenty-four months whereas newer programs are experimenting with stays of three to six months.

Treatment Modalities. Treatment interventions can be categorized in terms of behavioral, self-help, psychological, or pharmacological approaches. Although a specific treatment setting may emphasize one type of intervention, additional modalities are often employed. Generally, programs proficient in using diverse treatment methods are more likely to change their therapeutic interventions if the initial approaches appear ineffective.

Behavioral Approaches. Various behavioral treatments, using the psychological theories of operant and respondent conditioning, have been designed to treat substance abuse. Experimental psychologists found that behavior could be shaped if positive consequences occurred as a result of the changed behavior. Used with drug abusers, operant conditioning is complicated since many positive and negative reinforcers may promote continued drug use. These reinforcers include: (1) the positive sensations related to the drug itself, (2) the avoidance of actual or conditioned withdrawal symptoms, (3) the perceived reduction of distressing psychologic symptoms, (4) the fear of losing a social network centered on drug use, and (5) the anxiety associated with having to confront painful issues once drug use ceases.

Several clinicians have attempted to counter drug-promoting reinforcers with other reinforcers that were contingent on non–drug taking behavior. Higgins et al. (1993) developed a voucher system in which negative urine screens were rewarded with vouchers that could be used to purchase a variety of community-based items viewed as prosocial and consistent with a drug-free lifestyle. When compared to a control group that had received standard drug counseling, it was found that the behavioral group remained in treatment longer and had more discrete periods of abstinence.

Operant techniques can be applied in various treatment settings by using fairly simple yet effective reinforcers. For example, methadone programs may offer drug abusers take-home doses for negative urine results. Because compliance is more likely to occur if the positive reinforcement is temporally

linked with the desired behavior, take-home doses immediately offered after two weeks of negative urine tests work better than if the take-home doses are delayed until a prolonged period of abstinence has been accomplished. CONTINGENCY CONTRACTING and respondent conditioning are two alternative behavioral interventions that are occasionally used for treating substance abuse. Contingency contracting applies negative contingencies to undesirable behavior. For example, patients who are concealing their drug use from their bosses, family members, or anyone else may be asked to sign a "contract" that allows their therapist to inform one or more specific individuals if their drug use resumes.

Respondent conditioning may involve the use of noxious stimuli. For example, individuals may be given a chemical that induces nausea (e.g., apomorphine) while receiving an injection of their drug of choice or while handling drug-related paraphernalia. The drug may come to induce unpleasant feelings as a result of its association with the noxious stimuli. Poor patient acceptance, ethical issues, and insufficient data regarding efficacy limit the use of these AVERSIVE TREATMENT approaches.

Self-Help Approaches. These interventions have evolved from the personal experiences and ideas generated by Bob Smith and Bill Wilson, two alcoholics who cofounded Alcoholics Anonymous. The organization has grown until, in the early 1990s, it numbered over 47,000 groups in the United States and Canada (Nace, 1992). Although AA's approach to gaining SOBRIETY (the Twelve Steps) and its principles (the Twelve Traditions) are commonly integrated into many treatment programs, it remains unclear which patients benefit most from self-help programs, particularly when they are used without other interventions. The concepts of AA have also been applied to other psychoactive-substance use disorders (e.g., in the programs of COCAINE ANONYMOUS and Narcotics Anonymous).

Psychological Approaches. Psychological approaches are used to try to understand the psychological or cognitive issues that promote drug use and, with this knowledge, to provide appropriate treatment interventions. As Zweben (1986) emphasized, the goals of recovery-oriented psychotherapy change as addicted individuals progress in their recovery. The manner in which recovery "progresses" has been clearly conceptualized by Gorski and Miller (1986) in their six-stage developmental model. Each of the stages has a primary goal, and different types of psychological interventions become appropriate, depending on the goal.

During the first two phases, pretreatment and stabilization, the focus is placed on challenging the denial of patients regarding the consequences of their disease and, subsequently, on addressing the symptoms of acute and post-acute withdrawal. For therapists to engage patients into treatment, they need to be skillful at both confrontational and supportive approaches. During the third and fourth stages of early and middle recovery, the patients' major goals are to learn to function without drugs or alcohol and to develop a healthy lifestyle. For these stages, a cognitive approach focused on RELAPSE PREVENTION is useful. Marlatt and Gordon (1985) stressed that drug relapse was often due to ineffective coping with high-risk situations. Although individuals have their own unique list of high-risk situations, the situations are usually related to interpersonal conflicts, social pressure, conditioned cues, or negative emotional states. The therapeutic work of this approach is to develop effective coping responses as well as learn to handle a "lapse" (i.e., a single drink or drug administration) such that it does not degenerate into a "relapse" (i.e., problem use).

The final stages, late recovery and maintenance, emphasize personal growth in areas such as self-esteem, spirituality, intimacy, and work while individuals are maintaining a drug-free lifestyle. When there are deficits in these areas, insight-oriented therapy may be helpful. The reasons for continued inadequate functioning can be extremely complex and may involve unresolved issues from childhood. Kaufman and Redoux (1988) emphasized that uncovering core conflicts and confronting maladaptive defenses might elicit intense anxiety. Unless patients were in the late recovery stage, they might revert to their former maladaptive mode of coping—namely, using drugs.

The developmental model should be used as a guideline in understanding the recovery process rather than as a paradigm that is directly applicable to all patients. Additionally, there may be exceptions to when certain psychological interventions should be utilized. For example, an individual with major depression might not benefit from relapse-prevention techniques until the depression has been treated. *Pharmacologic Approaches.* Medications can serve as useful adjuncts in a comprehensive treatment plan.

The appropriate use of these agents depends on the patient's medical and psychiatric status, prior treatment experience, and the clinical setting. Generally, the novel as well as established pharmacotherapies can be put into four classifications: (1) AGONISTS, (2) ANTAGONISTS, (3) antiwithdrawal agents, and (4) anticraving agents.

Agonists bind and activate receptors on cell membranes, and these operations then lead to a cascade of biologic activities. Drugs themselves are usually agonists and may generate strong physiologic responses (i.e., full agonists) or weak responses (i.e., partial agonists). The use of a specific agonist is limited to treatment of abuse of a drug from the same pharmacological class. Agonists are generally used for detoxification or for medication maintenance, and, when chosen for these purposes, they are likely to be well absorbed orally and slowly eliminated from the body. Slowly metabolized medications are less likely to produce a severe withdrawal syndrome but are more likely to produce a protracted, albeit less intense, one. Because agonists induce positive drug effects, they are well accepted. This, however, also means that they have the potential for abuse.

The most commonly used agonist for both maintenance and for opiate withdrawal is methadone, which itself is an opiate. BUPRENORPHINE, a partial opioid agonist, is being evaluated in the mid-1990s and may have less potential for abuse and be associated with fewer withdrawal symptoms than methadone when used for opiate detoxification. L-ALPHA-ACETYLMETHADOL (LAAM), also an opiate drug, has recently (1993) received FDA approval for use in treating opiate abuse. Unlike methadone, which must be taken daily, LAAM can be given three times a week, thereby decreasing the number of clinic visits for the patient as well as the risk of medication diversion. Few agonist drugs have been developed for other types of drug abuse, although NICOTINE, delivered transdermally, is being used with some success to treat tobacco dependence.

Antagonists prevent agonists (i.e., the abused drug) from producing their full physiologic response, either by blocking the receptor site or by disrupting the functioning of the receptor. Short-acting antagonists are most commonly used for treatment of acute intoxication or overdose and long-acting ones for rapid detoxification and relapse prevention. The benefits of antagonists are that they produce no euphorigenic effect, have no potential for abuse, and produce no withdrawal syndrome. Although generally only antagonists that block the specific receptor activated by the specific drug can be used for drug-abuse treatment, research is suggesting that the opiate antagonist NALTREXONE may play a role in diminishing alcohol drinking after a single drink.

Commonly used opioid antagonists include naloxone and naltrexone. Naloxone reverses the respiratory depression associated with opiate overdoses. Naltrexone is used after detoxification to maintain abstinence. Unfortunately, relatively few patients take an antagonist as prescribed because of its lack of pleasant effect, its lack of effect on withdrawal if the patient ceases taking the medication, and at times the persistence of craving (Kleber, 1989). Development of a monthly, long-acting injectable formulation may soon increase compliance when it reaches the market.

Antiwithdrawal medications are given to minimize the discomfort associated with detoxification from drugs that induce physiologic dependence. Agents used for opiate detoxification include methadone, CLONIDINE, and lofexidine; although effective for opiate detoxification, the latter two have not received FDA approval for this indication. The use of the dopamine agonists bromocriptine and AMANTADINE have been suggested for the manifestations of cocaine withdrawal, but their efficacy remains unclear. The most appropriate antiwithdrawal regimen for a particular clinical situation is not always the one chosen. This situation may be due to federal and state regulations, physician or patient bias, reimbursement issues, and the lack of available expertise within a community in the use of particular methods (Kleber, 1994).

The development of anticraving agents to treat drug dependence is a new treatment strategy. Earlier conceptualizations of craving focused on the physical aspects (i.e., the individual "craved" the drug because he or she was experiencing physical withdrawal symptoms). Thus the emphasis was placed on developing antiwithdrawal rather than anticraving drugs. During the last decade, as cocaine use soared, clinicians noted that craving could be psychologically based and be a significant relapse trigger (Gawin & Kleber, 1986). Much research was consequently done to find useful anticraving medications. Although desipramine remains promising, no medication has been unequivocally shown to be an effective anticraving agent for cocaine addiction.

ASSESSMENT OF TREATMENT OUTCOME

Although treatment for substance abuse can work, which treatment setting or modality will work best for each patient cannot invariably be predicted. Using a number of outcome studies, researchers at the Institute of Medicine (Gerstein & Harwood, 1990) reached several conclusions regarding the efficacy of various treatment modalities:

1. *Methadone Programs* Opiate-dependent individuals maintained on methadone exhibit less illicit drug use and other criminal behavior than do individuals discharged after being in the program for a period of time or not treated at all. For opiate-dependent individuals, there are higher retention rates in methadone programs as compared to other programs, and patients tend to do better if they are stabilized at higher doses. Problems include continued use of nonopiate drugs, especially cocaine, and difficulty withdrawing.

2. *Therapeutic Communities* The length of stay within these communities, even for those who do not complete the program, is the best predictor of treatment outcome measured by drug use, criminal behavior, and social functioning. Graduates from therapeutic communities have superior outcomes when compared to dropouts. Dropout rates are unfortunately as high as 75 percent, although data suggest that even those who do not graduate derive some benefit if they have stayed for a period of time.

3. *Outpatient Nonmethadone Programs* As with individuals in therapeutic communities, individuals who graduate from these programs have better outcomes than those who drop out, and individuals who enter the programs have better outcomes than those who were contacted but did not begin the programs. These programs tended to treat less severely dependent patients.

4. *Chemical Dependency Programs* There were inadequate data to evaluate the efficacy of residential or inpatient programs (so-called 28-day MINNESOTA MODEL programs) designed to treat drug problems, and there were no data regarding whether hospital or free-standing programs were more effective.

Hubbard (1992) found that individuals referred from the criminal justice system performed as well in treatment as did other patients entering without such pressure, and that drug-abuse treatment provides a favorable cost-benefit ratio to society within one year of completion of treatment.

Recognizing that treatment success is multifactorial, investigators have sought comprehensive yet practical ways to characterize both patients and treatment programs. One instrument increasingly used to assess patient functioning is the ADDICTION SEVERITY INDEX (ASI) (McLellan et al., 1980). Using the ASI, the interviewer rates the severity of the patient's problem across six domains: alcohol and drug use, medical status, employment and support status, family and social relationships, legal status, psychiatric status. By giving the ASI at admission and repeating it over time, treatment success can be assessed in a standardized manner. Using this instrument, McLellan et al. (1984) found that opiate-addicted patients with severe psychological problems did worse over time when placed in a therapeutic community compared to those placed in methadone programs. As this study illustrates, it is critically important to assess "nondrug" variables when evaluating treatment response, and to carry out a comprehensive assessment prior to, during, and after treatment.

In the past few years, there has been greater emphasis on understanding how the specific aspects of treatment programs (e.g., therapeutic skills of the counselors, treatment modalities used, psychosocial services offered) influence treatment outcome. In regard to treatment services, McLellan et al. (1992) developed a rapid interview, the Treatment Services Review (TSR), which provides an evaluation of the amount and type of psychosocial services provided to patients during treatment. The investigators have suggested that this type of review might be useful when comparing different programs or for determining if the needs of individual patients were met during treatment. A recent study by McLellan et al. (1993) found that methadone-maintained patients who received enhanced psychosocial services did significantly better than those who received standard or minimal services.

No single study, no matter how comprehensive, can address all of the factors that influence treatment outcome. Instead, studies will need to focus on specific subpopulations of patients when comparing various treatment interventions as well as the impact on treatment of factors often overlooked (e.g., the patient's stage of recovery and the extent of program hours).

RECOMMENDED TREATMENT POLICIES

Since many Americans are still in need of treatment for drug abuse problems, rational treatment policies need to be established on the basis of our current knowledge regarding the extent of the problem and what interventions work. Such policies should address the following issues (Kleber, 1993):

1. Available treatment needs to be expanded. Although there are approximately 6 million individuals in need of drug treatment, the current system can treat less than 2 million a year.

2. Patients need to have access to a wide variety of treatment modalities. Since no one treatment is suitable for all patients, a community with a diversity of treatment services can more likely offer appropriate interventions to its population.

3. For treatment improvement to occur, there must be more funds dedicated to research along with efficient dissemination of new technologies. Without new research, progress will not be achieved. Without training and education of staff regarding new research findings, treatment will not improve.

4. Pressure must be exerted to encourage drug-addicted individuals to enter treatment. As noted earlier, those who enter under pressure from the criminal justice system do as well as those entering voluntarily. The family, employer, or criminal justice system can all be instrumental in getting individuals to enter and remain in treatment. This pressure must be sustained since when it remits, the individual often drops out of treatment.

5. The treatment needs of special populations (e.g., prisoners, pregnant women, HIV-infected individuals) require greater attention. There are few programs designed to treat drug-addicted prisoners while they are incarcerated or newly released. For pregnant drug abusers to engage in treatment, programs need to be accessible, be affordable, include child care (for optimal results), and reflect a nonjudgmental view. For HIV-infected individuals, comprehensive medical care should be linked with the substance-abuse treatment, especially considering the rising incidence of tuberculosis in this group.

6. Rehabilitation and habilitation need to be integrated into substance-abuse treatment programs. Some drug-dependent individuals have the educational background or skills that allow them to

gain employment once their drug problem has been treated. Others may require job-seeking skills, job training, or additional schooling prior to seeking employment. A goal of treatment needs to be integration into society, not simply cessation of drug use.

When examining the different modalities of treatment the question is not, "Does treatment work?" but rather, "What works best for a particular individual?" and "What can be done to engage drug abusers in appropriate, well-organized treatment systems?" If these issues are successfully addressed, treatment strategies can be designed for each patient and yet remain affordable. Millions spent on effective treatment will save billions spent elsewhere.

(SEE ALSO: *Abuse Liability of Drugs; Coerced Treatment for Substance Offenders; Comorbidity and Vulnerability; Research; Substance Abuse and AIDS; Treatment; Treatment in the Federal Prison System*)

BIBLIOGRAPHY

GAWIN, F. H., & KLEBER, H. D. (1986). Abstinence symptomatology and psychiatric diagnosis in cocaine abusers: Clinical observations. *Archives of General Psychiatry, 43*, 107–113.

GERSTEIN, D. R., & HARWOOD, H. J. (1990). *Summary—Treating Drug Problems: A study of the evolution, effectiveness, and financing of public and private drug treatment systems*, Vol. 1. Institute of Medicine, Committee for the Substance Abuse Coverage Study Division of Health Care Services. Washington, DC: National Academy Press.

GORSKI, T., & MILLER, M. (1986). *Staying sober: A guide for relapse prevention.* Independence, MO: Independence Press.

HIGGINS, S. T., ET AL. (1993). Achieving cocaine abstinence with a behavioral approach. *American Journal of Psychiatry, 150*, 763–769.

HUBBARD, R. L. (1992). Evaluation and treatment outcome. In J. H. Lowinson et al. (Eds.), *Substance abuse: A comprehensive textbook* (2nd ed.). Baltimore: Williams & Wilkins.

KAUFMAN, E., & REDOUX, J. (1988). Guidelines for the successful psychotherapy of substance abusers. *American Journal of Drug Alcohol Abuse, 14*, 199–209.

KLEBER, H. D. (1994). Detoxification from opioid drugs. In M. Galanter & H. D. Kleber (Eds), *The American Psychiatric Press Textbook of Substance Abuse Treatment.* Washington, DC: American Psychiatric Press.

KLEBER, H. D. (1993). America's drug strategy: Lessons of the past . . . steps toward the future. Paper presented at a Senate Judiciary Committee Hearing, April, Washington, DC.

KLEBER, H. D. (1989). Treatment of drug dependence: What works. *International Review of Psychiatry, 1*, 81–100.

MARLATT, G. A., & GORDON, J. R. (1985). Relapse prevention: Maintenance strategies in the treatment of addictive behaviors. New York: Guilford Press.

McLELLAN, A. T., ET AL. (1993). The effects of psychosocial services in substance abuse treatment. *Journal of the American Medical Association, 269*, 1953–1959.

McLELLAN, A. T., ET AL. (1992). A new measure of substance abuse treatment: Initial studies of the treatment services review. *Journal of Nervous and Mental Disease, 180*, 101–110.

McLELLAN, A. T., ET AL. (1984). The psychiatrically severe drug abuse patient: Methadone maintenance or therapeutic community? *American Journal of Drug & Alcohol Abuse, 10*, 77–95.

McLELLAN, A. T., ET AL. (1980). An improved diagnostic instrument for substance abuse patients: The Addiction Severity Index. *Journal of Nervous and Mental Disease, 168*, 26–33.

NACE, E. P. (1992). Alcoholics anonymous. In J. H. Lowinson et al. (Eds.), *Substance abuse: A comprehensive textbook* (2nd ed.). Baltimore: Williams & Wilkins.

ROUNSAVILLE, B. J., WEISSMAN, M. M., & KLEBER, H. D. (1983). An evaluation of depression in opiate addicts. *Research in Community and Mental Health, 3*, 257–289.

ROUNSAVILLE, B. J., ET AL. (1991). Psychiatric diagnoses of treatment-seeking cocaine abusers. *Archives of General Psychiatry, 48*, 43–51.

WEISS, R. D., ET AL. (1986). Psychopathology in chronic cocaine abusers. *American Journal of Alcohol Abuse, 12*, 17–29.

ZWEBEN, J. E. (1986). Recovery oriented psychotherapy. *Journal of Substance Abuse Treatment, 3*, 255–262.

FRANCES R. LEVIN
HERBERT D. KLEBER

Art of Acupuncture The art of acupuncture is an ancient and integral part of the armamentarium used in China for the treatment of medical problems.

Acupuncture consists of the insertion of very fine needles into the skin at specific points intended, according to traditional Chinese medicine, to influence specific body functions or body parts. In the traditional Chinese view of the body, life energy, (chi), circulates through pathways; blockage of the pathways leads to deficiency of chi, or disease. The goal of the traditional acupuncturist is to open up the pathways and stimulate the movement of chi. The specific points for needle insertion are based on traditional anatomy maps that depict which pathways affect which body functions.

Following President Richard M. Nixon's historic trip to China in 1972, considerable public interest in acupuncture was generated when the media observed that acupuncture was not only effective in relieving pain, but could also be a substitute for general anesthesia. The following year, Dr. H. L. Wen, a neurosurgeon in Hong Kong, reported a serendipitous observation that acupuncture with electrical stimulation (AES) eliminated withdrawal symptoms in a narcotics addict on whom he had intended to perform brain surgery to treat drug addiction. The discovery occurred the day before the scheduled surgery while Dr. Wen was demonstrating to the patient that AES could relieve pain. Fifteen minutes after the AES had begun, the patient reported a significant reduction of his drug withdrawal symptoms, which disappeared altogether thirty minutes after AES was started. Dr. Wen followed this patient, noting that AES had to be administered every eight hours for the first three days, and gradually the intervals could be increased. Within a week there were no further signs or symptoms of withdrawal. This led Dr. Wen to conduct a study of AES in 40 narcotics addicts experiencing withdrawal. All but one (who required medication for severe pain and was dropped from the study) were successfully detoxified. It is noteworthy that Dr. Wen's initial observations occurred prior to the discovery, in 1975, of endogenous opioid substances in the brain (also called endorphins).

In a later study, in 1977, Dr. Wen noted that AES increased endorphin levels and relieved abstinence syndromes while simultaneously inhibiting the autonomic nervous system, primarily the parasympathetic nervous system. The findings by Dr. Wen and several other scientific groups that peripheral stimulation could release endogenous opioid substances in the central nervous system (CNS) gave scientific credibility to the possibility that this traditional Chinese therapy could help to deal with a contemporary problem. Chronic or repeated exposure to opioids leads to adaptive changes in the CNS; withdrawal symptoms occur when these drugs are abruptly discontinued. Since the administration of opioid drugs alleviates withdrawal, it was reasonable to believe that one's own endogenous opioids might do the same.

During the mid-1970s, the use of acupuncture became popular in the United States, despite the absence of the kind of rigorous clinical investigation typically required for new pharmacological treatments. There were probably a number of factors that contributed to its popularity. Because it involved no pharmacological agents, it was seen as being more compatible with the approach espoused by SELF-HELP groups, ranging from ALCOHOLICS ANONYMOUS (AA) to THERAPEUTIC COMMUNITIES. Also, acupuncture did not initially require medical personnel, so it was relatively inexpensive compared to either psychotherapy or pharmacotherapy. In addition, its popularity increased at a time when some people objected to using METHADONE for drug detoxification or for maintenance, on the grounds that such use made drug-dependent minority-group members dependent upon the medical establishment. A technique from a non-Western tradition seemed, therefore, to have special appeal for treatment programs that dealt predominantly with minorities.

One such program was the Division of Substance Abuse at Lincoln Hospital in the south Bronx, New York, under the leadership of Dr. Michael O. Smith. Smith was interested in alternatives to methadone for detoxification. Based on Wen's work, Smith first used electrical stimulation along with acupuncture, but he later discarded the use of electrical stimulation. Eventually, a standard protocol was developed which used four or five acupuncture points on each ear. By 1975, the use of acupuncture as a treatment for drug abuse was extended to alcohol patients, then later to cocaine and crack-cocaine patients.

In 1985 Smith founded the National Acupuncture Detoxification Association (NADA) at 3115 Broadway, #51, New York, New York 10027. By 1993, when the second international conference of NADA was held in Budapest, Hungary, there were participants from all over the world.

In the early 1990s, the use of acupuncture in addiction treatment had become popular with many people working in the criminal-justice system. Most of the funding for treatment programs using acupuncture at that time came initially from the criminal-justice system, rather than from the federal and state agencies that usually fund drug treatment programs. Although the scientific community had been unable to show the efficacy of acupuncture in properly controlled clinical studies, this relatively inexpensive and easily expanded procedure became the mainstay of a number of "drug courts," where judges involved themselves directly in managing the treatment of drug offenders.

At many clinics in the United States, acupuncture treatment is now offered as part of a broad psychosocial program that has elements of self-help and TWELVE-STEP programs, plus traditional medicine and alternative medicine (some clinics, for example, use a "sleep mix" tea brewed from a variety of herbs).

As practiced in the United States, several technical procedures broadly described as acupuncture have been used. *Standard bilateral acupuncture* is the application of five needles to the concha and cartilage ridge of each ear at defined points (*shen men*, lung, sympathetic, kidney, and liver) determined from traditional Chinese anatomy maps. With *unilateral acupuncture*, the needles are applied to one ear. *Acupressure* involves applying pressure by hand or by an object to the same areas. *Electroacupuncture* applies low level electric current to needles placed at the traditional points. With *moxibustion*, herbs are burned near the needles to add heat; and with *neuroelectric stimulation*, low dose electrical current is passed through surface electrodes. Some practitioners advocate the use of surface electrodes and special currents, designating this approach *neuroelectrical therapy* (NET). There is no more evidence for the efficacy of added electrical current in the acupuncture treatment of drug and alcohol problems than there is for acupuncture itself.

Many acupuncture practitioners in the United States belong to and are accredited by the American Association of Acupuncture and Oriental Medicine (AAAOM), founded in 1981. Others may be accredited by the National Acupuncture and Oriental Medicine Alliance (NAOMA), founded in 1992, which accepts a broader range of training for purposes of certification than AAAOM.

In 1991, the NATIONAL INSTITUTE ON DRUG ABUSE (NIDA) sponsored a technical review of the current state of knowledge about the use of acupuncture in

the treatment of alcoholism and other drug-dependence problems. One of the participants, Dr. George Ulett, noted that although there is some evidence that electrical stimulation through needles or electrodes placed at certain points on the body can release endogenous opioids and other neuropeptides in the central nervous system, there is little evidence that such release is caused by needles alone. He also asserted that the critical factor is the frequency characteristic of the current, not the specific placement site of needles or electrodes. This group of researchers concluded that part of the difficulty in deciding whether acupuncture is effective was the lack of standard terminology and standard methods. A number of procedures, all called acupuncture, were being applied to a variety of drug and alcohol problems, but in different ways, over varying periods of time, with results measured in differing ways. For example, different numbers of acupuncture needles could be used, at different sites, with or without electrical current. One study of acupuncture for alcohol detoxification, by Bullock and coworkers, which came closest to being scientifically valid, used appropriate controls (placement of needles in non-sites) and staff who were "blinded" as to which group was control and which was receiving acupuncture at specific body sites. This study found a far better outcome for patients in the specific body-site group than for controls—and that the difference persisted even when measured six months later. However, another research group using similar methodology could not replicate the findings and reported no difference between point-specific acupuncture, sham transdermal stimulation, or standard care (no acupuncture control).

Many practitioners who have used acupuncture, even those who are convinced of its efficacy, report that only a small proportion of people who start treatment actually complete the typical series of ten to twenty treatments. Those who have used the technique believe that the minimal amount of treatment required for benefit is at least one twenty-minute session per day of bilateral acupuncture for at least ten days. In general, among both opioid-dependent and cocaine-dependent patients, those with lighter habits seemed to fare best.

The NIDA technical review panel concluded that, at the time of the review (1991), there was no compelling evidence that acupuncture is an effective treatment for opiate or cocaine dependence. Nevertheless, they found no evidence that acupuncture is harmful.

BIBLIOGRAPHY

BRUMBAUGH, A. G. (1993). Acupuncture: New perspectives in chemical dependency treatment. *Journal of Substance Abuse Treatment, 10,* 35–43.

MCLELLAN, A. T., ET AL. (1993). Acupuncture treatment for drug abuse: A technical review. *Journal of Substance Abuse Treatment, 10,* 569–576.

JOYCE H. LOWINSON
JEROME H. JAFFE

Assertiveness Training Assertiveness training is one part of a behavioral approach to treatment (RELAPSE PREVENTION), which holds that poor coping skills contribute to problematic substance use. Cummings, Gordon, and Marlatt (1980) found that many people relapsed in situations that included frustration and anger resulting from social interaction. For example, a man who is trying to remain abstinent from ALCOHOL might drink shortly after getting into an argument. Assertiveness training teaches alternative styles of dealing with these high-risk situations crucial to successful abstinence.

Experimental evidence supports this notion. Marlatt, Kosturn, and Lang (1975) found that participants in an experiment—who were frustrated without the possibility of expressing their anger—drank significantly more than those who were able to express their anger before being given the opportunity to drink. This supported the hypothesis that heavy drinkers who express their anger would drink less than those who did not have an adequate COPING response. Relapse prevention holds that an inadequate coping response to a frustrating situation will result in a feeling of less personal control in one's life, and may contribute to relapse. If people do drink at this point, their sense of personal control further erodes, because of the violation of the goal of abstinence. Uncontrolled use of the substance is more likely to occur, coupled with fantasies of power and/or actual aggressive behavior as the continuation of a maladaptive means of coping with the initial frustration. The individual is then likely to attribute the problematic behavior (e.g., arguing with a spouse) not to themselves but to the results of using alcohol. This in turn adds to a sense of poor personal control and the increased probability of continued drinking.

Relapse prevention teaches the client assertiveness skills so as to provide a repertoire of successful coping strategies for high-risk situations. In particular, clients are taught drink-refusal skills and alternative ways to cope with anger and frustration. Drink-refusal skills center on clients generating as many possible ways as they can think of to decline a drink, and then to practice these refusal skills during the course of therapy until the response feels comfortable. Alternative coping skills center on planning responses that allow the individual to get through the situation without drinking. These can include both behavioral strategies, such as avoidance of a stressful stimulus (e.g., not attending a party at a friend's house) and cognitive strategies, such as self-talk that helps the individual regulate thoughts ("I can handle this"). Research has supported the importance of both of these areas of focus, having found that the faster an individual can respond to a problematic high-risk situation, the less likely they are to drink (Chaney, O'Leary, & Marlatt, 1978).

The techniques of teaching these coping responses include a combination of direct verbal instruction, modeling of appropriate skills through role play, and rehearsal of the skills within the therapy session. By coaching the client and providing feedback, the therapist can help the client gain a sense of personal control in situations that have historically led to drinking. Research with alcoholics has shown that skill training does lead to fewer days of excessive drinking, fewer days of continuous drinking, fewer total drinks consumed, and more days of controlled drinking when compared with drinkers who did not receive the same instruction (Chaney, O'Leary, & Marlatt, 1978).

Assertiveness training is also used in primary prevention by teaching children, for example, to "just say no" to friends who may try to tempt them with a substance.

(SEE ALSO: *Prevention; Vulnerability As Cause of Substance Abuse*)

BIBLIOGRAPHY

CHANEY, E. F., O'LEARY M. R., & MARLATT, G. A. (1978). Skill training with alcoholics. *Journal of Consulting and Clinical Psychology, 46,* 1092–1104.

Cummings, C., Gordon, J. R., & Marlatt, G. A. (1980). Relapse: Prevention and prediction. In W. R. Miller (Ed.), *The addictive behaviors.* New York: Pergamon.

Marlatt, G. A., & Gordon, J. R. (1985). *Relapse prevention: Maintenance strategies in the treatment of addictive behaviors.* New York: Guilford Press.

Marlatt, G. A., Kosturn, C. F., & Lang, A. R. (1975). Provocation to anger and opportunity for retaliation as determinants of alcohol consumption in social drinkers. *Journal of Abnormal Psychology, 84,* 652–659.

ALAN MARLATT
MOLLY CARNEY

Aversion Therapy For many years, attempts have been made to condition alcoholics to dislike alcohol. For example, alcoholics are asked to taste or smell alcohol just before a preadministered drug makes them nauseated. Repeated pairing of alcohol and nausea results in a conditioned response—after a while, alcohol alone makes them nauseated. Thereafter, it is hoped, the smell or taste of alcohol will cause nausea and discourage drinking.

Instead of pairing alcohol with nausea, other therapists have associated it with pain, shocking patients just after they drink, or they have associated it with panic from not being able to breathe by giving them a drug that causes very brief respiratory paralysis. Others have trained patients to imagine unpleasant effects from drinking, hoping to set up a conditioned response without causing so much physical distress.

Does it work? Some degree of conditioning is usually established, but it is uncertain how long the conditioning lasts. The largest study that involved conditioning alcoholics was conducted many years ago in Seattle, Washington (Lemere & Voegtlin, 1940). More than 34,000 patients conditioned to feel nauseated when exposed to alcohol were studied ten to fifteen years after treatment. Sixty-six percent were abstinent, an impressive recovery rate compared to other treatments. The patients who did best had had booster sessions—that is, they had come back to the clinic after the initial treatment to repeat the conditioning procedure. Of those who attended booster sessions, 90 percent were abstinent. Based on this study, the nausea treatment for alcoholism would seem an outstanding success. Why hasn't it been universally accepted?

One reason is that the results can be attributed to factors other than the conditioning. The patients in the study were a special group. Generally, they were well educated, had jobs, and were well off financially. They may not have received the treatment otherwise, since the clinic where they were treated was private and expensive. Studies of alcoholics have often shown that certain subject characteristics are more predictive of successful treatment outcome than the type of treatment administered. These factors include job stability, living with a relative, absence of a criminal record, and living in a rural community. In the Seattle study there was no control group that did not receive conditioning therapy. It is possible that this select group of patients, many having characteristics that favor a good outcome, would have done as well without conditioning.

Furthermore, in conditioning treatments, motivation is important. Treatment is voluntary and involves acute physical discomfort; presumably few would consent to undergo the therapy if they were not strongly motivated to stop drinking. The Seattle study makes this point graphically clear. Those who came back for booster sessions did better than those who didn't, but another group did better still: those who *wanted* to come back but couldn't because they lived too far from the hospital. All of these people remained abstinent.

For many years, chemically induced aversive conditioning of alcoholics was virtually ignored in the literature. Then, in 1990, Smith and Frawley published an outcome study of patients who received aversion therapy as part of their inpatient treatment. From a randomly selected sample of 200 patients, 80 percent were located and interviewed by telephone. Between thirteen and twenty-five months had passed since their discharges from the hospital. The overall abstinence rate for the first twelve months was 71 percent; it was 65 percent for the total period.

Follow-up studies of alcoholism treatment rarely report abstinence rates this high. How should these be interpreted?

As in the original Seattle study, in the Smith and Frawley study, the patients, by and large, had good prognostic features. At the time of admission, more than 50 percent were married and had some college education. Nearly 80 percent were employed. They could afford a private hospital. In short, with characteristics that favor a good outcome, they might have done as well without conditioning. Moreover, the inpatient program involved more than aversive conditioning. It included many ingredients found in other treatment programs, including counseling, a family program and aftercare plan, and ALCOHOLICS ANONYMOUS.

One finding in this report was similar to that of the original study—booster sessions are important. One month and three months after discharge, the patients were asked to return for reinforcement treatments. Just as in the original studies, those who returned for the booster sessions had a particularly good outcome. In fact, the most powerful predictor of abstinence was the number of reinforcement treatments utilized by each patient. Those taking two reinforcement treatments had a twelve-month abstinence rate of 70 percent; those who took only one had a 44 percent rate; and those who had no reinforcement had only a 27 percent rate. Seven percent took *more* than two reinforcement treatments and had a phenomenal twelve-month abstinence rate of 92 percent.

The importance of reinforcement sessions may reflect motivation on the part of the patient, actual Pavlovian conditioning, or both. The paper does not tell whether the patients developed a true conditioned response to alcohol at any time. Information about this would help separate nonspecific motivational factors from actual conditioning.

The study lacked a control group. This was remedied in a report (Smith, Frawley, & Polissar, 1991) that compared 249 alcoholic inpatients who received aversion therapy with patients from a national treatment registry who did not receive aversion therapy. The patients treated with aversion therapy had significantly higher abstinence rates at six and twelve months, suggesting that motivation and good prognostic features may not completely explain the success of this still rather unpopular treatment.

Frawley and Smith (1992) have also reported remarkably high abstinence rates from cocaine (current abstinence of at least six months, 68 percent) among a similar group of patients, with good prognostic features, treated with aversion therapy and follow-up at an average of fifteen months after treatment. Again there was no control group.

Aversion treatment for cigarette smoking has been studied by using appropriate controls. The technique involves encouraging the smoker to keep inhaling at rapid intervals over a period of five to ten minutes until he or she becomes sick, presumably because the nicotine levels exceed the smoker's tolerance levels. This approach has consistently produced higher levels of abstinence from smoking than have control groups.

(SEE ALSO: *Calcium Carbimide; Disulfiram*)

BIBLIOGRAPHY

FRAWLEY, P. J., & SMITH, J. W. (1992). One-year follow-up after multimodal inpatient treatment for cocaine and methamphetamine dependencies. *Journal of Substance Abuse Treatment, 9,* 271–286.

LEMERE, F., & VOEGTLIN, W. L. (1940). Conditioned reflex therapy of alcoholic addiction: Specificity of conditioning against chronic alcoholism. *California and Western Medicine, 53*(6), 1–4.

SMITH, J. W., & FRAWLEY, P. J. (1990). Long-term abstinence from alcohol in patients receiving aversion therapy as part of a multimodal inpatient program. *Journal of Substance Abuse Treatment, 7,* 77–82.

SMITH, J. W., FRAWLEY, P. J., & POLISSER (1991). Six- and twelve-month abstinence rates in inpatient alcoholics treated with aversion therapy compared with matched inpatients from a treatment registry. *Alcohol: Clinical and Experimental Research 5,* 862–870.

DONALD W. GOODWIN

Behavior Modification Within a behavioral model of drug dependence, drug use is considered a learned behavior that is directly influenced by antecedent and consequent events of drug use. Drug use is deemed a legitimate and primary target of assessment and intervention. The interventions are directed toward elimination of drug use and increasing prosocial behaviors incompatible with drug use and dependence.

Four types of behavioral interventions are covered in this article: (1) contingency-management procedures, (2) behavioral counseling/relapse prevention, (3) antecedent stimulus counterconditioning, and (4) multimodal behavioral interventions (for a more detailed review, see Stitzer, Bigelow & Gross, 1989).

CONTINGENCY-MANAGEMENT PROCEDURES

Contingency-management procedures are the most thoroughly researched of the behavioral interventions for drug abuse. The goal of these interventions is to alter drug use by systematically arranging consequences designed to weaken drug use and strengthen drug abstinence. Contingencies can be placed directly on drug use or on other behaviors deemed important in the treatment process, such as regular attendance at counseling sessions or compliance with adjunct medication regimens. Often, but not always, contingency-management procedures are implemented through written behavioral contracts.

Drug use in people dependent on any one of the major drugs of abuse—including ALCOHOL (Miller, 1975), AMPHETAMINES (Boudin, 1972), COCAINE (Anker & Crowley, 1982), MARIJUANA (Budney et al., 1991), NICOTINE (Stitzer & Bigelow, 1983), OPIOIDS (Higgins et al., 1986), and SEDATIVES (Stitzer et al., 1982)—has been demonstrated to be sensitive to direct reinforcement contingencies. Contingency-management procedures have been employed effectively in both inpatient and outpatient settings. Reinforcing and aversive consequences are effective in these interventions, although the former are preferred for ethical reasons and because they are likely to be associated with greater patient acceptability and treatment retention (Stitzer et al., 1986).

An example of contingency-management procedures designed to improve compliance with a treatment regimen comes from a study conducted in a methadone-maintenance clinic (Stitzer et al., 1977). In these clinics, patients generally visit the clinic daily to consume their medication under supervision and attend regularly scheduled counseling sessions on approximately a once-weekly basis. A common problem is that patients consume their medication but fail to attend the required counseling sessions. To counter this problem, patients in this study were provided an opportunity to take methadone doses home from the clinic (a break from the grind of daily clinic attendance) contingent on attending scheduled counseling sessions. Implementation of this contingency increased counseling attendance significantly.

An example of use of contingency-management procedures to directly alter drug use also comes from the METHADONE-MAINTENANCE clinic. It was a study conducted to decrease sedative use among methadone-maintenance patients (Stitzer et al., 1982). Chronic sedative users were offered a choice of reinforcers that included medication take-home privileges or monetary payment. These reinforcers were available contingent on providing sedative-free urine specimens during twice-weekly urinalysis monitoring. Sedative use was reduced in eight of the ten subjects during the intervention period.

Research has demonstrated convincingly that contingency management procedures are effective in drug-abuse treatment. The most noteable shortcoming associated with their use is that treatment gains are often lost at some point after the contingencies are removed. This is not a problem unique to these procedures. It is shared by most interventions for drug dependence. Nevertheless, it is an important problem to be surmounted. Use of the multimodal interventions described below may be helpful in meeting this challenge.

BEHAVIORAL COUNSELING/ RELAPSE PREVENTION

Behavioral counseling/RELAPSE-PREVENTION therapy emphasizes environmental restructuring and the acquisition of specific skills deemed necessary to achieve abstinence and avoid relapse. Patients are counseled to identify antecedents and consequences related to their drug use. For example, if drug use is likely when patients are in the company of certain individuals or settings, they are counseled to minimize contact with those people or settings. Regarding consequences, patients are counseled to make explicit the negative consequences of drug use and to identify alternatives to drug reinforcement. Patients are also provided drug-refusal skills.

Evaluation of this approach is still preliminary, but it appears promising. For example, in a recent study, forty-two outpatients who met diagnostic criteria for cocaine dependence were randomly assigned to relapse-prevention therapy or interpersonal psychotherapy (Carrol, Rounsaville, & Gawin, 1991). Treatment was for twelve weeks, with once-weekly counseling sessions and random urinalysis monitoring. The subjects assigned to relapse-prevention therapy achieved more continuous weeks of cocaine abstinence than those in interpersonal psychotherapy (57% vs. 33%) and were more likely to complete treatment (67% vs. 38%).

ANTECEDENT STIMULUS COUNTERCONDITIONING

A convincing body of basic research demonstrates that environmental events including the rituals associated with taking abused drugs and the physical characteristics of the drugs themselves come to elicit responses from the user that could increase the likelihood of drug use (O'Brien, Ehrman, & Ternes,

1986). Thus, treatments have been designed to alter these conditioned effects such that those same stimuli would come to be either aversive or neutral.

The most common form of counterconditioning is AVERSION THERAPY. Characteristic stimulus effects of the abused drug are paired with an aversive event; for example, a chemical agent that smells nauseating, electric shock, or unpleasant thoughts. Although it has been demonstrated to have efficacy in the treatment of alcohol dependence (Rimmele, Miller, & Dougher, 1989) and tobacco smoking, the efficacy of aversion therapy with other forms of drug dependence has not been supported by data from controlled clinical trials. High abstinence from cocaine one year after treatment has been reported from uncontrolled trials, but without controls with comparable prognostic factors, much about abstinence rates cannot be interpreted.

Another form of counterconditioning relies on extinction. That is, patients are repeatedly exposed to drug-related stimuli with the goal that such exposure to the conditioned stimuli, in the absence of any pairing with the pharmacological effects of the drug, will eventually render the stimuli neutral.

Research with this procedure is also preliminary, but initial results suggest that individuals enrolled in outpatient treatment for cocaine dependence who receive extinction procedures in combination with psychotherapy have better treatment retention and less cocaine use than patients treated with other treatment packages (Childress et al., 1988).

MULTIMODAL INTERVENTIONS

Treatment packages can be developed that utilize aspects of several of the above components in a comprehensive effort to eliminate use of the target drug and increase prosocial behaviors predictive of success in drug-abuse treatment. The seminal approach illustrating the efficacy of multimodal-behavioral intervention was the Community Reinforcement Approach to the treatment of alcoholism (Hunt & Azrin, 1973; Sisson & Azrin, 1989).

More recently, such an approach was used effectively in outpatient treatment of cocaine dependence (Higgins et al., 1991, 1993). For example, in one study, thirty-eight patients were randomly assigned to a multimodal behavioral-treatment package or standard outpatient drug-abuse counseling (Higgins et al., 1993). Treatment was twenty-four weeks in du-

ration, with twice-weekly counseling and thrice-weekly urinalysis testing. The behavioral treatment included a contingency contract wherein patients earned vouchers exchangeable for retail items for each cocaine-free urine specimen. Patients' significant others were also telephoned after each test and informed of the urinalysis results; they provided social reinforcement for cocaine abstinence. Additionally, patients received marital counseling, vocational, and social/recreational counseling. Those enrolled in standard counseling received group and individual therapy based on the TWELVE STEPS of recovery, as espoused by ALCOHOLICS ANONYMOUS. A significantly greater proportion of those who received the behavioral intervention completed treatment than those in standard counseling (58% vs. 11%). Additionally, more than 50 percent of patients in the behavioral treatment achieved nine or more weeks of continuous cocaine abstinence versus 11 percent in standard counseling.

There has not been sufficient research to determine the efficacy of these multimodal behavioral interventions in achieving long-term abstinence. However, they appear to hold promise for bringing about the many changes likely to be necessary to maintain drug abstinence.

(SEE ALSO: *Causes of Substance Abuse; Research, Animal Model; Toughlove; Wikler's Pharmacologic Theory of Drug Addiction*)

BIBLIOGRAPHY

ANKER, A. A., & CROWLEY, T. J. (1982). Use of contingency contracts in specialty clinics for cocaine abuse. In L. S. Harris (Ed.), *Problems of drug dependence 1981*, NIDA Research Monograph 41, pp. 452–459. Washington, DC: U.S. Government Printing Office.

BOUDIN, H. M. (1972). Contingency contracting as a therapeutic tool in the reduction of amphetamine use. *Behavior Therapy, 3*, 604–608.

BUDNEY, A. J., ET AL. (1991). Contingent reinforcement of abstinence with individuals abusing cocaine and marijuana. *Journal of Applied Behavior Analysis, 24*, 657–665.

CARROL, K. M., ROUNSAVILLE, B. J., & GAWIN, F. H. (1991). A comparative trial of psychotherapies for ambulatory cocaine abusers: Relapse prevention and interpersonal psychotherapy. *American Journal of Alcohol and Drug Abuse, 17*, 229–247.

CHILDRESS, A., ET AL. (1988). Update on behavioral treatments for substance abuse. In L. S. Harris (Ed.), *Problems of drug dependence 1988*, NIDA Research Monograph 90, pp. 183–192. Washington, DC: U.S. Government Printing Office.

HIGGINS, S. T., ET AL. (1993). Achieving cocaine abstinence with a behavioral approach. *American Journal of Psychiatry, 150*, 763–769.

HIGGINS, S. T., ET AL. (1991). A behavioral approach to achieving initial cocaine abstinence. *American Journal of Psychiatry, 148*, 1218–1224.

HIGGINS, S. T., ET AL. (1986). Contingent methadone delivery: Effects on illicit-opiate use. *Drug and Alcohol Dependence, 17*, 311–322.

HUNT, G. M., & AZRIN, N. H. (1973). A community-reinforcement approach to alcoholism. *Behavior Research and Therapy, 11*, 91–104.

MILLER, P. M. (1975). A behavioral intervention program for chronic public offenders. *Archives of General Psychiatry, 32*, 915–918.

O'BRIEN, C. P., EHRMAN, R. N., & TERNES, J. W. (1986). Classical conditioning in human opioid dependence. In S. R. Goldberg & I. P. Stolerman (Eds.), *Behavioral analysis of drug dependence*, pp. 329–356. Orlando, FL: Academic.

RIMMELE, C. T., MILLER, W. R., & DOUGHER, M. J. (1989). Aversion therapies. In R. K. Hester & W. R. Miller (Eds.), *Handbook of alcoholism treatment approaches: Effective alternatives*, pp. 128–140. New York: Pergamon.

SISSON, R. W., & AZRIN, N. H. (1989). The community reinforcement approach. In R. K. Hester & W. R. Miller (Eds.), *Handbook of alcoholism treatment approaches: Effective alternatives*, pp. 242–258. New York: Pergamon.

STITZER, M. L., & BIGELOW, G. E. (1983). Contingent reduction for carbon monoxide reduction: Effects of pay amount. *Behavior Therapy, 14*, 647–656.

STITZER, M. L., BIGELOW, G. E., & GROSS, J. (1989). Behavioral treatment of drug abuse. In T. B. Karasu (Ed.), *Treatment of psychiatric disorders: A task force report of the American Psychiatric Association*, vol. 2, pp. 1430–1447. Washington, DC: American Psychiatric Association.

STITZER, M. L., ET AL. (1986). Effects of methadone dose contingencies on urinalysis test results of poly-abusing methadone maintenance patients. *Drug and Alcohol Dependence, 18*, 341–348.

STITZER, M. L., ET AL. (1982). Contingent reinforcement for benzodiazepine-free urines: Evaluation of a drug abuse treatment intervention. *Journal of Applied Behavior Analysis, 15*, 493–503.

STITZER, M. L., ET AL. (1977). Medication take-home as a reinforcer in a methadone maintenance program. *Addictive Behaviors, 2*, 9–14.

STEPHEN T. HIGGINS
ALAN J. BUDNEY

Cognitive Therapy Cognitive treatment is based on the assumption that the way one thinks is a primary determinant of feelings and behavior. Developed from Beck's research (Beck et al., 1979, 1993), cognitive treatment is approached as a collaborative effort between the client and therapist to examine the client's errors and distortions in thinking that contribute to problematic behavior. This examination is fostered through a combination of verbal techniques and behavioral experiments to test the underlying assumptions the client holds about the problematic behavior.

Cognitive treatment in the substance-abuse field was a direct extension of Beck's work. Beck's catalog of distorted thoughts examined in depression were found to be applicable to cognitive distortions and errors that accompany addictive disorders. Various cognitive treatments for substance abuse focus on these distortions and vary primarily in the techniques used to change these thought processes.

In RELAPSE PREVENTION (Marlatt & Gordon, 1985), cognitive distortions are viewed as instrumental in the process that leads to relapse. By helping the client thoroughly examine the thoughts that accompany substance use, therapy can reduce the likelihood of a lapse (single use), as well as help prevent a lapse from becoming a relapse (return to uncontrolled use). This is accomplished by examining the following cognitive errors:

1. Overgeneralizing—this is one of the most frequently occurring cognitive errors that helps a single lapse become a full-blown relapse. By viewing the single use as a sign of total relapse, the client overgeneralizes the single use of a substance as a symptom of total failure, thereby allowing for increasing use over time and in a variety of situations. This is sometimes referred to as the ABSTINENCE VIOLATION EFFECT (AVE).

2. Selective abstraction—by excessively focusing on the immediate lapse, with an accompanying ne-

glect of all past accomplishments and learning, the client interprets a single slip as equivalent to total failure. The individual measures progress almost exclusively in terms of errors and weaknesses.

3. Excessive responsibility—by attributing the cause of a lapse to personal, internal weaknesses or lack of willpower, the client assumes total responsibility for the slip, which in turn makes reassuming control more difficult than when environmental factors are considered partially responsible for the slip.

4. Assuming temporal causality—here, the client views a slip as the first of many to come, thereby dooming all future attempts at self-control.

5. Self-reference—when the client thinks that a lapse becomes the focus of everyone else's attention, believing that others will attribute blame for the event to the client, this adds to feelings of guilt and shame that may already be present within the person.

6. Catastrophizing—the client believes the worst possible outcome will occur from a single use of the substance instead of thinking about how to cope successfully with the initial lapse.

7. Dichotomous thinking—by viewing events in "black and white," clients view their addictive behavior exclusively in terms of abstinence or relapse and leave no logical room for "gray" areas, where they can get back on track once a slip has occurred.

8. Absolute willpower breakdown—here, the client assumes that once willpower has failed, loss of control is inevitable, never to be regained.

9. Body over mind—the cognitive error here is assuming that once a single lapse has occurred, the physiological process of addiction has exclusive control over subsequent behavior, making continued use inevitable.

These errors in thinking are targeted for change in relapse prevention by helping the client learn how to reattribute the cause of a lapse from internal, stable, personal causes to mistakes or errors in the learning process. To facilitate the client's sense of personal control, lapses are viewed as opportunities for corrective learning, instead of indications of total failure. Congruent with the research in the area (Shiffman, 1991), the therapist presents a lapse as a frequently occurring event in the journey toward recovery. The therapist therefore encourages the client to examine the thoughts and expectancies that surround the lapse closely, with the aim of learning alternative coping skills for similar situations that may arise in the future. By reframing a lapse as a learning opportunity, the client is encouraged to view the event as a chance to hone the skills required for abstinence, thereby countering the cognitive errors of selective abstraction.

To intervene with the errors of overgeneralization and temporal causality, the client is taught to view a lapse as a specific, unique event in time and space, instead of as a symptom with greater significance attached to it (e.g., the beginning of the inevitable end). The errors of self-reference and willpower breakdown can be countered by teaching the client to reattribute a lapse to external, specific, and controllable factors. By examining the difficulty of the high-risk situation, the appropriateness of the coping response employed, and any motivational deficits (fatigue or excessive stress), the client can maintain a sense of control over the event and the process of recovery.

Each of these techniques is aimed at conveying the idea that abstinence is the result of a learning process, requiring an acquisition of skills similar to many other skills one learns. This general metaphor can help the client reverse catastrophizing, by reframing a relapse as a "prolapse," as a fall forward rather than backward. This view, combined with viewing a lapse as a unique event in time, helps client maintain a sense of personal control, since abstinence or control is framed as just a moment away if use is discontinued.

Several skills are taught to the client in relapse prevention to facilitate these cognitive changes and prevent future lapses. Identifying specific sources of stress that contribute to urges, cravings, or lapses helps isolate the event in time as well as identify other distortions that may be present. For example, clients may identify discussing money with one's spouse as the high-risk situation that preceded a lapse. While discussing the lapse with a therapist, clients can learn to anticipate that discussing money in the marriage may trigger an urge or craving to drink. Teaching clients to use visual imagery, such as viewing the urge as a wave that they can surf, can help manage the feeling that urges will continue to build until they must inevitably be given in to. Self-talk is encouraged if a client believes this will help gain a sense of personal control (such as reciting a phrase to oneself about the goal of abstinence or remembering who can be telephoned when an urge is experienced). In addition, clients are taught to be alert for "apparently irrelevant decisions," which can inadvertently lead to relapse. For example, an abstinent gambler may decide to take a scenic drive through Reno, only to find a situation that would be extremely difficult for many to ignore, thus in this case causing a relapse.

Other theorists have developed treatments based exclusively on changing irrational thinking. Ellis and colleagues (1988) founded a self-help group network called RATIONAL RECOVERY (RR), based on the principles of rational emotive therapy. Developed as an alternative to the ALCOHOLICS ANONYMOUS network, RR focuses on "addictive thinking" and views abstinence as possible—purely as a result of changing these thought processes. This differs from the relapse prevention model described above, which in its entirety combines cognitive and behavioral techniques. Ellis's RR movement teaches addicts how to identify their own faulty thinking through a self-help manual (Trimpey, 1989) and the attendance at support groups.

(SEE ALSO: *Alcoholism; Causes of Substance Abuse; Disease Concept of Alcoholism and Drug Abuse*)

BIBLIOGRAPHY

BECK, A. T., ET AL. (1993). *Cognitive therapy of substance abuse.* New York: Guilford Press.

BECK, A. T., ET AL. (1979). *Cognitive therapy of depression.* New York: Guilford Press.

ELLIS, A., ET AL. (1988). *Rational-emotive therapy with alcoholics and substance abusers.* New York: Pergamon.

MARLATT, G. A., & GORDON, J. R. (1985). *Relapse prevention: Maintenance strategies in the treatment of addictive behaviors.* New York: Guild Press.

SHIFFMAN, S. (1992). Relapse process and relapse prevention in addictive behaviors. *Behavior Therapist*, January 8–11.

TRIMPEY, J. (1989). *The small book*, 3rd ed. Lotus, CA: Lotus Press.

MOLLY CARNEY
ALAN MARLATT

Group and Family Therapy The illnesses of drug addiction and alcoholism are so severe that they pervade every aspect of an individual's existence. It is rare that so extensive an illness can be reversed by individual therapy alone. Thus therapists are espousing an integration of individual, TWELVE-STEP, group, and family treatment, with specific combinations of treatments tailored to each individual's needs.

Dealing with the family is one more involvement with the patient's ecosystem, which includes working with the treatment team, twelve-step groups, sponsors, employers, EAPs (EMPLOYEE ASSISTANCE PROGRAM counselors), managed-care workers, parole officers, and other members of the legal system. However, family work is most critical to the success of treatment.

Group therapy has frequently been designated as the treatment of choice for addicted patients. This article views group therapy as an essential component of the integrated, individualized approach to addicts and alcoholics.

FAMILY THERAPY

The family treatment of substance abuse begins with developing a system to achieve and maintain abstinence. This system, together with specific family therapeutic techniques and knowledge of patterns commonly seen in families with a substance-abusing member, provides a workable, therapeutic approach to substance abuse.

Family treatment of substance abuse must begin with an assessment of the extent of substance dependence as well as the difficulties it presents for the individual and the family. The quantification of substance-abuse history can take place with the entire family present; substance abusers often will be honest in this setting, and "confession" is a helpful way to begin communication. Moreover, other family members can often provide more accurate information than the substance abusers (also known as the identified patient, IP). However, some IPs will give an accurate history only when interviewed alone.

In taking a drug-abuse history, it is important to know current and past use of every type of abusable drug as well as of ALCOHOL: quantity, quality, duration, expense, how intake was supported and prevented, physical effects, tolerance, withdrawal, and medical complications. At times, other past and present substance abusers within the family are identified; their own use and its consequences should be quantified without putting the family on the defensive. It is also essential to document the family's patterns of reactivity to drug use and abuse. Previous attempts at abstinence and treatment are reviewed to determine components of success and failure. The specific method necessary to achieve abstinence can be decided only after the extent and nature of substance abuse are quantified.

Establishing a System to Achieve a Substance-Free State. It is critical first to establish a system for enabling the substance abuser to become drug-free, so that family therapy can be effective. The specific methods employed to achieve abstinence vary according to the extent of use, abuse, and dependence. Mild-to-moderate abuse in adolescents can often be controlled if both parents agree on clear limits and expectations, and how to enforce them. Older abusers may stop if they are aware of the medical or psychological consequences to themselves or the effects on their family.

If substance abuse is moderately severe or intermittent and without physical dependence, such as intermittent use of HALLUCINOGENS or weekend CO-CAINE abuse, the family is offered a variety of measures, such as regular attendance at ALCOHOLICS ANONYMOUS (AA), NARCOTICS ANONYMOUS (NA), or COCAINE ANONYMOUS (CA) for the IP and Al-Anon or Naranon for family members.

If these methods fail, short-term hospitalization or treatment in an intensive outpatient program (20 hours or more per week) may be necessary to establish a substance-free state and to begin effective treatment even with nondependent patients. In more severe cases of drug abuse and dependence, more aggressive methods are necessary to establish a substance-free state.

Family Education. A substantial amount of family education is generally very helpful in the early stages of the family's involvement in therapy. In many inpatient addiction treatment programs, the family spends several days or more receiving appropriate education. If this is not available, the therapist should include this education process in early sessions.

Some of the issues covered by this educational emphasis are: (1) the physiological and psychological effects of drugs and alcohol; (2) the disease concept; (3) cross addiction (which helps families learn that a recovering cocaine addict should not drink or vice versa); (4) common family systems—emphasizing the family's roles in addiction and recovery, including enabling, scapegoating, and CODEPENDENCY; (5) the phases of treatment, with an emphasis on the

deceptiveness of the "honeymoon" period in early recovery; and (6) the importance of twelve-step family support groups (AL-ANON, ALATEEN).

Working with Families with Continued Drug Abuse. The family therapist is in a unique position with regard to continued substance abuse and other manifestations of the IP's resistance to treatment, including total nonparticipation. The family therapist still has a workable and highly motivated patient(s): the family. One technique that can be used with an absent or highly resistant patient is the intervention, which was developed for use with alcoholics but can be readily adapted to work with drug abusers, particularly those who are middle class, involved with their nuclear families, and employed.

In this technique, the family (excluding the abuser) and significant network members (e.g., employer, fellow employees, friends, and neighbors) are coached to confront the substance abuser with concern, but without hostility, about the destructiveness of his or her drug abuse and behavior. They agree in advance about what treatment is necessary and then insist on it. As many family members as possible should be included, because the breakthrough for acceptance of treatment may come from an apparently uninvolved family member, such as a grandchild or cousin. The involvement of the employer is crucial, and in some cases may be sufficient in and of itself to motivate the drug abuser to seek treatment. The employer who clearly makes treatment a condition of continued employment, who supports time off for treatment, and who guarantees a job on completion of the initial treatment course is a very valuable ally. The employer's model is also a very helpful one for the family, who need to be able to say "We love you, and because we love you, we will not continue to live with you if you continue to abuse drugs and alcohol. If you accept the treatment being offered to you and continue to stay off drugs, we will renew our lifetime commitment to you."

If substance abusers do not meet the above criteria for an intervention or if the intervention has failed, we are left with the problems of dealing with a substance-abusing family. Berenson (1976) offers a workable, three-step therapeutic strategy for dealing with the spouses or other family members of individuals who continue to abuse substances or who are substance dependent. Step one is to calm down the family by explaining problems, solutions, and coping mechanisms. Step two is to create an external support network for family members so that the emotional intensity is not all in the relationship with the substance abuser or redirected to the therapist. There are two types of support systems available to these spouses. One is a self-help group on the Al-Anon, Naranon, or Coanon model; the other is a significant others (SO) group led by a trained therapist. In the former, the group and sponsor provide emotional support, reinforce detachment, and help calm the family. An SO group may provide more insight and less support for remaining with a substance-abusing spouse.

Step three involves giving the client three choices: (1) keep doing exactly what you are doing; (2) detach or emotionally distance yourself from the drug abuser; or (3) separate or physically distance yourself. When the client does not change, it is labeled an overt choice 1. When a client does not choose 2 or 3, the therapist can point out that he or she is in effect choosing not to change. If not changing becomes a choice, then the SO can be helped to choose to make a change. In choice 2, SOs are helped to avoid overreacting emotionally to drug abuse and related behavior, and they are taught strategies for emotional detachment. Leaving, choice 3, is often difficult when the family is emotionally or financially dependent on the substance abuser.

Each of these choices seems impossible to carry out at first. The problem of choosing may be resolved by experiencing the helplessness and powerlessness in pursuing each choice.

As part of the initial contract with a family, it is suggested that the abuser's partner continue individual treatment, Al-Anon, Coanon, or an SO group even if the abuser drops out. Other family members are also encouraged to continue in family therapy and support groups. It should be reemphasized that whenever therapy is maintained with a family in which serious drug abuse continues, the therapist has the responsibility of not maintaining the illusion that the family is resolving problems, when in fact they are really reinforcing them. Even when the substance abuser does not participate in treatment, however, therapy may be quite helpful to the rest of the family.

The concept of the family as a multigenerational system necessitates that the entire family be involved in treatment. The family members for optimum treatment consist of the entire household and any relatives who maintain regular (approximately weekly) contact with the family. In addition, relatively emancipated family members who have less

than weekly contact may be very helpful to these families.

The utilization of a multigenerational approach involving grandparents, parents, spouse, and children at the beginning, as well as certain key points throughout, family therapy is advised. However, the key unit with substance abusers younger than about age 24 is the IP with siblings and parents. The critical unit with married substance abusers older than 24 is the IP and spouse. However, the more dependent the IP is on the parents, the more critical is family work with these parents. The majority of sessions should be held with these family units; the participation of other family members is essential to more thorough understanding and permanent change in the family.

Family therapy limited to any dyad is most difficult. The mother-addicted-son dyad is almost impossible to treat as a sole entity; some other significant person, such as a lover, grandparent, aunt, or uncle should be brought in if treatment is to succeed. If there is absolutely no one else available from the natural family network, then surrogate family members in multiple-family therapy groups can provide support and leverage to facilitate restructuring maneuvers.

AN INTEGRATED APPROACH TO A WORKABLE SYSTEM OF FAMILY TREATMENT

Family Diagnosis. Accurate diagnosis is as important a cornerstone of family therapy as it is in individual therapy. Family diagnosis looks at family interaction and communication patterns and relationships. In assessing a family, it is helpful to construct a map of the basic alliances and roles, as well as to examine the family rules, boundaries, and adaptability.

Family Treatment Techniques. Each system of family therapy presently in use is briefly summarized below, with an emphasis on the application of these techniques to substance abusers. They are classified into four schools: structural-strategic, psychodynamic, Bowen's systems theory, and behavioral. Any of these types can be applied to substance abusers if their common family patterns are kept in mind and if a method to control substance abuse is implemented.

Structural-Strategic Therapy. These two types are combined because they were developed by many of the same practitioners, and shifts between the two are frequently made by the therapist, depending on the family's needs. The thrust of structural family therapy is to restructure the system by creating interactional change within the session. The therapist actively becomes a part of the family, yet retains sufficient autonomy to restructure it. The techniques of structural therapy have been described in detail by Kaufman (1985). They include the contract, joining, actualization, marking boundaries, assigning tasks, reframing, the paradox, balancing and unbalancing, and creating intensity.

According to strategic therapists, symptoms are maladaptive attempts to deal with difficulties, which develop a homeostatic life of their own and continue to regulate family transactions. The strategic therapist works to substitute new behavior patterns for the destructive repetitive cycles. The techniques used by strategic therapists include the following:

1. Using tasks with the therapist responsible for planning a strategy to solve the family's problems.
2. Putting the problem in solvable form.
3. Placing considerable emphasis on change outside the sessions.
4. Learning to take the path of least resistance, so that the family's existing behaviors are used positively.
5. Using paradox, including restraining change and exaggerating family roles.
6. Allowing the change to occur in stages; the family hierarchy may be shifted to a different, abnormal one before it is reorganized into a new functional hierarchy.
7. Using metaphorical directives in which the family members do not know they have received a directive.

Stanton et al. (1982) successfully utilized an integrated structural-strategic approach with heroin addicts on METHADONE MAINTENANCE treatment.

Psychodynamic Therapy. This approach has rarely been applied to substance abusers because they usually require a more active, limit-setting emphasis on the here and now than is generally associated with psychodynamic techniques. However, if certain basic limitations are kept in mind, psychodynamic principles can be extremely helpful in the family therapy of these patients.

There are two cornerstones for the implementation of psychodynamic techniques: the therapist's self-knowledge and a detailed history of the substance abuser's family.

Important elements of psychodynamic family therapy include the following:

countertransference—The therapist may have a countertransference problem toward the entire family or any individual member of the family, and may get into power struggles or overreact emotionally to affect, content, or personality. The IP's dependency, relationship suction and repulsion, manipulativeness, denial, impulsivity, and family role abandonment may readily provoke countertransference reactions in the therapist. However, family therapists view their emotional reactions to families in a systems framework as well as a countertransference context. Thus they must be aware of how families will replay their problems in therapy by attempting to detour or triangulate their problems onto the therapist. The therapist must be particularly sensitive to the possibility of becoming an enabler who, like the family, protects or rejects the substance abuser.

the role of interpretation—Interpretations can be extremely helpful if they are made in a complementary way, without blaming, guilt induction, or dwelling on the hopelessness of longstanding, fixed patterns. Repetitive patterns and their maladpative aspects for each family member can be pointed out, and tasks can be given to help change these patterns. Some families need interpretations before they can fulfill tasks. An emphasis on mutual responsibility when making any interpretation is an example of a beneficial fusion of structural and psychodynamic therapy.

overcoming resistance—Resistance is defined as behaviors, feelings, patterns, or styles that prevent change. In substance-abusing families, key resistance behaviors that must be dealt with involve the failure to perform functions that enable the abuser to stay "clean."

Every substance-abusing family has characteristic patterns of resistant behavior, in addition to individual resistances. This family style may contribute significantly by resistance; some families may need to deny all conflict and emotion, and are almost totally unable to tolerate any displays of anger or sadness; others may overreact to the slightest disagreement. It is important to recognize, emphasize, and interpret the circumstances that arouse resistance patterns.

Bowen's Systems Family Therapy. In Bowen's (1974) approach, the cognitive is emphasized and the use of affect is minimized. Systems theory focuses on triangulation, which implies that whenever there is emotional distance or conflict between two individuals, tensions will be displaced onto a third party, issue, or substance. Drugs are frequently the subject of triangulation.

Behavioral Family Therapy. This approach is commonly used with substance-abusing ADOLESCENTS. Its popularity may be attributed to the fact that it can be elaborated in clear, easily learned steps.

Noel and McCrady (1984) developed seven steps in the therapy of alcoholic couples that can readily be applied to married adult drug abusers and their families:

1. Functional analysis. Families are taught to understand the interactions that maintain drug abuse.
2. Stimulus control. Drug use is viewed "as a habit triggered by certain antecedents and maintained by certain consequences." The family is taught to avoid or change these triggers.
3. Rearranging contingencies. The family is taught techniques to provide reinforcement for efforts at achieving a drug-free state by frequent reviewing of positive and negative consequences of drug use and self-contracting for goals and specific rewards for achieving these goals.
4. Cognitive restructuring. IPs are taught to modify self-derogatory, retaliatory, or guilt-related thoughts. They question the logic of these "irrational" thoughts and replace them with more "rational" ideation.
5. Planning alternatives to drug use. IPs are taught techniques for refusing drugs through role-playing and covert reinforcement.
6. Problem solving and assertion. The IP and family are helped to decide if a situation calls for an assertive response and then, through role-playing, to develop effective assertive techniques. IPs are to perform these techniques twice daily and to utilize them in situations that would have previously triggered the urge to use drugs.
7. Maintenance planning. The entire course of therapy is reviewed, and the new armamentarium of skills is emphasized. IPs are encouraged to practice these skills regularly as well as to reread handout materials that explain and reinforce these skills.

Families can also be taught through behavioral techniques to become aware of their nonverbal communication, so as to make the nonverbal message concordant with the verbal and to learn to express interpersonal warmth nonverbally as well as verbally.

FAMILY READJUSTMENT AFTER CESSATION

Once the substance abuse has stopped, the family may enter a honeymoon phase in which major conflicts are denied. They may maintain a superficial harmony based on relief and suppression of negative feelings. When the drug-dependent person stops using drugs, however, other family problems may be uncovered, particularly in the parents' marriage or in other siblings. These problems, which were present all along but obscured by the IP's drug use, will be "resolved" by the IP's return to symptomatic behavior if they are not dealt with in family therapy. In the latter case, the family reunites around their problem person, according to their old, familiar pathological style.

Too many treatment programs in the substance-abuse field focus their efforts on brief, high-impact treatment, neglecting aftercare. Many of these programs include a brief, intensive family educational and therapeutic experience, but have even less focus on the family in aftercare than on the IP. These intensive, short-term programs have great impact on the family system, but only temporarily. The pull of the family homeostatic system will draw the IP and/or other family members back to symptomatic behavior. The family must be worked with for months, and often years, after substance abuse first abates if a drug-free state is to continue. In addition, ongoing family therapy is necessary for the emotional well-being of the IP and other family members.

GROUP THERAPY

Group therapy varies with each of the three phases in the psychotherapy of substance abusers: achieving abstinence, early SOBRIETY, and late sobriety (achieving intimacy).

Early Phase: Achieving Abstinence. In the first phase of psychotherapy, the type of group utilized will depend on the treatment setting: hospital, residential, intensive outpatient (also termed partial hospitalization), or limited outpatient.

In hospital settings, educational groups are an essential part of the early treatment process, and the subjects covered in these groups are quite similar to those in educational family groups (described in the first section of this article). The major difference of emphasis in patient educational groups is on the physiological aspects and risk factors of drugs and alcohol. Other important didactic groups cover in detail issues such as (1) ASSERTIVENESS TRAINING; (2) other compulsive behaviors, such as sexuality, eating, working, and GAMBLING; (3) RELAPSE PREVENTION; (4) the prolonged abstinence syndrome; (5) leisure skills; and (6) cross addiction. All educational groups include appropriate coping strategies, some of which are developed from the experiences of recovering members.

One advantage of 28-day residential programs (now more often 7 to 21 days, followed by an intensive 6-hours-a-day outpatient program) is that group therapy can be started immediately after drinking or drug use stops. In the first few sober days, the addict or alcoholic is so needy that his/her resistance to groups is low. At this stage, the therapist and the group should show the substance abuser how to borrow the confidence that life without alcohol or drugs is possible and better than life with it. This hope is best offered by a therapist or cotherapist who is a recovering substance abuser with solid sobriety. Therapeutic groups in these settings will also deal with appropriate expressions of feelings, relationships with significant others, childhood molestation and abuse, building self-esteem, and development of strategies for self-care.

A critical aspect of early group therapy is for the patient to experience the sharing of a group of individuals struggling against their addiction. This helps to overcome the feelings of isolation and shame that are so common in these patients. The formation of a helping, sober peer group that provides support for a lifetime, in and out of twelve-step groups, is very helpful and dramatic when it occurs.

In outpatient programs there is less of an opportunity to perform uncovering therapy in the early phases because there is less protection and less of a holding environment than in residential settings.

Others, particularly Woody et al. (1986), have developed detailed group therapy techniques for methadone patients. Also, Brown and Yalom (1977) and Vanicelli (1992), with alcoholics, and Khantzian et al. (1990), with cocaine addicts, have adapted psychodynamic techniques for group work.

Ex-addicts and recovering alcoholics are valuable as cotherapists, or even as primary or sole therapist, particularly in the early stages of groups. Commonality of experience with the client, by itself, does not qualify an individual to be a therapist. Recovering persons should have at least two years of sobriety before they are permitted to function as group therapists. The techniques that help ex-addicts become experienced therapists are best learned gradually and under close supervision, preferably by experienced paraprofessionals and professionals.

Also helpful in cotherapy is male-female pairing, which provides a balance of male and female role models and transference.

During the early sessions of group therapy with substance abusers, the focus is on the shared problem of drinking or drug use, and its meaning to each individual. The therapist should be more active in this phase, which should be instructional and informative as well as therapeutic.

Alcoholics tend toward confessionals and monologues about prior drinking. These can be politely interrupted or minimized by a ground rule of "no drunkalogues." Romanticizing past use of drugs or alcohol is strongly discouraged.

Outpatient Groups. The desire to drink or use drugs and the fear of slipping are pervasive, early concerns in outpatient groups. The patient's attitude is one of resistance and caution, combined with fear of open exploration. Members are encouraged to participate in AA and other relevant twelve-step groups, yet the "high support, low conflict, inspirational style" of AA may inhibit attempts at interactional therapy. Therapists should not be overly protective and prematurely relieve the group's anxiety because this fosters denial of emotions. On the other hand, the members' recognition of emotions and responsibility must proceed slowly because both are particularly threatening to substance abusers. Patients are superficially friendly, but do not show real warmth or tenderness. AA-type hugs are an easy way to begin to show physical support. They are afraid to express anger or to assert themselves. However, sudden irritation, antipathy, and anger toward the leaders and other members inevitably begin to become more overt as the group progresses.

Gradually, tentative overtures of friendship and understanding become manifest. There may be a conspiracy of silence about material that members fear could cause discomfort or lead to drug use or drinking. The therapists can point out to the members that they choose to remain static and within comfortable defenses rather than expose themselves to the discomfort associated with change. Patients usually drop out early if they are still committed to using drugs or drinking. Other patients who drop out early do so because they grow increasingly alarmed as they become aware of the degree of discomfort that any significant change requires.

Middle Phase: Early Sobriety. In the middle phase of group therapy, the emphasis is quite similar to that of individual therapy. Therapists should continue to focus on cognitive behavioral techniques to maintain sobriety. Intensive affects are abreacted toward significant persons outside of the group but are minimized and modulated between group members. In this stage there evolves a beginning awareness of the role of personality and social interactions in the use of drugs and alcohol. Alcoholics are ambivalent about positive feedback. They beg for it, yet reject it when it is given. They repeatedly ask for physical reassurance, such as a warm hug, but may panic when they receive it because of fear of intimacy and a reexperiencing of their unmet past needs. There is a fear of success and a dread of competing in life as well as in the group. Success means destroying the other group members (siblings) and loss of therapist (parent).

Alcoholics are reluctant to explore fantasies because the thought makes them feel as guilty as the act. They view emotions as black or white. This makes them withhold critical comments because they fear their criticism will provoke upset and the resumption of drinking in other members. This withholding may be conscious or unconscious. Rage has been expressed either explosively or not at all. Its expression in the middle phase of group should be encouraged, but gradually and under slowly releasing controls.

The other crucial affect that must be dealt with is depression. There is an initial severe depression, which occurs immediately after detoxification. It appears to be severe but usually remits rapidly, leaving the substance abuser with a chronic, low-grade depression—frequently expressed by silence, lack of energy, and vegetative signs. These patients should be drawn out slowly and patiently. Ultimately, they are encouraged to cry or mourn, and a distinction is made between helping them deal with despair as opposed to rushing to take it away from them.

The success of the middle phase of group therapy with substance abusers depends on the therapist's and the group's ability to relieve anxiety through support, insight, and the use of more adaptive, concrete ways of dealing with anxiety. Alcohol and drugs

must become unacceptable solutions to anxiety. In this vein, it is important not to end a session with members in a state of grossly unresolved conflict. This can be avoided by closure when excessively troubling issues are raised. Closure can be achieved by the group's concrete suggestions for problem solution. When this is not possible, group support, including extragroup contact by members, can be offered. Brown and Yalom (1977) utilize a summary of the content of each group that is mailed to members between sessions and helps provide closure and synthesis.

Final Phase: Late Sobriety. In the final phase of therapy, substance abusers express and work through feelings, responsibility for behavior, interpersonal interactions, and the functions and secondary gain of drugs and alcohol. In this phase, reconstructive group techniques as practiced by well-trained professionals are extremely helpful and essential if significant shifts in ego strength are to be accomplished. Here, the substance abuser will become able to analyze defenses, resistance, and transference. The multiple transferences that develop in the group are recognized as "old tapes" that are not relevant to the present. Problems of sibling rivalry, competition with authority, and separation anxiety become manifest in the group, and their transference aspects are developed and interpreted. Conflicts are analyzed on both the intrapsychic and interpersonal levels. Ventilation and catharsis take place, and may be enhanced by group support. Excessive reliance on fantasy is abandoned.

Alcoholics who survive a high initial dropout rate stay in groups longer than neurotic patients, and thus a substantial number of middle-phase alcoholics will reach this final phase. By the closing phase, the alcoholic has accepted sobriety without resentment and works to free himself or herself from unnecessary neurotic and character problems. He or she has developed a healthy self-concept, combined with empathy for others, and has scaled down inordinate demands on others for superego reassurance. He or she has become effectively assertive rather than destructively aggressive and has developed a reasonable sense of values. More fulfilling relationships with spouse, children, and friends can be achieved.

When members leave the group, the decision to leave should be discussed for several weeks before a final date is set. This permits the group to mourn the lost member and for the member to mourn the group. This is true regardless of the stage of the group, but the most intense work is done in the later phases. In open-ended groups, the leadership qualities of the graduating member are taken over by others, who then may apply these qualities to life outside the group.

By the time substance abusers have reached this phase, they act like patients in highly functioning neurotic groups. Other forms of group treatment combine the principles of group and family work, such as multiple family group treatment and couples groups.

Multiple Family Group Treatment (MFGT). This is a technique that can be used in any treatment setting for substance abusers but is most successful in hospital and residential settings, where family members are usually more available. In a residential setting, the group may be composed of all of the families or separated into several groups of three or four closely matched families. Most MGFTs now include the entire community because this provides a sense of the entire patient group as a supportive family. In residential settings these groups are held weekly for two or three hours. In hospitals, a family week or weekend is often offered as an alternative or adjunct to a weekly group.

Couples Groups. There are two types of couples groups: one for the parents of young substance abusers and one for the significant other and the substance abuser.

Couples often have difficulty dealing with the role of their own issues in family or other couple therapy dysfunction when the children are present. This boundary is generally appropriate, and thus ongoing couples groups should be an integral part of any family-based treatment program.

When the presenting problem of substance abuse is resolved, content shifts to marital problems. It is often at this point that parents want to leave the MFGT and attend a couples group. In a couples group, procedures are reversed. Couples should not speak about their children but, rather, focus on the relationship between themselves. If material is brought up about the children, it is allowed only if it is relevant to problems that the couples have.

Couples must support each other while learning the basic tools of communication. When one partner gives up substance misuse, the nonusing partner must adjust the way he or she relates to the formerly using partner. There are totally new expectations and demands. Sex may have been used for exploitation and pacification so often that both partners have given up hope of resuming sexual relations and have stopped serious efforts toward mutual satisfaction. In addition, drugs and alcohol may have physiologically diminished the sex drive. Sexual communication must be slowly redeveloped. Difficulties may arise because the recovering abuser has given up the most precious thing in his or her life (drugs or alcohol) and expects immediate rewards. The spouse has been "burned" too many times (and is unwilling to provide rewards when sobriety stabilizes the spouse) to trust one more time; at the same time the recovering abuser is asked to reevaluate expectations for trust.

Couples groups in an adult or an adolescent program provide a natural means for strengthening intimacy. Spouses are encouraged to attend Al-Anon, Naranon, Coanon, and Coda to help diminish their reactivity and enhance their coping and self-esteem.

Couples groups have been used even more widely with alcoholics than with drug abusers, and the techniques are similar to those described above. Spouses of alcoholics are encouraged to attend Al-Anon, which facilitates an attitude of loving detachment.

Many studies have demonstrated that spousal involvement facilitates the alcoholic's participation in treatment and aftercare. It also increases the incidence of sobriety and enhanced function after treatment. Further, the greater the involvement of the spouse in different group modalities (Al-Anon, spouse groups, etc.), the better the prognosis for treatment of the alcoholic.

(SEE ALSO: *Causes of Substance Abuse; Comorbidity and Vulnerability; Contingency Contracts; Families and Drug Use; Sobriety; Toughlove*)

BIBLIOGRAPHY

ABLON, J. (1974). Al-Anon family groups. *American Journal of Psychotherapy, 28*, 30–45.

ANDERSON, C. M., & STEWART, S. (1983). *Mastering resistance: A practical guide to family therapy.* New York: Guilford Press.

BERENSON, D. (1976). Alcohol and the family system. In P. J. Guerin (Ed.), *Family therapy.* New York: Gardner.

BOWEN, M. (1974). Alcoholism as viewed through family

systems therapy and family psychotherapy. *Annals of the New York Academy of Sciences, 233*, 114.

BROWN, S., & YALOM, I. D. (1977). Interactional group therapy with alcoholics. *Journal of Studies on Alcohol, 38*, 426–456.

CADOGAN, D. A. (1973). Marital group therapy in the treatment of alcoholism. *Quarterly Journal of Studies on Alcohol, 34*, 1187–1197.

CAHN, S. (1970). *The treatment of alcoholics: An evaluative study.* New York: Oxford University Press.

FOX, R. (1962). Group psychotherapy with alcoholics. *International Journal of Group Psychotherapy, 12*, 56–63.

HOFFMAN, H., NOEM, A. A., & PETERSEN, D. (1976). Treatment effectiveness as judged by successfully and unsuccessfully treated alcoholics. *Drug and Alcohol Dependence, 1*, 241–246.

JOHNSON, V. E. (1980). *I'll quit tomorrow* (rev. ed.). San Francisco: Harper & Row.

KAUFMAN, E. (1994). *Psychotherapy of addicted persons.* New York: Guilford Publications.

KAUFMAN, E. (1985). *Substance abuse and family therapy.* New York: Grune & Stratton.

KAUFMAN, E. (1982). Group therapy for substance abusers. In M. Grotjahn, C. Friedman, & F. Kline (Eds.), *A handbook of group therapy.* New York: Van Nostrand Reinhold.

KAUFMAN, E., & KAUFMAN, P. (1992). *Family therapy of drug and alcohol abuse* (2nd ed.). Boston: Allyn & Bacon.

KAUFMAN, E., & KAUFMAN, P. (1979). *Family therapy of drug and alcohol abuse.* New York: Gardner.

KHANTZIAN, E. J., HALLIDAY, D. S., & McAULIFFE, W. E. (1990). *Addiction and the vulnerable self.* New York: Guilford Press.

McCRADY, B., ET AL. (1986). Comparative effectiveness of three types of spousal involvement in outpatient behavioral alcoholism treatment. *Journal of the Studies of Alcohol, 14*(6), 459–467.

MINUCHIN, S. (1974). *Families and family therapy.* Cambridge, MA: Harvard University Press.

NOEL, N. E. & McCRADY, B. (1984). Behavioral treatment of an alcohol abuser with a spouse present. In E. Kaufman (Ed.), *Power to change: Family case studies in the treatment of alcoholism.* New York: Gardner.

STANTON, M. D., ET AL. (1982). *The family therapy of drug abuse and addiction.* New York: Guilford Press.

VANICELLI, M. (1992). *Removing the roadblocks.* New York: Guilford Press.

WOODY, G. E., ET AL. (1986). Psychotherapy for substance abuse. *Psychiatric Clinics North America, 9*, 547–562.

WRIGHT, K. D., & SCOTT, T. B. (1978). The relationship of wives' treatment to the drinking status of alcoholics. *Journal of Studies on Alcohol, 39*, 1577–1581.

YALOM, I. D., ET AL. (1978). Alcoholics in interactional group therapy. *Archives of General Psychiatry, 35*, 419–425.

EDWARD KAUFMAN

Hypnosis Hypnosis is a normal state of attentive, focused concentration with a relative suspension of peripheral awareness, a shift in attention mechanisms in the direction of focus at the expense of the periphery. Being hypnotized is something like looking through a telephoto lens. What is seen, is seen in great detail, but at the expense of context. The use of hypnosis has been associated with inducing a state of relaxation and comfort, with enhanced ability to attend to a therapeutic task, with the capacity to reduce pain and anxiety, and with heightened control over somatic function. For these reasons, hypnosis has been used with some benefit as an adjunct to the treatment of certain kinds of DRUG and ALCOHOL ABUSE and ADDICTION.

Therapeutic approaches involving hypnosis include using it as a substitute for the pleasure-inducing substance, taking a few minutes to induce a self-hypnotic state of relaxation (for example, by imaging oneself floating in a bathtub or a lake, or visualizing pleasant surroundings on an imaginary screen). In this strategy the hypnosis is a safe substitute for the pleasure-inducing effects of the drug. A second approach involves ego-enhancing techniques, providing the subject with encouragement, picturing himself or herself living well without the substance and able to control the desire for it. A third approach involves instructing subjects to reduce or eliminate their craving for the drug. A fourth involves cognitive restructuring, diminishing the importance of the craving for the drug by focusing instead on a commitment to respect and protect the body by eliminating the damaging drug. One widely used technique for smoking control, for example, has people in hypnosis repeat to themselves three points: (1) For my body, smoking is a poison; (2) I need my body to live; (3) I owe my body respect and protection. This approach places an emphasis on a positive commitment to what the person is for, rather than paying attention to being against the drug, thereby keeping attention on protection rather than on abstinence.

Hypnosis has been most widely used in the treatment of NICOTINE dependence, and although the results vary, a number of large-scale studies indicate that even a single session of training in self-hypnosis can result in complete abstinence of six months or more by approximately one out of four smokers.

There are fewer systematic data regarding use of hypnosis with COCAINE, OPIATE, or alcohol addiction. The success of the approach is complicated by the fact that the acute effects of substance intoxication and/or the chronic effects on cognitive function of alcohol and other drug abuse hampers hypnotic responsiveness, thereby diminishing the potential of addicted individuals to enter this state and benefit from it. Nonetheless, there may be occasional individuals who are sufficiently hypnotizable and motivated to use this approach as an adjunct to other treatment, diminishing the dysphoria and discomfort that can accompany WITHDRAWAL and abstinence while enhancing and supporting their commitment to a behavior change. Hypnosis can be used by licensed and trained physicians, psychologists, dentists, and other health-care professionals who have special training in its use. The treatment is employed in offices and clinics as well as in hospital settings. It should always be used as an adjunct to a broader treatment strategy.

Hypnosis is a naturally occurring mental state that can be tapped in a matter of seconds and mobilized as a means of enhancing control over behavior, as well as the effects of withdrawal and abstinence, in motivated patients supervised by appropriately trained professionals.

DAVID SPIEGEL

Long-Term Approaches The various ways of helping people with alcohol and other substance-abuse disorders to achieve and maintain long-term behavior changes are described here. Treatment approaches can be considered from several perspectives: (1) the substances at which they are directed (e.g., alcohol, cocaine, opiates, tobacco); (2) the population(s) at which they are targeted (e.g., youth, the elderly, women, IV-drug users, the severely dependent); (3) the setting in which services are delivered

(e.g., general health practitioners' offices; specialized addiction services including outpatient counseling, residential programs, halfway house programs, inpatient hospital programs, and therapeutic communities). They also can be classified by their underlying rationale, which is usually related to their procedures. Pharmacological approaches may be based on hypotheses that drug problems either occur because of, or result in, biochemical imbalances, with medication intended to address the imbalance. Approaches grounded in psychological learning theory may postulate that people have learned to use drugs as an inappropriate way of coping with problems; treatment focuses on how to handle the problems without drugs. In this article approaches are considered from the perspective of how they are thought to result in behavior change.

Different approaches may be appropriate at different stages of the recovery process. Recovery involves at least three phases: readiness for change, achieving change, and maintaining change. A long-term treatment may need to start by increasing the individual's motivation for change. Motivational interviewing is one approach that helps people decide that change is necessary. If the person is committed to change but has not yet changed, then an approach that helps accomplish change would be appropriate (e.g., social-skills training). Once change is achieved, relapse frequently occurs among substance-use disorders. Approaches such as RELAPSE PREVENTION are aimed at maintaining recovery.

APPROACHES INTENDED TO INCREASE MOTIVATION TO CHANGE

Approaches intended to increase a person's desire to change first became a focus of research in the 1980s. The best-known approach, motivational interviewing, is used with persons who are in conflict about whether they should change or who are not strongly committed to change. First developed for use with alcohol abusers, it is basically a set of guidelines instructing the therapist how to use a nonthreatening, supportive interview style to help people examine and evaluate their drinking or drug use and related consequences, and to guide the person to a decision that stopping or reducing use is essential. The objective is for the client to recognize and become committed to the need for change. Other procedures may then be used to help the person change, but many people change their behavior without help. This means that in some cases simply increasing motivation might be sufficient to produce change. Behavioral treatments that emphasizes self-change (i.e., the client's taking major responsibility for formulating and implementing the treatment plan) often include motivational procedures such as allowing clients to select their treatment goals.

A different motivational approach, used by programs such as ALCOHOLICS ANONYMOUS and similar groups, is based on a traditional disease approach. The approach is aimed at persuading the individual that he or she lacks any potential for control over substance use and must accept the need for a permanently drug-free lifestyle, with social support from similarly afflicted individuals.

Some pharmacotherapies can be viewed as intended to increase motivation—for example, by decreasing the perceived need to drink or use drugs or by attaching aversive consequences to alcohol or drug consumption. At present, medications to decrease the desire to drink or use drugs are a topic of research. It has been speculated that individuals in the presence of strong drinking or drug-use cues (e.g., HEROIN addicts in the presence of other addicts who are injecting) may experience a desire to use based on a learned association between the cues and substance use in the past. Since learned associations presumably have a neurochemical basis, medication theoretically could disrupt the association and thereby avoid the increased desire to use. A way of disrupting the association without using medication is by cue-exposure (i.e., extinction) procedures in which the cues are presented in the absence of substance use. This might be thought of as teaching the individual to resist temptation. It is undetermined, however, whether medications or cue exposure would be useful with severely dependent individuals who have a strong learned response to drug cues or with persons whose problems are not severe and for whom the effects of drug cues are weak and might be more easily disrupted. Another way medications might eventually be useful is if compounds are found that reduce or block the neurological reward properties of alcohol. Much research is presently seeking to discover medications that can be effective treatments for substance use disorders.

Other pharmacotherapies that might be considered as affecting motivation are agents that produce an unpleasant reaction if the target substance is ingested. DISULFIRAM (Antabuse) and citrated calcium carbamide (Temposil, not available in the U.S.) make a person ill if alcohol is ingested. Since people who effectively use antialcohol drugs rarely consume alcohol, the medications' effectiveness appears to be in their deterrent function rather than in the aversive reactions. Presumably, a person's desire to drink is reduced if he or she knows that drinking will make him or her ill. NALTREXONE, a pharmacological antagonist to heroin, blocks the reward effects of the drug. Although naltrexone works differently than antialcohol drugs, it shares a practical problem in that abusers often discontinue use after a short time. Presently, no blocking agents for stimulant drugs have been discovered.

While some might not think of Alcoholics Anonymous (AA) as a "treatment," its goal is to help people accomplish long-term behavior change. A central tenet of AA is that for people to overcome a drinking problem, they must first admit to being powerless over their drinking (i.e., they cannot drink like a "normal" drinker). This and other aspects of AA (e.g., making a list of wrongs done to others by one's drinking) can be considered as intended to increase commitment to change, and thus to increase motivation. Similar self-help groups (e.g., NARCOTICS ANONYMOUS, COCAINE ANONYMOUS) take a similar approach.

Once change has occurred, maintaining the change can be considered a motivational issue. Relapse prevention is a cognitive approach (i.e., it focuses on thought processes) to maintaining change that encourages people to take a long-term perspective on recovery and to stay committed to their long-term objective even if they experience problems. This is important because alcohol and drug problems tend to be recurrent. This approach is intended to help people deal constructively with relapses. When people who have not used drugs for some time experience a relapse, they may interpret the relapse as an indication that they are unable to change, and consequently continue abusing drugs. This can be viewed as a self-fulfilling prophecy. Relapse prevention teaches clients that although relapses are unfortunate, they are not uncommon among persons trying to recover from alcohol or drug problems. Relapse prevention encourages (1) viewing relapse as regrettable events from which something can be learned, (2) overcoming the relapse as soon as possible, and (3) continuing to strive toward the long-term objective of recovery.

APPROACHES THAT ENCOURAGE PEOPLE TO TAKE RESPONSIBILITY FOR GUIDING THEIR OWN CHANGE

A recent development has been self-change approaches aimed at helping people help themselves. The hallmark of such approaches is the idea that people can solve their own problems. Treatment which tends to be short-term and outpatient, typically uses reading materials and homework assignments, and provides clients with considerable choice in how they deal with their problems. For alcohol, self-change approaches for persons whose problems are not severe have typically included allowing clients to choose their drinking goal (i.e., reduced drinking or abstinence). Although there are elements of this approach in other treatments, this discussion relates to approaches in which empowerment of the individual is the central treatment focus.

APPROACHES INTENDED TO SUBSTITUTE FOR DRUG EFFECTS

Some approaches are based on the hypothesis that drug use serves various purposes for the individual besides producing an intoxicated state. For example, it has long been thought that people learn to drink excessively as a way of coping with stressful situations; alcohol is thought to lessen bad feelings and to disinhibit the individual so that behaviors which would otherwise be held in check (e.g., aggression) can be expressed. Consequently, treatment focuses on providing users with alternative ways of accomplishing the functions served by drugs. Often these treatments involve (1) identifying factors or situations that set the stage for substance use, (2) identifying the outcomes achieved by substance use, and (3) providing training in alternative ways of successfully dealing with those situations. Thus, people who say they use drugs in order to feel comfortable in social situations might be given social-skills or relaxation training. This may involve talking about the situations and practicing behaviors in treatment sessions, as well as carrying out assignments outside of treatment.

A pharmacological approach that seeks to substitute for drug effects is relevant to cases in which both substance use and psychiatric disorders are present and there is reason to think the individual is using drugs to self-medicate. In such cases medications, such as antidepressants, can be used as treatment for both disorders. When the psychiatric disorder is one in which medications can induce dependence (e.g., tranquilizers for anxiety disorders), medication use must be carefully monitored.

Another type of substitution approach is demonstrated by the use of methadone, an opiate that is longer-acting than heroin and interferes less with a person's daily functioning. Heroin addicts who are transferred to methadone can obtain their opiates legally (by prescription), which is thought to diminish criminal activities related to acquiring money to buy heroin. Although use of methadone was first proposed as a "maintenance" treatment that would be used indefinitely, many programs have disregarded this notion and have used decreasing doses of methadone with the objective of totally withdrawing the individuals from opiate use.

An approach that is similar to substituting for drug effects is the use of NICOTINE GUM or nicotine patches to treat nicotine dependence. These methods change how nicotine enters the body (orally or through the skin rather than smoking), with a long-term aim of withdrawing the individual from nicotine altogether. Hypothetically, these alternative forms of nicotine delivery could be used indefinitely as a maintenance treatment, since the main risks from smoking are from factors other than nicotine (e.g., tar).

APPROACHES INTENDED TO REDUCE OR STOP DRUG USE INDIRECTLY

Psychotherapeutic approaches to treating substance-use disorders are based on the notion that substance use occurs as a symptom of an individual's unresolved conflicts associated with his or her personality development. The basic assumption is that symptoms will not disappear until the underlying conflicts are identified and resolved, although the type of symptoms could change. Approaches based on resolution of underlying conflicts typically involve therapy sessions conducted weekly or more frequently over several months or even years. One reason they have not been particularly popular as treatments for alcohol or drug problems may be that the substance use is not dealt with directly.

Treatment of persons having coexisting psychiatric and substance-use disorders can be considered as intended to affect the substance use indirectly if it concentrates on resolving the psychiatric disorder, with an assumption that the substance-use disorder will stop if the psychiatric problem is resolved. This might occur when there is a functional relationship between the two disorders (e.g., a person drinks excessively to deal with anxiety). Whether substance use and other psychiatric disorders are causally related, it has been found that serious psychiatric problems are associated with a poor response to treatment for substance abuse.

APPROACHES THAT REARRANGE THE ENVIRONMENT TO PREVENT OR DISCOURAGE SUBSTANCE USE

If environmental factors, such as a person's social group, living conditions, leisure activities, or work, play a role in maintaining substance use, then changing those factors may enable the individual to recover. Community reinforcement treatment seeks to get significant others in the client's environment to establish rules concerning how they will act when the client uses drugs. For example, they may refuse to spend time with the client if he or she is intoxicated. If a client's environment is not supportive of recovery, it may be helpful for the client to move or to enter a residential treatment program. For alcohol abusers the latter tend to be 28-day MINNESOTA MODEL programs that are based on a disease model of alcoholism and use educational and counseling activities and AA. Such programs usually involve aftercare meetings as well as attendance at AA meetings for several months after the residential stay. The most popular residential treatment programs for drug abuse other than alcohol are therapeutic communities, such as Synanon, which began in 1958. These programs typically require a one-to-two-year residence or longer, use confrontational and group therapies, enforce rigid codes of behavior, and require acceptance of the need for a drug-free lifestyle. Given the extent to which participating in these programs will intrude into an individual's life, the programs are generally seen as most appropriate for persons with severe drug problems.

Another approach that can be considered rearranging the individual's environment is through family interventions, which typically means involving family members in the user's treatment sessions. Family approaches have especially been used with adolescents.

TREATMENT AS IT TYPICALLY IS PROVIDED

This article has reviewed a variety of treatment approaches to achieving long-term behavior change by persons with alcohol or drug problems. To do this, it has drawn largely on the scientific and clinical research literature. Unfortunately, in actual practice many alcohol and drug treatment programs lack a basis in any theory of behavior change; also, some treatments combine multiple approaches. As in seeking other types of health care, it is valuable for prospective clients to inquire about a program's approach and its basis.

(SEE ALSO: *Addiction: Concepts and Definitions; Alcoholism; Comorbidity and Vulnerability. Disease Concept of Alcoholism and Drug Abuse; Methadone Maintenance Programs; Wikler's Pharmacologic Theory of Drug Addiction*)

BIBLIOGRAPHY

COOK, C. C. H. (1988). The Minnesota model in the management of drug and alcohol dependency: Miracle, method or myth? Part I. The philosophy and the programme. *British Journal of Addiction, 83*, 625–634.

HEATHER, N., & ROBERTSON, I. (1983). *Controlled drinking.* New York: Methuen.

HESTER, R. K., & MILLER, W. R. (EDS.). (1990). *Handbook of alcoholism treatment approaches: Alternative approaches.* New York: Pergamon.

KHANTZIAN, E. J. (1980). The alcoholic patient: An overview and perspective. *American Journal of Psychotherapy, 34*, 4–19.

KOSTEN, T. R., & KOSTEN, R. T. (1991). Pharmacological blocking agents for treating substance abuse. *Journal of Nervous and Mental Disease, 179*, 583–592.

LITTEN, R. Z., & ALLEN, J. P. (1991). Pharmacotherapies for alcoholism: Promising agents and clinical issues. *Alcoholism: Clinical and Experimental Research, 15*, 620–623.

MARLATT, G. A., & GORDON, J. R. (EDS.). (1985). *Relapse prevention.* New York: Guilford Press.

McLELLAN, A. T. (1986). "Psychiatric severity" as a predictor of outcome from substance abuse treatments. In R. E. Meyer (Ed.), *Psychopathology and addictive disorders.* New York: Guilford Press.

MILLER, W. R., & ROLLNICK, S. (1991). *Motivational interviewing: Preparing people to change addictive behavior.* New York: Guilford Press.

MONTI, P. M., ET AL. (1989). *Treating alcohol dependence.* New York: Guilford Press.

PLATT, J. J., HUSBAND, S. D., & TAUBE, D. (1990–1991). Major psychotherapeutic modalities for heroin addiction: A brief overview. *International Journal of the Addictions, 25,* 1453–1477.

RAWSON, R. A. (1991). Chemical dependency treatment: The integration of the alcoholism and drug addiction/use treatment systems. *International Journal of the Addictions, 25,* 1515–1536.

SOBELL, M. B., & SOBELL, L. C. (1993). *Problem drinkers: Guided self-change treatment.* New York: Guilford Press.

MARK B. SOBELL
LINDA C. SOBELL

Minnesota Model   Origins of the Minnesota Model of drug abuse treatment are found in three independent Minnesota treatment programs: Pioneer House in 1948, HAZELDEN in 1949, and Wilmar State Hospital in 1950. The Hazelden Clinics are still in existence and are located in Minnesota and Florida. The original treatment programs recognized ALCOHOLICS ANONYMOUS (AA) as having success in bringing about recovery from ALCOHOLISM. Unique to this early stage of the Minnesota Model was the blending of professional behavioral science understandings with AA's principles. Important in the development of the Minnesota Model is the way treatment procedures emerged from listening to alcoholics, from trial and error, from acknowledgment of the mutual help approach of AA, and from the use of elementary assumptions rather than either a well-developed theoretical position or a generally accepted therapeutic protocol. In many ways, the Minnesota Model may be seen as having come about in a grassroots, pragmatic manner.

Because of its evolutionary, noncentralized development, the Minnesota Model is not a standardized set of procedures but an approach organized around a shared set of assumptions. These assumptions have been articulated by Dan Anderson, the former president of Hazelden Foundation and one of the early professionals working with the Minnesota Model at Wilmar State Hospital. They are the following: (1) Alcoholism exists in a consolidation of symptoms; (2) alcoholism is an illness characterized by an inability to determine time, frequency, or quantity of consumption; (3) alcoholism is nonvolitional—alcoholics should not be blamed for their inability to drink ethanol (alcohol); (4) alcoholism is a physical, psychological, social, and spiritual illness; and (5) alcoholism is a chronic primary illness—meaning, that once manifest, a return to nonproblem drinking is not possible. Although these assumptions are phrased as pertaining to alcoholism, early experience with the Minnesota Model demonstrated that drug abuse other than alcoholism can also be understood and treated within these assumptions. *Chemical dependency* is the term generally used by clients and treatment providers when referring to substance abuse. The Minnesota Model provides treatment for chemical dependency—for both alcohol and other drugs.

A twenty-four to twenty-eight day inpatient treatment stay, or approximately eighty-five hours in outpatient rehabilitation, characterizes the Minnesota Model treatment. Inpatient treatment may occur in hospital settings or free-standing facilities and may be run by for-profit or nonprofit organizations. Different treatment settings have different mixes of staff positions, but the multidisciplinary team of medical and psychological professionals plus clergy and focal counselors are frequently found—either in a close interacting network or a more diffuse working arrangement.

Primary focal counselors have either received specific training in the Minnesota Model approach to treatment or have learned their counseling skills in an apprenticelike placement. Most counselors are neither mental-health-degreed professionals nor holders of medically related degrees, but they are commonly working on their own twelve-step programs because of life experience with chemical dependency or other addictions. As in AA, this shared personal experience of both clients and counselors is important for the client/counselor relationship and the behavior modeling the counselor provides for the client.

Minnesota Model treatment programs vary in the centrality of counseling staff and the programmed autonomy of the treatment experience. Some treatment programs have the counselor facilitating the majority of the groups and visibly directing the treatment experience. Other programs have the treatment groups carrying out the treatment experience where the activity follows a prescribed format, but the group members are the visible actors while the counseling staff maintains a low profile as they seek to empower clients to acquire the insights and resources necessary for their recovery. Treatment also varies in the amount of confrontation, the presence of a family program requirement, the extent of assigned reading, the detail of client record documentation, and other attributes.

What Minnesota Model treatment has without exception is the use of AA principles and understandings (steps and traditions) as primary adjuncts in the treatment experience. Clients are provided with the AA "Big Book" (*Alcoholics Anonymous*) and *The TWELVE STEPS and Twelve Traditions*. Both of these books are required reading. Spirituality is emphasized as important to recovery, which is consistent with the AA understanding. AA group meetings occur in the schedule of rehabilitation activities, and clients may visit a community AA meeting as part of their treatment experience. Clients will work on AA steps during their treatment experience; some programs focus on the first five steps while others emphasize all twelve steps.

Treatment it not just an intensive exposure to AA. It motivates treatment participants to develop mutual trust and to share and be open about how the use of chemicals has come to control their lives. Clients are told that they have the disease of chemical dependency. Their behavior has been directed by the disease, but they have been unable to see the reality of their behavior and the consequences because of the disease characteristic of denial. Treatment plans are individualized based on assessments by the multidisciplinary staff. Generally, the first goal of treatment is to break the client's denial and the second goal is for the client to accept the disease concept. Because treatment has clients ranging from new admissions to those ready to complete their program, senior peers are very influential in helping clients who are in the early stages of treatment to understand denial and the DISEASE CONCEPT.

Acceptance and awareness that they are able to change if they take appropriate action to deal with their chronic condition is the message in the final treatment stage. The rehabilitation staff develops an aftercare plan with the client that will continue to support some of the changes that have taken place during treatment and it encourages changes that will promote ongoing recovery. Characteristically, clients comment on their increased awareness of simple pleasures and being with other people without trying

to manipulate them. They are told that they must continue to work the AA steps, attend AA meetings, and address other problems of living if they are going to experience recovery because primary treatment is just one part of an ongoing continuum of care. Recovery is hard work made even more difficult by possible bouts of depression, problems of regaining trust from their family, and establishing new friends and activities not tied to alcohol and drug use.

Treatment outcome studies carried out by Hazelden for their treatment clients and for ten treatment programs in the Hazelden Evaluation Consortium are in general agreement with outcome evaluation findings reported by Comprehensive Assessment and Treatment Outcome Research for approximately one hundred hospital and freestanding treatment programs throughout the United States. About 50 percent of all clients treated, including noncompleters, are abstinent for one year following treatment discharge. This percentage is higher for treatment completers and for clients having fewer complications and more stability in their lives. Thirty-three percent of the clients have returned to heavy use patterns within the year, and the remainder have had slips or a period of resumed drinking/use but also have sustained periods of abstinence. Abstinent clients have fewer legal, health, interpersonal, and job-related problems, and about 75 percent attend AA and/or continuing care.

The Minnesota Model is a label that is applied to a broad range of programming. Nevertheless, it represents a highly visible treatment modality serving a large number of clients throughout the United States, although it is more dominant in certain regions. It has a counterpart known as the Icelandic Model, and both of these treatment models have influenced treatment in SWEDEN and other parts of Scandinavia. International interest in adopting the Minnesota Model appears to be growing, with scattered treatment programs appearing in many countries. Little research has been done on the diffusion of this treatment model to other cultures.

(SEE ALSO: *Alcoholism; Treatment, History of*)

BIBLIOGRAPHY

ANDERSON, D. J. (1981). *Perspectives on treatment: The Minnesota experience.* Center City, MN: Hazelden Educational Services.

Cook, C. C. H. (1988). The Minnesota Model in the management of drug and alcohol dependency: Miracle, method or myth? Part I. The philosophy and programme. *British Journal of Addiction, 83,* 625–634.

Cook, C. C. H. (1988). The Minnesota Model in the management of drug and alcohol dependency: Miracle, method or myth? Part II. Evidence and conclusions. *British Journal of Addiction, 83,* 735–748.

Laundergan, J. C. (1982). *Easy does it: Alcoholism treatment outcomes, Hazelden and the Minnesota Model.* Center City, MN: Hazelden Educational Services.

J. Clark Laundergan

Nonmedical Detoxification The term *detoxification* is used to refer to the management of two distinct types of problem resulting from excessive ALCOHOL or other drug use. These are the symptoms and behavioral changes associated with extreme intoxication on the one hand and of WITHDRAWAL following extended use on the other. Although both involve recovering from the toxic effects of a drug while refraining from further use, the problems associated with each are quite different and require different methods to tackle them. In relation to Western society's favorite drug, alcohol, these problems are so common that the challenge is to develop methods that can be widely used without excessive cost. This requirement tends to rule out an exclusive reliance on expensive medical settings, medical personnel, and medication—even though both problems carry with them a small but significant risk of death or serious injury. Despite this restriction, human ingenuity has devised a number of relatively safe and cost-effective alternatives to hospital care, which are frequently preferred by the clients in need of "sobering up" or "drying out." These innovative services have usually been developed for people who run into problems with their use of alcohol; they have later been emulated by services for people who use other types of mood-altering drug.

The most visible problems associated with extreme intoxication concern public order, particularly in relation to the use of alcohol. Drunkenness is associated with violence both to the self and to others as well as with "public nuisance" offenses. The habitual drunken offender, who may otherwise be quite harmless, and the potentially dangerous disorderly "drunk" present themselves in huge numbers to police forces the world over and, typically, then clog up already overburdened court and penal systems. In the past two decades, several countries have experimented with having drunkenness decriminalized—that is, made no longer a criminal offense. The aim of this has been to free up the courts and the police so that they can concentrate on more serious crimes. Another impetus for decriminalization of drunkenness has been a growing awareness that locking up drunk people in police cells puts them at risk of serious harm. In Australia, for example, the tragic deaths of many Aboriginal people while in police custody are thought to have been caused by the combined effects of alcohol and confinement. In the United States, where holding cells have several occupants, drunks may be abused, involved in fights, or even killed by cellmates.

Historically, the setting up of nonmedical detoxification services occurred hand-in-hand with the decriminalization of drunkenness. Early experiments in the 1970s were undertaken by the ADDICTION RESEARCH FOUNDATION in the Canadian province of Ontario and St. Vincent's Hospital in New South Wales, Australia. In both cases, services were set up with the principal aim of diverting drunken offenders from the criminal-justice system to a more humane setting, where they might also be counseled to seek help for their drinking problems. Both utilized a residential social setting staffed by nonmedical personnel and provided no medical care or medication. To this day, they successfully supervise thousands of problem drinkers, mainly self-referred, through sobering-up and/or alcohol withdrawal, with an impressive record of safety. For example, in more than 10 years of operation, the New South Wales facility has dealt with nearly 14,000 admissions and recorded only 2 fatalities among this high-risk population. Only 1 percent have required transfer to a nearby hospital for specialized medical care, often for reasons unrelated to ALCOHOL WITHDRAWAL. These facilities have not been successful, however, in terms of attracting referrals from the police. In New South Wales, for example, the police have accounted for only 0.2 percent of referrals. These facilities may be diverting some potential offenders before they come to police attention.

SOBERING-UP SHELTERS

In an excellent review of detoxification services worldwide, Orford and Wawman (1986) suggest that the design of the above services has confused the problems of intoxication and withdrawal. They may be seen as highly successful and cost-effective alternatives to hospital care for alcohol withdrawal but *not* the solution for how society should treat the habitual drunken offender. Australia's continuing concern to prevent Aboriginal deaths in custody has also prompted an increasing use of what have come to be called *sobering-up shelters*. These provide supportive nonmedical settings, where people can stay a few hours or, if necessary, overnight—until they have literally sobered up. They have been found to provide an inexpensive alternative to prison and have succeeded in gaining the necessary support of the local police. Experience suggests that close liaison between shelter staff and police officers is necessary so that all concerned are clear about the specific aims of the project and how each can help the other. It is important that specialist treatment facilities be available to the sobering-up shelters, so that people requiring urgent medical attention or longer-term help with a drinking problem can be referred to them.

It should be noted that there are also potentially serious medical emergencies associated with extreme levels of drug intoxication. Poisoning through overdose, accidental or otherwise, is a common cause of admission to hospital emergency rooms, which all too frequently may result in death. The most common of such instances are deliberate acts of self-poisoning, usually with prescribed medication, closely followed by cases of accidental alcohol poisoning. Overdosing on HEROIN is also common, especially as a result of users having lost tolerance to the drug's effects after a period of abstinence or if they are unknowingly sold an unusually pure supply. For this reason, the staff of sobering-up shelters, or of any facility that also caters to drug users, should be trained to identify the warning signs of overdose—so that the sufferer may be taken to the hospital with as little delay as possible. Similarly, there is a great educational need among the general drug-using and drinking public who all too often abandon their friends to "sleep it off" and later find them asphyxiated.

DEALING WITH ALCOHOL AND OTHER DRUG WITHDRAWAL

Since the pioneering Canadian and Australian development of "social-setting" detoxification services to assist people safely through alcohol withdrawal, a variety of other nonmedical approaches have been developed. Really, detoxification services should be seen as being on a continuum ranging from supervision by an informed lay person—a relative, a recovered problem drinker or user, or nonmedical professionals—all the way to twenty-four-hour nursing and medical care in a specialist hospital unit. Even in the latter case, substantial variations exist regarding the amount of medication used during withdrawal—or even whether any medication is used at all. Detoxification services designed to minimize discomfort and the possibility of actual harm occurring during withdrawal may be nonmedical in several senses: by, variously, using nonmedical settings (e.g., hostels, the client's home), nonmedical personnel (e.g., relatives, ex-problem drinkers), or nonmedical procedures. There is wide consensus that medical assistance needs to be available if required, but the responsibility for accessing this need not be left only with medical personnel.

The Ontario model of nonmedical detoxification was created following the results of a study reported in 1970. It found in the relative safety of an alcoholism treatment unit that only 5 percent of admissions required any form of medical assistance. In addition to the residential social-setting model of detoxification, ambulatory (or outpatient) detoxification procedures were developed that relied on drinkers stopping in daily to a clinic to collect their medication and receive a brief checkup. Evaluations of these types of service conducted in several countries have demonstrated that their success rates in terms of both safety and effectiveness is at least the equal of inpatient care—and it is considerably cheaper.

A variation of this approach is home detoxification, an approach that has been most extensively researched in Great Britain with problem drinkers. This involves a community alcohol worker (e.g., nurse, counselor, or psychologist) assisting a family practitioner to assess a drinker who wishes to stop drinking alcohol but who may experience severe withdrawal symptoms in the process. Providing that the home environment is deemed to be supportive and that the client is sufficiently motivated to stop drinking, the detoxification then occurs in the patient's home with supportive visits from the alcohol worker. The family doctor's telephone number is provided to the client and any close relative or partner in case of emergency. A particular effort is made to screen out drinkers with a history of withdrawal seizures, DELIRIUM TREMENS (DTs), or KORSAKOFF'S SYNDROME.

To reduce the real risk of overdose with some types of medication (notably chlormethiazole), either the alcohol worker or a relative holds the medication. An important reason for developing this service in Britain was the discovery that many family doctors were already prescribing chlormethiazole to cover alcohol withdrawal but in the absence of any supervision and frequently longer than the recommended maximum period—sometimes even indefinitely. It was found that this was the single most common method of managing alcohol withdrawal among a group of patients who, for many reasons, were loathe to attend a psychiatric hospital or specialized treatment unit. Home detoxification may therefore offer a safe alternative to completely unsupervised withdrawal as well as a cost-effective alternative to inpatient hospital care. The cost of home detoxification per client has been estimated to be approximately a quarter that of inpatient hospital care. Formal evaluations of the British service suggest that not only is there no loss in terms of either safety or efficacy but that clients prefer to be treated at home and that many would refuse to attend a hospital facility.

CONCLUSIONS

Nonmedical detoxification services have been developed to cope with the problems associated with alcohol withdrawal in chronic heavy drinkers and also with episodes of alcohol-induced intoxication. While such services are being developed for users of other mood-altering drugs, there is, as yet, little published research concerning their efficacy. Nonmedical detoxification services need clear aims and objectives and should be part of a comprehensive range of services for people with alcohol problems. Both intoxication and alcohol withdrawal are so common in Western society that, although they carry a small but significant risk of serious injury or death, it is too costly to attempt to provide specialist medical care in every instance. In a number of countries, safe and inexpensive alternatives have been developed that are to be highly recommended over a laissez-faire or a punitive approach.

(SEE ALSO: *Alcohol; Public Intoxication; Temperance Movement; Treatment, History of, Treatment; Withdrawal*)

BIBLIOGRAPHY

ANNIS, H. (1985). Is inpatient rehabilitation of the alcoholic cost effective? *Advances in Alcohol and Substance Abuse, 5,* 175–190.

ORFORD, J., & WAWMAN, T. (1986). *Alcohol detoxification services: A review.* London: DHSS, HMSO.

PEDERSON, C. (1986). Hospital admissions from a nonmedical alcohol detoxification unit. *Alcohol and Drug Review, 5,* 133–137.

STOCKWELL, T., BOLT, E., & HOOPER, J. (1986). Detoxification from alcohol at home managed by general practitioners. *British Medical Journal, 292,* 733–735.

STOCKWELL, T., ET AL. (1991). Home detoxification for problem drinkers: Its safety and efficacy in comparison with inpatient care. *Alcohol and Alcoholism, 26*(2), 207–214.

TIM STOCKWELL

Outpatient versus Inpatient Although inpatient treatment (within a hospital or residential facility) continues to be a very common form of treatment for substance abuse, a number of studies show that outpatient treatment can be equally effective. For example, with regard to DETOXIFICATION from ALCOHOL (ethanol), research indicates that detoxification can be accomplished in most patients in (1) a hospital setting with minimal medical supervision (Whitfield et al., 1978), (2) in a nonhospital social setting (Feldman et al., 1975), or (3) in an outpatient medical clinic (Stinnett, 1982). None of these studies have, however, compared inpatient versus outpatient alcohol detoxification systematically. Unless participants are assigned on a random basis to the alternative forms of treatments, we have no way of knowing whether the findings were due to different kinds of patients volunteering for the different types of treatment. Later in this article a random-assignment study will be described, conducted at the Addiction Research Center of the University of Pennsylvania, which compared the effectiveness and costs of inpatient and outpatient medical detoxification (Hayashida et al., 1989).

Several studies have randomly assigned some middle socioeconomic status alcoholic patients to receive postdetoxification rehabilitation in either an inpatient or an outpatient setting (Fink et al., 1985; McLachlan & Stein, 1982). In the two settings, treatment was not found to differ in effectiveness, and outpatient treatment costs were significantly lower than those of inpatient treatment. Indeed, in the study by Fink and associates, both inpatient and outpatient participants received the same behaviorally based treatments from the same treatment team. Later in this article we will discribe two random assignment studies recently conducted at the Addiction Research Center of the University of Pennsylvania on lower socioeconomic men receiving either inpatient or intensive day treatment. In one study the participants were primarily COCAINE dependent (Alterman et al., 1994); in the second study, the participants were primarily alcohol dependent (O'Brien et al., 1992).

REVIEW OF RANDOM ASSIGNMENT STUDIES

Alcohol Detoxification. In a study by Hayashida and colleagues (1989), 87 patients applying for treatment at a Veterans' Administration Medical Center were randomly assigned to outpatient medical treatment; 77 were assigned to inpatient treatment. The primary form of treatment was prescription of oxazepam, a benzodiazepine, to relieve withdrawal symptoms. More inpatients completed treatment than outpatients, 95 percent to 72 percent respectively; however, the two groups were not found to differ one and six months after entry into the study. Both groups showed significant improvements in substance related, medical, and psychosocial functioning at both of these follow-up assessments. Costs were ten to twenty times greater for inpatient detoxification, however.

Cocaine and Alcohol Dependence. Two studies were conducted that compared the effectiveness and costs of intensive (27 hours per week) day rehabililtation to inpatient rehabilitation for either cocaine-dependent (study 1) or alcohol-dependent (study 2) men seeking treatment at a Veterans' Administration Center (Alterman et al., 1994; O'Brien et al., 1992). Both treatment programs were conventional in content and approximately one month in duration. The 58 cocaine-dependent patients were randomly assigned to each treatment in study 1; 23 alcohol-dependent men were randomly assigned to each program in study 2. In both studies, significantly greater proportions of inpatients completed the one month regime of treatment. However, the short-term (4 months) and longer-term (7 and 13 months) outcomes of the two treatment groups were

similar. In study 1, there was some indication of fewer alcohol-related problems in the day-treatment group. Both groups showed significant improvements at all follow-up assessments. The costs of intensive day treatment were about 40 to 60 percent those of inpatient treatment.

CONCLUSIONS

Most studies that have compared outpatient and inpatient treatment have shown few differences in longer term effectiveness, although inpatient treatment is clearly more effective in retaining patients in treatment. This may be important in some circumstances. Both treatment settings appear to result in substantial improvements. Outpatient treatment is far less costly than inpatient treatment.

These conclusions need to be qualified. First, up to 10 percent of people seeking treatment are clearly in need of acute medical and psychiatric treatments that are most readily available and applicable in an inpatient setting. Such patients were not included in the comparative studies we reviewed. Second, inpatient treatment resources are particularly important for outpatient detoxification failures and for patients who repeatedly fail to benefit from outpatient rehabilitation. Third, these conclusions may not apply to women, adolescent patients, or patients with serious psychiatric impairment. Finally, the type of treatment given appears to be the most important factor, not the setting in which the treatment is provided.

(SEE ALSO: *Comorbidity and Vulnerability; Treatment; Treatment Outcome Prospective Study*)

BIBLIOGRAPHY

ALTERMAN, A. I., ET AL. (1994). Effectiveness and costs of inpatient vs. day hospital treatment for cocaine dependence. *Journal of Nervous and Mental Disorders, 182,* 157–163.

ALTERMAN, A. I., O'BRIEN, C. P., & DROBA, M. (1994). Day hospital versus inpatient rehabilitation of cocaine abuse: An interim report. *NIDA Technical Reviews.* Rockville, MD.

FELDMAN, D. J., ET AL. (1975). Outpatient alcohol detoxification: Initial findings on 564 patients. *American Journal of Psychiatry, 132,* 407–412.

FINK, E. B., ET AL. (1985). Effectiveness of alcoholism treatment in partial versus inpatient settings: Twenty-four month outcomes. *Addictive Behaviors, 10,* 235–248.

HAYASHIDA, M., ET AL. (1989). Comparative effectiveness and costs of inpatient and outpatient detoxification of patients with mild-to-moderate alcohol withdrawal syndrome. *New England Journal of Medicine, 320,* 358–365.

McLACHLAN, J. F. C., & STEIN, R. L. (1982). Evaluation of a day clinic for alcoholics. *Journal of Studies of Alcohol, 43,* 261–272.

O'BRIEN, C. P., ALTERMAN, A. I., & McLELLAN, A. T. (1992). Developing and evaluating new treatments for alcoholism and cocaine dependence. In M. Galanter (Ed.), *Recent developments in alcoholism,* Vol. 10, 303–325. New York: Plenum.

STINNETT, J. (1982). Outpatient detoxification of the alcoholic. *International Journal of the Addictions, 17,* 1031–1046.

WHITFIELD, C. ET AL. (1978). Detoxification of 1,024 alcoholic patients without psychoactive drugs. *Journal of the American Medical Association, 239,* 1409–1410.

ARTHUR I. ALTERMAN

Pharmacotherapy, An Overview   Pharmacological agents may be used for several purposes in the treatment of drug and alcohol addiction. These include the alleviation of acute withdrawal symptoms, the prevention of relapse to drug or alcohol use, and the blocking of the euphorigenic effects of drugs of abuse. The various medications are used in the treatment of addiction to alcohol, opiates, cocaine, tobacco, and sedatives.

ALCOHOLISM

Detoxification.   The use and abuse of ALCOHOL has been known to humankind for centuries, and alcohol is currently one of the most widely used of the mood-altering substances. Habitual alcohol use is associated with the development of TOLERANCE and physiological (PHYSICAL) DEPENDENCE. Tolerance refers to a decrease in susceptibility to the effects of alcohol following chronic alcohol use, which results in the user consuming increasing amounts of alcohol over time. Physical dependence may be conceptualized as a physiological state in which the recurrent administration of alcohol is required to prevent the onset of withdrawal symptoms. Symptoms of alcohol withdrawal include irritability, tremulousness, anxi-

ety, sweating, chills, fluctuations in pulse and blood pressure, diarrhea, and, in severe cases, seizure. These symptoms generally begin within twenty-four hours following the last use of alcohol, peak within forty-eight hours, and subside over several days.

Pharmacotherapy for alcohol withdrawal includes the use of agents, such as BENZODIAZEPINES and BARBITURATES, that are cross-tolerant with alcohol. These agents attenuate the symptoms of withdrawal and result in decreased arousal, agitation, and potential for seizure development. Medication is provided in doses that are sufficient to produce mild sedation and physiological stabilization early in the withdrawal period; this is followed by a gradual dose reduction and then discontinuation over the next one to two weeks. Currently, benzodiazepines are the agents of choice for the treatment of alcohol withdrawal, because of the relatively high therapeutic safety index of these medications, their ability to be administered both orally and intravenously, and because of their anticonvulsant properties. Barbiturates may be used in a similar fashion, but they have a lower therapeutic index of safety than do benzodiazepines.

Recent additions to the pharmacotherapy of alcohol withdrawal include clonidine and carbamazepine. Clonidine is an antihypertensive agent (i.e., it lowers blood pressure) that has recently been used in the treatment of drug withdrawal states and chronic pain. This medication decreases autonomic hyperactivity (i.e., it lowers an increased pulse and blood pressure), but it does not have the anticonvulsant properties of the benzodiazepines or barbiturates. Carbamazepine has also been employed in the treatment of alcohol withdrawal and does have anticonvulsant properties. Neither medication is habit forming and thus may have potential in the treatment of alcohol withdrawal.

Maintenance Lithium.   Lithium is primarily employed in the treatment of bipolar mood disorder (previously termed manic-depressive disorder), but it may be beneficial in the treatment of other psychiatric disorders. It has received much attention in the investigation of pharmacologic agents for the treatment of alcohol dependence, and several studies have reported that its use had favorable effects on alcohol consumption. For example, after receiving doses of lithium comparable to those administered to human beings, laboratory animals demonstrated a significant reduction in alcohol consumption. In recovering alcoholics, lithium treatment has been associated with a decreased desire to continue drinking after alcohol use and, in several studies, with a higher rate of abstinence for those alcoholic patients who were compliant with therapy. Although these small studies on the efficacy of lithium for alcohol dependence appeared promising, a recent large placebo-controlled study failed to demonstrate a beneficial effect of lithium. At the present time, although lithium certainly has a place in the treatment of alcoholic patients with bipolar disorder, the indications for its use in other patients with alcohol dependence are less clear.

Antidepressants. Depressive symptoms are noted in many alcoholics at the time that they enter treatment. Because of the frequent co-occurrence of depression and alcoholism, the use of antidepressants would appear to be potentially useful in this population. Several studies have demonstrated favorable effects of antidepressants on alcohol consumption. Tricyclic antidepressants such as imipramine and desipramine inhibit the re-uptake of norepinephrine and serotonin in nerve terminals. These medications have been associated with decreased ethanol consumption in laboratory animals and in human alcoholic subjects. The serotonin re-uptake inhibitors (blockers) zimelidine, viqualine, fluvoxamine, and fluoxetine (Prozac) have also demonstrated favorable short-term results in the treatment of alcohol dependence. Although these medications are not routinely administered to all recovering alcoholics, many physicians consider the use of antidepressants in alcoholic patients if depressive symptoms do not resolve after several weeks of abstinence, or if a mood disorder was present prior to the onset of ethanol abuse.

Anxiolytics. Used to decrease anxiety, anxiolytics include benzodiazepines, such as chlordiazepoxide (Librium) and diazepam (Valium), and azaspirodecadiones, such as buspirone. Both classes of medication have been investigated for use in alcohol dependence. Early studies supported the use of benzodiazepines in recovering alcoholics with claims of decreased alcohol craving and consumption after chlordiazepoxide administration. Other controlled trials refuted this, however, and many physicians would question the use of benzodiazepines in this population. The azaspirodecadiones such as buspirone are nonaddictive medications that have been marketed for the treatment of anxiety. Although few controlled trials have been conducted that evaluated the effect of buspirone on human al-

cohol use, animal studies have demonstrated decreased alcohol consumption after treatment with this agent. Unlike benzodiazepines, buspirone is not known to be habit forming and thus may be a promising agent for additional controlled studies in human subjects.

Dopaminergic Agents. The effects of dopaminergic agents on the consumption of alcohol in animal studies have been conflicting, since both agents that augment dopaminergic activity and those that diminish it have been noted to decrease alcohol consumption. In humans, controlled studies with apomorphine and bromocriptine, both of which increase dopaminergic activity, have revealed decreases in alcohol craving, anxiety, and depression, and increased abstinence among alcoholic depressed patients.

Opioid Antagonists. Opioid antagonists are competitive antagonists of OPIOIDS at opiate receptors. They include NALOXONE, which may be used intramuscularly or intravenously to rapidly reverse opiate intoxication, and NALTREXONE, which is prescribed orally to prevent or reverse intoxication from opioids. Unlike opioids, these medications are not habit forming and may have a place in the treatment of alcohol-dependent patients. A variety of studies have demonstrated a reduction of alcohol consumption or self-administration by experimental animals treated with these agents. In human subjects, naltrexone administered as an adjunct to substance-abuse treatment has resulted in a decreased rate of alcohol consumption. In addition, those patients who did experience a "slip" were less likely than those who were not treated with naltrexone to suffer a complete relapse to alcohol use.

Antidipsotropics. Antidipsotropics are medications that are used to decrease alcohol consumption by creating an adverse reaction following alcohol use. They include DISULFIRAM, CALCIUM CARBIMIDE, and Flagyl. Disulfiram use results in an accumulation of acetaldehyde following the consumption of alcohol. Acetaldehyde levels accumulate if patients who are receiving disulfiram ingest alcohol, with the result that the patients may experience symptoms of acetaldehyde toxicity. These include sweating, chest pain, palpitations, flushing, thirst, nausea, vomiting, headache, difficulty breathing, hypotension, dizziness, weakness, blurred vision, and confusion. Symptoms may begin within five to fifteen minutes following alcohol ingestion and may last from thirty minutes to several hours.

The use of disulfiram is based upon the premise that the fear or actual experience of this adverse event may serve as a deterrent to alcohol use. Despite its toxicity, disulfiram has been used safely by thousands of recovering alcoholics since its introduction in 1948. Supervised voluntary use of the medication as an adjunct to other rehabilitative therapy has resulted in reduced alcohol consumption and decreased alcohol-related criminal behavior among alcohol-dependent patients.

Compliance is the key to successful use of disulfiram in alcohol dependence, since patients need only discontinue using disulfiram if they wish to resume drinking. Indeed, in an unsupervised setting, disulfiram administration shows no superiority over placebo on outcome measures related to alcohol use. Methods that have been investigated to improve compliance include surgical implants of disulfiram, reinforcement by providing a reward for compliance, and contingency management techniques. Although surgical implants have met with little success, the other two methods have demonstrated various degrees of efficacy.

OPIOID DEPENDENCE

The opioids include opiates, drugs derived from the opium poppy (*Papaver somniferum*), as well as those synthesized to produce similar narcotic effects. Opium has been used as a medicinal substance for at least 6,000 years. Widespread abuse of opiates was noted by the eighteenth century, with the smoking of opium in Asia; currently, HEROIN is a major opiate of abuse in the United States. Pharmacotherapy for opiate dependence may be employed both during the acute withdrawal syndrome and later to maintain abstinence from illicit opioids (e.g., heroin).

Acute Opioid Withdrawal. The syndrome of acute withdrawal from opiates varies in regard to the opiate of abuse. The time of onset, intensity, and duration of withdrawal symptoms depend on several factors, including the half-life of the drug, the dose, and the chronicity of use. Heroin is a relatively short-acting agent; symptoms of withdrawal often begin within eight to twelve hours after the last use. Early symptoms include craving, anxiety, yawning, tearing, runny nose, restlessness, and poor sleep. Symptoms may progress to include pupil dilation, irritability, muscle and bone aches, piloerection (the goose bumps—thus the term *cold turkey*), and hot

and cold flashes. Peak severity occurs 48 to 72 hours after the last dose and includes nausea and vomiting, diarrhea, low-grade fever, increased blood pressure, pulse, and respiration, muscle twitching, and occasional jerking of the lower extremities (which explains the term *kicking the habit*). The opiate withdrawal syndrome following chronic heroin use may last seven to ten days, but with longer-acting agents such as METHADONE, a similar constellation of symptoms may occur; they begin later, peak on the third to eighth day, and persist for several weeks.

A variety of medications may be used in the treatment of acute opiate withdrawal. The most common method is to use opiates alone. A dose high enough to stabilize the patient is administered on the first day and then gradually tapered over one to two weeks. Generally, long-acting opiates such as methadone are employed, but any opiate may be used.

Other medications used for opiate withdrawal are CLONIDINE and BUPRENORPHINE. Clonidine is an alpha-2 adrenergic agonist that is commonly employed as an antihypertensive medication. It is active on central nervous system (CNS) locus coeruleus neurons in the same areas at which opiates exert their effects. Clonidine appears most effective in decreasing symptoms such as elevation of pulse and blood pressure and may be less effective in relieving other symptoms of withdrawal. The major side effects of clonidine are orthostatic hypotension and sedation. A recent development in the pharmacotherapy of opiate withdrawal is rapid detoxification through the combined use of clonidine with opiate antagonists such as naltrexone. This treatment may decrease the time required for the detoxification process to two to three days. Opiate addicts may be stabilized on buprenorphine, a mixed opioid agonist/antagonist, with minimal discomfort and then withdrawn over five to seven days with less severe withdrawal symptoms than those associated with methadone withdrawal.

Antagonists. Opiate antagonists such as naloxone and naltrexone compete with opiates for CNS opioid receptors. Naloxone has a short half-life (two to three hours) and is generally employed on a short-term basis to reverse acute opiate intoxication. Naltrexone has a longer duration of action (approximately twenty-four hours) and is used as a long-term maintenance medication to inhibit euphoria in opioid addicts. Both medications have been used with relative safety for several years, and maltrexone has been successfully employed as an adjunct to other therapies in the treatment of opioid addicts. Clinically, side effects of naltrexone may include mild dysphoria and elevation in cortisol and beta-endorphin levels; no withdrawal syndrome has been noted following its discontinuation. Naltrexone is generally administered three to four times a week at an average dose of 50 milligrams per day. Despite its advantages, many opioid addicts resist therapy with this medication, and even in the most successful of programs, six-month retention rates may range from only 20 to 30 percent. The addition of psychosocial interventions such as counseling and contingency-management programs is helpful. When these interventions are added, naltrexone has been noted to be particularly effective in selected groups, such as those made up of health care professionals, business people, and prisoners on work-release programs.

Methadone Maintenance. Methadone has been used as a safe and effective treatment for opioid dependence for over twenty years. Heroin addicts easily adapt to using this long-acting opiate that possesses all of the physiological characteristics of heroin. When taken orally, methadone may have less abuse potential than heroin, but the onset of its CNS effects are slower and its tendency to induce euphoria is generally less than that of intravenous or inhaled heroin. In addition, it has a longer half-life than heroin and if it is administered daily, tissue levels accumulate, thereby decreasing interdose withdrawal symptoms that may lead to repeated opiate use. Methadone maintenance may be helpful for addicts who have difficulty adjusting to a drug-free lifestyle or for those who have been unsuccessful with other forms of treatment.

During maintenance therapy, methadone is initiated at a low dose and then gradually increased to higher doses, which are associated with decreased opiate craving and secondary illicit opiate use. With methadone maintenance treatment, many patients show significant decreases in illicit drug use, depression, and criminal activity, and they demonstrate increased employment. Therapy that is provided for extended periods of time and in the context of other psychosocial services has been associated with the highest success rates.

Another maintenance medication currently under investigation is levo-alpha-acetylmethadol (LAAM). LAAM is a long-acting form of methadone that requires administration three times per week instead of daily as with methadone. Although LAAM has been associated with a reduction in illicit opioid use,

its slower onset of action may lead to decreases in treatment retention compared to the use of methadone. The initiation of treatment with methadone and subsequent conversion to LAAM therapy may improve compliance with this medication. LAAM is not yet routinely used in the treatment of opioid dependence, and additional studies will be necessary to determine the appropriate use of this agent.

Buprenorphine. Buprenorphine is a mixed opioid agonist/antagonist that has been used for several years as a possible maintenance medication for opioid dependence. Although it has only recently been available within the United States, preliminary studies indicate that it may be a promising agent for the treatment of opioid dependence. As with methadone, maintenance treatment consists of daily administration of buprenorphine, but the optimal daily dose of medication remains under investigation. At low doses, buprenorphine has agonist effects at opioid receptors, but at higher doses antagonistic effects may occur. Buprenorphine maintenance has been associated with good treatment retention, decreased illicit opiate use, and a relatively mild withdrawal syndrome. On the basis of early studies, buprenorphine was thought to be a promising agent in the treatment of both cocaine and opioid dependence, but significant benefits have not been confirmed by better-controlled studies.

COCAINE DEPENDENCE

Cocaine abuse has increased markedly since the 1970s, and by 1984, more than 20 million Americans reported that they had tried cocaine. In addition to psychotherapy and other traditional approaches to substance-abuse treatment, a variety of pharmacotherapeutic interventions may be of benefit to cocaine abusers.

Pharmacotherapy for cocaine abuse may be employed to address specific symptoms that occur during the cocaine-withdrawal syndrome. Gawin and Kleber identified three phases in the cocaine abstinence syndrome. The crash phase generally begins soon after cocaine use ends and may last up to four days. Symptoms experienced at this time may include depression, suicidal ideation, irritability, anxiety, and intense cocaine craving. Sedatives such as alcohol and heroin may be used by addicts to alleviate these symptoms. The second or withdrawal phase may last two to ten weeks and is characterized by anxiety, depression, inability to experience pleasure, and increased cocaine craving. The third or extinction phase may last three to twelve months; during this phase, cocaine craving may continue as well as increased susceptibility to relapse in response to environmental cues.

Pharmacotherapy for cocaine dependence may be used to alleviate symptoms experienced during the cocaine abstinence syndrome. During the crash period, early symptoms such as anxiety and insomnia may be relieved by benzodiazepines such as CHLORDIAZEPOXIDE. Neuroleptics (ANTIPSYCHOTICS) may also be helpful during this period to alleviate psychotic symptoms such as paranoia.

Other agents that may be used on a short-term basis include dopaminergic agents such as bromocriptine and AMANTADINE. Some investigators postulate that CNS dopamine may be depleted by chronic cocaine use. Dopaminergic agents may be used to augment CNS dopaminergic function, and various dopaminergic agents such as amantadine, bromocriptine, and L-dopa have been employed for this purpose. Although few long-term, double-blind, placebo-controlled studies have been conducted, several studies have supported the use of dopaminergic agents such as amantadine as anticraving medications during withdrawal.

Antidepressants may be helpful during the withdrawal and extinction stages of cocaine abstinence. One controlled and several uncontrolled studies in recovering cocaine addicts suggested that the tricyclic antidepressant desipramine might decrease cocaine use and craving. Other antidepressants investigated in pilot studies include fluoxetine, imipramine, doxepin, and trazodone. Antidepressants may take several weeks to begin to alleviate symptoms of depression or craving, however, and some cocaine addicts may drop out of treatment during this period. These patients may benefit from initiation of treatment with a short-term agent (such as a dopaminergic agent) followed by long-term treatment with an antidepressant. As with every treatment, however, no firm conclusions are warranted about any agent until it has been tested in a controlled clinical trial that has been replicated at least once.

Pharmacotherapy may also be helpful for patients with psychiatric diagnoses other than cocaine dependence. In some patients, cocaine abuse may be an attempt at self-medication to address the discomfort of depression or other psychiatric disorders. Patients with major depressive disorder and bipolar disorder may respond to therapy with antidepressants or lithium, and those with attention deficit disorder may benefit from the cautious use of low doses of stimulant medication.

In summary, antipsychotics and benzodiazepines may be used to alleviate symptoms of acute cocaine withdrawal, whereas tricyclic antidepressants and dopaminergic agents may be helpful in the long-term treatment of cocaine withdrawal. Pharmacotherapy should be considered an adjunct to other forms of rehabilitative therapy during the long-term treatment of the cocaine-dependent patient.

TOBACCO DEPENDENCE

One commonly used pharmacological treatment for tobacco dependence is a nicotine-containing gum called Nicorette. The main reason to quit smoking cigarettes is its powerful association with lung cancer, emphysema, and other medical problems. Yet nicotine, the active ingredient in cigarettes, is another drug that is associated with pleasant effects and with withdrawal discomfort, thereby making it an extremely addicting drug. Providing cigarette smokers with nicotine replacement in the form of a gum will help them avoid the health risks associated with smoking cigarettes. One problem with Nicorette is that it is difficult to chew correctly and therefore people need to be trained in how to chew it in order to derive the therapeutic effect. Recently, a patch has been developed that is placed on the arm and automatically releases nicotine. A method that shows good potential as a treatment, the patch was made available in the early 1990s. Detoxification from nicotine may also be facilitated with the medication clonidine, the same agent used to help alleviate opiate withdrawal symptoms.

SEDATIVE DEPENDENCE

Current treatments for sedative dependence include detoxification agents rather than anticraving agents. Detoxification is accomplished by tapering the dosage of benzodiazepines over two to three weeks. More recently, carbamazepine, an antiseizure medication, was shown to relieve alcohol and sedative withdrawal symptoms, including seizures and delirium tremens. Future work with agents that block the actions of benzodiazepines may hold promise as a maintenance or anticraving agent used to help the sedative abuser abstain from drug abuse.

CONCLUSIONS

Medications must be accompanied by psychological and social treatments and support; they do not work on their own. Moreover, medications to block illicit-drug effects in the brain may be of little use if the patient does not take them. More research in many fields is needed to identify potential medications, but this research must recognize the psychosocial as well as the neurobiological areas of therapy. Without this integration, the work to develop more effective treatments for the difficult problem of drug abuse and dependence cannot begin.

(SEE ALSO: *Causes of Substance Abuse; Comorbidity and Vulnerability; Complications; Disease Concept of Alcoholism and Drug Abuse; Nicotine Delivery Systems for Smoking Cessation*)

BIBLIOGRAPHY

FRANCES, R. J., & FRANKLIN, J. E. (1990). Alcohol and other psychoactive substance use disorders. In J. A. Talbott, R. E. Hales & S. C. Yudofsky (Eds.), *The American Psychiatric Press textbook of psychiatry.* Washington, DC: American Psychiatric Press.

GAWIN, F. H., & KLEBER, H. D. (1986). Abstinence symptomatology and psychiatric diagnosis in chronic cocaine abusers. *Archives of General Psychiatry, 43,* 107–113.

JAFFE, J. H. (1989). Drug dependence: Opioids, nonnarcotics, nicotine (tobacco) and caffeine. In H. I. Kaplan & B. J. Sadock (Eds.), *Comprehensive textbook of psychiatry* (5th ed.). Baltimore: Williams & Wilkins.

JAFFE, J. H. (1985). Drug addiction and drug abuse. In A. G. Gilman et al. (Eds.), *Goodman and Gilman's the pharmacological basis of therapeutics,* 7th ed. New York: Macmillan.

KOSTEN, T. R., & KLEBER, H. D. (EDS.). (1992). *Clinician's guide to cocaine addiction.* New York: Guilford Press.

LOWINSON, J. H., RUIZ, P., & MILLMAN, R. B. (EDS.). (1992). Baltimore, MD: Williams & Wilkins.

SCHUCKIT, M. A., & SEGAL, D. S. (1987). Opioid drug use. In E. Braunwald et al. (Eds.), *Harrison's principles of internal medicine,* 11th ed. New York: McGraw-Hill.

ELIZABETH WALLACE
THOMAS R. KOSTEN

Psychological Approaches Psychological treatments of drug dependence assume that drug abuse is a learned behavior. As such, it is not different from other less controversial and more healthful behaviors in its development. That is, a psychological perspective suggests that drug abuse is, for the most part, learned in many of the same ways as behaviors such as reading or driving a car. This perspective also suggests that drug abuse can be changed in the ways that other behaviors are changed. Forces for change include rewards (reinforcers) and unpleasant events (punishments); cues that signal the need for specific actions (discriminative stimuli); and training in new ways of thinking about oneself and the world that lead to ways of living that do not involve drugs.

Operant Learning Models. Psychological treatments for drug abuse can be grouped into three categories, based on the models of behavior that they represent. The first are those that draw from operant learning models. These models suggest that many important behaviors, including those many behaviors that end with the use of an illegal drug are controlled by environmental events, rather than events inside the individual. Internal events may come into play but, ultimately, these are caused by external events. These models suggest that the important factor in determining drug use is the balance between the rewards and the punishments of use. CONTINGENCY CONTRACTING, a system of rewards for abstinence and punishment for drug use, is an example of an operant-based treatment.

Classical Conditioning. A second model used is classical conditioning. A neutral event is paired repeatedly with another important event, one that usually evokes a response for the organism. A man who has experienced heroin withdrawal many times may eventually find that certain rooms of his apartment itself have come to cause him to crave drugs, because the apartment itself has become associated with withdrawal. A treatment based on classical conditioning, for example, is an attempt to remove the craving induced by the sight of drug paraphernalia, by repeatedly presenting pictures of those paraphernalia with no drugs, and therefore with lack of a reinforcing response.

Social Learning Models. Other treatments draw from social learning models. These assume that behaviors, such as drug abuse, are learned in many ways, including operant conditioning, classical conditioning, imitation (learning by watching someone else), and learning certain ways of thinking. These models also usually assume that imitation and learning new ways of thinking are more important for humans than other ways of learning. An example of a treatment based on a social learning model is cognitive behavioral psychotherapy, where the drug abuser is taught new ways of viewing old situations, as well as new social skills, in the hope that these new thoughts and skills will lead to a less troubled life, which does not demand drug abuse to make it tolerable.

OPERANT MODELS: CONTINGENCY MANAGEMENT

Contingency management has been incorporated into many drug-treatment programs as a way of assisting people in reducing drug use. In contingency management, reinforcers or punishers are applied depending on the patient's behavior. Often, contingencies are formalized in a contract. In contingency contracting, a treatment plan is developed and agreed to by treatment staff and patient. As part of the contract, both agree that certain consequences will occur as a result of certain behaviors on the part of the patient.

Early work indicating the usefulness of contingencies was completed largely at Johns Hopkins University. Working in a methadone-maintenance program, investigators at Johns Hopkins found that money and the opportunity to raise dose levels all served to decrease drug abuse. Work at the University of California in detoxification treatment programs also indicated that payment for drug abstinence was an effective adjunct to short-term detoxification treatments, where methadone is used for only about three weeks, to help drug abusers in their transition from heroin use to a drug-free state. Both of these experimental programs focused on rewards for desired behavior, rather than punishments for drug use. Contingencies also have been used to help clients conform to other treatment demands, including attending counseling sessions (Stitzer & Kirby, 1991).

Even though early work focused on providing positive reinforcers for desired behavior, the adaptations of this work in most clinics around the country has involved negative consequences. For reasons not clear, most clinical sites that have adopted the

contingency contracting procedures use punishers, not reinforcers. A common example is the use of a detoxification contract in methadone-maintenance treatment. Frequently, patients who are using illegal drugs sign a contract with treatment staff indicating that if they do not terminate all unapproved drug use within a certain period of time, their methadone dose will be reduced. If they continue to use drugs, their dose is incrementally reduced until they are no longer receiving methadone. At any point in the sequence, however, that the patient shows evidence of discontinuing drug use, the methadone dose can be raised and the person continued on the treatment program. Usually, the contract indicates that patients are given a certain amount of time to decrease the number of drug-positive urines or they are gradually detoxified from the program.

Contingency management has been used with practically every addiction, both by itself and in conjunction with other treatments. The evidence is now convincing that contingencies, especially positive contingencies, are effective in decreasing drug abuse. Work is needed to train clinic staff in using contingency programs, especially those employing positive contingencies (Stitzer & Kirby, 1991).

CLASSICAL CONDITIONING: AVERSIVE CONDITIONING

A form of behavioral therapy once widely used is AVERSION THERAPY. Here, the drug or the cues that remind drug users of it are paired with unpleasant events. The notion is that by pairing this very desirable substance with an unpleasant event, the association with the substance will become negative. The most successful of these has been rapid smoking, a treatment for tobacco dependence. In rapid smoking, the smoker smokes and inhales at a rate about 6 times that of normal. During this process, the therapist points out negative things about smoking, including the smell of the smoke, burning eyes, racing heart, and pounding head. Over time, the poisonous elements of the smoke itself (usually an amount of NICOTINE that exceeds the smoker's tolerance) may make the smoker nauseated. Thus, the cues associated with a cigarette (its appearance and smell) rather than calling forth pleasant reactions in the smoker, come to call forth unpleasant ones. Aversive-conditioning treatments have been attempted with other drugs, most notably ALCOHOL and CO-CAINE. Usually, for example, a chemical that induces vomiting is given so that nausea and vomiting occur at about the same time the patient is drinking in a controlled setting. However, aversion treatments for drug abuse other than TOBACCO abuse have had limited success or, at least, limited popularity. There are at least two reasons for this. First, with other drugs, the dose of the problem drug needed to produce unpleasant reactions may be physiologically dangerous. Second, rapid smoking is unique in that it is the actual drug, tobacco smoke, that is used to form the aversion. There is evidence in the psychological literature that such aversions are especially potent.

Aversive smoking has been evaluated in several well-controlled studies. It appears that when it is done correctly, abstinence rates can be as high as 60 percent after one year—a very high abstinence rate indeed—since the average abstinence rate after treatment for cigarette smoking is about 20 percent. The data for aversion for alcoholics using chemicals is not so clear. There are few comparisons with other treatments or with no treatment. Individuals who choose aversion treatment may be especially motivated to change, and they might have achieved high abstinence rates even without treatment.

One variant of aversion conditioning is covert conditioning. In covert conditioning, the drug abuser, with the help of a therapist, imagines both the drug use and the unpleasant consequences of it. For example, alcoholics might picture a cold beer, prepare to savor it, look at it, sip it, then slowly feel increasingly nauseated until they become violently ill. Thus, both the aversive events and the unpleasant consequences are imagined, rather than real. This has advantages if the drug of choice is illegal or quite dangerous, because it avoids drug use at all. Also, patients who might refuse to participate in actual aversive conditioning may feel able to do so when the aversion experienced is imagined. Unfortunately, however, there is not a great deal of evidence to support the usefulness of this approach (Council on Scientific Affairs, 1987).

The use of aversion conditioning has decreased recently, except in a limited number of private psychiatric hospitals. There are several reasons contributing to its demise. The first is the lack of demonstrated efficacy in controlled clinical trials with drugs other than tobacco. The second is its ex-

pense when compared with other treatments. Last, because of its intrinsically unpleasant nature, it has low acceptability.

SOCIAL LEARNING MODELS

Skill Training. In skill training, drug abusers, and others at risk for drug abuse, are taught skills that will help them not to use drugs. These can be simple and direct; for example, teaching junior high school students effective ways to refuse a cigarette. The skills learned may also be complex. Consider, for example, a smoker who knows the temptation to smoke when angry, because in the past anger-provoking situations have resulted in relapse. A therapist working with such a person in skill training would first review the situations that produce anger. These might be as diverse as incorrect charges on a credit card bill to a fight with the boss. After identifying the situations, the smoker and therapist would then discuss the details of the situation. For example, they might imagine what the boss would say to smokers to elicit anger. They would attempt to find ways of handling the situation that would leave the smokers feeling satisfied after it was over. They would discuss the usual response that would culminate in smoking. They would then identify alternative responses. Finally, they would role-play the alternative responses. The therapist would play the role of both the boss and the smoker, to give the smoker a model of different ways to handle the situation. In this way, the smokers would learn to handle anger in a better way, would be satisfied with the new responses, and be less likely to smoke. The smoker would also have ready responses other than smoking. Skill-training programs have been studied with smokers, alcoholics, cocaine abusers, and abusers of multiple drugs. Skill training is closely related to the recovery training and self-help that is discussed below. Recent data indicate skill training may be an especially useful treatment for heroin and/or cocaine abusers and alcoholics when used in the context of a large therapy program (Carroll, Rounsaville, & Gawin, 1991).

Skill training has been shown to be especially useful as an ancillary to other treatments. For example, one program developed a workshop to train drug-treatment patients in job-finding skills. There was a great deal of practice in new ways to interview for jobs. Patients were taught how to fill out a job application to maximize their strengths—also how to handle the existence of prison records or long lapses in employment. They practiced their interviews and saw themselves in practice interviews on videotape. The rationale was that if drug abusers could be taught to present themselves positively in a job interview, they would be more likely to get jobs. And, were they to become employed, they would be less likely to use drugs, for several reasons. These reasons include increased general life satisfaction and making new friends and social contacts who are not drug abusers. Studies using this technique found that it was helpful in increasing employment rates in both METHADONE-MAINTENANCE clients and former addicts recruited from the criminal-justice system. These studies did not address the length of time the job was held, however. It may be that a separate set of skills is needed to maintain employment. This set should be the object of further study (Hall et al., 1981).

Some programs have attempted to combine several approaches, so that abstinence is supported in multiple ways. Among the most successful of these is the community-reinforcement approach to alcoholism treatment developed by Azrin (1976). The original community-reinforcement approach incorporated (1) placement in jobs that interfered with drinking; (2) marriage and family counseling; (3) a self-governing social club; and (4) encouragement to engage in hobbies and recreational activities that could substitute for drinking. This procedure was found to decrease time spent drinking alcohol, increase rates of employment, increase time spent with families, and decrease the time spent in the hospital being treated for alcoholism. A later revision of the program also encouraged patients to take DISULFIRAM, a drug which produces unpleasant reactions if one drinks after taking it; taught alcoholics how to identify and handle danger signals so that they did not lead to drinking; provided patients with a "buddy" in the client's neighborhood; and switched from individual to group counseling. This procedure produced even more strikingly positive results than the original program. It can be argued that subjects in these studies had resources available to them that many drug abusers and alcoholics do not have, including the opportunity to receive inpatient treatment, a local economy that provides a choice of job opportunities, and supportive families. Recent work with cocaine abusers has replicated these positive re-

sults. The finding is especially impressive because the cocaine abusers were treated on an outpatient basis, and they traditionally have fewer resources than alcoholics.

Psychotherapy. Psychotherapy has also been useful in treating drug addicts, especially those with social and psychological problems that complicate their drug abuse. The assumption behind providing psychotherapy to drug abusers is that drug abuse is motivated by the problems that abusers have with other people, as well as their feelings about themselves. Early workers in the field attempted to provide psychotherapy as the sole treatment for drug abuse. Most found that it was not successful; they assumed that this was because the personality characteristics of addicts were not those that allowed people to succeed in psychotherapy—that is, addicts are often distrustful of nonaddicts and may not easily reveal their feelings to professionals. Also, they may not be especially reliable and often appear to have shaky to no motivation to change. Nevertheless, a large-scale study at the University of Pennsylvania—using clients who were already in methadone maintenance—found that, in the context of a larger treatment program, drug-treatment clients with other or extensive psychological problems do benefit from the addition of psychotherapy. The forms of psychotherapy available included one focusing on feelings and emotions (supportive–expressive) and one focusing on thought and behaviors (cognitive–behavioral). These researchers found that the type of therapy was not important, just participating in therapy was important (Woody et al., 1983).

The Recovery Training and Self-Help Model. Researchers at Harvard University studied a model that combined skill training in RELAPSE PREVENTION with SELF-HELP GROUPS. In their study, opiate addicts attended a recovery-training session once a week and a self-help group led by a former addict. Members also met informally outside the treatment meetings and in group-sponsored recreation and community activities. In the professionally led recovery meetings, leaders addressed a variety of topics, including high-risk situations, friendships, physical illness, and relations with family; they developed new ways of handling these situations that would be less likely to lead to drug use. The self-help groups supported these changes and further reinforced them. In two studies, one in the United States and one in Hong Kong, this treatment led to higher rates of abstinence or infrequent use than was found in a control condition, to increases in employment, and to fewer reports of criminal behavior. These differences were quite long-lasting—occurring six months to one year after entrance into treatment (McAuliffe & Ch'ien, 1986).

Twelve-Step Programs. The most well-known TWELVE-STEP program for helping substance abusers is ALCOHOLICS ANONYMOUS (AA). AA, founded in 1935 by a group of recovering alcoholics, is a fellowship of men and women who are committed to helping other alcoholics. NARCOTICS ANONYMOUS (NA), founded in 1953, was adapted from AA principles to include all substance abusers, not only alcoholics.

AA and NA programs focus on alcoholism and substance abuse as a disease for which there is no cure—therefore recovery becomes a lifetime commitment. These programs emphasize the personal powerlessness of individuals in combating their illness and get individuals to recognize that they must give themselves to a greater power so that they may be saved.

The guiding tenets of AA and NA programs are called the Twelve Steps. Each step is a passage through recovery, combining self-discovery with spiritual guidance. They involve five psychological tasks: (1) recognition and admission of powerlessness over alcohol; (2) acceptance of a high power as a source of strength and guidance during recovery; (3) self-help appraisal and self-disclosure in the service of personal change; (4) making amends for past wrongs; and (5) carrying the AA message to others (Anderson & Gilbert, 1989).

One can argue that aspects of AA parallel psychological approaches. For example, similar to psychotherapy, AA and NA members are encouraged to "work through" problems and to change the attitudes and actions associated with an alcohol- or drug-using lifestyle. These programs also use principles common to other self-help groups. Members are encouraged to attend meetings on a daily or weekly basis, at which the steps are discussed and made relevant, speakers recount their lives, and connections with support networks and role models are made.

Nevertheless, despite the facility with which psychological models might explain such approaches, they are not psychological approaches. They were developed from a spiritual approach, not from psychological principles.

SUMMARY

There are many psychological treatments that appear to be useful in aiding drug abusers to stop using drugs, no matter whether the drug be an illegal one, or alcohol or nicotine. Positive results come from contingency-contracting programs and multifaceted-reinforcement programs that are offered in the context of complex treatment programs or from skill-training programs that address several facets of the drug abuser's life. Also, there is evidence for the usefulness of different forms of psychotherapy for drug abusers, especially for those who have psychological and social problems. Drug abuse is increasingly becoming identified as a complicated problem that involves both biological and psychological factors. Because of this and the clear usefulness of psychological intervention, we can expect to see the development of new psychological treatments for drug abuse.

(SEE ALSO: *Addiction: Concepts and Definitions; Adjunctive Drug Taking; Causes of Substance Abuse; Disease Concept of Alcoholism and Drug Abuse; Prevention; Vulnerability; Wikler's Pharmacologic Theory of Drug Addiction*)

BIBLIOGRAPHY

ANDERSON, J. G., & GILBERT, F. S. (1989). Communication skills training with alcoholics for improving performance of two of the Alcoholics Anonymous recovery steps. *Journal of Studies on Alcohol, 50,* 361–367.

AZRIN, N. H. (1976). Improvements in the community-reinforcement approach to alcoholism. *Behavior Research and Therapy, 14,* 339–348.

CARROLL, K. M., ROUNSAVILLE, B. J., & GAWIN, F. H. (1991). A comparative trial of psychotherapies for ambulatory cocaine abusers: relapse prevention and interpersonal psychotherapy. *American Journal of Drug and Alcohol Abuse, 17*(3), 229–247.

COUNCIL ON SCIENTIFIC AFFAIRS. (1987). Aversion therapy. *Journal of the American Medical Association, 258,* 2562–2566.

EMRICK, C. D. (1987). Alcoholics Anonymous: Affiliation processes and effectiveness as treatment. *Alcoholism, 11,* 416–423.

HALL, S. M., ET AL. (1981). Increasing employment in ex-heroin addicts II: Criminal justice sample. *Behavior Therapy, 12,* 453–460.

McAuliffe, W. E., & Ch'ien, J. M. (1986). Recovery training and self-help. *Journal of Substance Abuse Treatment, 3,* 9–20.

Ogborne, A. C., & Glaser, F. B. (1981). Characteristics of affiliates of Alcoholics Anonymous: A review of the literature. *Journal of Studies on Alcohol, 42,* 661–675.

Sheeren, M. (1988). The relationship between relapse and involvement in Alcoholics Anonymous. *Journal of Studies on Alcohol, 49,* 104–106.

Stitzer, M. L., & Kirby, K. C. (1991). Reducing illicit drug use among methadone patients. In *Improving Drug Abuse Treatment* (National Institute on Drug Abuse Research Monograph 106). Rockville, MD: National Institute on Drug Abuse.

Woody, G. E., et al. (1983). Psychotherapy for opiate addicts. *Archives of General Psychiatry, 40,* 639–645.

Sharon Hall
Meryle Weinstein

Self-Help and Anonymous Groups Self-help groups for drug and alcohol abuse, often called mutual-help groups, are of two basic types. First are the long-standing anonymous groups closely patterned after ALCOHOLICS ANONYMOUS (AA). An alternative type also has a group context, but rejects the spiritual aspects (such as reliance on "higher power") of AA and urges members instead to take personal responsibility for gaining sobriety. The AA-like anonymous groups embrace the TWELVE STEPS, applying them to their own particular disorder. In some instances, they also adapt the AA Twelve Traditions. NARCOTICS ANONYMOUS, Emotions Anonymous, Overeaters Anonymous, Gamblers Anonymous, AL-ANON, COCAINE ANONYMOUS, and Nicotine Anonymous are prominent examples. Examples of the alternatives to AA are RATIONAL RECOVERY (RR), SECULAR ORGANIZATION FOR SOBRIETY (SOS), and WOMEN FOR SOBRIETY (WFS). Numerous members of these groups have been dropouts from AA.

In embracing AA's Twelve Steps, the first type of organization teaches powerlessness over their malady, reliance on the group or on some entity as a "higher power," catharsis via self-inventory, confession and amends, and a commitment to search out and tell others suffering from the same disorder about their programs for recovery. The rationale is that members have deep-seated denials that must be blunted by admitting helplessness and invoking the

The Twelve Steps of Alcoholics Anonymous

1. We admitted we were powerless over alcohol—that our lives had become unmanageable.
2. Came to believe that a Power greater than ourselves could restore us to sanity.
3. Made a decision to turn our will and our lives over to the care of God *as we understood Him.*
4. Made a searching and fearless moral inventory of ourselves.
5. Admitted to God, to ourselves, and to another human being the exact nature of our wrongs.
6. Were entirely ready to have God remove all these defects of character.
7. Humbly asked Him to remove our shortcomings.
8. Made a list of all persons we had harmed, and became willing to make amends to them all.
9. Made direct amends to such people wherever possible, except when to do so would injure them or others.
10. Continued to take personal inventory, and when we were wrong, promptly admitted it.
11. Sought through prayer and meditation to improve our conscious contact with God *as we understood Him,* praying only for knowledge of His will for us and the power to carry that out.
12. Having had a spiritual awakening as the result of these steps, we tried to carry this message to others, and to practice these principles in all our affairs.

SOURCE: The Twelve Steps are reprinted with permission of Alcoholics Anonymous World Services, Inc. Permission to reprint this material does not mean that AA has reviewed or approved the contents of this publication, nor that AA agrees with the views expressed herein. AA is a program of recovery from alcoholism *only*—use of the Twelve Steps in connection with programs and activities patterned after AA, but which address other problems, does not imply otherwise.

group and a higher power to help them. Moreover, this powerlessness is seen as a lifetime condition and the Twelve Steps are seen as providing a mechanism for ensuring a lifetime cessation of the compulsive behavior. The steps were devised in the late 1930s by Bill W., the major cofounder of AA, in conjunction with a small group of his earlier followers.

The second type of organization emphasizes that individuals, as individuals, must use their own resources and, in effect, "Save Our Selves" (SOS). The founder of WFS has written Thirteen Statements of Acceptance around which meetings are anchored: For example, number 5 is "I am what I think," and number 13 is "I am responsible for myself and my actions." The other statements encourage in women alcoholics a strong feeling of self-worth even though they have symptoms of a serious disease (Kirkpatrick, 1989).

The two types of organizations differ on basic treatment strategies. One difference is their divergent views of the permanency of their obsessive behavior. AA, and the many AA-like groups, view their problems as lifetime conditions over which they are powerless. In short, they will never recover; they are permanently "recovering" from a disease. In contrast, RR, for example, plays down the disease concept, and the higher-power notion that goes with it, and appeals to forces within a member's own intellect and willpower. Self-reliance is taught. WFS targets the development of self-value, self-esteem, and self-confidence as a way to meet the emotional needs of modern women, thereby, members believe, reducing significantly the basic roots of alcohol abuse for them.

The success rates of the AA fellowship have been assessed at two points in time. Of those initially attracted to AA, a large proportion drop out—somewhere between 35 and 65 percent. Of those who become active members, 65 to 70 percent "improve to some extent, drinking less or not at all during A.A. participation" (Emrick, 1989:45). Membership in AA seems to be associated with relatively high abstinence rates, but with fairly typical improvement rates (Emrick, Lassen, & Edwards, 1977). It appears that AA is effective only with some 25 to 30 percent of the population with alcohol-related problems. AA, then, is a highly selective treatment source—attracting and holding those alcohol-troubled persons with severe alcohol problems who have high affiliative needs, conformist tendencies, proneness to guilt, and need for external controls (Trice & Roman, 1970; Ogborne & Glaser, 1981).

Unfortunately, the alternative type of organization has yet to be scrutinized by objective researchers. But subjective estimates of the number of groups and members have been put forward. SOS claims 1,000 groups with 2,000 members (Christopher, 1992); Hall (1990:1,46) has estimated that RR has meetings in 100 cities, "with perhaps two thousand members at any one time," and Hall (1990) estimated 5,000 members in 32 groups for WFS. Assuming that, like AA, there are dropouts and misfits for each type of group, these numbers must be sharply discounted. Nevertheless all three have demonstrated some staying power. SOS even publicizes it-

self as a demonstrated and proven alternative to AA. As yet no reliable data support this contention, but the fact that sizable numbers have been attracted to it suggests that it, or groups like it, are realistic contenders for some of AA's approximately 1 million members.

(SEE ALSO: *Alcoholism; Disease Concept of Alcoholism and Drug Addiction; Ethnic Issues and Cultural Relevance in Treatment; Women and Substance Abuse*)

BIBLIOGRAPHY

CHRISTOPHER, J. (1992). The S.O.S. story. *S.O.S. National Newsletter, 5*(1), 1, 2.

EMRICK, C. (1989). Alcoholics Anonymous: Membership characteristics and effectiveness as treatment. In M. Galanter (Ed.), *Recent developments in alcoholism: Treatment and research* (pp. 37–53). New York: Plenum Press.

EMRICK, C. D., LASSEN, C. L, & EDWARDS, M. T. (1977). Nonprofessional peers as therapeutic agents. In A. S. German & A. M. Razin (Eds.), *Effective psychotherapy: A handbook of research* (pp. 120–161). New York: Pergamon Press.

HALL, T. (1990). New way to treat alcoholism shuns spirituality. *New York Times*, December, 4, 1, 46.

KIRKPATRICK, J. (1989). Women for sobriety. *The Counselor*, January/February: 9.

OGBORNE, A. C., & GLASER, F. B. (1981). Characteristics of affiliates of Alcoholics Anonymous: A review of the literature. *Journal of Studies on Alcohol, 42*(7), 661–675.

TRICE, H. M., & ROMAN, P. M. (1970). Sociopsychological predictors of affiliation with Alcoholics Anonymous: A longitudinal study of "treatment success." *Social Psychiatry, 5,* 51–59.

HARRISON M. TRICE

Therapeutic Communities Therapeutic Communities (TCs) are drug-free residential treatment facilities for drug and/or alcohol addiction. TCs emerged in the 1960s as a self-help alternative to the conventional medical and psychiatric approaches being used at that time. Antecedents of the TC include the communities pioneered in the 1950s by Maxwell Jones and others in Britain, for treating psychiatric patients, and the 1930s organization of ALCOHOLICS ANONYMOUS (AA). SYNANON, established in California in 1958 by Charles (Chuck) Dederich, a former alcoholic, was the first TC devoted to the treatment of the drug addicted. Although now highly evolved into sophisticated human-services institutions, the TC prototype can be found in all forms of communal healing and group support.

TCs aim to completely change a drug abuser's lifestyle, promoting ABSTINENCE from illicit substances, eliminating antisocial attitudes and activity, developing employable skills, and inculcating prosocial attitudes and values. Although TCs originally attracted NARCOTIC (OPIOID) addicts, by the 1990s, TC clients mostly abused other drugs. This reflects changing trends in drug-use patterns, as well as the ability of this treatment modality to respond to clients with drug problems of varying severity, different lifestyles, and various social, economic, and ethnic/cultural backgrounds. Therapeutic communities now encompass a variety of short- and long-term residential programs, as well as day treatment and/or ambulatory programs. The effectiveness of the traditional long-term residential TC has been well-documented, and it is this model that will be described here (De Leon & Ziegenfuss, 1986; De Leon & Rosenthal, 1989; De Leon, 1994).

THE TRADITIONAL TC

Most traditional TCs have similar features, including their organizational structures, staffing patterns, perspectives, rehabilitative regimes, and a fifteen- to twenty-four-month duration of stay. They differ greatly, however, in size (30 to 600 beds) and client demography. Primary clinical staff are usually former substance abusers who were rehabilitated and trained in TC programs. Other staff are the professionals who provide medical, mental-health, vocational, educational, family-counseling, fiscal, administrative, and legal services.

Traditional TCs also share a defining view of substance abuse as a deviant behavior that reflects impeded personality development or chronic deficits in social, educational, and economic skills, which may be attributed to psychological factors, poor family effectiveness and, frequently, to socioeconomic disadvantage. Drug abuse is thus seen as a disorder of the whole person, which leads to cognitive and behavioral problems, as well as to mood disturbances. Thinking may be unrealistic or disorganized; values are confused, nonexistent, or antisocial. Fre-

quently, there are deficits in verbal, reading, writing, and marketable skills. Rehabilitation requires multidimensional influence and training which, for many drug abusers, can only occur in a long-term 24-hour residential setting.

Within the TC, comprehensive interventions and services are coordinated within a single treatment setting. These include vocational counseling, work therapy, recreation, group and individual therapy, educational, medical, family, legal, and social services. The community itself—consisting of peers and staff (who are role models of successful personal change)—serves as the primary "therapist" and teacher. Staff members also serve as rational authorities and guides in the recovery process. Thus, the community as a whole provides a crucial twenty-four-hour context for continued learning.

VIEW OF THE PERSON

The TC distinguishes individuals along dimensions of psychological dysfunction and social deficit, rather than by drug-use patterns. Because many TC clients have never acquired conventional lifestyles or mainstream values, the goal of the TC experience is to *habilitate*—to develop socially productive, conventional lifestyles. For clients from more advantaged backgrounds, whose drug abuse is more likely to be seen as a manifestation of psychological disorder, the TC aims to *rehabilitate*, emphasizing a return to a socially approved lifestyle. Despite individual differences, all residents in the TC must follow the same regime, but specific treatment plans modify the emphasis of each one's TC experience.

VIEW OF RECOVERY

The aim of the TC is to change both lifestyle and personal identity. The primary psychological goal is to change the negative behavior patterns that predispose to drug use; the main social goal is to develop the skills, attitudes, and values of a responsible drug-free lifestyle. Stable recovery requires successfully integrating these psychological goals through commitment to the values of abstinence, achievement, and self-reliance. Such behavioral change requires the development of insight and the integration of emotions, conduct, skills, cognitive attitudes, and values. The rehabilitative regime in a traditional TC is shaped by several broad assumptions about recovery.

Motivation. Recovery depends on both positive and negative pressures to change. Some clients have been coerced into treatment; others come voluntarily. For all, however, remaining in treatment requires continued motivation to change; therefore, elements of the TC rehabilitation approach are designed to sustain motivation.

Self-Help and Mutual Self-Help. Strictly speaking, treatment is not *provided* in the TC but *made available* through staff and peers, plus the daily regime of work, groups, meetings, seminars, and recreation. The effectiveness of these elements depends on constant and full participation in the treatment regime. The patient must make the main contribution to the change process (*self-help*). The main messages of recovery, personal growth, and right living are mediated by peers through confrontation and sharing in groups, by example as role models, and as supportive, encouraging friends in daily interactions (*mutual self-help*).

Social Learning. Because negative behavior patterns, attitudes, and roles were not acquired in isolation, they cannot be changed in isolation. Recovery, therefore, depends not only on what is learned but also on how and where learning occurs. This is the rationale for the community itself serving as teacher. Socially responsible roles are acquired by acting the roles. Sustained recovery requires a perspective on self, society, and life which must be continually affirmed by a positive social network of others within and beyond residency in the TC.

Treatment As an Episode. A person's relatively brief residency in a TC must compete with the influences of all the years before and after treatment. For this reason, unhealthy "outside" influences are minimized until the patient is better prepared to confront them. Thus, the treatment regimen is designed for high impact: Life in the TC is necessarily intense; its daily regime demanding; and its therapeutic confrontations unmoderated.

VIEW OF RIGHT LIVING

TCs adhere to certain precepts that constitute a view of healthy personal and social living. Moral positions regarding social and sexual conduct are unambiguous: Explicit right and wrong behaviors—

such as exploitative sexual conduct and other antisocial behaviors and attitudes, and the negative values of the streets, jails, or negative peers—are identified and paired with appropriate rewards and sanctions.

A central issue in the recovery process is guilt—toward self, significant others, and the community outside the TC. It is believed that if these guilts are not addressed, the self-acceptance necessary for personal change will be blocked. Personal growth and social learning are stressed, with emphasis placed on acquiring the positive values of the work ethic, self-reliance, earned rewards and achievement, personal accountability, responsible concern ("brother's/sister's keeper"), honesty in word and deed, social manners, and community involvement. Past behavior and past circumstances are explored only to illustrate current dysfunctional patterns. The personal present (here and now) is the focus of treatment and individuals are encouraged and trained to assume personal responsibility for their present reality and their destinies.

WHO COMES FOR TREATMENT?

Most people entering TCs have used multiple drugs—including TOBACCO, MARIJUANA, ALCOHOL, OPIOIDS, pills, and, recently, COCAINE and CRACK-cocaine. In addition to their substance abuse, most TC clients also have a considerable degree of psychosocial dysfunction. In traditional TCs, 70 to 75 percent of clients are men, but admissions for women are increasing. Most community-based TCs are integrated across gender, race/ethnicity, and age. In general, Hispanics, Native Americans and clients under twenty-one years of age have represented the smallest proportion of TC admissions. About half of all adult admissions are from disrupted or dysfunctional families. Most have poor work histories and have engaged in criminal activity at some time in their lives. A substantial number have already been in treatment for drug use at least once before. A large proportion of the adolescents who come to TCs are school DROPOUTS and have been arrested or involved with the criminal-justice system at least once. Many have histories of family deviance; some have been treated for psychological problems; some have had prior treatment for drug use.

Several studies have shown that the psychological profiles of people admitted to TCs are remarkably similar, reflecting in part the large proportion of residents referred from the criminal-justice system. These psychological profiles vary little across age, sex, race, primary drug, or admission year, and are not significantly different from drug abusers in other treatment modalities. Typically, symptom measures of depression and anxiety are high, socialization scores are poor, and IQ is in the dull-normal range.

In the few studies where psychiatric diagnoses of TC admissions were obtained using the DIAGNOSTIC INTERVIEW SCHEDULE [DIS], over 70 percent of the admission sample revealed a lifetime nondrug psychiatric disorder in addition to substance abuse or dependence (COMORBIDITY). One third of those admitted had a current history of mental disorder. The most frequent nondrug diagnoses were phobias, generalized ANXIETY, affective disorders, psychosexual dysfunction, and ANTISOCIAL PERSONALITY (De Leon 1988a, 1993; Jainchill, 1989, 1994; Jainchill, De Leon & Pinkham 1986).

Guiding the approach to treatment in the TC is the view, based on long experience, that clients admitted have problems with authority and low tolerance for delay of gratification for all forms of discomfort. They are unable to manage feelings (particularly hostility, guilt, and anxiety); they have poor impulse control (particularly sexual or aggressive); and they have poor judgment and reality testing concerning the consequences of their actions. They are given to unrealistic self-appraisal; tend to use lying, manipulation, and deception as COPING behaviors; and are personally and socially irresponsible. TCs regard positive changes in these negative characteristics as essential for stable recovery.

Contact and Referral. Clients come to TCs from several sources—social agencies, treatment providers, self-referral, and through active recruitment by the programs. Outreach teams—consisting usually of trained graduates of TCs and selected staff—recruit in hospitals, jails, courtrooms, social agencies, and the street, by conducting brief orientations or face-to-face interviews to determine receptivity to the TC. Although most people come into TCs voluntarily, many are actually being pressured to do so by family or significant others, by difficulties in the workplace, or by anticipated legal problems. Approximately 30 percent of TC admissions are mandated to treatment by the courts (De Leon, 1988b); the percentage is higher for adolescents (40–50%) than adults (25–30%).

COMMON CHARACTERISTICS

Admission Interview. Structured admission interviews last about one hour and may be followed by a second interview, which often includes family members or other significant persons. Previous legal, medical, psychiatric, and drug-treatment histories are also evaluated. Interviews are conducted by trained paraprofessionals and form the basis for program placement within the TC treatment system or referral to an oustide source.

Detoxification. Because most primary abusers of opioids, cocaine, alcohol, BARBITURATES, and AMPHETAMINES have undergone self or medical DETOXIFICATION prior to seeking admission to a TC, traditional TCs do not provide this service. Barbiturate users are routinely referred for medically supervised detoxification. HALLUCINOGEN or PHENCYCLIDINE (PCP) users are referred for psychiatric service if they appear to be having psychiatric complications.

Criteria for Residential Treatment. Traditional TCs maintain an "open-door" policy with respect to admission to residential treatment, thereby attracting people who may not be equally motivated or suited to the regime of such treatment. There are, however, some exclusionary criteria having to do with the patient's ability to meet the demands of the TC regime and with potential community risk. A person coming into treatment is expected to be able to participate in groups, fulfill work assignments, and live in an open community with minimal privacy (usually a dormitory). Risk refers to the extent to which a person poses a threat to the community (such as a history of arson or violence) or a management burden to the staff (such as serious psychiatric disorder). While those on medication for psychiatric disorder may be excluded, other chronic conditions requiring medication do not necessarily result in exclusion, providing they do not interfere with the ability of the client to participate fully in the program. This is true as well for those physically disabled.

All those entering a TC have complete physical examinations and laboratory workups after admission. Most TCs have established policy and practice concerning testing for HUMAN IMMUNODEFICIENCY VIRUS (HIV), for management of AIDS (ACQUIRED IMMUNODEFICIENCY SYNDROME) or AIDS-related complex (ARC), for hepatitis and tuberculosis.

Suitability for the TC. A number of those seeking admission to the TC may not be ready for treatment nor for the demands of a long-term residential regime. Assessment of these factors at admission provides a basis for appropriate referral. Some indicators of motivation, readiness, and suitability for TC treatment are the acceptance of the severity of one's drug problem; the acceptance of the need for treatment ("can't do it alone"); willingness to sever ties with family, friends, and current lifestyle while in treatment; willingness to meet the expectations of a structured community. Although motivation, readiness, and suitability are not criteria for admission to the TC, the importance of these factors often emerges after entry to treatment and, if not identified and addressed, lead to early dropout (De Leon & Jainchill, 1986; De Leon et al., 1993; De Leon, 1994).

THE TC APPROACH

The diverse elements and activities of the therapeutic community are used to foster rehabilitative change. The TC & is stratified into junior, intermediate, and senior peer levels; they constitute the community, the family. This structure serves to strengthen the individual's identification with an ordered network of others. More importantly, it arranges relationships of mutual responsibility with others at various levels in the program.

The TC is managed by staff who monitor and evaluate client status, supervise resident groups, assign and supervise resident job functions, and oversee house operations. Staff members conduct therapeutic groups (other than peer encounters), provide individual counseling, organize social and recreational projects, and confer with significant others. They decide matters of resident status, discipline, promotion, transfer, discharge, furlough, and treatment planning.

Working together under staff supervision, the residents are responsible for the daily operation of the community. Job assignments include providing all house services (e.g., cooking, cleaning, kitchen service, minor repair), serving as apprentices, running all departments, and conducting house meetings, certain seminars, and peer-encounter groups.

New clients enter a setting of upward mobility. The resident job functions are arranged in a hierarchy, according to seniority, clinical progress, and productivity. Job assignments begin with limited responsibility and lead vertically to levels of coordi-

nation and management. Indeed, clients come in as patients and can become staff. This social organization of the TC reflects the fundamental aspects of its rehabilitative approach, work as education and therapy, mutual self-help, peers as role models, and staff as rational authorities.

Work. Vertical job movements carry the obvious rewards of status and privilege, although the more frequent lateral job changes provide exposure to all aspects of the community. This increased involvement also heightens the residents' sense of belonging and strengthens their commitment to the community. Conversely, lateral or downward job movements also create situations that require demonstrations of personal growth. A resident may be removed from one job to a lateral position in another department or dropped back to a lower status position for clinical reasons. These movements are designed to teach new ways of coping with reversals and change that appear to be unfair or arbitrary.

Mutual Self-help. The essential dynamic in the TC is mutual self-help. The day-to-day activities of a TC are conducted by the residents themselves. In their jobs, groups, meetings, recreation, and personal and social time, residents continually transmit to each other the main messages and expectations of the community.

Peers As Role Models. In the mutual self-help context, peers as role models and staff as role models and rational authorities become the primary mediators of the recovery process. All members of the community are expected to be role models: roommates; older and younger residents; junior, senior, and directorial staff. All members who demonstrate the expected behaviors and reflect the values and teachings of the community are viewed as role models. This is illustrated in two main attributes.

Resident role models "*act as if*"; that is, they behave as the person they should be, rather than as the person they have been. Despite resistances, perceptions, or feelings to the contrary, they engage in the expected behaviors and consistently maintain the attitudes and values of the community. These include self-motivation, commitment to work and striving, positive regard for staff as authority, and an optimistic outlook toward the future. *Acting as if* is viewed as an essential mechanism for more complete psychological change, since altered self-perceptions, feelings, and insights often follow, rather than precede, behavior change.

Role models display *responsible concern*. This concept is closely akin to the notion of "I am my brother's/sister's keeper." Showing responsible concern involves willingness to confront others whose behavior is not in keeping with the rules of the TC, the spirit of the community, or the knowledge that is consistent with growth and rehabilitation. Role models have the obligation to be aware of the appearance, attitude, moods, and performances of their peers—and to confront negative signs in these. They must in particular be aware of their own behavior within the community and in the process prescribed for personal growth.

Staff as Rational Authorities. Through their managerial and clinical functions, and in their psychological relationships with the residents as role models, parental surrogates, and rational authorities, staff members foster the self-help learning process. In order to gain personal autonomy, residents need a successful experience with an authority figure who is viewed as supportive, corrective, protective, and credible (recovered). Staff members provide the reasons for their decisions and explain the meaning of consequences; they use their powers to teach and guide, to facilitate and correct, rather than to punish, control, or exploit.

THE TC PROCESS:
BASIC PROGRAM ELEMENTS

Therapeutic–Educative Element. Individual counseling and various group processes provide settings for expressing feelings, diverting negative acting-out, examining and confronting behavior and attitudes, and resolving personal and social issues. These activities increase communication and interpersonal skills, and provide guidance in alternate modes of behavior.

The four main forms of group activity are encounters, probes, marathons, and tutorials. While differing somewhat in format, objectives, and method, they all aim to foster trust, personal disclosure, intimacy, and peer solidarity to facilitate therapeutic change. The focus of the encounter is behavioral. Its approach is confrontation, and its objective is the direct modification of negative behavior and attitudes. Probes and marathons have as their primary objective significant emotional change and psychological insight. Tutorial groups stress learning concepts and specific skills.

Encounters are the cornerstone of group process in the TC. The term encounter is generic, describing a variety of forms that utilize confrontational procedures as their main approach. The basic encounter is a peer-led group composed of twelve to twenty residents meeting at least three times weekly, usually for two hours in the evening, followed by thirty minutes for snacks and socializing. Although often intense and profoundly therapeutic, the basic objective of each encounter is modest and limited: To heighten individual awareness of specific attitudes or behavioral patterns that should be modified.

Probes are staff-led group sessions composed of ten to fifteen residents, conducted as needed, to obtain in-depth clinical information on clients during the first two to six months of their residency. The main objectives of probes are to increase staff understanding of a client's background for purposes of treatment planning and to increase openness, trust, and mutual identification. Unlike the encounter, which stresses confrontation and focuses on the behavior of here and now, the probe, which may last from four to eight hours, emphasizes the use of support, understanding, and the empathy of the other group members; it explores past events and experiences.

Marathons are extended group sessions intended to initiate resolution of life experiences that have impeded the person's development. They are usually composed of large groups of selected residents and meet for eighteen to thirty-six hours. Every resident participates in several marathons during a typical eighteen-month stay. These groups are conducted by all staff, assisted by senior residents with marathon experience ("shepherds"). They employ a wide variety of techniques, including elements from primal therapy, theater, and psychodrama. Considerable personal and professional experience is required to assure safe and effective marathons.

The aim of the marathon is to break down defensiveness and resistance to painful but meaningful experiences. The supportiveness and safety of the setting promotes intimacy and bonding among the participants, facilitates emotional processing ("working through") of significant life events, and encourages the continued examination of the importance of life-altering past issues, which have been identified in counseling, probes, and other groups. These may include abandonment, sexual abuse, death of significant others, and other traumatic or catastrophic events.

Tutorial groups aim primarily to train and teach, rather than to promote psychological insight or correct behavior. They are scheduled as needed, are usually led by staff, and address specific themes, such as recovery and right living (self-reliance, maturity, relationships); job skills; and clinical skill training (e.g., use of encounter techniques).

Other Groups that vary in composition, focus, and format are convened as needed to supplement the main groups. They may, for example, be gender, ethnic, or age-specific groups and may use any of the major group techniques to focus on particular issues.

Counseling on a one-to-one basis aims to balance the needs of the individual with those of the community. Both formal and informal staff counseling sessions are scheduled as needed. The main features of TC staff counseling are transpersonal sharing, giving both direct support and instruction, providing minimal interpretation, and engaging in concerned confrontation. Since open peer exchange is ongoing within the community, in daily life informal counseling is always taking place.

COMMUNITY ENHANCEMENT

Activities designed to facilitate assimilation into the community include the morning meeting, seminars, and the house meeting, which are held every day, and the general meeting, called when needed.

Morning meeting is convened after breakfast and usually lasts about thirty minutes; it includes all residents and staff. Teams of residents conduct the meeting which usually consists of a planned program of readings, songs, and skits, and a recitation of the TC philosophy. The meeting is intended to motivate residents, strengthen unity, and begin the day's activities with a positive attitude. *Seminars*, conducted by residents, staff, or outside speakers, meet in the afternoon and usually last about one hour. Seminars emphasize listening, speaking, and conceptual skills. *House meetings* are mainly to transact community business. They are held every night after dinner, usually last for one hour, and are coordinated by a senior resident. *General meetings* are called only when needed, usually to address special circumstances or events, such as incidents in the community, or negative behavior and attitudes. They are led by multiple staff members who may use a variety of techniques (such as lecturing, public testimony, dispensing sanctions, or guilt-relieving special sessions). They deal with problem people or conditions,

reaffirm motivation, and reinforce positive behavior and attitudes.

COMMUNITY AND CLINICAL MANAGEMENT

Activities intended to maintain the psychological and physical safety of the community environment, ensure orderly and productive life for the residents, and strengthen the community as a context for social learning include the use of privileges, disciplinary sanctions, and surveillance.

Privileges. These are explicit rewards that reinforce the value of achievement. They are earned by demonstrating clinical progress in the program, may be lost through displays of inappropriate behavior or negative attitude, and can be regained by showing improvement. Privileges may range from being permitted to write letters or make telephone calls early in treatment to overnight furloughs at later stages. Specific privileges are linked to the degree of personal autonomy achieved by successfully moving through different stages in the program. Loss of privileges constitutes a drop in status in the vertical social system of the TC and can be a painful experience. Since many substance abusers have never learned to distinguish privilege from entitlement, TC residents learn through this reward system that productive membership in a family or community is earned by personal effort. The fact that privileges are gained through investing time and energy, working on self-modification, and risking failure and disappointment gives them social and psychological relevance and enhances their potency as behavioral reinforcers. They are tangible rewards contingent on evidence of individual change.

Discipline and Sanctions. Since social and physical safety are prerequisites for psychological trust, sanctions are invoked against any behavior that threatens the safety of the therapeutic environment. Violence or the threat of violence can result in immediate expulsion from the community. Even minor infractions of rules are regarded as threats to community well-being that must be addressed. Disciplinary actions ("contracts") are related to the severity of the infraction, the history of infractions, and the time in the program. They may range from verbal reprimands to loss of privileges, speaking bans, job demotions, or expulsion. Contracts are intended to provide a learning experience by compelling residents to attend to their conduct, reflect on their motivation, experience consequences of their behavior, and consider alternate forms of acting under similar situations. Since the entire facility is made aware of all disciplinary actions, they serve to deter violations and provide vicarious learning experiences for others.

Surveillance: The House Run. The house run is the TC's most comprehensive method for monitoring the physical and psychological status of the community. Staff and senior residents walk through the entire facility several times a day. They examine its overall condition and assess cleanliness, routines, safety procedures, and general morale. During house run they can observe psychological and social functioning of individual residents and peer groups, including evidence of self-management skills, attitudes toward self and the program, mood and emotional status, and the general level of awareness of self, and the physical and social environment of residents and staff. When infractions of house regulations or expectations are noted during house run, action (such as reprimand or instruction) may be taken immediately or the issue may be deferred to later encounter groups or house meetings.

Urine Testing. Most TCs utilize unannounced weekly random urine testing or, more commonly, incident-related urine-testing based on suspicion or observation of drug use. Residents who refuse urine testing on request are rejecting a fundamental expectation in the TC, which is to trust staff and peers enough to disclose undesirable behavior. The voluntary admission of drug use initiates a learning experience, which includes a review of the reasons or "triggers" precipitating the infraction. Denial of actual drug use, either before or after urine testing, is viewed as blocking the learning process and may lead to termination or drop-out. When evidence of drug use (positive urine) is detected, the action taken depends on the drug used, time in the program, previous history of drug and other infractions, place and condition of drug use, and the precipitating factors. Actions may range from demotion in community status to expulsion.

THE CHANGE PROCESS

Rehabilitation and recovery in the TC is a developmental process taking place in a social-learning setting. The goals are to establish a positive personal–social identity and a new lifestyle. Values, conduct, emotions, and cognitive understanding (in-

sight) must be integrated toward these goals. Achieving the goals reflects the individual's relationship to the community and acceptance of its teachings. This is a changing relationship which can be characterized as *compliance, conformity* and *commitment* (De Leon, 1994).

Compliance refers to adherence to norms and expectations of the community, primarily to *avoid* negative consequences, such as disciplinary sanctions, or undesirable alternatives, such as discharge to the street, a return to jail or to an unpleasant home situation. *Conformity* refers to adherence to the expectations and norms primarily to *maintain* affiliation with the community, personal relationships, or the acceptance of peers and staff. *Commitment* refers to adherence to a *personal resolve* to remain in the change process, which reflects internalization of the therapeutic and educational teachings of the TC.

PROGRAM STAGES

The three major program stages that characterize change in long-term residential TCs are orientation–induction, primary treatment, and reentry. There are also substages or phases in the process. Typical profiles of each stage are briefly outlined.

Stage I (Orientation–Induction, 0 to 60 Days). The main objectives of this initial phase of residency are further assessment and orientation to the TC. The admission evaluation does not always yield a complete picture of the client, as is evident from the high dropout rates at this phase. Thus, clinical assessment continues during the first two months to clarify specific treatment needs and overall suitability for the long-term residential TC.

The individual is quickly assimilated into the community through full participation and involvement in all activities. Group process aims to facilitate involvement in the community and acceptance of the regime, and to train clients to use the group itself. Formal seminars and informal peer instruction focus on reducing anxiety by conveying information about cardinal rules (no drugs, no violence, no threat of violence); house regulations (maintaining manners, no leaving the facility, no stealing, borrowing or lending, etc.); expected conduct regarding speaking, dressing, punctuality, attendance, etc.; program essentials (organization, job functions, privilege system, process stages, philosophy and perspective); therapeutic community tools (encounter and other groups).

Successful passage through this initial stage is reflected mainly by client retention; the fact that clients remain thirty to sixty days indicates that they have adhered to the rules of the program enough to meet the orientation objectives of this stage and have passed the period of highest vulnerability to early dropout. At the end of this period, new reisdents are expected to know the structure and to work at maintaining compliance.

Stage II (Primary Treatment, 2 to 12 Months). Primary treatment consists of three phases that roughly correlate with time in the program (2–4 months; 5–8 months; and 9–12 months). These phases are marked by plateaus of stable behavior that signal the need or readiness for further change. The daily therapeutic-educational regimen, meetings, groups, job functions, peer and staff counseling, remains the same throughout the year of primary treatment. Progress is reflected at the end of each phase in terms of three interrelated dimensions of change: community member development, maturity, and overall psychological adjustment.

Four-month residents have junior status in the TC, limited freedom, and lower-level jobs. They display a general knowledge of the TC approach and can *"act as if"* they understand and accept the perspective and regime. They comply with the program, follow directions, participate fully in daily activities, and engage in expected behaviors (such as getting up in the morning, making the bed, keeping the area clean, attending all meetings). They adhere to the cardinal and house rules and accept disciplinary contracts. At this stage, they are able to acknowledge the seriousness of their drug use and demonstrate awareness of personal-growth issues and other life problems. They show some separation from the language and attitudes of the drug culture or street code. Their participation in groups increases, and they display limited personal disclosure in groups and in one-to-one sessions.

Eight-month residents "set an example." Their earned privileges give them greater personal freedom, including the right to leave the facility briefly, without escort, when going to specified places for specific reasons. They hold higher positions in the job hierarchy, some of which may pay a small stipend. Their attitudes reflect their acceptance of the values of "right living": honesty, responsibility, and the importance of role models. They adapt to job changes, accept staff as rational authorities, and have learned to cope with and contain negative

thoughts and emotions. The TC precept to *"act as if"* has been internalized as a personal mode of learning, and a personal commitment to remain in the change process is evident. They generally reveal elevated self-esteem, particularly through positive and optimistic assertions about next steps. They are more self-aware and accept full responsibility for their behavior, problems, and solutions. They are less defensive when confronted and more open about personal disclosure. They have acquired group skills and are able to assist facilitators in encounter groups.

Twelve-month residents are established role models in the TC. They have more personal autonomy, more privacy, and can regularly obtain furloughs. They accept responsibility for themselves and for other members in the community. They hold responsible positions and can effectively run the house. As senior coordinators under staff supervision they may be responsible for arranging seminars, trips, or the movement of residents. They are eligible to be staff-in-training, in executive management offices or as junior counselors. They assist staff in monitoring the facility overnight and on weekends. Some may be enrolled in vocational-educational programs. Job performance is consistent; goal-setting and self-assessment are realistic. Social interactions with staff are comfortable, and during furlough or recreation they socialize with a network of positive peers. They can adapt to new situations and can teach TC concepts to others. They are fully trained participants in the groups process and can serve as facilitators. They display insight into their own drug problems and personalities and offer a high level of personal disclosure in groups, peer exchange, and staff counseling sessions. Although eager to move forward, the twelve-month resident may express anxiety and uncertainty about the future.

Stage III (Reentry, 13 to 24 Months). Reentry is a two-phase stage at which the client must strengthen skills for autonomous decision making and the capacity for self-management, with less reliance on rational authorities or a well-formed peer network.

The main objective of the *early reentry phase* (13 to 18 months) is to prepare for healthy separation from the community. There is less emphasis on rational authority, and more individual decision making about privileges, social plans, and life design. The group process involves fewer leaders, fewer encounters, and more shared decision making. Life-skills seminars are emphasized, with mandatory sessions on budgeting, job seeking, alcohol use, sexuality, parenting, use of leisure time, etc. In collaboration with key staff members and peers, clients make plans for long-term personal, social, psychological, educational, and vocational goals. During this phase, clients may be attending school or holding full-time jobs, either in or outside the TC, but they are still expected to participate in house activities whenever possible and to assume some community responsibilities.

The objective of the *later reentry phase* (18 to 24 months) is to successfully separate from residency in the TC. Clients are on "live-out" status, involved in full-time jobs or education, maintaining their own households, usually with live-out peers. They may attend such aftercare services as ALCOHOLICS ANONYMOUS (AA) or NARCOTICS ANONYMOUS (NA), or take part in family or individual therapy. This phase is viewed as the end of residency, but not of program participation. Contact with the program is frequent at first and only gradually reduced to weekly phone calls and monthly visits with a primary counselor.

Completion marks the end of active program involvement, and *graduation*, an annual event conducted in the facility for completers, usually takes place a year beyond their residency. While residence in the program facilitates a process of change, this change must continue throughout life. What is learned in treatment are the tools to guide the individual on a steady path of continued change. In the TCs' view, completion, or graduation, therefore, is not an end but a beginning.

AFTERCARE

Although aftercare is not a stage of the residential treatment process, it does underscore the continuing, perhaps lifelong process of recovery. Until recently, long-term TCs have not formally acknowledged aftercare as a definable period following program involvement (De Leon, 1991). Nevertheless, TCs have always appreciated the clients' efforts to maintain sobriety and a positive lifestyle beyond graduation. In most TCs, key clinical and life-adjustment issues of aftercare are addressed during the reentry stages of the two-year program. As discussed in the later section on modifications, however, many TCs now offer explicit aftercare components within their systems or through linkages with outside agencies.

CURRENT MODIFICATIONS OF THE TC MODEL

Most community-based traditional TCs have expanded their social services or have incorporated new interventions to address the needs of diverse client populations. In some case these additions enhance but do not alter the basic TC regime; in others they significantly modify the TC model itself. Some therapeutic communities are experimenting with substantially shorter periods of residence (6 months, for example), followed by a period of six to eight months during which participants live in shared apartments while maintaining contact and participating in the activities of the residential center.

Some current TC innovations actually represent the reintroduction of activities and philosophies of some first-generation TCs, which were dropped because of cost or pressure from outside regulators. For example, in the late 1960s, Synanon, Odyssey House, and GATEWAY FOUNDATION allowed women to move into therapeutic communities with their children. This not only reduced the conflict that women experienced when faced with leaving their children for a year to get treatment, but it also reduced their tendency to leave the TC prematurely. In the opinion of some TC pioneers, the presence of children had beneficial effects for the children, who enjoyed the attention of a number of sober adults (sometimes for the first time in their lives), and also for the residents, who learned parenting skills while learning to manage their own lives. For various reasons, the practice of including children in the TC setting was discontinued in the 1970s, but it is now being rediscovered.

Some early innovations have rarely been replicated. For example, in 1970, Tinley Park was established as part of the Illinois Drug Abuse Programs (IDAP) to provide the therapeutic potential of the TC to those who had opted for methadone maintenance treatment but who were not able to control their alcohol use, cease illicit drug use entirely, or stabilize their lives. Tinley Park admitted any IDAP participant who needed an extended residential program, not just those taking methadone, and provided a back-up for outpatient programs. As part of the IDAP system, Tinley Park was able to draw on the talents of Gateway staff, which included David Deitch (IDAP's Director of Training) and other former Daytop Village staff. Tinley Park demonstrated the possibilities of mixing treatment modalities—TC approaches, methadone, and outpatient groups—in one facility. Like the first-generation TCs, women with children were welcome (Glasscote et al., 1972).

THE TC MOVEMENT

Most of the first-generation TCs (those that began in the 1960s) are still active and many have expanded to multiple sites, some of them international—these include DAYTOP VILLAGE, PHOENIX HOUSE, Odyssey House, GATEWAY, and MARATHON. Among the active well-known second-generation TCs, begun in the 1970s, are Gaudenzia, Samaritan, AMITY, WALDEN HOUSE, ABRAXIS, Integrity, and CURA.

Therapeutic communities are now a key element in the U.S. response to drug dependence. There are more than 400 TC programs in the United States and Canada, serving more than 100,000 people each year. Representatives of most of these programs regularly share experiences at the annual meetings of the Therapeutic Communities of America (TCA), an organization founded in 1975. In 1994, the thirty-one TCs listed as Board or Corresponding Members of TCA were from twelve states and Canada. These included: (from Arizona) Amity; (California) Asian American Recovery Services, Inc., Center Point, Inc., G.R.O.U.P., Inc., House of Metamorphosis, Inc., Tarzana Treatment Center, and Walden House, Inc.; (Florida) OPERATION PAR, INC.; (Illinois) Gateway Foundation, Inc.; (Maryland) SECOND GENESIS, Inc.; (Massachusetts) Spectrum, Inc.; (Michigan) SHAR, Inc.; (Nevada) WestCare, Inc.; (New Jersey) CURA, Inc., Discovery, Inc., Integrity, Inc., and Newark Renaissance House, Inc.; (New York) A-Way Out, Inc., APPLE, Inc., Daytop Village, Inc., H.E.L.P./Project Samaritan, Inc., Outreach Project, Inc., Phoenix House Foundation, Inc., PROJECT RETURN FOUNDATION, Inc., Samaritan Village, Inc., and Stay'n Out; (Pennsylvania) Abraxas Group, Inc., Gaudenzia, Inc., and House of the Crossroads; (Rhode Island) MARATHON, Inc.

Further information on these and other TCs is available from Therapeutic Communities of America, 1555 Wilson Boulevard, Suite 300, Arlington, VA 22209. Telephone: 703/875-8636; FAX: 703/812-8875. (See Appendix for more information.)

By the 1970s, the therapeutic-community movement had already expanded beyond the United States and Canada; in the 1990s, TCs were well established in at least thirty countries. Many belong to the World Federation of Therapeutic Communities. There are TCs in Spain, England, Sweden, and Germany, and a religious variation of the TC concept in Hungary. Italy has several TCs developed and supported by Roman Catholic priests (C.E.I.S.), although religious observance is not a required element of the program, as well as TCs that are not affiliated with the Roman Catholic church. There are several TCs in Argentina and Brazil, and a few in Peru and Colombia. TCs have also been established in Malaysia and the Philippines. There are a number of sites in France and other parts of Europe under the general designation of LE PATRIARCHE, after the charismatic founder of the program; although these programs are self-help communities, they do not incorporate the educational and therapeutic elements of the TC model as described above. TC research has expanded considerably in recent years, illustrated in the establishment of a national Center for Therapeutic Community Research.

(SEE ALSO: *Civil Commitment; Coerced Treatment for Substance Offenders; Prisons and Jails: Drug Treatment in; Shock Incarceration and Boot Camp Prisons*)

BIBLIOGRAPHY

DE LEON, G. (1994). Therapeutic communities: Toward a general theory and model. In F. M. Tims, G. De Leon, and N. Jainchill (Eds.), *Therapeutic Community: Advances in research and application*. National Institute on Drug Abuse Research Monograph 144. Publication no. 94-3633. Rockville, MD: National Institute on Drug Abuse.

DE LEON, G. (1993). Cocaine abusers in therapeutic community treatment. In F. M. Tims & C. G. Leukefeld (Eds.), *Cocaine Treatment: Research and clinical perspectives*. National Institute on Drug Abuse Research Monograph, Publication no. 93-3639. Rockville, MD: National Institute on Drug Abuse.

DE LEON, G. (1991). Aftercare in therapeutic communities. *International Journal of Addictions*, 25(9A + 10A), 1229–1241.

DE LEON, G. (1989). Psychopathology and substance abuse: What we are learning from research in therapeutic communities. *Journal of Psychoactive Drugs*, 21(2), 177–188.

DE LEON, G. (1988). Legal pressure in therapeutic communities. In C. G. Leukefeld & F. M. Tims (Eds.), *Compulsory treatment of drug abuse: Research and clinical practice*. National Institute on Drug Abuse Research Monograph. Publication (ADM)88-1578. Rockville, MD: National Institute on Drug Abuse.

DE LEON, G., & JAINCHILL, N. (1986). Circumstances, motivation, readiness and suitability (CMRS) as correlates of treatment tenure. *Journal of Psychoactive Drugs, 8*(3), 209–213.

DE LEON, G., & ROSENTHAL, M. S. (1989). Treatment in residential therapeutic communities. In T. B. Karasu (Ed.), *Treatments of Psychiatric Disorders*. Washington, DC: American Psychiatric Press.

DE LEON, G., & ZIEGENFUSS, J. (Eds.) (1986). *Therapeutic communities for addictions: Readings in theory, research and practice*, Springfield, IL: Charles C. Thomas.

DE LEON, G., ET AL. (1994). Circumstances, motivation, readiness and suitability (the CMRS scales): Predicting retention in therapeutic community treatment. *American Journal of Drug and Alcohol Abuse, 20*(4), 495–575.

DE LEON, G., ET AL. (1993). Is the therapeutic community culturally relevant? Findings on race/ethnic differences in retention in treatment. *Journal of Psychoactive Drugs, 25*(1), 77–86.

GLASSCOTE, R., ET AL. (1972). *The treatment of drug abuse: Programs, problems, prospects*. Washington, DC: Joint Information Service.

JAINCHILL, N. (1994). Co-morbidity and therapeutic community treatment. In F. M. Tims, G. De Leon, & N. Jainchill (Eds.), *Therapeutic community research*. National Institute on Drug Abuse Research Monograph 144. Publication no. 94-3633. Rockville, MD: National Institute on Drug Abuse.

JAINCHILL, N. (1989). *The relationship between psychiatric disorder, retention in treatment, and client progress among admissions to a residential drug-free modality*. Unpublished Ph.D. dissertation, New York University.

JAINCHILL, N., DE LEON, G., & PINKHAM, L. (1986). Psychiatric diagnosis among substance abusers in therapeutic community treatment. *Journal of Psychoactive Drugs, 18*, 209–312.

GEORGE DE LEON
JEROME H. JAFFE

Traditional Dynamic Psychotherapy *Dynamic psychotherapy* is the term for the various psychological treatments, primarily talking treatments, intended to modify and ameliorate behaviors based on inner conflicts (e.g., "Should I study for the test or cheat?") and/or interpersonal conflicts (difficulties with others). These techniques range from those intended primarily to support individuals, lending them the therapist's strength or understanding ("If you do that you'll get in trouble. Have you thought of handling it this way?"), to helping patients reach their own understanding of the origins and implications of their behaviors. The application of these techniques to the treatment of alcoholics and substance abusers is supported by the high incidence of cooccurrence of psychiatric illness—in several studies, 70 percent—some of which may play a role in initiating or maintaining the behavior. It has been suggested that for some substance abusers, the use of illicit compounds is a misguided attempt at self-medication. Often, psychotherapy must be provided in conjunction with other treatments—pharmacologic, such as DISULFIRAM for alcoholics or METHADONE for HEROIN abusers; SELF-HELP groups, such as ALCOHOLICS ANONYMOUS; or family or group psychotherapy.

Psychotherapy is based on the assumption that the patient will think and talk about ideas and feelings rather than acting upon them. This may prove particularly difficult for substance abusers who often have little sense of what they feel, other than generalized pain, and who are used to action and immediate gratification. Therefore, treatment, particularly at the beginning, must take place within a structure that both supports and helps control impulsive behavior. Sometimes, treatment starts in a hospital or other residential setting; often, it is accompanied by regular drug testing. After the agreement to start therapy and setting goals, therapist and patient meet once to several times a week. As trust is developed between patient and therapist, the therapist can expect less lying and less denial of difficulties; treatment can, if indicated, begin to move from support toward expression of feelings—toward identification of conflicts and the understanding of their origins. Initially the therapist listens, struggling to understand the patient's inner experience and its meaning. The therapist then attempts to help patients to understand what they have presented, with appropriate changes and qualifications based on further information provided by the patient. Important issues to be explored in treatment include current relationships (with spouse, children, friends, coworkers), past relationships (with parents and other family), and the relationship within the treatment between the patient and the therapist. Often, the difficulties and distortions within this relationship mirror past and current relationships and may be used to help the patient see the nature and impact of the past on current behaviors.

Treating substance abusers can be frustrating for therapists; there are many slips with return to drug use, and patient behavior is often calculated to make the therapist angry and to give up. It is essential that therapists who make the attempt carefully monitor their own feelings so that they do not interfere with the treatment itself. It is also important to remember that when properly done, treatment can make the difference between suffering with chronic problems and successful adaptation. This is particularly true when substance abuse is accompanied by other psychiatric disease and/or disability.

(SEE ALSO: *Causes of Substance Abuse: Psychological (Psychoanalytic) Perspective; Comorbidity and Vulnerability; Disease Concept of Alcoholism and Drug Abuse; Epidemiology*)

BIBLIOGRAPHY

AMERICAN PSYCHIATRIC ASSOCIATION. (1989). *Treatments of psychiatric disorders: A task force report of the American Psychiatric Association.* Washington, DC: Author.

WILLIAM A. FROSCH

Treatment Strategies Although drug and alcohol problems have been recognized for millennia, the treatment of such problems is a recent phenomenon. Most treatment approaches were developed in the 1960s or later. Although not considered formal treatment, an exception is Alcoholics Anonymous (AA), which started in the 1930s. In this article AA will be considered a treatment approach because it is directed at helping people refrain from drinking.

Treatment can have several objectives. It can be directed at helping an individual stop using alcohol or drugs (i.e., treating withdrawal symptoms). It can be directed at eliminating substance use problems over the long term. Finally, it can concern dealing with complications of substance abuse (e.g., liver disease) rather than the substance use itself. This article presents an overview of approaches to treating

withdrawal and accomplishing long-term behavior change.

STRATEGIES FOR FACILITATING WITHDRAWAL

If people use certain drugs, such as alcohol and heroin, frequently and at sufficient doses, they may experience withdrawal symptoms when they stop their use. Generally, the symptoms are opposite to drug effects (e.g., for depressant drugs, withdrawal symptoms tend to reflect high levels of arousal) and are unpleasant for the user. It has been speculated that once a person has used enough of a drug so that withdrawal symptoms will occur upon stopping use, continued use is motivated partly by a desire to avoid withdrawal. Nicotine withdrawal symptoms, for example, are thought to be one of the main reasons why smokers find it very difficult to stop smoking.

Since stopping use is often necessary to achieve long-term change in substance use, treatments have been developed explicitly to avoid or lessen withdrawal symptoms. Two main approaches have been used: medical and nonmedical detoxification. Medical treatments involve the use of medications, usually for a few days. Sometimes these medications substitute for the abused substance while avoiding adverse side effects. For example, an alcohol-dependent person who stops drinking might be given diazepam (Valium) and then be tapered off the diazepam. In the case of smokers, nicotine, is the medication of choice and the method of drug delivery (smoking) is replaced by nicotine gum or a nicotine skin patch. In other cases medication is used to prevent or reduce a particular withdrawal symptom (e.g., anticonvulsant medication to prevent seizures).

Nonmedical detoxification involves providing social support and monitoring the condition of the person withdrawing. Withdrawal symptoms decrease in severity over time and eventually disappear, usually within a few days. The vast majority of alcohol-dependent individuals can withdraw successfully on an outpatient basis and without the use of medication. For certain drugs (e.g., BARBITURATES) or when an individual is at risk of life-threatening withdrawal symptoms (e.g., delirium tremens resulting from alcohol withdrawal), withdrawal may be conducted in an inpatient setting. Whatever the procedure, the objective is to get the individual to the point where withdrawal symptoms are no longer present or are reduced to tolerable levels. Thus, withdrawal is only the first step in treatment intended to produce long-term behavior change. The use of NICOTINE GUM or METHADONE can sometimes be an exception to this rule, because these substances can also be used as maintenance medications. In a few cases withdrawal may be all that is needed, such as for an individual who has become dependent on a prescription drug and the medical reason for drug use is no longer present, but the person is unable to stop use because of withdrawal. Finally, some drugs do not produce a withdrawal syndrome (e.g., hallucinogens), and for drugs that can produce a syndrome, many users do not use sufficient amounts or use them frequently enough to experience withdrawal upon stopping use.

STRATEGIES FOR OVERCOMING ALCOHOL OR OTHER DRUG PROBLEMS

Treatment approaches are intended to produce and maintain change. In this article, approaches will be considered from the standpoint of how they are thought to work. Often more than one approach is used in treating the same individual. Treatment objectives are based on an assessment of the individual's presenting problems, history, and needs. For example, a heavy heroin user who lives with other users, has a poor job history, and reports difficulty in coping with life problems might be assessed as needing medical detoxification, coping-skills training, vocational counseling, and a living environment that does not support drug use. A problem drinker who now and then drinks to excess but has not experienced serious adverse consequences may be assessed as needing brief outpatient counseling with an objective of avoiding excessive drinking. As with other health problems, treatment should be tailored to the individual.

Strategies Providing Alternative Ways to Serve Drug Functions. The substitution of alternative behaviors or other effects is a common treatment approach that is based on the notion that excessive drinking or drug use is a learned but inappropriate way of accomplishing certain purposes. Usually treatment involves identifying situations and circumstances associated with substance use and the purposes served by use on those occasions. For example, a person may report drinking heavily to reduce anxiety in a social situation (e.g., meeting new

people at a party), using amphetamines (stimulants) for dieting, or using benzodiazepines (e.g., Valium) as a way of dealing with feelings of depression. Plans are then developed for looking at ways of dealing with the situation other than by using drugs. For example, treatment may help the person learn ways of socializing effectively (social-skills training) or ways to reduce anxiety (e.g., relaxation training). Sometimes the alternative may involve pharmacological treatment, such as the use of antidepressant or anti-anxiety medication.

At times these options may involve continued use of the drug, but in a way designed to avoid serious negative consequences. With smokers, an alternative method of nicotine administration (i.e., gum or a skin patch) can avoid the main risks to health (e.g., tar in cigarettes, which is associated with lung cancer). Use of methadone as a treatment for heroin abusers is perhaps the best-known example of this sort of harm-reduction strategy. Methadone is an opiate that has a longer-acting effect than heroin (it need be taken only once, rather than several times, per day), does not produce as much behavioral impairment, and can be medically prescribed. Although theory holds that its use should be as a "maintenance" medication (i.e., taken indefinitely), it is most often used as a substitute for heroin from which the user is then gradually withdrawn. Benefits from methadone use include the user's being less likely to engage in illegal behavior or to risk health consequences (e.g., AIDS, overdose).

Strategies Changing the User's Environment. Considerable research indicates that factors in the environment influence the risk of drug abuse. These factors range from ease of obtaining drugs, to social support for drug use, to whether alternative activities (e.g., recreation) are readily available. Some treatments are intended to rearrange the user's environment so as to discourage or prevent drug abuse. Perhaps the best-known approach involves residential treatment programs for drug abusers. The idea is that treatment will be facilitated by placing individuals in a social environment that supports recovery rather than continued drug use. Although research suggests that residential treatment confers no special benefits in most cases, for alcohol abusers it is a very common approach, especially in North America. The most popular version is the 28-day MINNESOTA MODEL, a multicomponent treatment program including education, AA, group therapy, and a disease philosophy. Residential treatment programs have a goal of abstinence.

For other drugs, residential treatment is typically conducted in THERAPEUTIC COMMUNITIES; however, outpatient treatment is a more common approach. The therapeutic community approach originated with SYNANON in the late 1950s and depends heavily on confrontation, group therapy, Narcotics Anonymous, self-reliance, and personal responsibility through example. Residential programs for drug abusers usually involve a much longer period of time (up to a year) than residential alcohol treatment programs.

An environmental-change strategy that has received considerable research support is contingency management. Based on learning theory, the environment is arranged so as to reward the desired behaviors (e.g., nonuse of drugs, alternatives to drug use) and punish undesired behaviors (e.g., use of drugs results in loss of privileges). This approach, sometimes called community reinforcement, has produced impressive results. Unfortunately, it requires a very large investment of time and effort by the treatment program and is highly dependent upon features of the environment that are difficult to control (e.g., actions of the abuser's friends). An exception to this is the use of take-home methadone as a reinforcer for persons being treated for heroin abuse. Persons on methadone come to the treatment program daily to receive their methadone and consume it in front of treatment staff (an opiate drug, methadone can be sold on the street). As trust develops, they may be given take-home methadone privileges, receiving a few days' supply at a time. Persons who use heroin sometimes also abuse alcohol and may increase their alcohol use when they stop using heroin. A contingency-management approach that has been applied to both problems makes take-home methadone privileges contingent upon the person's taking medication that will make him or her ill if alcohol is consumed.

Other approaches to environmental change are often part of an overall treatment plan. Such components may include changing social situations (e.g., avoiding friends who are drug abusers), changing recreational activities (e.g., working out at a gym instead of going to a bar), or even changing jobs or moving.

Strategies for Changing the Effects of Psychoactive Drugs. Much effort is being spent in developing treatments thought to operate through counteracting the effects of drugs. One major line of research is pharmacological treatments. There are various ways that medications can interfere with or moderate drug effects (e.g., directly counteract reward effects in the brain; decrease desire to use drugs; produce adverse reactions when drugs are ingested). Examples of medications that can be effective include disulfiram (Antabuse), which produces an allergic-type reaction when alcohol is ingested, and methadone and naltrexone, which block the "high" normally produced by heroin. A major problem with medications to counteract drug effects is exemplified by research investigating the effectiveness of Antabuse. Although Antabuse may deter some alcoholics from drinking, the problem is that they will not reliably take the medication. This highlights the fact that pharmacological treatments may need to be accompanied by psychosocial strategies—what use is a medication that clients will not take? Several medications are being investigated, and some have been found to reduce drug consumption (e.g., SEROTONIN-UPTAKE INHIBITORS have been found to reduce alcohol consumption by about 10% to 20%). How these medications work is little understood.

Nonpharmacological treatments aimed at changing drug effects are also being evaluated. These procedures are aimed at modifying conditioned (learned) reactions to drugs and to cues associated with drug use. Research indicates that various learned responses become associated with drug use. Some of these are opposite in direction to drug effects and can be thought of as preparing the body to receive drugs. For example, the sight of heroin (a depressant) injection paraphernalia can elicit in a heroin user an agitated (stimulated) reaction. Since these responses become associated with cues that signal drug use and can be aversive, it is thought that such cues can lead an individual to relapse (i.e., consuming a depressant drug can help alleviate the agitation). Borrowing from the effective treatment of obsessive/compulsive disorders, studies are investigating the value of cue exposure as a way of counteracting these learned responses. Such treatment puts the person in the presence of cues but instructs him or her not to engage in drug use. With sufficient exposure and no drug use, the cues lose their capacity to elicit a reaction.

Strategies for Increasing Commitment to Change. A recent development in treatments is the use of motivational interventions, procedures designed to increase a person's commitment to change his or her behavior. For many years drug abusers' motivation to change received little attention from clinicians other than as a convenient excuse for treatment failures (people did not recover because they were "not motivated"). This circular reasoning has been replaced by the recognition that motivation is a state which can change and can be influenced by treatment. A drug abuser can be considered to be either comfortable in the drug-using pattern and not thinking about changing; aware that change may be necessary and considering whether to change; in the process of changing; or having changed and taking actions to maintain the change. The value of this approach is that it points out that if an individual is not firmly committed to change, it makes little sense to invest much effort in effecting change. Rather, the first objective of treatment should be to increase the person's conviction that change is necessary. A major tactic of this sort is motivational interviewing, a non-confrontational style of interviewing intended to help people examine their drug use and, it is hoped, lead them to decide that change is necessary. Once a person is solidly committed to changing, treatment strategies aimed at helping him or her change can be used.

Self-change approaches to treatment employ some motivational components. For example, clients may be asked to select their own treatment goals and to develop their own strategies for change. Although self-change treatments often embody other procedures (e.g., learning alternatives to using drugs), here they are considered as motivational treatments because although they involve identifying appropriate options, the procedures typically do not include actual training in implementing the options. Because self-change treatments seldom involve more than a few outpatient sessions and often provide reading and homework assignments to clients, they have been used largely with clients who are educated and whose problems are not severe.

After change is accomplished, the goal is to maintain that change. The best-known procedures for this purpose are RELAPSE-PREVENTION procedures. They help people anticipate situations that might provoke a return to earlier drug-use patterns and plan alternative ways of dealing with those situations. Another focus of relapse prevention is on individuals' thoughts about relapses that might occur. People

who are committed to abstinence yet use drugs have to explain that contradiction to themselves. One explanation could be that the person simply is not capable of refraining from use, but this can be self-defeating and lead to further drug use. Using the relapse-prevention approach, people are counseled to think of maintaining recovery as a difficult task that sometimes may be interrupted by lapses. Although the lapses are setbacks and signify that more work is needed to prevent a future similar situation from leading to a lapse, they do not mean that the individual is incapable of preventing lapses. The individual is encouraged to stop a lapse as quickly as possible (i.e., not allow it to become a relapse), and to construe it as an unfortunate event from which something can be learned but that should not end attempts at recovery. There is a particular emphasis on recognizing that one's reaction to a lapse should not become self-defeating (e.g., "This proves I just can't do it").

Self-Help Approaches. It has been known for some time that some people are able to recover from alcohol and drug problems on their own (without treatment). Others find joining self-help groups such as AA, NARCOTICS ANONYMOUS, and spin-off groups to be helpful. Although it is debatable whether these approaches should be considered treatments, it is clear that many use them as if they are treatments, and for that reason they will be discussed. AA and similar approaches might be considered motivational interventions. Although they require recognition by users of their dependency on a substance, they place the responsibility for avoiding substance use squarely on the user. The driving force of the approach is to avoid an initial lapse (use), since that is viewed as placing the individual under control of the substance effects. Although the effectiveness of the various self-help approaches, including AA, has not been established by research, they have considerable popular support and are very cost-effective.

Public-Health Strategies. Historically, the bulk of substance-abuse countermeasures have been focused on intensive treatment of very chronic abusers. When treatment was first introduced, this was understandable, because care was scarce and those individuals were most at risk. As clinical services became established, it became apparent that a great deal could be accomplished by general health professionals (e.g., primary-care physicians, health educators) as opposed to specialized addictions workers. Secondary prevention (catching problems early in their development) also can have substantial benefits for society, because nonintensive interventions are often effective. Conducting secondary prevention in general health-care settings in the community helps make treatment available to a much larger population than can be served by specialized addictions agencies.

It has also been recognized that many smokers and alcohol abusers recover without formal help, and that it may be possible to facilitate self-change in the community. The recent decrease in smoking attributed to public-health campaigns (such as the surgeon general's reports) can be viewed as an example of this sort of facilitated self-change. Although community mass-media campaigns probably should not be considered treatments, other aspects of the public-health approach do involve treatments. It has been found that brief interventions by primary-care physicians can identify patients who are smokers or problem drinkers and encourage them to give up smoking or to reduce or stop their drinking. Such interventions have not yet been evaluated for substances other than TOBACCO and ALCOHOL, and they may be most appropriate for persons whose problems are not severe. They are highly cost-effective, however, and they reach a population far larger than persons who seek out treatment.

TREATMENT AS IT TYPICALLY IS CONDUCTED

The various treatment approaches have been discussed as they are reported in the research literature. Unfortunately, treatment as typically offered often bears little relation to the research literature. Despite considerable evidence that most people with alcohol problems benefit as much from less costly outpatient treatment as from inpatient treatment, inpatient treatment remains the most popular approach in North America. Treatment for alcohol problems is further plagued by controversy about treatment goals. Abstinence is the goal mandated by the vast majority of treatment providers, but research indicates that if secondary prevention initiatives are to succeed, moderation or reduced use must be recognized as a legitimate goal. Similarly, in clinical practice methadone tends to be prescribed in much lower doses and for much shorter intervals than research has demonstrated is necessary.

Finally, the most effective approaches may involve more than one type of intervention. It has been found, for example, that in smoking cessation nicotine gum combined with a behavioral treatment is more effective than either approach alone.

(SEE ALSO: *Alcohol; Alcoholism*)

BIBLIOGRAPHY

DELEON, G., & ZIEGENFUSS, J. T., JR. (1986). *Therapeutic communities for addictions.* Springfield, IL: Charles C. Thomas.

HAYASHIDA, M., ET AL. (1989). Comparative effectiveness and costs of inpatient and outpatient detoxification with mild-to-moderate alcohol withdrawal syndrome. *New England Journal of Medicine, 320,* 358–365.

HESTER, R. K., & MILLER, W. R. (EDS.). (1990). *Handbook of alcoholism treatment approaches: Effective alternatives.* New York: Pergamon.

KLEBER, H. D., & GAWIN, F. H. (1987). Pharmacological treatments of cocaine abuse. In A. W. Washton & M. S. Gold (Eds.), *Cocaine: A clinician's handbook.* New York: Guilford.

LITTEN, R. Z., & ALLEN, J. P. (1991). Pharmacotherapies for alcoholism: Promising agents and clinical issues. *Alcoholism: Clinical and Experimental Research, 15,* 620–633.

MARLATT, G. A., & GORDON, J. R. (EDS.). (1985). *Relapse prevention.* New York: Guilford.

MILLER, W. R., & HESTER, R. K. (1986). The effectiveness of alcoholism treatment: What research reveals. In W. R. Miller & Heather (Eds.), *Treating addictive behaviors: Processes of change.* New York: Plenum.

MILLER, W. R., & ROLLNICK, S. (1991). *Motivational interviewing: Preparing people to change addictive behavior.* New York: Guilford.

PATTISON, E. M.; SOBELL, M. B.; & SOBELL, L. C. (1977). *Emerging concepts of alcohol dependence.* New York: Springer.

PROCHASKA, J. O., & DICLEMENTE, C. C. (1986). Toward a comprehensive model of change. In W. R. Miller & N. Heather (Eds.), *Treating addictive behaviors: Process of change.* New York: Plenum.

RAWSON, R. A. (1991). Chemical dependency treatment: The integration of the alcoholism and drug addiction/use treatment systems. *International Journal of the Addictions, 25,* 1515–1536.

SOBELL, M. B., & SOBELL, L. C. (In press). Treatment for problem drinkers: A public health priority. In J. S. Baer,

G. A. Marlatt, & R. J. McMahon (Eds.), *Addictive behaviors across the lifespan: Prevention, treatment, and policy issues.* Beverly Hills, CA: Sage.

SOBELL, M. B., & SOBELL, L. C. (1993). *Problem drinkers: Guided self-change treatment.* New York: Guilford.

MARK B. SOBELL
LINDA C. SOBELL

Twelve Steps, The   The heart of the ALCO-HOLICS ANONYMOUS (AA) is a program called the Twelve Steps set forth by cofounder Bill W. and his early followers. The Twelve Steps establish a suggested, unfolding process for becoming, and remaining, sober. The process begins with an admission of powerlessness over alcohol, along with unmanageable lives, and builds momentum gradually into a commitment to carry the AA program via the Twelve Steps to active alcoholics. Newcomers are not pressed to follow all the steps if they feel unwilling or unable to do so. This suggested policy seems to be followed. Thus, Madsen (1974) found that 41 of the 100 AA members he studied had gone through all the Twelve Steps. And Rudy (1986:10) reports that "in Mideast City, A.A. members talk about and emphasize steps 1, 2, 3, 4, and 12 more than others." This pragmatic view of the Twelve Steps can be heard in an AA saying—"Take the best and leave the rest." The steps are:

1. We admitted we were powerless over alcohol—that our lives had become unmanageable.
2. Came to believe that a Power greater than ourselves could restore us to sanity.
3. Made a decision to turn our will and our lives over to the care of God *as we understood Him.*
4. Made a searching and fearless moral inventory of ourselves.
5. Admitted to God, to ourselves, and to another human being the exact nature of our wrongs.
6. Were entirely ready to have God remove all these defects of character.
7. Humbly asked Him to remove our shortcomings.
8. Made a list of all persons we had harmed, and became willing to make amends to them all.
9. Made direct amends to such people wherever possible, except when to do so would injure them or others.
10. Continued to take personal inventory, and when we were wrong, promptly admitted it.
11. Sought through prayer and meditation to improve our conscious contact with God *as we understood Him,* praying only for knowledge of His will for us and the power to carry that out.
12. Having had a spiritual awakening as the result of these steps, we tried to carry this message to alcoholics and to practice these principles in all our affairs [Alcoholics Anonymous World Services, 1976:59].

Step one meant for Bill W., the founder of AA, "the destruction of self centeredness" (Alcoholics Anonymous, 1939:16). In informal talk, AA members often urge everyone "to leave their egos at the door." Trice (1957:45) found that affiliation with AA was initially encouraged among those newcomers who reported that they had no willpower models among their friends or relatives for quitting alcohol abuse. Many observers have noted the strong tendency among alcoholics toward an "exaggerated belief in the ability to control their impulses, especially the impulse to use alcohol ... that they are in charge of themselves, that they are autonomous and able to govern themselves" (Khantzian & Mack, 1989:74). AA teaches that until alcoholics accept the first step they will continue to believe a fiction—that they are clever enough and strong enough to control their drinking. In any event, by taking the first step, newcomers to AA dramatically change their conception of self from believing they can control their drinking to believing they cannot ever do so.

In step one, AA taps into the repentant role in U.S. tradition. Redemptive religions emphasize that one can correct a moral lapse, even one of long duration, by public admission of guilt and repentance. AA members can assume this repentant role, beginning with step one, and it becomes, along with the other steps, a social vehicle whereby they can reenter the community (Trice & Roman, 1970).

This role is strengthened by step two and step three, wherein alcoholics agree there is a power greater than themselves who will help and agree to turn their destiny over to this higher power as they conceive of it. In essence, members believe that one does not have to stand alone against alcohol abuse and the strains of life; AA offers the group itself and its collective notion of a higher power to help the powerless.

By accepting and executing step four and step five, AA members believe they are engaging in a realistic self-examination of the factors of fear, guilt, and resentment that cause their drinking. In step four, new members list all people they now resent or have resented in the past. Along with this list, newcomers note what they believe to be the substance of the resentment. Following this exercise, new members work out ways to try to alter conceptions of these resented persons. They also attempt an inventory of their own behaviors that have contributed to their fears, guilts, and resentments. In step five, alcoholics acknowledge these inventories to a higher power and confess them to some other individual, for example, a friend, pastor, therapist, or sponsor. Members believe that this moral inventory and its reduction in resentments enable them to live through emotional experiences that in the past were managed by the abuse of alcohol.

Steps six and seven are reinforcements of the changes produced by acting out steps four and five. In step six, members indicate and reaffirm a readiness to respond to help from a higher power. In step seven, with as much humility as possible, members actually request that the higher power help them eliminate the inventory of "shortcomings" assembled by the member. In steps eight and nine, members seek to make further changes and reinforce past changes by providing restitution to those they have hurt in the past. Members list those actually harmed by their past behaviors and then do as much as they can to make amends and try to cancel out the harm caused. Most members agree that some amends might actually do harm to either themselves or others and caution against them. For example, the member might grievously damage a spouse by confessing in detail sexual infidelities. Step ten is a repetition and a reinforcement of steps four and five. In this step, members continue to "take my moral inventory" and admit their wrongs to themselves, others, and the Greater Power. Step eleven also acts as an implementer, but this time for step three, in which through meditation and prayer they again decide to turn over their willpower and their lives to a higher power.

Step twelve is the culmination of all these steps. Members are urged to carry their experiences and stories to active alcoholics in treatment centers, hospitals, even homes—in effect, to offer the redemptive model of AA sobriety to them. AA participants argue that, by becoming helpers, they help themselves at the same time and that they derive new commitments to the truths believed to be manifest in the other eleven steps. Furthermore, in twelfth-step work, there is a one-on-one, often a two-on-one (two AA members and one active alcoholic) meeting that often results in a sponsor-sponsoree relationship between a newcomer and older (in AA "birthdays") members. The group wisdom of AA teaches that new members are more likely to join during a crisis. Consequently, twelfth-step workers do not press for an admission of alcoholism during initial contacts. Rather, they try to be nonjudgmental, accepting, and reassuring, while nevertheless trying to help the prospect define the problem and what he or she will do about it. Members do, however, briefly describe their recovery via AA and invite the prospect to come to their meetings. If there is a positive response, they will promise to take the prospective member. According to Bales (1962:575), the sponsor-sponsoree relationship, along with the actual twelfth-step work itself, is "the heart of the therapeutic process" in AA.

The use of these steps is supported by basic assumptions: that intense self-examination and confession are cathartic; that alcoholics cannot control even moderate drinking and therefore are incapable of drinking at all. In other words, "once an alcoholic, always an alcoholic." According to the first step, "We admitted we were powerless over alcohol." The assumption of being powerless has been the focus of considerable controversy outside AA. The controversy centers around a follow-up study of 11,000 alcoholics whose drinking patterns were obtained 6 months and 18 months after experiencing one of a variety of treatment programs. The study, which contained numerous flaws (e.g., short follow-up time), showed that the majority of former alcoholics (who drank, on average, more than 8 ounces a day of ethanol [alcohol]) who had experienced a treatment program could drink moderately (2.5 ounces per day) at levels that many believe to be no problem (Armor, Polich, & Stambul, 1976).

A competing assumption is that ALCOHOLISM is a disease—that alcoholics suffer from an "allergy." This belief has also been controversial. An alternative has been the concept of the "problem drinker," the heavy drinker who gets into trouble, directly or indirectly, because of drinking alcohol. This bypasses the debate about alcoholism being a disease and about the amount drunk; it focuses instead on the "problem" correlates of drinking, that is, a role-impairment definition—financial problems and problems with family, police, friends, and neighbors. For example, Trice (1966:29) suggests that role impairment—such as job impairment—would be one of the performance criteria for the definition of alcoholism: alcoholics differ from those around them because the performace of their adult roles becomes clearly impaired by their recurrent use of alcohol. In the United States, most alcoholics are very poor husbands and fathers or wives and mothers; on the job, they falter and disappoint their coworkers. In addition, their unreliable behavior makes for doubts and confusion in intimate friendships. In sum, drinking behavior that significantly damages the performance of basic roles is the phenomenon, and it is not necessarily a disease as AA claims. Calahan and Room (1974) reported significant correlations between heavy drinking and impairments in the performance of these elementary roles. Such a definition opens the door for other therapies that assume that moderate drinking is possible. It even assumes that there may be "spontaneous recovery," that no therapy of any kind may be involved in some recoveries.

Finally, it should be noted that the Twelve Steps of AA are, in many members' minds, inevitably associated with AA's Twelve Traditions, which are aphorisms for the maintenance and continuity of AA itself at the group level. Examples are: Tradition 1—Our common welfare should come first; personal recovery depends upon AA unity. Tradition 10—We need always maintain personal anonymity at level of press, radio, and films (Alcoholics Anonymous World Services, 1965).

(SEE ALSO: *Alcoholism; Disease Concept of Alcoholism and Drug Abuse; Rational Recovery; Relapse Prevention; Sobriety; Treatment, History of; Vulnerability As Cause of Substance Abuse*)

BIBLIOGRAPHY

ALCOHOLICS ANONYMOUS WORLD SERVICES. (1976). *Alcoholics Anonymous: The story of how thousands of men and women have recovered from alcoholism* (3rd ed.). New York: A.A. Publishing.

ALCOHOLICS ANONYMOUS WORLD SERVICES. (1965). *Twelve steps and twelve traditions.* New York: Author.

ALCOHOLICS ANONYMOUS WORLD SERVICES. (1939). *Alcoholics Anonymous* (1st ed.). New York: Author.

ARMOR, D. J., POLICH, J. M., & STAMBUL, H. B. (1976). *Alcoholism and treatment.* Santa Monica, CA: Rand.

BALES, R. F. (1962). The therapeutic role of Alcoholics Anonymous as seen by a sociologist. In D. Pittman & C. R. Snyder (Eds.), *Society, culture, and drinking patterns,* pp. 573–578. New York: Wiley.

KHANTZIAN, E. J., & MACK, J. E. (1989). Alcoholics Anonymous and contemporary psychodynamic theory. In M. Galanter (Ed.), *Recent developments in alcoholism: Treatment research,* Vol. 7, pp. 67–89. New York: Plenum Press.

MADSEN, W. (1974). *The American alcoholic.* Springfield, IL: Charles C. Thomas.

RUDY, D. R. (1986). *Becoming alcoholic: Alcoholics Anonymous and the reality of alcoholism.* Carbondale: Southern Illinois University Press.

TRICE, H. M. (1966). *Alcoholism in America.* New York: McGraw-Hill.

TRICE, H. M. (1957). A study of the process of affiliation with Alcoholics Anonymous. *Quarterly Journal of Studies on Alcohol, 18,* 39–54.

TRICE, H. M., & ROMAN, P. M. (1970). Delabeling, relabeling and Alcoholics Anonymous. *Social Problems, 17*(4), 538–546.

HARRISON M. TRICE

TRIPLICATE PRESCRIPTION

It is estimated that hundreds of millions of prescribed medication doses are diverted to the street each year. Triplicate-prescription programs were developed as an effort to decrease the diversion of prescription medications to illicit markets. There are currently seven states (California, Idaho, Illinois, Indiana, Michigan, New York, and Texas) that require physicians to write prescriptions on special triplicate forms for all Schedule II drugs, including narcotic analgesics, BARBITURATES, and stimulants. In 1989, New York State passed legislation requiring triplicate prescribing for the BENZODIAZEPINES (Schedule IV substances).

In triplicate prescribing, the physician keeps one copy of the prescription for five years and sends two copies with the patient to the pharmacist. The pharmacist also keeps one copy and forwards the third to a specified state agency. Here the prescription is used to track the physician's prescribing practices and the patient's use of the controlled substances. With some exceptions, refills are not permitted for medications prescribed under this system. Opponents of the triplicate-prescription system claim that although it is effective in decreasing diversion, it does so at the expense of some patients who are unjustly denied analgesics, anxiolytics, or sedative-hypnotics. The New York experience with triplicate prescribing of benzodiazepines is often considered an example of this. Although benzodiazepine prescriptions were reduced by up to 60 percent, the number of prescriptions for the older and potentially more hazardous sedatives (such as MEPROBAMATE, methyprylon, ETHCHLORVYNOL, butalbital, and CHLORAL HYDRATE) increased markedly—in contrast to continued decreases in prescribing them in the rest of the United States.

In 1990, an attempt to federally legislate triplicate prescriptions for Schedule II medications for all states was unsuccessful in the House of Representatives.

(SEE ALSO: *Controls: Scheduled Drugs/Drug Schedules, U.S.; Iatrogenic Addiction; Legal Regulation of Drugs and Alcohol; Multidoctoring*)

BIBLIOGRAPHY

AMERICAN MEDICAL ASSOCIATION COUNCIL ON SCIENTIFIC AFFAIRS. (1982). Drug abuse related to prescribing practices. *Journal of the American Medical Association, 247*(6), 864–866.

BRAHAMS, D. (1990). Benzodiazepine overprescribing: Successful initiative in New York State. *Lancet, 336,* 1372–1373.

WEINTRAUB, M., ET AL. (1991). Consequences of the 1989 New York State triplicate benzodiazepine prescription regulations. *Journal of the American Medical Association, 266*(17), 2392–2397.

WILFORD, B. (1991). Prescription drug abuse: Some considerations in evaluating policy responses. *Journal of Psychoactive Drugs, 23*(4), 343–348.

MYROSLAVA ROMACH
KAREN PARKER

TWELVE STEPS, THE *See* Treatment/ Treatment Types.

U

UNITED NATIONS CONVENTION AGAINST ILLICIT TRAFFIC IN NARCOTIC DRUGS AND PSYCHOTROPIC SUBSTANCES, 1988

This international treaty was intended to extend and augment the agreements among the signatories that were contained in the 1961 SINGLE CONVENTION ON NARCOTIC DRUGS and the 1971 CONVENTION ON PSYCHOTROPIC SUBSTANCES. The 1988 Convention came into force in November 1990. By November 1994, 103 governments and the European Economic Community had been parties to the Convention. Included among the provisions are arrangements and agreements to legalize seizure of drug-related assets; criminalize MONEY LAUNDERING; relax bank-secrecy rules; permit extradition of individuals charged with drug-law violations; control shipments of precursor and essential chemicals; continue to support CROP CONTROL and eradication; and share evidence with law enforcement and prosecuting agencies of governments who are party to the conventions.

BIBLIOGRAPHY

U.S. DEPARTMENT OF JUSTICE, OFFICE OF JUSTICE PROGRAMS, BUREAU OF JUSTICE STATISTICS. (1992). *Drugs, crime and the justice system* (December NCJ-133652). Washington, DC: U.S. Government Printing Office.

JEROME H. JAFFE

URINE TESTING   *See* Drug Testing.

U.S. DRUG POLICY   *See* Anslinger, Harry J., and U.S. Drug Policy; U.S. Government/U.S. Government Agencies.

U.S. DRUG UNDERCOVER OPERATIONS   *See* Drug Interdiction.

U.S. GOVERNMENT/U.S. GOVERNMENT AGENCIES

It is generally accepted that the terms *War on Drugs*, *National Strategy*, and *Drug Czar* (a federal official with overall responsibility to coordinate the many federal agencies involved in implementing federal drug strategy) began with the initiatives of the Nixon administration during the period 1969–1973. In general, the articles on government activities that are in this section use this period as a starting point. However, history clearly shows that drug and alcohol problems have been high priority concerns in the United States for many years. Appendix IV (Volume 4) presents the milestones in U.S. drug policy that began in the late nineteenth century, when several of the western states prohibited opium dens and the first federal laws were enacted to prohibit the importation of opium by Chinese nationals living in the United States. At the turn of the century, concerns about total prohibition of alcohol had a major influence on national political life. Public concern about illicit drugs did not then reach the high levels experienced since the 1960s, but the federal government has long been involved with control of drugs other than alcohol. There are at least a few people who would assert that the first "drug czar" was Harry J. ANSLINGER of the Bureau of Narcotics in the U.S. Treasury Department.

The articles in *U.S. Government/U.S. Government Agencies* describe the origins and functions of selected U.S. government agencies that have (or had) important roles in dealing with alcohol- and drug-abuse issues. In addition to individual articles on agencies, offices, centers, and institutes there are general articles: *Agencies in Drug Law Enforcement and Supply Control; Agencies Supporting Substance Abuse Prevention and Treatment; Agencies Supporting Substance Abuse Research; Drug Policy Offices in the Executive Office of the President;* and *The Organization of U.S. Drug Policy.* Articles on individual agencies describe the current (or past) activities of the *Bureau of Narcotics and Dangerous Drugs;* the *Center for Substance Abuse Prevention;* the *Center for Substance Abuse Treatment;* the *National Institute on Alcoholism and Alcohol Abuse;* the *National Institute on Drug Abuse;* the *Office for Drug Abuse Law Enforcement;* the *Office of Drug Abuse Policy;* the *Office of National Drug Control Policy;* the *Special Action Office for Drug Abuse Prevention;* the *Substance Abuse and Mental Health Services Administration;* the *U.S. Customs Service;* and the *U.S. Public Health Service Hospitals.*

This is certainly not a totally inclusive list of agencies involved in substance-abuse treatment or research, prevention, law enforcement, and policy—but the reader should nonetheless be able to learn a great deal about the efforts of the U.S. federal government in these endeavors.

U.S. GOVERNMENT The following articles appear in this section:

Agencies in Drug Law Enforcement and Supply Control;

Agencies Supporting Substance Abuse Prevention and Treatment;

Agencies Supporting Substance Abuse Research;

Drug Policy Offices in the Executive Office of the President;

The Organization of U.S. Drug Policy

Agencies in Drug Law Enforcement and Supply Control So many agencies are involved in drug law-enforcement and supply-control activities that none are discussed here in detail. Except for the Drug Enforcement Agency (DEA), the order in which these descriptions appear is not necessarily related to the importance of an agency's role in the overall supply-control effort: Their functions frequently fit together like parts of an intricate puzzle.

The DEA was created in 1973 as a result of a reorganization that merged the activities and personnel from four federal drug law-enforcement programs into one agency within the Department of Justice (DOJ). John Bartels, Jr., was the first director. The offices and programs merged into DEA were the Bureau of Narcotics and Dangerous Drugs (BNDD), the Office for Drug Abuse Law Enforcement (ODALE), the Office for National Narcotic Intelligence, and U.S. Customs Service activities primarily directed to drug law enforcement. Since that time, DEA has been the lead federal agency for enforcement of drug laws.

DEA operates domestically and in foreign countries with the agreement of the government in each country. Its legal authority stems primarily from the CONTROLLED SUBSTANCES ACT and other laws directed at control of essential chemicals and precursors. DEA's efforts are directed against illicit drug production and high level drug-smuggling and drug-trafficking organizations operating within the United States or abroad. This agency is responsible for working with foreign governments to identify and disrupt the cultivation, processing, smuggling, and distribution of illicit substances, and the diversion of legally manufactured pharmaceuticals to illicit traffic in the United States. It maintains formal relationships with INTERPOL and the United Nations and works with them on international narcotics-control programs. The U.S. Department of State also has major responsibilities in working with foreign governments in this aspect of drug-traffic control. In carrying out these activities, DEA works closely with the state department, the Coast Guard, the Internal Revenue Service, and the U.S. Customs Service, and also with state and local law-enforcement agencies.

One of DEA's major domestic responsibilities is the enforcement of regulations concerning importation, manufacture, storage, and dispensing of all drugs scheduled under the Controlled Substances Act. Related to this function is the oversight, authorized by the Drug Treatment Act of 1974, of drug treatment programs using such drugs as LAAM or METHADONE (in METHADONE MAINTENANCE). DEA employs approximately 400 administration compliance officers to enforce regulations dealing with production and distribution of PRESCRIPTION DRUGS and supports a training program for narcotics officers at state and local levels. Virtually all state legislatures have passed a version of a prototype law, the Uniform Controlled Substances Act, which places legal CONTROLS on drugs at the state level similar to those at the federal level and establishes penalties under state law for violation of those laws. The Uniform Controlled Substances Act promotes uniformity in the way drugs are regulated, but individual states may schedule drugs not included in federal schedules and may place any drug at a different level of scheduling.

Because of similar laws at the federal and state levels, and overlapping responsibilities among federal agencies, several law-enforcement agencies may have jurisdiction with respect to any single drug offense or group of offenders. The decision about which of the cooperating agencies takes the lead and under which law a case will be tried depends on mutual assessment among enforcement agencies and prosecutors of their capabilities and procedures, and of which jurisdiction is most likely to obtain a conviction, since rules of evidence and procedures differ between federal and local courts. Generally, federal agencies will focus on high level drug traffickers and networks. Local police are empowered only to enforce state and local drug laws and are not permitted to arrest people for breaking a federal drug law. Federal agents may not enforce state and local drug laws unless specifically authorized to do so. The DEA also has enforcement responsibilities under the Chemical Diversion and Trafficking Act of 1988. This law was designed to control the availability of chemicals and precursors used by clandestine laboratories to produce DESIGNER DRUGS or to further process plant products such as COCA leaf into pure COCAINE. Since at least thirty-seven states have passed similar laws, this is another area where federal and local enforcement agencies may have concurrent jurisdiction.

Other major responsibilities of DEA include investigation of major drug traffickers operating at interstate and international levels; personnel training; scientific research related to control or prevention of illicit trafficking; management of a narcotics intelligence system; seizure and forfeiture of assets derived from or traceable to illicit drug trafficking.

Forfeiture is the loss of ownership of property used in connection with drug-related criminal activity or property derived from its income. Such forfeiture was authorized in the Comprehensive Drug Prevention Control Act of 1970 and the Racketeering Influenced and Corrupt Organization (RICO) Statute also passed in 1970. In 1990, DEA seized assets valued at more than one billion dollars, although not all of this property was ultimately forfeited. Forfeited property is usually sold at public auction and the proceeds are used for government activities and shared with cooperating state governments. States have used these funds for drug treatment and education programs as well as for drug law enforcement. Some goes into a special forfeiture fund within the Office of National Drug Control Policy (ONDCP), which in turn transfers it to other federal agencies. For example, significant amounts were transferred to the Center for Substance Abuse Treatment (CSAT) to support treatment programs for pregnant addicts.

In addition to DEA, several other organizations within the DOJ and other Cabinet departments have responsibility in areas concerning drug laws and related matters. The Office of Justice Programs (OJP) in the DOJ, established by the Justice Assistance Act of 1984, contains several bureaus involved with these issues. Three having significant roles at the present time are the Bureau of Justice Assistance (BJA), the Bureau of Justice Statistics (BJS), and the National Institute of Justice (NIJ). The BJA provides technical and financial assistance to state and local government for controlling drug trafficking and violent crime. Under the terms of the Anti-Drug Abuse Act of 1988, states may apply for grants to assist them in enforcing local and state laws against offenses comparable to those included in the Controlled Substances Act. Part of the application for these "formula grant" funds requires devising a statewide anti-drug and -violent crime strategy. The BJS collects, analyzes, and disseminates information on crime, its victims, and its perpetrators. Its 1992 report, *Drugs, Crime, and the Justice System*, the source for much of the material in this article, may be the best written and most comprehensive summary on the topic ever produced by the federal government. BJS also manages the Drugs and Crime Data Center and Clearinghouse (tel. 1-800-666-3332), which gathers and evaluates existing data on drugs and the justice system. The NIJ is the major research and development entity within the DOJ. Among its other activities, NIJ evaluates the effectiveness of programs supported by BJA, such as community anti-drug initiatives, and SHOCK INCARCERATION AND BOOT-CAMP PRISONS.

Other drug law-enforcement entities within the DOJ include the Federal Bureau of Investigation (FBI); the U.S. Attorneys, who are the chief federal law-enforcement officers in their districts and are responsible for prosecuting cases in federal court; the Immigration and Naturalization Services (INS); and the U.S. Marshals Service, which manages the Asset Forfeiture Fund. The FBI became more prominently involved in antidrug activities when its resources were significantly expanded in 1982 under President Ronald W. Reagan's reinvigoration of the "war on drugs." At that time it was given concurrent jurisdiction with DEA to investigate drug offenses, with the FBI concentrating primarily on drug trafficking by organized crime, electronic surveillance techniques, and drug-related financial activities such as investigations of international MONEY LAUNDERING.

Treasury Department agencies that play a role in controlling illicit drugs include the U.S. Customs Service, which stops and seizes illegal drugs as well as other contraband being smuggled into the United States; The Bureau of Alcohol, Tobacco, and Firearms (BATF), which investigates violations of laws dealing with weapons, particularly federal drug offenses involving weapons; and the Internal Revenue Service (IRS), which assists in financial investigations, particularly money laundering.

Two agencies in the Department of Transportation, the Federal Aviation Administration (FAA) and the U.S. Coast Guard, are significantly involved in drug-control activities. The FAA uses its radar systems to assist in detecting smuggling by air; the Coast Guard is involved in interdiction of drugs being smuggled into the U.S. by water.

The Postal Inspection Service of the U.S. Postal Service is also involved in the antidrug effort. This agency enforces laws against using the mail to transport drug paraphernalia and illegal drugs.

The Department of State's role in international drug policy is to coordinate drug-control efforts with foreign governments. Within State, the Bureau of International Narcotics Matters (INM) is responsible for international antidrug policy. This bureau provides technical assistance, money, and equipment to foreign governments for local law enforcement, transportation of personnel, and equipment for crop eradication. It also monitors worldwide drug production. Each U.S. Embassy abroad has a designated narcotics coordinator. In countries where there is considerable drug-related activity, there may be an entire narcotics-assistance section at the embassy. The state department also helps selected foreign governments with demand-reduction activities. Helping countries adversely affected economically by drug CROP CONTROL and eradication is a responsibility of the Agency for International Development. The U.S. Information Agency provides information about drug policy and relevant laws to U.S. officials serving in foreign countries.

The Department of Defense (DOD) is involved in detecting and monitoring aircraft and ships that might be involved in smuggling drugs into the United States. Until the 1980s, the military was prohibited from exercising police power over U.S. civilians by the Possae Comitatus Act of 1876. Changes in the act allow the military to share resources with civilian law-enforcement agencies, although military personnel are still not permitted to arrest civilians. The National Guard also assists federal agencies in border surveillance and in marijuana eradication.

Eleven agencies are involved in the Intelligence Center at El Paso, Texas (EPIC), operated by the DEA. EPIC is designed to target, track, and interdict drugs, aliens, and weapons moving across U.S. borders. The participating agencies, in addition to the DEA, are the Federal Bureau of Investigation (FBI); the Immigration and Naturalization Service (INS); the Customs Service; the U.S. Marshals Service; the U.S. Coast Guard; the Federal Aviation Administration (FAA); the Secret Service; the Department of State Diplomatic Service; the Bureau of Alcohol, Tobacco and Firearms (BATF); and the Internal Revenue Service (IRS). There is also a Counternarcotics Center developed by the Central Intelligence Agency (CIA) that coordinates international intelligence on narcotics trafficking. This effort involves personnel from the National Security Agency (NSA), the Customs Service, the DEA, and the Coast Guard.

(SEE ALSO: *Crime and Drugs; Drug Interdiction; International Drug Supply Systems; Terrorism and Drugs*)

BIBLIOGRAPHY

BUREAU OF JUSTICE STATISTICS, OFFICE OF JUSTICE PROGRAM, U.S. DEPARTMENT OF JUSTICE. (1992). *Drugs, crime, and the justice system.* Washington, DC: U.S. Government Printing Office.

DRUG ABUSE POLICY OFFICE, OFFICE OF POLICY DEVELOPMENT, THE WHITE HOUSE. (1984). *National strategy for prevention of drug abuse and drug trafficking.* Washington, DC: U.S. Government Printing Office.

EXECUTIVE OFFICE OF THE PRESIDENT, THE WHITE HOUSE. (1995). *National drug control strategy.* Washington, DC: U.S. Government Printing Office.

OFFICE OF THE FEDERAL REGISTER, NATIONAL ARCHIVES AND RECORDS ADMINISTRATION. (1993). *United States government manual 1993/1994.* Washington, DC: U.S. Government Printing Office.

JEROME H. JAFFE

Agencies Supporting Substance Abuse Prevention and Treatment

Within the U.S. Department of Health and Human Services (DHHS), originally established in 1953 as the Department of Health, Education, and Welfare (DHEW), a number of Public Health Service (PHS) agencies have been involved in reducing drug abuse. From 1974 to 1992, many demand-reduction activities have related to increasing, through research, the scientific foundations for a better understanding of how drugs of abuse interact with individuals, so as to prevent drug abuse and effectively treat those who do abuse drugs. Included among these agencies are the National Institute on Drug Abuse (NIDA) and the National Institute on Alcohol Abuse and Alcoholism (NIAAA), both components of the National Institutes of Health (NIH), as well as the Center for Substance Abuse Prevention (CSAP) and the Center for Substance Abuse Treatment (CSAT), components of the Substance Abuse and Mental Health Services Administration (SAMHSA). In addition, the Health Resources and Services Administration (HRSA) and the National Institute of Child Health and Human Development (NICHD), another NIH component, play a role in the department's anti-drug abuse mission. Although not all inclusive, the chart below

shows the organizational hierarchy of these agencies within the department.

From its creation in 1974 by statute, the National Institute on Drug Abuse has conducted RESEARCH on drugs of abuse and their effects on individuals. In its early days, NIDA supported PREVENTION and TREATMENT programs and conducted clinical training programs for professional health-care workers (particularly in schools of medicine, nursing, and social work) and counselor and other paraprofessional training. With the advent of the Alcohol and Drug Abuse and Mental Health Services block grant, enacted into statute in 1981, the direct provision of treatment and prevention services became a state responsibility. Enactment of the block grant that is currently administered within SAMHSA served to refocus NIDA's role on the generation of knowledge through scientific research, so that more could be learned about strategies and programs to help prevent and treat drug abuse.

The National Institute on Alcohol Abuse and Alcoholism (NIAAA) conducts research on alcohol abuse and alcoholism. Because a comprehensive approach to prevention and treatment of drug abuse requires attention to alcohol as well as to illicit drugs, and because individuals who abuse illicit drugs often abuse alcohol as well, the research programs of NIDA and NIAAA are symbiotic. Furthermore, the genetic, environmental, and social influences important to the initiation of drug and alcohol use are similar, and research in one area suggests researchable hypotheses in the other.

The Center for Substance Abuse Prevention (CSAP), established in 1986 as the Office for Substance Abuse Prevention (OSAP), has led the nation's efforts to prevent alcohol and other drug use, with a special emphasis on youth and FAMILIES at particularly high risk for drug abuse. Youth considered to be at high risk include school DROPOUTS, economically disadvantaged youth, or children of parents who abuse drugs or alcohol or who are at high risk of becoming drug or alcohol abusers. CSAP administers a variety of programs, including Prevention demonstration grants targeting youth at high risk and projects for pregnant and postpartum women and their infants.

The Center for Substance Abuse Treatment (CSAT), formerly the Office of Treatment Improvement (OTI), was established administratively in 1990 with a focus on improving treatment services and expanding the capacity for delilvering treatment services. In addition to administering the Alcohol and Drug Abuse block grant, CSAT administers a number of demonstration grant programs such as the Target Cities, Critical Populations, and Criminal Justice treatment programs.

Drug and alcohol abuse are complex behaviors that often result in a multitude of adverse consequences. Thus, to understand them necessitates multifaceted, often crosscutting areas of research. Because many individuals who suffer from alcohol or drug abuse also suffer from mental illness, NIAAA and NIDA, as well as the National Institute of Mental Health (NIMH) of the NIH, are engaged in initiatives to learn more about individuals who are dually diagnosed.

Acquired immunodeficiency syndrome (AIDS) has become a growing health program among intravenous drug users, and an increased risk of human immunodeficiency virus (HIV) infection in those who share drug paraphernalia with other drug users has been clearly demonstrated (Chaisson et al., 1987; Schoenbaum et al., 1989). Accordingly, NIDA collaborates with the Centers for Disease Control (CDC) on AIDS prevention programs and with the National Institute of Allergy and Infectious Diseases (NIAID) to provide HIV therapeutics to intravenous drug abusers with HIV.

The study of maternal and fetal effects of drug abuse is another high-priority focus within the department. Research and demonstration programs have been undertaken by NIDA and CSAP, and the NICHD is also conducting studies in this area.

Recent research has shown that the most effective treatment for drug abusers is a comprehensive array of services that address not only their drug-abuse problems but also other health problems and their potential need for education and vocational rehabilitation, as well as a host of ancillary services. Accordingly, NIDA, the centers within SAMHSA, and HRSA are exploring the effectiveness of providing a comprehensive range of drug-abuse and other primary-care services, both in drug-abuse settings and primary-care settings.

Besides the DHHS, there are many other agencies involved in prevention and treatment efforts. For example, the Food and Drug Administration (FDA), plays a determining role in deciding when new pharmacological treatment agents can be marketed for clinical use, and it is one of the key agencies setting policies and standards for the use of OPIOID drugs in the treatment of opioid dependence. Both the Department of Education and the Department of Justice (through the Drug Enforcement Agency [DEA]) have significant programs aimed at prevention; the Department of Veterans Affairs and the Department of Defense (U.S. MILITARY) have also made major commitments to treatment.

(SEE ALSO: *Education and Prevention; Prevention Movement; Research; Substance Abuse and HIV/AIDS*)

BIBLIOGRAPHY

CHAISSON, R. E., ET AL. (1987). Human immunodeficiency virus infection in heterosexual intravenous drug users in San Francisco. *American Journal of Public Health, 77*(2), 169–172.

SCHOENBAUM, E. E., ET AL. (1989). Risk factors for human immunodeficiency virus infection in intravenous drug users. *New England Journal of Medicine, 321*(13), 874–879.

RICHARD A. MILLSTEIN

Agencies Supporting Substance Abuse Research In the United States, federal support of drug-abuse research began in the 1920s with the work of Lawrence Kolb. It became more formalized with the establishment of the Addiction Research Center in 1935. A small research unit was formed with only fifteen employees in a U.S. Public Health Service Hospital in Lexington, Kentucky, by 1944. The Addiction Research Center was designed for federal prisoners who were narcotics addicts. This research group became part of the National Institute of Mental Health (NIMH) in 1948, the year the institute was established. In 1979, the Addiction Research Center moved to Baltimore, Maryland, and became the in-house (intramural) research program of the National Institute on Drug Abuse (NIDA), which was itself established by Congress in 1974.

In the early 1990s, it was estimated that NIDA funded 88 percent of the drug-abuse research in the world. In 1992, the NIDA budget for the almost 1,000 research grants awarded to universities and other research institutions (i.e., extramural research) totaled 338 million dollars. NIDA's 1992 intramural research budget for the Addiction Research Center was 24 million dollars. The research thus funded includes studies in practically every basic and clinical science, both biomedical and social. The National Institute on Alcohol Abuse and Alcoholism (NIAAA), established in 1970, conducts parallel efforts in the area of alcohol-abuse research. In 1992, its budget for extramural research was 155 million dollars for over 600 research projects. NIAAA's intramural research arm, located in Bethesda, Maryland, had a budget of nearly 20 million dollars.

Both NIDA and NIAAA became part of the National Institutes of Health (NIH) in October 1992. They had previously been part of the Alcohol, Drug Abuse, and Mental Health Administration (ADAMHA), which included both research and services components. By separating these two components, the Congress indicated its intention to give proper emphasis to both. Now treatment and prevention services for alcohol and drug abuse are under the direction of the Substance Abuse and Mental Health Services Administration (SAMHSA).

NIDA and NIAAA are the two largest federal research institutes dedicated to drug abuse and alcohol research, but there are many other agencies that have a stake in these areas. They include other institutes in the National Institutes of Health; for example, the National Institute of Child Health and Development centers its research on the effects of drugs and alcohol on fetal development and on the consequences for the neonate of exposure to drugs and alcohol during pregnancy. The National Institute of Mental Health conducts research on the high coincidence of mental illness and substance-abuse disorders. Some of the other institutes have similarly targeted interests, as, for example, the National Cancer Institute, which played an important role in support of research on tobacco dependence and the adverse health effects of tobacco.

Other parts of the Public Health Service also play a role in substance abuse research. The Centers for Disease Control (CDC) use their epidemiological expertise to resolve certain questions about the nature and extent of the abuse of drugs and alcohol. The Agency for Health Care Policy and Research conducts research on the costs associated with medical care and health insurance for drug and alcohol abusers seeking treatment.

Beyond the Public Health Service and the Department of Health and Human Services, many other federal agencies and departments are concerned with and conduct research on the social problems caused by drug and alcohol abuse: the departments of education, labor, transportation, treasury, justice, state, veterans affairs and even defense—each has a stake in drug-abuse research. The Department of Education is concerned primarily with drug and alcohol prevention; the departments of labor and transportation with workplace performance impaired by drugs and alcohol.

The Department of Veterans Affairs has played an important role in both basic and clinical research. Some of the most important work on the treatment of opioid dependence and on alcoholism and the toxic effects of alcohol have been conducted by researchers based at Veterans Administration (VA) hospitals and funded in part by research funds from the Department of Veterans Affairs. Other federal agencies have a regulatory role in certain types of drug-abuse research. Many of the drugs that are studied in animals and volunteer human subjects are included under the CONTROLLED SUBSTANCES ACT of 1970. In order to obtain and store the drugs, researchers must be properly registered with the Drug Enforcement Agency (DEA). The DEA is also responsible for ensuring that the drugs are properly stored and the records of their use are properly kept by the researchers. In addition, researchers who are interested in studying any drug not yet approved for clinical use, or studying an approved drug for a new use (such as using the antihypertensive agent, CLONIDINE, to control alcohol, tobacco, or opioid withdrawal), must obtain permission obtaining an Investigational New Drug (IND) authorization from the Food and Drug Administration (FDA). Further, when a new agent seems promising, a sponsor (usually a pharmaceutical company) must submit the data supporting its safety and effectiveness to the FDA before it can be approved for marketing and general use.

Both the Department of Justice and the Department of the Treasury are concerned with law enforcement issues surrounding drug and alcohol use, and they have funded research on detection of clandestine laboratories and the nature of DESIGNER DRUGS. The 1994 National Strategy showed that of the entire federal drug-abuse research budget, some 500 million dollars, approximately 67 million was allocated to domestic law-enforcement research.

The Department of State and the Department of Defense are involved in matters relating to international narcotics control. The U.S. Information Agency (USIA) and the Agency for International Development sponsor small drug-abuse research programs, mostly epidemiological in nature, in various countries. The Office of National Drug Control Policy (ONDCP) was given the mandate by Congress in 1988 to coordinate the federal antidrug-abuse effort. It does this through its budgetary oversight and through the Research, Data, and Evaluation Committee. The ONDCP for several years has had a Science and Technology subcommittee, which oversees the Counter-Drug Technology Assessment Center (CTAC). CTAC is involved in both medical research and supply-related counter-drug technology development. The latter includes activities such as the use of satellites for wide area surveillance, non-intrusive inspections, and development of information systems to permit sharing of data among criminal justice data bases. All of these policy-related organizations rely on facts based on the biomedical, epidemiological, and behavioral research funded by NIDA, NIAAA, and NIMH.

(SEE ALSO: *Addiction Research Foundation (Canada); Addiction Research Unit (U.K.); Education and Prevention; Prevention Movement; Wikler's Pharmacologic Theory of Drug Addiction*)

BIBLIOGRAPHY

EXECUTIVE OFFICE OF THE PRESIDENT. (1994). *National Drug Control Strategy.* Washington, DC: U.S. Government Printing Office.

GORDIS, E. (1988). Milestones. *Alcohol Health and Research World, 12*(4), 236–239.

History of NIDA. (1991). *NIDA Notes, 5*(5), 2–4.

CHRISTINE R. HARTEL

Drug Policy Offices in the Executive Office of the President

The Executive Office of the President (EOP) is an administrative group of key advisors and agencies supporting the president and the White House staff. Changes to the organization and functions of the EOP reflect the priorities and interests of each president. The organization of the EOP can be modified by executive order, by reorganization plan (when authorized), or by legislation.

Since 1970, several drug-policy activities have been established in the EOP. The list includes three separate EOP agencies, authorized and funded by statute, and three drug-policy offices, authorized by

the president and located within a larger EOP agency. The drug-policy offices are listed immediately below, followed by a general description of each activity.

Separate Agencies. Special Action Office for Drug Abuse Prevention (SAODAP), 1971–1975. Office of Drug Abuse Policy (ODAP), 1977–1978. Office of National Drug Control Policy (ONDCP), 1989–present.

Offices. Federal Drug Management (Office of Management & Budget), 1973–1977. Drug Policy Office (Domestic Policy Staff), 1978–1980. Drug Abuse Policy Office (Office of Policy Development), 1981–1989.

SPECIAL ACTION OFFICE FOR DRUG ABUSE PREVENTION (SAODAP)

A separate agency in the EOP from 1971 to 1975, SAODAP was responsible for providing leadership and coordination of all federal drug-abuse prevention activities (demand related) and to coordinate the demand-related activities with the supply-related efforts of law enforcement agencies.

Directors.
Jerome H. Jaffe, 1971–1973 (also Consultant to the President for Narcotics and Dangerous Drugs) Robert L. Dupont 1973–1975.

Authorization and Role.
Established by President Richard M. Nixon (E. O. 11599, June 17, 1971). Legislative authorization: Public Law 92–255, March 21, 1972; the "Drug Abuse Office and Treatment Act of 1972." The director reported to the president, working through the Domestic Council and the White House staff. SAODAP had a staff of over 100 and an annual budget of approximately $50 million. About 50 percent of the budget was in a "Special Fund for Drug Abuse" to be transferred to other federal agencies as an incentive to develop more effective prevention programs.

SAODAP provided oversight of all categories of "Demand Reduction" functions and made recommendations to the Office of Management and Budget (OMB) on funding for drug-abuse programs. SAODAP published three federal strategies under the auspices of the relatively inactive Strategy Council on Drug Abuse.

When the authorizing statute expired on June 30, 1975, SAODAP's treatment, rehabilitation, and prevention functions were moved from the EOP to the National Institute on Drug Abuse in the Department of Health, Education, and Welfare.

Bibliography of Associated Major Policy Publications (SAODAP):

U.S. Executive Office of the President. Strategy Council on Drug Abuse. *Federal Strategy for Drug Abuse and Drug Traffic Prevention. 1973.* Washington, DC: Government Printing Office, 1973.

U.S. Executive Office of the President. Strategy Council on Drug Abuse. *Federal Strategy for Drug Abuse and Drug Traffic Prevention. 1974.* Washington, DC: Government Printing Office, 1974.

U.S. Executive Office of the President. Strategy Council on Drug Abuse. *Federal Strategy for Drug Abuse and Drug Traffic Prevention. 1975.* Washington, DC: Government Printing Office, 1975.

FEDERAL DRUG MANAGEMENT, OFFICE OF MANAGEMENT AND BUDGET

Opened in 1973 as a unique office within OMB, Federal Drug Management (FDM) was designed to manage federal activities directed at illegal drugs during a time of rapid expansion and major reorganization. FDM continued in operation until early 1977.

FDM Chiefs.

Walter C. Minnick, 1973–1974

Edward E. Johnson, 1974–1977.

Authorization and Role.

Established by OMB memorandum, the authority of the staff office and the budget for operating expenses were derived from OMB. Initially, FDM was responsible for coordinating the implementation of drug policy, resolving interagency disputes, assisting drug agencies with reorganization and management, and working closely with other inter-agency drug-coordinating structures. In August 1974, FDM's budget and management responsibilities reverted to the normal OMB divisions and FDM continued to provide Executive Office oversight of the domestic and international drug abuse programs, interdepartmental coordination, and staff support to the cabinet councils on drug abuse.

Located in the Old Executive Office Building, FDM's five-person staff functioned with little public visibility. Working with other OMB staff, FDM guided the implementation of Reorganization Plan No. 2 of 1973, including union negotiations. FDM continued through the Ford Administration, providing staff assistance and policy advice to OMB, the Domestic Council, and the National Security Council. FDM was eliminated in early 1977 during the transition to the Carter Administration.

Bibliography of Associated Major Policy Publications (FDM):

U.S. Executive Office of the President. The Domestic Council Drug Abuse Task Force. *White Paper on Drug Abuse, September 1975.* Washington, DC: Government Printing Office, 1975.

U.S. Executive Office of the President. Strategy Council on Drug Abuse. *Federal Strategy. Drug Abuse Prevention. 1976.* Washington, DC: Government Printing Office, 1976.

OFFICE OF DRUG ABUSE POLICY (ODAP)

In March 1976, Congress authorized the Office of Drug Abuse Policy, located in the EOP and intended to be the successor agency to SAODAP. President Gerald R. Ford did not activate the new agency, choosing instead to continue with the existing FDM staff. President Jimmy Carter opened ODAP in March of 1977 and abolished it one year later. The director's office was located in the West Wing of the White House and the staff offices were in the Old Executive Office Building.

Director.

Dr. Peter G. Bourne, 1977–1978 (also Special Assistant to the President for Health Issues).

Authorization and Role.

Congress established ODAP in Public Law 94–237 and provided an annual budget of $1.2 million. The director was the principal advisor to the president on policies, objectives, and priorities for federal drug-abuse functions. The director coordinated the performance of drug-abuse functions by federal departments and agencies.

ODAP, with a staff of approximately fifteen, conducted a comprehensive set of drug-policy reviews using interagency study teams. The director and staff sought a close cooperative relationship with Congress and testified when requested before various congressional committes. The director was required to prepare an annual report on the activities of ODAP and to oversee the preparation of a drug-abuse strategy.

In mid-1977, the President's Reorganization Project prepared a reorganization of the EOP that included abolishing ODAP. Congress objected to the loss of ODAP. After spirited congressional hearings emphasizing the continuing need for executive coordination of the drug program, ODAP was abolished in March 1978 and its responsibilities transferred to the Domestic Policy Staff.

Bibliography of Associated Major Policy Publications (ODAP):

U.S. Executive Office of the President. Office of Drug Abuse Policy. *Border Management and Interdiction—An Interagency Review,* September 1977.

U.S. Executive Office of the President. Office of Drug Abuse Policy. *Supply Control: Drug Law Enforcement—An Interagency Review,* December 1977.

U.S. Executive Office of the President. Office of Drug Abuse Policy. *International Narcotics Control Policy, March 1978.*

U.S. Executive Office of the President. Office of Drug Abuse Policy. *Narcotics Intelligence* (Classified), 1978.

U.S. Executive Office of the President. Office of Drug Abuse Policy. *Drug Use Patterns, Consequences and the Federal Response: A Policy Review,* March 1978.

U.S. Executive Office of the President. Office of Drug Abuse Policy. *Drug Abuse Assessment in the Department of Defense: A Policy Review,* November 1977.

U.S. Executive Office of the President. Office of Drug Abuse Policy. *1978 Annual Report.* Washington, DC: Government Printing Office, 1978.

DRUG POLICY OFFICE (DPO), DOMESTIC POLICY STAFF

The Drug Policy Office (DPO) opened March 26, 1978, as an integral part of the White House Domestic Policy Staff. Six people were transferred from ODAP, and the DPO provided direction and over-

sight of federal drug-program activities through 1980.

Director.
Lee I. Dogoloff, 1978–1980 (Associate Director for Drug Policy in the Domestic Policy Staff).

Authorization and Role.
Reorganization Plan No. 1 of 1977 transferred the ODAP responsibilities to the Domestic Policy Staff in the EOP. President Carter signed Executive Order No. 12133 on May 9, 1979, formally designating the associate director for Drug Policy in the Domestic Policy Staff as

primarily responsible for assisting the President in the performance of all those functions transferred from the Office of Drug Abuse Policy and its Director . . . in formulating policy for and in coordinating and overseeing, international as well as domestic drug abuse functions by all Executive Agencies.

DPO continued to report to Dr. Bourne as special assistant to the president for health issues. On numerous occasions, the associate director testified before Congress on drug-policy matters.

DPO published a 1979 federal strategy under the auspices of the Strategy Council on Drug Abuse, an annual report in 1980, and an annual budget crosscut of all drug-abuse prevention and control activities. Both the Domestic Policy Staff and DPO were eliminated during the transition to the Reagan Administration.

Bibliography of Associated Major Policy Publications (DPO):

U.S. Executive Office of the President. Strategy Council on Drug Abuse. *Federal Strategy for Drug Abuse and Drug Traffic Prevention. 1979.* Washington DC: Government Printing Office, 1979.

U.S. Executive Office of the President. Domestic Policy Staff. *Annual Report on the Federal Drug Program. 1980.* Washington, DC: Government Printing Office, 1980.

DRUG ABUSE POLICY OFFICE (DAPO), OFFICE OF POLICY DEVELOPMENT

Similar in organization and responsibilities to the preceding DPO, the Drug Abuse Policy Office (DAPO) was the principal EOP drug-abuse staff during the eight years of President Ronald W. Reagan's administration. In 1981, DAPO was established within the White House Office of Policy Development.

Directors.

Carlton E. Turner, 1981–1986 (also Special Assistant to the President; promoted in March 1985 to Deputy assistant to the President).

Dr. Donald Ian MacDonald, 1987–1989, (Special Assistant to the President; promoted in August 1988 to Deputy Assistant to the President).

Authorization and Role.

The statutory basis for the office (21 USC 1111 & 1112) required the president to establish a system to assist with drug abuse policy functions and to designate a single officer to direct the drug functions. Presidential Executive Order 12368, signed on June 24, 1982, assigned the Office of Policy Development (OPD) to assist the president with drug-abuse policy functions, including international and domestic drug-abuse functions by all executive agencies. The director of ODAP was responsible for advising the president on drug-abuse matters and assisting Nancy D. Reagan and her staff in developing the First Lady's drug-abuse prevention program.

The director and staff developed policies regarding all aspects of drug abuse, including drug law enforcement, international control, and health-related prevention and treatment activities for both government and the private sector. DAPO coordinated the development and publication of 1982 and 1984 drug-abuse strategies.

In October 1984, Public Law 98–473, which created the National Drug Enforcement Policy Board to oversee drug law enforcement, also included a new statutory duty for DAPO; "to insure coordination between the National Drug Enforcement Policy Board and the health issues associated with drug abuse."

In March 1987, Executive Order 12590 established a National Drug Policy Board (NDPB) to assist the president in formulating all drug-abuse policy, replacing the director of DAPO in that role. The new executive order made the director a member of the NDPB and assigned DAPO to assist both the president and the NDPB in the performance of drug-policy functions. The DAPO director assisted in developing the health-related aspects of the national drug strategy published in the board's 1988 report *Toward a Drug-Free America—The National Drug Strategy and Implementation Plans.*

DAPO was terminated early in the administration of President George H. Bush by Public Law 100–

690, which created the Office of National Drug Control Policy.

Bibliography of Associated Major Policy Publications (DAPO):

> U.S. Executive Office of the President. Drug Abuse Policy Office, Office of Policy Development, The White House. *Federal Strategy for Prevention of Drug Abuse and Drug Trafficking 1982.* Washington, DC: Government Printing Office, 1982.
>
> U.S. Executive Office of the President. Drug Abuse Policy Office, Office of Policy Development, The White House. *1984 National Strategy for Prevention of Drug Abuse and Drug Trafficking.* Washington, DC: Government Printing Office, 1984.

OFFICE OF NATIONAL DRUG CONTROL POLICY (ONDCP)

In January 1989, the Office of National Drug Control Policy (ONDCP) was established as an agency in the EOP to oversee all national drug-control functions and to advise the president on drug-control matters. Functioning as the so-called drug czar, the director of ONDCP had the broadest combination of staff, funding, and authority of any previous EOP drug agency or office.

Directors.

William J. Bennett, 1989–1990.

Bob Martinez, 1991–1992.

Lee P. Brown 1993–

Authorization and Role.

Established by Public Law 100–690 (21 USC 1504) with a five-year authorization, ONDCP had a staff of approximately 130 and a Fiscal Year 1993 budget of $59 million for salaries, expenses, and support for High Intensity Drug Trafficking Areas. The fiscal year 1994 budget request reduces the ONDCP staff to 25 positions. The director controls a Special Forfeiture Fund with over $75 million appropriated in Fiscal Year 1993 to provide added funding for high-priority drug-control programs.

ONDCP was responsible for national drug control policies, objectives and priorities, and annual strategy, and a consolidated budget. ONDCP was also required to make recommendations to the president regarding changes in the organization, management, personnel, and budgets of the federal departments and agencies engaged in the antidrug effort.

ONDCP was required to promulgate an annual national drug control strategy and to coordinate and oversee the implementation of the strategy. The director had to consult with and assist state and local governments regarding drug-control matters.

Bibliography of Associated Major Policy Publications (ONDCP):

> U.S. Executive Office of the President. Office of National Drug Control Policy. *National Drug Control Strategy, September 1989.* Washington, DC: Government Printing Office, 1989.
>
> U.S. Executive Office of the President. Office of National Drug Control Policy. *National Drug Control Strategy, January 1990.* Washington DC: Government Printing Office, 1990.
>
> U.S. Executive Office of the President. Office of National Drug Control Policy. *National Drug Control Strategy, February 1991.* Washington, DC: Government Printing Office, 1991.
>
> U.S. Executive Office of the President. Office of National Drug Control Policy. *National Drug Control Strategy, January 1992.* Washington, DC: Government Printing Office, 1992.
>
> U.S. Executive Office of the President. Office of National Drug Control Policy. *National Drug Control Strategy, February, 1994.* Washington, DC: Government Printing Office.

(SEE ALSO: *Anslinger, Harry J., and U.S. Drug Policy*)

BIBLIOGRAPHY

BONAFEDE, D. (1971). White House Report/Nixon's offensive on drugs treads on array of special interests. *National Journal, 3*(27), 1417–1423.

HAVEMANN, J. (1973). White House Report/Drug agency reorganization establishes unusual management group. *National Journal, 5*(18), 653–659.

HOGAN, H. (1989). *Drug control at the federal level: Coordination and direction.* Washington, DC: Congressional Research Service, the Library of Congress. Report 87–780 GOV.

U.S. CONGRESS, HOUSE, SELECT COMMITTEE ON NARCOTICS ABUSE AND CONTROL. (1978). *Congressional resource guide to the federal effort on narcotics abuse and control, 1969–76, Part 1.* A Report of the Select Committee on Narcotics Abuse and Control. 95th Congress, 2nd sess. Washington, DC: U.S. Government Printing Office.

U.S. NATIONAL ARCHIVES AND RECORDS ADMINISTRATION, OFFICE OF THE FEDERAL REGISTER. *The United States Government Manual.* Washington, DC: U.S. Government Printing Office.

RICHARD L. WILLIAMS

The Organization of U.S. Drug Policy

Reducing drug abuse has been a priority for the U.S. government since the late 1960s, with continuing expansion of management attention and federal budgets. In 1969, eight agencies and four cabinet departments received drug-program funding; in 1975, seventeen agencies in seven cabinet departments were included; the federal drug control program for 1993 involves forty-five agencies and twelve cabinet departments. In 1969, the total budtet for federal drug-abuse programs was $81 million; for 1993, the budget was approximately $12.7 billion.

WHY IS IT DIFFICULT TO ORGANIZE DRUG POLICY?

Drug-policy issues are complex. The organization for drug-policy development must be able to handle the complexity of the drug problem and of the government's response.

Illegal drugs come from both international and domestic sources; they include a wide variety of substances; they involve many different forms of transportation, geographical areas, criminal activities, use patterns, and social effects. All these elements are dynamic—constantly adjusting to changes in supply and demand. Drug traffickers and continuing users immediately react to drug law enforcement pressures by shifting to areas or techniques that have less risk. Federal managers and policymakers must recognize the complex changes (and the probable causes) and be capable of adjusting the federal effort promptly and effectively.

National leadership, including an accepted strategy and a process to ensure implementation, is essential to real progress in eliminating illegal drugs and their use. The president must have congressional cooperation in authorizing and funding the strategy. The cabinet departments and agencies must be willing participants, with an effective procedure for resolving interdepartmental differences of opinion.

The complex drug issue, however, does not fit the usual organization of the federal government: There is no cabinet department with line authority over all drug-program resources; and only a few federal agencies are organized around a single drug-related function (e.g., the Drug Enforcement Agency and the National Institute on Drug Abuse). Most of the drug control agencies and all the departments have various other important roles, so they must balance their drug and nondrug responsibilities.

Every step in the policy-determination and -implementation process is complex and subject to bureaucratic, political, and technical differences of opinion. Two of the most difficult aspects of the drug problem are (1) seeking agreement on the extent and nature of the problem, and (2) attempting to assess the impact of the federal effort on the ever changing situation.

During the past two decades, the federal organization for determining drug policy and implementing drug programs has expanded to involve a significant portion of the federal government. The following list of cabinet departments and agencies that execute drug policy reflects the breadth of implementation activities.

NATIONAL DRUG CONTROL AGENCIES

The 1992 National Drug Control Strategy lists over forty-five agencies and several activities in twelve cabinet departments involved in drug-control efforts:

ACTION
Agency for International Development
Department of Agriculture
    Agricultural Research Service
    U.S. Forest Service
Central Intelligence Agency
Department of Defense
Department of Education
Department of Health and Human Services
    Administration for Children and Families
    Alcohol, Drug Abuse, and Mental Health Administration (includes the National Institute of Mental Health, the National Institute on Drug Abuse, the National Institute on Alcohol Abuse and Alcoholism, the Office for Substance Abuse Prevention and the Office for Treatment Improvement)
    Centers for Disease Control
    Food and Drug Administration
    Health Care Financing Administration
    Indian Health Service
Department of Housing and Urban Development
Department of the Interior
    Bureau of Indian Affairs
    Bureau of Land Management
    Fish and Wildlife Service
    National Park Service
    Office of Territorial and International Affairs
The Judiciary
Department of Justice
    Assets Forfeiture Fund
    U.S. Attorneys
    Bureau of Prisons
    Criminal Division
    Drug Enforcement Administration
    Federal Bureau of Investigation
    Immigration and Naturalization Service
    INTERPOL/U.S. National Central Bureau
    U.S. Marshals Service
    Office of Justice Programs
    Organized Crime Drug Enforcement Task Forces
    Support of U.S. Prisoners
    Tax Division
Department of Labor
Office of National Drug Control Policy
    Counter-Narcotics Technology Assessment Center
    High Intensity Drug Trafficking Areas
    Special Forfeiture Fund
Small Business Administration
Department of State
    Bureau of International Narcotics Matters
    Bureau of Politico/Military Affairs
    Diplomatic and Consular Service
Department of Transportation
    U.S. Coast Guard
    Federal Aviation Administration

National Highway Traffic Safety Administration

Department of the Treasury

Bureau of Alcohol, Tobacco, and Firearms

U.S. Customs Service

Federal Law Enforcement Training Center

Financial Crimes Enforcement Network

Internal Revenue Service

U.S. Secret Service

U.S. Information Agency

Department of Veterans Affairs

Weed and Seed Program

COORDINATING MECHANISM FOR DRUG POLICY

In reviewing historical drug-policy coordinating systems since the late 1960s, each system reflects a complex set of considerations. Two elements seem to differentiate between the various approaches: Either a drug-policy adviser and supporting drug staff is fully integrated into the regular policy processes at the White House, or a high-priority cabinet-level activity or agency is established with its own special policy process but with less participation in White House internal staff activity.

Each president selects his own White House staff and establishes a policy-development process to meet his needs. Therefore, any policy-coordinating mechanism that is closely related to a president must be expected to change with each new administration.

Congress has repeatedly attempted to establish a "drug czar" in the Executive Office of the President (EOP)—one person to oversee drug policy and to advise both the president and Congress.

HISTORY

A chronological summary of drug-policy coordinating mechanism is presented here, beginning with 1971—first from the perspective of the Executive Branch, then from the perspective of Congress.

Executive Drug Policy 1971–1976. On the demand side, President Richard M. Nixon created the Special Action Office for Drug Abuse Prevention (SAODAP) in the EOP in June 1971—to lead and coordinate all federal drug-abuse prevention activi-

ties. The first director, Dr. Jerome H. Jaffe, was given the added title of Consultant to the President for Narcotics and Dangerous Drugs. SAODAP then monitored the annual budget process and prepared budget analyses of all federal drug-abuse programs, by agency and by activity.

Also in 1971, President Nixon called for "an all out global war on the international drug traffic" (1973 Federal Strategy, p. 112), and his organization for policy reflected the international perspective. International efforts were coordinated by the Cabinet Committee on International Narcotics Control (CCINC), chaired by the secretary of state. Established in August 1971, CCINC was responsible for developing a strategy to stop the flow of illegal narcotics into the United States and to coordinate federal efforts to implement that strategy. Domestic drug-law enforcement had a high priority within the normal cabinet-management system.

In January 1972, President Nixon created the Office of Drug Abuse Law Enforcement (ODALE) in the Department of Justice and gave the ODALE director, Myles J. Ambrose, the added title of Consultant to the President for Drug Abuse Law Enforcement. The directors of both SAODAP and ODALE had a policyoversight role in advising the president.

The 1972 legislation authorizing SAODAP also created the Strategy Council on Drug Abuse (known as "The Strategy Council") and directed the "development and promulgation of a comprehensive, coordinated, long-term Federal strategy for all drug abuse prevention and drug traffic functions conducted, sponsored, or supported by the Federal government." The cabinet-level strategy council, with the directors of SAODAP and ODALE as cochairmen, prepared the 1973 Federal Strategy for Prevention of Drug Abuse and Drug Trafficking, the first explicit strategy document.

During 1973, the drug program and drug-policy organizations underwent major change. The Office of Management and Budget (OMB) established a special management office called Federal Drug Management (FDM), which supported OMB's senior officials, the CCINC, and the White House Domestic Council. Given unusually wide latitude in providing direct management assistance to the drug-related operating agencies, FDM assisted in implementation of President Nixon's Reorganization Plan No. 2 of 1973. Also in 1973, Dr. Jaffe was succeeded at SAODAP by Dr. Robert Dupont who in 1975 became the first director of the newly established National Institute on Drug Abuse. FDM also assumed oversight of the demand-related drug activities as SAODAP was phased out of the EOP. Before terminating in mid-1975, SAODAP published the 1974 and 1975 federal strategies, under the auspices of a relatively inactive Strategy Council.

In early 1975, President Gerald R. Ford directed the White House Domestic Council to review the federal drug effort. Vice-President Nelson A. Rockefeller chaired an interagency task force called the Domestic Council Drug Abuse Task Force, with the chief of FDM as study director. The task force, with advice from community organizations, prepared a comprehensive White Paper on Drug Abuse. The 1975 white paper recommended assigning responsibility for overall policy guidance to the Strategy Council on Drug Abuse; creating an EOP Cabinet Committee to coordinate prevention and treatment activities; and continuing a small staff in OMB to assist the Strategy Council and the EOP. In April 1976, President Ford announced two new cabinet committees, the Cabinet Committee on Drug Law Enforcement and the Cabinet Committee on Drug Abuse Prevention "to ensure the coordination of all government resources which bear on the problem of drug abuse" (1976 Strategy, p. 26). The cabinet committee structure, supported by the FDM staff, worked to the satisfaction of President Ford but did not satisfy Congress.

Congress enacted legislation establishing an Office of Drug Abuse Policy (ODAP) in March 1976, seeking a single individual in the EOP who had responsibility for the overall drug program. President Ford did not activate the new agency but continued with the three cabinet committees, supported by the FDM staff.

Executive Drug Policy 1977–1980. In March 1977, President Jimmy Carter revised the drug-policy structure, activating ODAP and abolishing the three drug-related cabinet committees. Also, he revitalized the strategy council, with the director of ODAP as executive director, to serve as the governmentwide advisory committee for all drug-abuse matters. ODAP worked particularly well with the White House staff, partially because Director Peter Bourne was also special assistant to the president for health issues and had an excellent relationship with President Carter and the White House staff. ODAP aggressively pursued a wide range of policy and coordination activities, including a major review of all federal drug programs.

The President's Reorganization Project reviewed the organization of the Executive Branch and recommended abolishing ODAP in mid-1977. Within the EOP, ODAP was an unusual federal agency, with a strong presence and authority for a single issue, somewhat contrary to the normal EOP structure. Thus, ODAP was a logical target in efforts to streamline the EOP. Congress disagreed strongly with the elimination of ODAP, however. After congressional hearings and negotiations, the Carter Administration compromised by continuing part of the ODAP staff and all the ODAP functions as part of the White House Domestic Policy Staff (DPS).

In March 1978, six members of ODAP's staff were transferred to DPS and became the Drug Policy Office (DPO). DPO continued to perform the ODAP functions, including responding to congressional interests and reporting directly to Peter Bourne. After Bourne departed the White House staff in 1978, the drug staff worked through the director of the DPS. In May 1979, the president affirmed the head of DPO (Lee Dogoloff, the associate director for drug policy)—as the individual primarily responsible for the federal government's drug-abuse prevention and control programs. DPO published the 1979 Federal Strategy and a 1980 Annual Report. A major policy-coordinating mechanism was the monthly meetings held by DPO with the heads of the major operating agencies (called the Principals Group). DPO also supported another policy-coordinating mechanism called the National Narcotics Intelligence Consumers Committee, established in April 1978. DPO also initiated efforts to increase military support for drug-interdiction activities. During the transition to the Reagan Administration in early 1981, most of President Carter's DPO staff departed.

Executive Drug Policy 1981–1988. In 1981, President Ronald W. Reagan's Office of Policy Development (OPD) included a Drug Abuse Policy Office (DAPO) similar in organization and role to the preceding DPO. President Reagan charged DAPO with (1) a full range of policy-development and -coordination activities, (2) international negotiations, and (3) assisting First Lady Nancy Reagan's drug-abuse prevention efforts. In addition to overseeing the efforts of the federal drug agencies, DAPO emphasized the use of all opportunities for the federal government to encourage a wide range of nongovernment antidrug activities. DAPO was directed by Carlton Turner, a pharmacologist, who was succeeded in 1987 by Dr. Donald Ian Macdonald, a pediatrician. DAPO published the 1982 Federal Strategy and, reflecting the broader policy direction, published the first "National" Strategy in 1984.

DAPO continued the coordination meetings with the agency heads (the previous Principals Group, renamed the Oversight Working Group) and assisted in the design and implementation of the National Narcotics Border Interdiction System (NNBIS), headed by Vice-President George H. Bush. DPO assisted the Cabinet Council on Legal Policy and the Cabinet Council on Human Resources with drug matters until the cabinet councils were replaced by the Domestic Policy Council in April 1985. The Domestic Policy Council Working Group on Drug Abuse Policy prepared a major presidential drug initiative in 1986, with assistance from DAPO.

During this period, the oversight of drug law enforcement moved away from the White House.

In 1984, Congress had established a federal drug law-enforcement czar to "facilitate coordination of U.S. operations and policy on illegal drug law enforcement." The attorney general was chairman of the new cabinet-level National Drug Enforcement Policy Board (NDEPB) with staff offices in the Department of Justice. DAPO was charged with ensuring "coordination between the NDEPB and the health issues associated with drug abuse," in addition to supporting the president and the White House staff. In January 1987, the NDEPB published the *National and International Drug Law Enforcement Strategy*, which expanded on the sections of the 1984 National Strategy involving drug law enforcement and international controls. DAPO continued to provide Executive Office oversight of the entire drug program.

In 1987, President Reagan replaced the NDEPB by creating a National Drug Policy Board (NDPB) to coordinate all drug-abuse policy functions. The director of the White House DAPO was a member and assisted the NDPB in developing the health-related drug policy. The NDPB published *Toward a Drug-Free America—The National Drug Strategy and Implementation Plans* in 1988.

The White House Conference for a Drug Free America was opened in 1987 with DAPO assistance; it was charged with reviewing a wide range of drug programs, policies, and informational activities—including focusing "public attention on the importance of fostering a widespread attitude of intolerance for illegal drugs and their use throughout all segments of our society" (Executive Order No.

12595, Section 1(c)). The conference, chaired by Lois Haight Herrington, published a final report in 1988 with 107 wide-ranging recommendations, including a "Cabinet-rank position of National Drug Director."

In late 1988, Congress again passed drug czar legislation, authorizing a new agency named the Office of National Drug Control Policy (ONDCP) in the EOP.

Executive Drug Policy 1989–1990s. ONDCP began operation in the EOP in early 1989, absorbing the NDPB, and terminating the two existing White House drug activities, DAPO and NNBIS. Although never actually a member of the cabinet, the first two cabinet-level directors were given broad responsibilities for developing and guiding a National Drug Control Program, including developing an annual strategy and overseeing its implementation. The first director, William Bennett, had been secretary of education in the Reagan administration; he was succeeded by Bob Martinez, a former governor of Florida. ONDCP had oversight of organization, management, budget, and personnel allocations of all departments and agencies engaged in drugcontrol activities. ONDCP used a complex set of interagency coordinating committees under a Supply Reduction Working Group, a Demand Reduction Working Group, and a Research and Development Committee. The director chaired the NSC's Policy Coordinating Committee for Narcotics which ensured coordination between drug law enforcement and national security activities. The director also provided administrative support to the President's Drug Advisory Council, which in turn assisted ONDCP in supporting national drug-control objectives through private sector initiatives. ONDCP was also required to establish realistic and attainable goals for the following two years and the following ten years and to monitor progress toward the goals. Following the election of President Bill Clinton, Lee Brown, a criminologist and former New York police commissioner, was appointed director of ONDCP and was also given membership in the cabinet.

CONGRESSIONAL DRUG-POLICY OVERSIGHT

Various legislative committees and subcommittees oversee the drug-control activities of the Executive Branch departments and agencies. In addition to the various standing committees, Congress had special drug-oversight activities, including the Senate Caucus for International Narcotics Control and the House Select Committee on Narcotics Abuse and Control. Special audits and evaluations by the General Accounting Office and support from the Congressional Research Service also assisted Congress in its oversight role.

The continuing congressional interest in establishing an effective drug-policy oversight mechanism reflected the difficulties of the various committees in attempting to address the drug activities of a single agency within the context of the overall federal effort. The frustration was reflected in the repeated legislative efforts to establish a drug czar in the EOP to oversee federal drug policy and to advise both the president and Congress.

For example, the Senate Committee on Government Operations had a long-term interest in drug-program oversight. Senator Charles H. Percy, responding to the plan to abolish ODAP in 1977, summarized the congressional view. Reiterating the programmatic needs for a single, high-level coordinating body with broad statutory authority over federal drug-abuse policy and its implementation, Senator Percy stated:

> My concerns are not limited to the question of whether the Federal drug abuse effort can function effectively under this proposal (to abolish ODAP). Indeed, my greatest opposition ... is that Congressional participation in the formulation and execution of Federal drug policy will be seriously impaired with the demise of ODAP. ... Although Congress has jurisdiction over the individual offices and agencies, this authority is meaningless without corresponding jurisdiction over those responsible for coordinating the line agencies' programs—the point where policy differences must be reconciled. [Congressional Record, September 30, 1977; S–16071–16072].

In the House of Representatives, the Select Committee on Narcotics Abuse and Control, headed by Representative Charles Rangel, played an important role in Congressional oversight of drug programs and policy. The select committee was formed in July 1976 "to oversee all facets of the Federal narcotics effort and coordinate the response of the seven legislative committees in the House which have jurisdiction over some aspect of the narcotics problem." Without legislative jurisdiction, the select committee was primarily a fact-finding activity to support the seven standing committees in the House of Representatives. The select committee also was a focal point for congressional pressure for a legislatively based federal drug czar. In early 1993, the select committee on Narcotics Abuse and Control was discontinued.

DRUG-POLICY LEGISLATION

In 1972, Congress passed legislation authorizing the Special Action Office for Drug Abuse Prevention, as requested by President Nixon. After SAODAP expired in 1975, Congress authorized a replacement drug-policy agency (ODAP), in early 1976, and was critical of President Ford's decision to not open the new agency.

When President Carter decided to activate ODAP in early 1977, Congress applauded the decision and confirmed the director and deputy director; but ODAP was abolished in early 1978 despite congressional objections, ending their successful relationship with ODAP. The resulting executive/congressional negotiations required the Drug Policy Office of the DPS to carry out the functions previously assigned to ODAP and to allow congressional access to the drug-policy staff.

In late 1979, Congress followed up with legislation requiring the president to establish a drug-abuse policy coordination system and to designate a single officer to direct the activities (21 USC 1111 & 1112). A system was established by President Carter (Executive Order 12133, 1979-Drug Policy Office) and by President Reagan (Executive Order 12368, 1982-Drug Abuse Policy Office).

In late 1982, Congress enacted a strong drug czar, in an Office of National and International Drug Operations and Policy, with a cabinet-level director. The director was granted broad powers to develop, review, implement, and enforce government policy and to direct departments and agencies involved. The explicit power to direct other departments and agencies was seen as too strong and in conflict with the principles of cabinet government. President Reagan did not accept the legislation.

In 1984, the Congress and the administration agreed to establish a cabinet-level NDEPB with a limited charter to coordinate drug law enforcement. The legislation designated the attorney general as chairman and primary adviser to the president and to Congress—on both national and international law enforcement.

In 1987, President Reagan signed Executive Order 12590, broadened the charter of the attorney general and the NDEPB to include the entire federal drug program and named the new activity the National Drug Policy Board.

In late 1988, Congress passed new drug czar legislation, creating the Office of National Drug Control Policy in the EOP, with a cabinet-level director and funding provisions for both operating expenses and program activities. President Bush accepted the new agency and appointed a cabinet-level director, but he did not include the first director or his successor in his immediate cabinet.

Thus, Congress achieved the drug czar objectives that it pursued for two decades—a cabinet-level drug-policy manager with broad oversight of policy and budgets, responsible both to Congress and the president.

(SEE ALSO: *Anslinger, Harry J., and U.S. Drug Policy International Drug Supply Systems; Opioids and Opioid Control, History; Prevention Movement; Treatment, History of*)

BIBLIOGRAPHY

HOGAN, H. (1989). Congressional Research Service, the Library of Congress. *Drug control at the federal level: Coordination and direction.* Report 87–780 GOV.

U.S. CONGRESS, HOUSE. Select Committee on Narcotics Abuse and Control. (1978). *Congressional resource guide to the federal effort on narcotics abuse and control, 1969–76, Part 1.* A Report of the Select Committee on Narcotics Abuse and Control. 95th Congress, 2nd sess. Washington, DC: Government Printing Office.

U.S. CONGRESS, HOUSE. Select Committee on Narcotics Abuse and Control. (1980). *Recommendation for continued house oversight of drug abuse problems.* A Report of the Select Committee on Narcotics Abuse and Control. Report No. 96–1380. 96th Congress, 2nd sess. Washington, DC: Government Printing Office.

U.S. EXECUTIVE OFFICE OF THE PRESIDENT. Domestic Council Drug Abuse Task Force.(1975). *White paper on drug abuse September. 1975.* Washington, DC: Government Printing Office.

U.S. EXECUTIVE OFFICE OF THE PRESIDENT. (1980). Domestic Policy Staff. *Annual report on the federal drug program. 1980.* Washington, DC: Government Printing Office.

U.S. EXECUTIVE OFFICE OF THE PRESIDENT. Drug Abuse Policy Office, Office of Policy Development, The White House. (1984). *1984 national strategy for prevention of drug abuse and drug trafficking.* Washington, DC: Government Printing Office.

U.S. EXECUTIVE OFFICE OF THE PRESIDENT. Office of Drug Abuse Policy. (1978). *1978 annual report.* Washington, DC: Government Printing Office.

U.S. EXECUTIVE OFFICE OF THE PRESIDENT. Office of National Drug Control Policy. (1990). *National drug control strategy. January 1990.* Washington, DC: Government Printing Office.

U.S. EXECUTIVE OFFICE OF THE PRESIDENT. Office of National Drug Control Policy. (1992). *National drug control strategy. January 1992.* Washington, DC: Government Printing Office.

U.S. EXECUTIVE OFFICE OF THE PRESIDENT. President's Advisory Commission on Narcotic and Drug Abuse. (1963). *Final report.* Washington, DC: Government Printing Office.

U.S. EXECUTIVE OFFICE OF THE PRESIDENT. Strategy Council on Drug Abuse. (1973) *Federal strategy for drug abuse and drug traffic prevention. 1973.* Washington, DC: Government Printing Office.

U.S. EXECUTIVE OFFICE OF THE PRESIDENT. Strategy Council on Drug Abuse. (1976). *Federal strategy. Drug abuse prevention. 1976.* Washington, DC: Government Printing Office.

U.S. EXECUTIVE OFFICE OF THE PRESIDENT. Strategy Council on Drug Abuse. (1979). *Federal strategy for drug abuse and drug traffic prevention. 1979.* Washington, DC: Government Printing Office.

RICHARD L. WILLIAMS

U.S. GOVERNMENT AGENCIES

The following articles appear in this section:

Bureau of Narcotics and Dangerous Drugs (BNDD);
Center for Substance Abuse Prevention (CSAP);
Center for Substance Abuse Treatment (CSAT);
National Institute on Alcoholism and Alcohol Abuse (NIAAA);
National Institute on Drug Abuse (NIDA);
Office of Drug Abuse Law Enforcement (ODALE);
Office of Drug Abuse Policy (ODAP);
Office of National Drug Control Policy (ONDCP);
Special Action Office for Drug Abuse Prevention (SAODAP);

*Substance Abuse and Mental Health Services
    Administration (SAMHSA);*
U.S. Customs Service;
U.S. Public Health Service Hospitals

Bureau of Narcotics and Dangerous Drugs
Presidential Reorganization Plan No. 1 of 1968 created the Bureau of Narcotics and Dangerous Drugs (BNDD) in the U.S. Department of Justice. The new agency combined the drug law enforcement functions of two predecessor organizations—the Federal Bureau of Narcotics (FBN) in the Department of the Treasury and the Bureau of Drug Abuse Control in the Food and Drug Administration, Department of Health and Human Services. Long-standing conflicts between two Department of the Treasury agencies that shared drug-enforcement responsibilities—the Federal Bureau of Narcotics and the Bureau of Customs—led to the decision to move the FBN functions into a new agency (BNDD) in a different cabinet department (Justice).

MISSION AND EXPERIENCE

BNDD's role was to suppress illicit narcotics trafficking and to control the diversion of legally manufactured drugs. BNDD was responsible for working with foreign governments to halt international drug traffic, immobilizing domestic illegal drug-distribution networks, providing a wide range of technical assistance and training to state and local officers, and preparing drug cases for prosecution.

BNDD emphasized investigations of high-level drug trafficking to identify and target major national and international violators. Director John E. Ingersoll described the success of BNDD as being "able to apprehend scores of illicit drug traffickers who were previously immune to the feeble efforts which law enforcement was formerly able to mount." In 1968 and 1969, BNDD contributed to major international success in stopping heroin traffic originating in Turkey.

The Bureau of Customs continued interdiction of drug smuggling at the borders and ports of entry. Customs special agents investigated drug cases based on seizures made by Customs inspectors and on antismuggling intelligence. Conflict between BNDD and Customs continued, with allegations of lack of

cooperation and failure to share intelligence with each other.

The White House and Office of Management and Budget (OMB) tried to resolve the conflict and, in early 1970, President Richard M. Nixon directed BNDD and Customs to work out a set of operating guidelines. After considerable interagency discussion, formal guidelines were prepared to give to BNDD full jurisdiction over drug-enforcement operations both within the United States and overseas. Customs was to be limited to border operations. The president approved the guidelines, but the conflicts continued. Neither Congress nor the White House was satisfied. Senator Abraham Ribicoff described the detailed guidelines as "more reminiscent of a cease-fire agreement between combatants than a working agreement between supposedly cooperative agencies."

ADDITIONAL DRUG ENFORCEMENT COMPLICATIONS

The "war against drugs" continued to expand. In 1972, President Nixon established two new drug agencies in the Department of Justice—the Office of Drug Abuse Law Enforcement (ODALE) and the Office of National Narcotics Intelligence (ONNI). ODALE's operational involvement with state and local law enforcement against local drug dealers was intended to complement BNDD's focus on high level traffickers. ODALE, however, depended on existing federal agencies for agents and attorneys, and BNDD was required to lend over 200 narcotics agents to ODALE. The additional antidrug agencies, combined with sensational reporting of conflicts between special agents from BNDD and Customs, added to the public perception of fragmentation and disorder in federal drug law enforcement.

In early 1973, another presidential reorganization plan was designed to eliminate the overlap and duplication of effort in drug enforcement. A factual assessment of the BNDD/Customs situation, provided to the Congress by the chief of OMB's Federal Drug Management Division, Walter C. Minnick, reported "Having attempted formal guidelines, informal cooperation and specific Cabinet-level mediation, all without success, the President concluded in March of 1972 that merging the drug investigative and intelligence responsibilities of Customs and BNDD into a single new agency was the only way to put a permanent end to the problem." Under Reorganization Plan No. 2 of 1973, BNDD, ODALE, and ONNI were eliminated; their functions and resources, along with 500 Customs special agents (those previously involved in drug investigations), were consolidated in the new Drug Enforcement Administration (DEA) in the Department of Justice.

(SEE ALSO: *Anslinger, Harry J., and U.S. Drug Policy*)

BIBLIOGRAPHY

BONAFEDE, D. (1970). Nixon seeks to heal top-level feud between customs, narcotics units. *National Journal, 2*(15), 750–751.

BONAFEDE, D. (1970). Nixon approves drug guidelines, gives role to Narcotic Bureau. *National Journal, 2*(29), 1532–1534.

FINLATOR, J. (1973). *The drugged nation.* New York: Simon & Schuster.

MOORE, M. H. (1978). Reorganization Plan #2 reviewed: Problems in implementing a strategy to reduce the supply of drugs to illicit markets in the United States. *Public Policy, 26*(2), 229–262.

RACHAL, P. (1982). *Federal narcotics enforcement.* Boston: Auburn House.

U.S. CONGRESS, SENATE, Committee on Government Operations. (1973). *Reorganization Plan No. 2 of 1973, Hearings before the Subcommittee on Reorganization, Research, and International Organizations.* 93rd Congress, 1st sess., Part 1. April 12, 13, and 26, 1973. Washington, DC.

RICHARD L. WILLIAMS

Center for Substance Abuse Prevention (CSAP) This agency was originally established as the Office for Substance Abuse Prevention (OSAP). It was created by the Anti-Drug Abuse Act of 1986 for the prevention of alcohol and other drug (AOD) problems among U.S. citizens, with special emphasis on youth and families living in high-risk environments. Dr. Elaine Johnson was appointed as the first director of the office. From 1986 to 1992, OSAP operated as a unit of the Alcohol, Drug Abuse, and Mental Health Administration (ADAMHA), one of the eight Public Health Service agencies within the U.S. Department of Health and Human Services.

In 1992, Public Law 102-321 reorganized ADAMHA and renamed it the Substance Abuse and Mental Health Services Administration (SAMHSA); it also created CSAP to replace OSAP.

The goal of CSAP is to promote the concepts of no use of any illicit drug and no illegal or high-risk use of alcohol or other legal drugs. (High-risk alcohol use includes drinking and driving; drinking while pregnant; drinking while recovering from alcoholism and/or when using certain medications; having more than two drinks a day for men and more than one for women, or to intoxication).

These are the principles that guide the prevention work of CSAP:

1. The earlier PREVENTION is started in a person's life, the more likely it is to succeed.
2. PREVENTION PROGRAMS should be knowledge based and should incorporate state-of-the-art findings and practices drawn from scientific research and field expertise.
3. Prevention programs should be comprehensive.
4. Programs should include both process and outcome evaluations.
5. The most successful programs are likely to be those initiated and conducted at the community level.

To utilize these principles and achieve its goals, CSAP performs the following functions:

1. Carries out demonstration projects targeting specific groups and individuals in high-risk environments.
2. Assists communities in developing long-term, comprehensive AOD-use prevention programs and early intervention programs.
3. Operates a national clearinghouse for publications on prevention and treatment and other materials and services, including the operation of the Electronic Communication System and the Regional Alcohol and Drug Awareness Resource (RADAR) Network.
4. Supports the National Training System, which develops new drug-use prevention materials and delivers training.
5. Supports field development.
6. Conducts an evaluation strategy consisting of individual grantee evaluations, contractual program-wide evaluations, and the National Evaluation Project.
7. Provides technical assistance for capacity building and promotes collaborations to help states, communities, and organizations develop and

implement communications, drug-use prevention, and early intervention efforts.

8. Develops and implements public information and educational media campaigns and other special-outreach and knowledge-transfer prevention programs.
9. Maintains a national drug-use prevention database to provide information on substance-abuse prevention programs.
10. Provides technical assistance and materials to small businesses for the development of EMPLOYEE-ASSISTANCE PROGRAMS.
11. Operates the National Volunteer Training Center for Substance Abuse Prevention.

To promote interagency cooperation and facilitate jointly sponsored prevention activities, CSAP's staff meets routinely with various federal organizations, including the departments of defense, justice, education, transportation, labor, housing and urban development, the Bureau of Indian Affairs, and others.

CSAP also develops partnerships with the research community, parent groups, foundations, policymakers, health-care practitioners, state and community leaders, educators, law enforcement officials, and others to enhance opportunities for comprehensive approaches to prevention and early intervention.

(SEE ALSO: *Education and Prevention; Parents Movement; Prevention Movement*)

ELAINE JOHNSON

Center for Substance Abuse Treatment (CSAT)

The Center for Substance Abuse Treatment (CSAT) was established in January 1990 as the Office for Treatment Improvement (OTI) of the Alcohol, Drug Abuse, and Mental Health Administration (ADAMHA) in the Department of Health and Human Services (DHHS). Dr. Beny J. Primm, a physician who had spent more than twenty years developing a major treatment program in New York City, was appointed its first director. Following reorganization of ADAMHA in 1992, the agency was renamed and is now part of the Substance Abuse and Mental Health Services Administration (SAMHSA), which replaced ADAMHA.

The congressional mandate of CSAT is to expand the availability of effective treatment and recovery services for people with drug and alcohol problems.

One of its goals is to ensure that new treatment technology is absorbed by the addiction-treatment infrastructure—that is, the system of state and local government agencies and public and private treatment programs providing addiction-treatment services. In carrying out this responsibility, CSAT collaborates with states, communities, and treatment providers to upgrade the quality and effectiveness of treatment and enhance coordination among drug-treatment providers, human-services, educational and vocational services, the criminal-justice system, and a variety of related services. CSAT provides financial and technical assistance for this purpose to targeted geographic areas and patient populations, with emphasis on assistance to minority racial and ethnic groups, ADOLESCENTS, HOMELESS people, WOMEN of childbearing age, and people in rural areas.

CSAT also collaborates with other government agencies, such as the National Institute on Drug Abuse (NIDA), the National Institute on Alcohol Abuse and Alcoholism (NIAAA), the National Institute of Mental Health (NIMH), the Center for Substance Abuse Prevention (CSAP), and state and local governments to promote the utilization of effective means of treatment and to develop treatment standards. In addition, CSAT has interagency agreements with the Department of Labor and the Department of Education that are designed to improve the coordination of health and human services, education, and vocational training. CSAT also promotes the mainstreaming of alcohol-, drug-abuse, and mental-health treatment into the primary health care system, and it is responsible for administering the Substance Abuse Prevention and Treatment (SAPT) Block Grant program, which provides federal support to state substance-abuse prevention and treatment programs (funded at $1.13 billion in fiscal year 1993).

Research has generated a vast body of knowledge regarding the nature of chemical dependency and about what works in the treatment of addiction and addiction-related primary health and mental-health disorders. From this research, three key observations formed the basis for CSAT's initial treatment philosophy. First, addiction is a complex phenomenon; people's addiction cannot be treated in isolation from addressing their primary health, mental health, or socioeconomic deficits. Second, addiction is frequently a chronic, relapsing disorder; the gains made during treatment often are lost following a person's return to the community. CSAT therefore tried to foster programs that provided those treated for chemical dependency with a series of interventions along a sustained continuum. These two observations constituted the basis for CSAT's Comprehensive Treatment Model, which was a central principle in all of its demonstration grant programs and technical-assistance initiatives. During its first few years of existence, CSAT targeted resources to the people it perceived as most adversely affected by extreme socioeconomic problems and at highest risk for addiction because of exposure to CRIME, abuse, POVERTY, and HOMELESSNESS, and also because of lack of access to primary health and mental health care, social services, and vocational training and education. For this reason, the early CSAT Comprehensive Treatment Model demonstration grants fostered a wide array of primary interventions geared to addressing each patient's health and human service needs, coupled with a readily accessible, intensive aftercare component.

At the core of CSAT's overall approach is, quite simply, the conviction that treatment works. Treatment has proved effective in reducing the use of illicit drugs and alcohol, improving rates of employment, reducing rates of HUMAN IMMUNODEFICIENCY VIRUS (HIV) seroconversion, reducing criminal activity, and reducing overall patient morbidity.

In addition to the SAPT Block Grant, CSAT awarded grants for a variety of demonstration and service programs: The treatment-capacity expansion program provided resources to the states to expand capacity in areas of demonstrated shortage; Target Cities assists metropolitan areas with particularly high-risk populations in providing treatment services and in developing systems to coordinate and improve the infrastructure of the programs. Critical Populations is a demonstration project for treatment program enhancement aimed at particularly at-risk groups—ADOLESCENTS; racial and ethnic minorities; residents of public housing; women and their infants and children; rural populations; drug and alcohol abusers who are homeless; patients with HIV or AIDS. Criminal justice-related programs include drug-abuse treatment programs in PRISONS AND JAILS; diversion to treatment; special services for probation or parole clients; screening, testing, referral, and treatment services for HIV/AIDS, TB, and other communicable diseases; literacy, education, job training, and job placement services; and case management and DRUG TESTING. CSAT also sup-

ported demonstration treatment campus programs; several programs aimed specifically at WOMEN and their infants and children; AIDS outreach for substance abusers; linkage of primary care and substance abuse model programs; state systems development programs; professional training and education; and collaborative efforts with other federal agencies.

After Dr. Primm's return to New York in 1992 and following Mr. David Mactas's appointment to head the agency in 1994, and as part of the Clinton administration's effort to reinvent government (redefine and refine its functions), CSAT's demonstration grant program emphasis shifted from improvement of services for the populations in greatest need to the development of knowledge about the effectiveness of treatment for different subgroups of the drug-using population.

Information regarding CSAT's current programs and technical initiatives is available from the CSAT Public Affairs Office, Center for Substance Abuse Treatment, Substance Abuse and Mental Health Services Administration, 5600 Fishers Lane, Rockville, MD 20857.

(SEE ALSO: *Ethnic and Cultural Relevance in Treatment; Treatment Types; Vulnerability As Cause of Substance Abuse*)

BENY J. PRIMM

National Institute on Alcohol Abuse and Alcoholism (NIAAA)   This is the leading federal entity for biomedical and behavioral research focused on combatting alcohol abuse and ALCOHOLISM. Alcohol abuse and alcoholism are major health problems in the United States. Approximately 15.3 million Americans experience serious alcohol-related problems, and the annual cost to society from alcoholism and alcohol abuse is estimated at 98.6 billion dollars.

LEGISLATIVE HIGHLIGHTS

December 31, 1970—Passage of Public Law (PL) 91-616 (Comprehensive Alcohol Abuse and Alcoholism Prevention, Treatment, and Rehabilitation Act of 1970), to develop and conduct comprehensive health, education, training, research, and planning programs for the prevention and treatment of alcohol abuse and alcoholism.

May 14, 1974—Passage of PL 93-282 (Comprehensive Alcohol Abuse and Alcoholism Prevention, Treatment, and Rehabilitation Act Amendments of 1974), establishing NIAAA, the National Institute of Mental Health (NIMH), and a new National Institute on Drug Abuse (NIDA) as coequal institutes within a newly created Alcohol, Drug Abuse, and Mental Health Administration (ADAMHA).

July 26, 1976—Expansion of NIAAA's research authority to include behavioral and biomedical etiology of the social and economic consequences of alcohol abuse and alcoholism under authority of the Comprehensive Alcohol Abuse and Alcoholism Prevention, Treatment, and Rehabilitation Act Amendments of 1976 (PL 94-371).

August, 1981—Passage of the Omnibus Budget Reconciliation Act of 1981 (PL 97-35), which transferred responsibility and funding for alcoholism treatment services to the states through the creation of an Alcohol, Drug Abuse, and Mental Health Services Block Grant administered by ADAMHA and which strengthened NIAAA's research mission.

October 27, 1986—Creation of a new Office for Substance Abuse Prevention (OSAP) in ADAMHA by the Anti-Drug Abuse Act of 1986 (PL 99-570), which consolidated the remainder of NIAAA's nonresearch prevention activities with those of NIDA and permitted NIAAA's total commitment to alcohol research.

July 10, 1992—NIAAA transferred to the National Institutes of Health by the ADAMHA Reorganization Act of 1992 (PL 102-321).

ORGANIZATION

The NIAAA is one of eighteen research institutes that comprise the U.S. National Institutes of Health (NIH). The NIH is a part of the U.S. Public Health Service, a major operating component of the U.S. Department of Health and Human Services. NIAAA's activities are carried out by three principal staff offices (Office of Scientific Affairs, Office of Policy Analysis, and Office of Planning and Resource Management), which together are responsible for overall institute management, planning, policy development, legislative activities, scientific information dissemination, and extramural grant review activities; and four research divisions (Division of Basic Research, Division of Clinical and Prevention Research, Division of Biometry and Epidemiology, and Division of Intramural Clinical and Biological Research), which are responsible for managing the institute's research portfolio.

HISTORY

Prior to the establishment of NIAAA, alcohol research within the federal government was supported and carried out by grants from the NIMH, and within the intramural programs of NIMH by units such as St. Elizabeth's Hospital in Washington, D.C.

MAJOR PROGRAM AND ACTIVITIES

In support of its mission, NIAAA conducts and supports biomedical and behavioral research aimed at understanding alcohol's effects on the human body and on society; conducts epidemiological studies and national and community surveys to assess the risks for and magnitude of alcohol-related problems; conducts and supports policy studies that have broad implications for alcohol problem prevention, treatment, and rehabilitation; and supports programs for training and developing scientists for careers in alcohol research. Research findings are shared with interested parties through bulletins, monographs and reports, an on-line bibliographic data base, and a professional journal distributed quarterly. The institute also collaborates with other national and international research institutes working in the alcohol field.

NIAAA supports research through extramural grant support to scientists at leading U.S. research institutions, through interdisciplinary National Alcohol Research Centers Program grants, and through an intramural research program. NIAAA also is involved in a number of important international research collaborations.

EXTRAMURAL RESEARCH

NIAAA supports research in ten broadly defined areas of investigation (see listing). Scientists from a variety of disciplines—including biology, medicine, and the social and behavioral sciences—participate in the extramural program.

Genetic and Molecular Biology: Research exploring genetic influence on alcoholism; how genes and the environment interact to precipitate alcoholism; biochemical and physiological markers of high risk; selective breeding of animal models of alcoholism; and studies of alcohol metabolizing enzymes.

Biochemistry and Metabolism: Investigations of how alcohol is metabolized by the body, alcohol's effects on membrane structure and function, and how alcohol affects the metabolism of nutrients, vitamins, and drugs.

Neuroscience and Behavior: Studies of how alcohol affects the central nervous system and on the biological mechanisms underlying the development of intoxication, tolerance, and dependence.

Biological and Environmental Determinants of Drinking: Studies of the factors in drinkers' environments that lead to abuse and disease.

Treatment: Research aimed at improving diagnostic criteria; controlled studies of existing treatments for alcohol abuse and alcoholism; studies on new treatment techniques including new medications and behavioral therapies; and studies measuring cost-effectiveness of specific types of treatments and settings.

Prevention: Investigations aimed at developing effective measures to reduce alcohol-related problems among children, adolescents, young adults, and other at-risk populations and studies of community, environment, and workplace issues; drinking and driving deterrence; and alcohol and the economy.

Alcohol-related Performance: Research on how alcohol affects the ability to perform important tasks, including the development of new methods for measuring alcohol impairment.

Alcohol and Pregnancy: Investigations of the adverse effects of fetal alcohol exposure, including alcohol's impact on fetal brain development and on fetal growth and development; determination of genetic factors contributing to fetal vulnerability to alcohol; and studies to improve the health-care system's ability to identify and intervene with pregnant women at high risk for having a child with alcohol-related birth defects.

Alcohol-related Medical Disorders: Studies of alcohol's toxic effects on human organs, including alcohol-related disorders of the liver, heart, pancreas, and brain; alcohol's effects on the immune system, including alcohol's role, if any, in the acquisition and clinical course of acquired immunodeficiency disease (AIDS).

Epidemiology and Biometry: Research examining the distribution of alcohol-related illness and death by place, by person (age, sex, ethnic and socioeconomic group) over time.

NATIONAL ALCOHOL RESEARCH CENTERS PROGRAM

NIAAA administers fourteen Alcohol Research Centers nationwide. The program provides long-term (typically 5 years) support for interdisciplinary research that focuses on particular aspects of alcohol abuse, alcoholism, or other alcohol-related problems, and it encourages outstanding scientists from many disciplines to provide a full range of expertise, approaches, and advanced technologies for developing knowledge in these areas. A primary goal of any NIAAA-funded center is to become a significant regional or national research resource. Each center also affords research training opportunities for persons from various disciplines and professions.

Specific areas of alcohol-center focus are—the genetic determinants of alcohol ingestion; social epidemiology of alcohol problems; pathological consequences of alcohol abuse; environmental approaches to prevention; effects of alcohol on cellular neurobiology; alcohol and the cell; etiology and treatment of alcohol dependence; alcohol and the aged; alcohol and immunology; genetic approaches to the neuropharmacology of alcohol; biobehavioral manifestations of adolescent alcohol abuse; neurobiology, genetics, epidemiology, and alcoholism;

neurobehavioral aspects of fetal alcohol exposure; and integrating basic and applied research in alcohol treatment.

INTRAMURAL RESEARCH

The goal of the NIAAA Intramural Research Program is to understand the mechanisms by which alcohol produces intoxication, dependence, and damage to vital body organs, and to develop tools to prevent and treat those biochemical and behavioral processes. Major areas of study include identification and assessment of genetic and environmental risk factors for the development of alcoholism; the effects of alcohol on the central nervous system, including how alcohol modifies brain activity and behavior; metabolic and biochemical effects of alcohol on various organs and systems of the body; noninvasive imaging of brain structure and activity related to alcohol use; development of animal models of alcoholism; and the diagnosis, prevention, and treatment of alcoholism and associated disorders.

Studies on the effects of alcohol on cell-membrane receptors, ion channels, and the expression of genes that code for these important proteins are yielding intriguing insights into the basic mechanisms of alcohol's action. Combined with studies on region-specific effects of alcohol on the release of neurotransmitters, these investigations will shed light on how alcohol produces reward, dependence, tolerance, and brain damage. Behavioral studies using primarily mice and monkeys, combined with molecular genetics and behavioral manipulations during development, examine important protective and causal factors for alcohol abuse and dependence.

NIAAA utilizes a combination of clinical and basic research facilities, which enables a coordinated interaction between basic research findings and clinical applications in pursuit of these goals. Four laboratories, an inpatient ward and a large outpatient program are located at the NIH Clinical Center.

INTERNATIONAL ACTIVITIES

The goals of the NIAAA International Activities Program are to further NIAAA domestic goals through international cooperation and collaboration with other countries and with international organiza-

tions—as well as to support U.S. foreign-policy objectives in relation to health and medical sciences, including working with other countries in their health endeavors to deal more effectively with their alcohol-related public-health problems. These goals are accomplished primarily through information exchange, exchange of scientists, technical cooperation and collaboration, and collaborative research projects.

(SEE ALSO: *Alcohol: Complications; Addiction Research Foundation (Canada); Complications: Alcohol (Liver); U.S. Government Agencies Supporting Substance Abuse Research*)

ENOCH GORDIS

National Institute on Drug Abuse (NIDA)

The National Institute on Drug Abuse provides national leadership for research on drug abuse, dependence, and addiction. Through its extramural program and its intramural program at the Addiction Research Center in Baltimore, Maryland, NIDA has implemented a comprehensive program to combat the illicit or nonmedical use and abuse of drugs. This program includes studies on the causes and consequences, the prevention and treatment, and the biological, social, behavioral, and neuroscientific bases of drug abuse and addiction. NIDA is charged with the development of medications to treat drug addiction. The Institute is also responsible for supporting research on the relationship between drug use and AIDS, tuberculosis, and other medical diseases. In addition, NIDA supports research training and career development, science and public education, and research dissemination. NIDA is the largest institution devoted to drug-abuse research in the world. Almost 90 percent of all U.S. drug-abuse research is supported by awards made by NIDA to scientists, primarily at major research facilities in the United States, abroad, and at NIDA's own Addiction Research Center (ARC).

HISTORY

Drug-abuse research and treatment have been a concern of the U.S. Public Health Service since the early 1930s. The Public Health Service Hospitals at

Lexington, Kentucky, and at Fort Worth, Texas, were established in 1929—and the research laboratories were established at Lexington in 1935. When the National Institute of Mental Health (NIMH) was established in 1946, the research laboratory at Lexington—the Addiction Research Center—became an intramural laboratory of the National Institute of Mental Health within the National Institutes of Health (NIH). In the late 1960s, NIMH became a separate bureau parallel to NIH, and the ARC was transferred to NIMH.

In the mid-1960s, a Center for Drug Addiction was established within NIMH, with responsibilities for making grants to universities for research and training. When the Narcotic Addict Rehabilitation Act was passed in 1966, giving NIMH responsibility for providing treatment to civilly committed addicts and for establishing community-based treatment programs, the Center for Drug Addiction was incorporated into a Division on Narcotic Addiction, still within NIMH. In 1972, the Drug Abuse Office and Treatment Act was passed, which created for the first time a special White House office to coordinate drug-abuse treatment, research, and prevention; this same legislation also created an Institute on Drug Abuse (still within NIMH) to come into existence when the White House office (Special Action Office for Drug Abuse Prevention) terminated in 1975.

The National Institute on Drug Abuse was established in 1974, and on the basis of additional legislation it was given a status comparable to that of NIMH and the National Institute on Alcohol Abuse and Alcoholism within a newly created coordinating structure—the Alcohol, Drug Abuse, and Mental Health Administration (ADAMHA). Dr. Robert Dupont became NIDA's first director in 1974, when it became one of three institutes comprising ADAMHA, a Public Health Service agency within the Department of Health and Human Services. NIDA's mandate was to collect information on the incidence, prevalence, and consequences of drug abuse, to improve the understanding of drugs of abuse and their effects on individuals, and to expand the ability to prevent and treat drug abuse. Through scientific research, NIDA has built a base of information on how drugs affect us—what they do to our bodies; to our behavior, thoughts, and emotions; to our relationships; and to our society. This understanding of the biological, social, behavioral, and environmental influences that place individuals at risk for turning to drug abuse is of great importance to prevention and treatment practitioners and to educators.

In October of 1992, the drug, alcohol, and mental-health activities within the Department of Health and Human Services were reorganized. In this reorganization, NIDA (along with the National Institute on Alcohol Abuse and Alcoholism and NIMH) were transferred from ADAMHA to NIH. The remaining service activities of ADAMHA were augmented and changed; the resulting organization, consisting of three service-oriented centers, was designated the Substance Abuse and Mental Health Services Administration (SAMHSA).

NIDA's mission can best be understood by considering the magnitude of the public-health problem of drug abuse. Studies supported by NIDA and by others show that in the United States millions have tried some illicit drug in the past year and in the past month. This includes both sexes, all races, all ages, and all economic levels and demographic regions. Data from 1985, projected to 1988 for inflation and other factors, indicated that the annual economic cost to society from drug abuse is approximately $58.3 billion. A further burden involves the increased risk for drug abusers of contracting the human immunodeficiency virus (HIV) that causes acquired immunodeficiency syndrome (AIDS). Nearly 35 percent of individuals newly diagnosed with AIDS are intravenous drug abusers. In addition to increasing the risk for AIDS, drug abuse also increasese the risks of exposure to and spread of other diseases, such as other sexually transmitted diseases (STDs), serum hepatitis (hep B), and tuberculosis (TB). The magnitude of the public-health problems associated with drug abuse, the research opportunities presented, and the cost to society have all had an impact on the research priorities supported by NIDA.

Prior to the 1992 reorganization, NIDA's priorities included an epidemiologic research program that monitors changing patterns of drug abuse, characteristics of drug users (see NATIONAL HOUSEHOLD SURVEY), and populations at highest risk of drug abuse. NIDA also collected data on the medical consequences of drug abuse, such as individuals in high-school (see HIGH SCHOOL SENIOR SURVEY), as well as data from hospital emergency-room visits caused by drug abuse and data from medical examiner cases. Through these activities, NIDA has developed a program that has been used to guide U.S. policy and to assist the states and communities to better understand the nature and extent of drug abuse; based on that understanding, they can better determine their own program priorities. After the 1992 reorganization that transferred NIDA to NIH, several of NIDA's epidemiological activities were transferred to SAMHSA. NIDA maintains responsibility for major epidemiological data collection activities such as the High School Senior Survey (Monitoring the Future Study) and the Community Epidemiology Workgroup program. It also continues to sustain a strong research portfolio through its Division of Epidemiology and Prevention Research.

FUNCTIONS

To improve the ability to prevent drug abuse, NIDA is concentrating on the variety of biological, behavioral, social, and environmental factors involved in vulnerability to drug abuse. This information enables NIDA to improve both prevention and treatment approaches—which are the keys to overcoming the demand for drugs—and to contribute to effective U.S. demand-reduction policies.

Successful treatment offers the best means for overcoming a life cycle revolving around drug-seeking behaviors and also reduces the spread of AIDS among intravenous drug abusers. Accordingly, NIDA is researching ways to improve the effectiveness of treatment—by matching the best treatment for each individual—and by increasing retention rates and reducing relapse rates (e.g., see DARP, TOPS). Drug dependence is a chronic, relapsing disorder, but treatment has been an effective tool in helping some to break the addiction cycle.

Through an understanding of the effects of drugs on the brain, NIDA is developing more effective treatments—including medications—for specific drugs of abuse, such as COCAINE and HEROIN, and for the toxic effects on the BRAIN and other organs that drugs of abuse produce. NIDA has engaged in a major effort to improve research on, and its application to, services for drug-abusing pregnant and postpartum women. NIDA also seeks to develop strategies to prevent or ameliorate the consequences of drugs of abuse on the children of drug-abusing parents.

To support this array of research programs, the research community needs an adequate supply of scientists with up-to-date skills and knowledge. Ac-

cordingly, NIDA sponsors drug-abuse research programs in the biomedical and behavioral sciences. These programs include support of pre- and post-doctoral training in medical schools, universities, and other institutions of higher education in basic, clinical, behavioral, and epidemiological research, to assure the steady supply of trained scientists.

A final important function of NIDA is to make available research findings. These are published for other researchers, prevention practitioners, treatment practitioners, young people, parents, policy-makers, and others. In addition, NIDA's Technology Transfer Program identifies research results ready for transfer to clinicians and translates those results into forms useful to practitioners. Through NIDA's research dissemination and technology transfer programs, science-based information can then be used to educate, prevent, treat, and rehabilitate.

CONCLUSION

NIDA conducts and supports RESEARCH that has as its underlying principles the goals of eliminating drug abuse, treating those whom prevention fails, increasing retention and decreasing relapse, and improving the health and well-being of all Americans, their families, their communities, and their nation.

NIDA collaborates with other research institutes, including the National Institute on Alcohol Abuse and Alcoholism, the other institutes within the National Institute of Mental Health, and the National Institutes of Health; with the Substance Abuse and Mental Health Services Administration; and with other agencies and departments of the U.S. government.

(SEE ALSO: *Comorbidity and Vulnerability; Substance Abuse and AIDS; Treatment; Vulnerability As Cause of Substance Abuse*)

RICHARD A. MILLSTEIN

Office of Drug Abuse Law Enforcement (ODALE) Located within the U.S. Department of Justice, the Office of Drug Abuse Law Enforcement (ODALE) was established by President Richard M. Nixon with Executive Order 11641 in January 1972.

Myles J. Ambrose was appointed director of ODALE and held two other concurrent titles: special consultant to the president for drug abuse law enforcement and special assistant attorney general.

FEDERAL, STATE, AND LOCAL TEAMWORK

Complementing federal efforts directed at "high-level drug traffickers," ODALE was charged with attacking the heroin-distribution system at the street level to reduce the drug's availability there. Patterned after the justice department's Organized Crime Strike Forces, the ODALE program included task forces of federal, state, and local law-enforcement officers and attorneys. The full use of federal, state, and local narcotics laws, the availability of assigned attorneys, and the use of the investigative grand jury made possible a wide range of approaches in pursuing violators.

ODALE established task forces in thirty-four cities in 1972 and encouraged citizens to "report information regarding alleged narcotics law violators in strict confidence." The federal government paid for task force equipment and operational expenses, including payments for a portion of the salaries and overtime of state and local officers. ODALE was credited with more than 8,000 narcotics arrests with a conviction rate of more than 90 percent during its 17 months of operation. Nevertheless, ODALE agents were widely criticized for conducting several drug raids involving unauthorized forcible entries into private homes and failures in identifying themselves as law officers during drug raids.

REORGANIZATION

ODALE was abolished on July 1, 1973, by Presidential Reorganization Plan No. 2 of 1973 and "those Federal operations designed to attack narcotics traffic at the street level in cooperation with local authorities" were transferred to the newly established Drug Enforcement Administration (DEA). The ODALE program was redesignated as DEA's State and Local Task Force program. ODALE's Deputy Director John R. Bartels, Jr., became the first administrator of the DEA.

(SEE ALSO: *Anslinger, Harry J., and U.S. Drug Policy*)

BIBLIOGRAPHY

RACHAL, P. (1982). *Federal narcotics enforcement*. Boston: Auburn House.

U.S. CONGRESS, SENATE, Committee on Government Operations. (1973). *Reorganization Plan No. 2 of 1973, Establishing a Drug Enforcement Administration in the Department of Justice*. Report of the Subcommittee on Reorganization, Research, and International Organizations, 93rd Congress, 1st sess., Report No. 93–469. Washington, DC.

U.S. GENERAL ACCOUNTING OFFICE. (1975). *Federal drug enforcement: Strong guidance needed*. Report No. GGD–76–32. Washington, DC.

RICHARD L. WILLIAMS

Office of Drug Abuse Policy   In March 1976, Congress authorized the creation of the Office of Drug Abuse Policy (ODAP) in the Executive Office of the President, with an annual budget of $1.2 million. President Jimmy Carter opened the office in March 1977 and appointed Dr. Peter G. Bourne as director.

The director of ODAP was given wide responsibilities in assisting the president with all federal drug-abuse matters, including providing "policy direction and coordination among the law enforcement, international and treatment/prevention programs to assure a cohesive and effective strategy that both responds to immediate issues and provides a framework for longer-term resolution of problems." The statutory authority included setting objectives, establishing priorities, coordinating performance, and recommending changes in organization.

During the first year of operation, ODAP conducted several international missions and worked closely with United Nations narcotics organizations. In coordinating federal drug activities, ODAP relied on biweekly discussion meetings with the heads of the principal drug agencies. Policy determination was executed through cooperative interagency study efforts. ODAP completed six comprehensive interagency policy reviews: border management, drug law enforcement, international narcotics control, narcotics intelligence, demand reduction, and drug abuse in the armed forces.

The ODAP staff coordinated preparation of President Carter's August 1977 Message to the Congress on Drug Abuse and initiated the planning for a comprehensive federal strategy to be published by the revitalized Strategy Council.

REORGANIZATION

After one year of successful operation, ODAP was abolished by Reorganization Plan No. 1 of 1977, effective March 31, 1978. Six ODAP staff members were transferred to a special drug-policy unit (Drug Policy Office) within the White House Domestic Policy Staff. The drug-policy staff continued to report to Dr. Bourne who became special assistant to the president for health issues.

(SEE ALSO: *Anslinger, Harry J. and U.S. Drug Policy*)

BIBLIOGRAPHY

HAVEMANN, J. (1978). Carter's reorganization plans—Scrambling for turf. *National Journal, 10*(20), 788–794.
U.S. EXECUTIVE OFFICE OF THE PRESIDENT. Office of Drug Abuse Policy. (1978). *1978 Annual Report*. Washington, DC: Government Printing Office.

RICHARD L. WILLIAMS

Office of National Drug Control Policy

The Office of National Drug Control Policy (ONDCP) was established on January 29, 1989, by Public Law 100–690 (21 USC 1504) as the drug-coordination agency for the Executive Office of the President (EOP) under President George H. Bush. ONDCP is responsible for coordinating federal efforts to control illegal drug abuse. It is the product of almost two decades of congressional efforts to mandate a so-called drug czar—the law providing for cabinet-level status and congressional involvement in drug-control policy. Its initial five-year authorization, to expire November 17, 1993, was extended.

ONDCP oversees international and domestic antidrug functions of all executive agencies and ensures that such functions sustain and complement the government's overall antidrug efforts.

THE DIRECTOR

ONDCP is led by a director (commonly referred to as the drug czar) with cabinet-level rank (Executive Level 1), two deputies (supply reduction and demand reduction), and one associate director (state and local affairs), all appointed by the president with the advice and consent of the Senate.

The director has a broad mandate for establishing policies, objectives, and priorities for the National Drug Control Program. Serving as the president's drug-control adviser and as a principal adviser to the National Security Council (NSC), the director has extraordinary management tools available to influence the national drug-control efforts.

ONDCP is required to produce an annual National Drug Control Strategy for the president and the Congress and for overseeing its implementation by the federal departments and agencies. Included is an annual consolidated National Drug Control Program budget and the director's certification that the budget is adequate to implement the objectives of the strategy. In addition to the strategy and program oversight, the director has two other legislated management tools—(1) approval of reprogramming of each agency's drug funds and (2) formal notification to the involved agency and the president when a drug-program agency's policy does not comply with the strategy. The director also recommends changes in organization, management, and budgets of departments and agencies engaged in the drug effort, including personnel allocations.

Reflecting congressional desire to participate in drug policy, the director must represent the administration's drug policies and proposals before Congress. Additionally, the authorizing legislation specifically allows Congress access to "information, documents, and studies in the possession of, or conducted by or at the direction of the Director" and to personnel of the office.

The first director of ONDCP was William J. Bennett, 1989–1990, previously the secretary of education during the administration of President Ronald W. Reagan. Director Bennett had the difficult job of starting the new agency from scratch and developing a new national drug-control strategy within the first year of operation. Reagan's successor, President Bush, declined to include the cabinet-level ONDCP director in his immediate cabinet, bringing congressional criticism. Bob Martinez (the former governor of Florida) was the next director, 1991–1992. The third director, Lee P. Brown, a criminologist and a former New York City police commissioner, was appointed by President Bill Clinton in 1993 and was given cabinet status.

ORGANIZATION AND AUTHORITY

Initially, ONDCP had approximately 127 staff positions and 40 additional members detailed from other federal agencies. ONDCP's Fiscal Year (FY) 1992 appropriation of $105 million included $86 million to be transferred to support the High Intensity Drug Trafficking Areas (HIDTA). The HIDTA funding provides $50 million for federal law-enforcement agencies and $36 million for state and local drug-control activities. President Clinton drastically reduced the size of the ONDCP staff soon after his election.

The director is responsible for a Special Forfeiture Fund, funded by the department of Justice Assets Forfeiture Fund, "to supplement program resources used to fight the war on drugs." For FY 1992, this fund included over $50 million for transfer to federal program agencies.

Additionally, ONDCP reviews and recommends funding priorities for the annual budget requests for over fifty federal agencies and accounts involved in the drug program (more than $12 billion in FY 1993).

ONDCP's authority to provide direction to diverse federal departments and agencies is based on a program-management structure known as the National Drug Control Program. The ONDCP program and budget authority coexists with the line authority of the cabinet departments and with the president's annual budget process (directed by the Office of Management and Budget). The structure for the parallel drug-control system is created by designating National Drug Control Program agencies, defined as "any department or agency and all dedicated units thereof, with responsibilities under the National Drug Control Strategy." The designated federal departments and agencies have special program and budget responsibilities to the director of ONDCP.

ONDCP's broad coordination authority over budgets and program activity also presents extraordinary opportunities for conflict with the existing line authority in the departments and agencies. Simultaneously, ONDCP receives congressional and press criticism regarding lack of influence over the operating activities.

POLICY DEVELOPMENT AND COORDINATION

The continued success of the complex drug-policy system depends on a continuing high priority for the drug programs, preventing bureaucratic turf battles, and seeking widespread understanding and endorsement of the goals and objectives of the national program. An essential element in communicating is a public document that explains the strategy, goals, and responsibilities—including a dynamic process of evaluating results and updating the strategy.

The annual National Drug Control Strategy, with accompanying Budget Summary (the January 1992 strategy was the fourth in the series) contains a description of the drug-abuse situation, an assessment of progress, and national priorities—with two-year and ten-year objectives and a federal budget "crosscut" and analysis. ONDCP has brought together a complex set of drug-control program functions and budgets in an understandable way; by function in the strategy and by agency in the budget summary. Under Lee P. Brown the office produced an interim strategy for 1993 and a fully developed strategy in February 1994.

The National Drug Control Strategy acknowledges that no single tactic will solve the drug problem. Therefore, the annual strategies call for improved and expanded treatment, prevention and education; increased international cooperation; aggressive law enforcement and interdiction; expanded use of the military; expanded drug intelligence; and more research.

ORGANIZATION FOR COORDINATION

ONDCP has established a drug-control management agenda, including federal coordinating mechanisms and senior-level management committees and working groups. The organization of ONDCP includes staff for supply reduction, demand reduction, and state and local affairs. ONDCP working groups and committees coordinate the implementation of the policies, objectives, and priorities established in the National Drug Control Strategy.

The federal drug-control agencies and departments are represented on the various working groups and committees, along with ONDCP staff. The organizational structure described in the 1992 strategy includes the following coordinating mechanism:

ONDCP Supply Reduction Working Group

Chaired by the ONDCP deputy director for supply reduction, the working group includes three committees:

The Border Interdiction Committee. Coordinates strategies and operations aimed at interdicting drugs between source and transit countries and at U.S. borders.

The Public Land Drug Control Committee. Coordinates federal, state, and local drug control programs (primarily marijuana eradication efforts) on federal lands.

Southwest Border and Metropolitan HIDTA Committees. Coordinates drug law enforcement activities in designated areas, including federal, state, and local enforcement task forces and intelligence activities. Four metropolitan HIDTAs have been designated: New York City, Miami, Houston, and Los Angeles.

ONDCP Demand Reduction Working Group

Chaired by the ONDCP deputy director for demand reduction, the working group coordinates policies, objectives, and outreach activities for treatment, education and prevention, workplace, and international demand reduction.

Research and Development Committee

Chaired by the director of ONDCP, the committee provides policy guidance for R&D activities of all federal drug control agencies, including the following R&D working committees—

The Data Committee. To improve the relevance, timeliness, and usefulness of drug-related data collection, research studies, and evaluations of both demand-related and supply-related activities.

The Medical Research Committee. Coordinates policy and general objectives on medical research by federal drug-control agencies and promotes the dissemination of research findings.

The ONDCP Science and Technology Committee. Chaired by the ONDCP chief scientist, the committee is responsible for oversight of counterdrug research and development throughout the federal government.

RELATED POLICY ACTIVITIES

The Counter-Narcotics Technology Assessment Center, established by Public Law 101–509 in 1991, provides oversight of the federal government's coun-ternarcotics research and development activities. ONDCP's chief scientist is responsible for defining scientific and technological needs for federal, state, and local law-enforcement agencies, and for determining feasibility and priorities. The chief scientist also coordinates the technology initiatives of federal civilian and military departments, including research on substance-abuse addiction and rehabilitation.

ONDCP works with the NSC, chairing the Policy Coordinating Committee for Narcotics to oversee coordination among agencies with law-enforcement and national-security responsibilities. The director also participates in meetings of the Domestic Policy council, which reviews the annual drug control strategy before it goes to the president.

ONDCP's state and local affairs staff sought wide public involvement in developing and implementing drug policy at all levels of government. Several National Conferences on State and Local Drug Policy were sponsored by ONDCP during 1990 and 1991 to highlight successful state and local programs, seek input to the national strategy, and inform participants of funding and initiatives available to them. ONDCP staff coordinated with both the White House Office of National Service and the president's Drug Advisory Council in encouraging private-sector and state-and-local initiatives for drug prevention and control.

ONDCP also provides administrative support to the president's Drug Advisory Council. With thirty-two private citizens as members, the Drug Advisory Council focuses on private-sector initiatives to support national drug-control objectives, and it assists the ONDCP. The advisory council is financed by private gifts.

(SEE ALSO: *Anslinger, Harry J., and U.S. Drug Policy; Opioids and Opioid Control, History of*)

BIBLIOGRAPHY

U.S. EXECUTIVE OFFICE OF THE PRESIDENT. Office of National Drug Control Policy. (1989). *National drug control strategy, September 1989.* Washington, DC: Government Printing Office.

U.S. EXECUTIVE OFFICE OF THE PRESIDENT. Office of National Drug Control Policy. (1990). *National drug control strategy, January 1990.* Washington, DC: Government Printing Office.

U.S. EXECUTIVE OFFICE OF THE PRESIDENT. Office of National Drug Control Policy. (1991). *National drug control strategy, February 1991.* Washington, DC: Government Printing Office.

U.S. EXECUTIVE OFFICE OF THE PRESIDENT. Office of National Drug Control Policy. (1992). *National drug control strategy, January 1992.* Washington, DC: Government Printing Office.

RICHARD L. WILLIAMS

Special Action Office for Drug Abuse Prevention (SAODAP)

The Special Action Office for Drug Abuse Prevention (SAODAP) was created by Executive Order of President Richard M. Nixon on June 17, 1971, as a response to public concern about drug abuse, particularly heroin addiction. SAODAP was given legislative authority by the Drug Abuse Office and Treatment Act on March 21, 1972. The formation of SAODAP represented the first attempt to establish a stable focus within the federal government for the coordination of the many facets of U.S. drug policy, including law enforcement, border control, control of selected medicines, treatment, prevention, education, and research.

More than twenty agencies, offices, and bureaus within the U.S. government were responsible for activities relating to drug problems. Yet there was no evident central authority other than the president. Congress and the public seemed eager to be able to hold accountable the head of one agency who, unlike the president, could be asked to testify before congress—a "drug czar." Although the term "drug czar" was popularly used, and it was expected that the person holding the office would exert power over the various agencies dealing with both law enforcement (supply side) and treatment and prevention (demand side) aspects of the problem, neither the president nor the Congress were entirely comfortable with delegating such broad authority to only one individual.

The legislation submitted to Congress by the White House, which finally emerged from debate, gave SAODAP unprecedented authority over demand-side activity—treatment, prevention, education, research—wherever these were carried out within the federal government. However, its mandate with respect to drug-control agencies such as the U.S. Customs Bureau, which reported to the secretary of the treasury, and the Bureau of Narcotics and Dangerous Drugs, which reported to the attorney general, was limited to coordination. SAODAP was also charged with developing a formal, written, national strategy for drug-abuse prevention. To head the new office, President Nixon appointed Dr. Jerome H. Jaffe, then a professor of psychiatry at the University of Chicago and director of the Illinois Drug Abuse Programs. Dr. Jaffe, who had helped the White House develop its response to HEROIN use in VIETNAM, was also appointed special consultant to the president on narcotics and dangerous drugs.

A primary goal of SAODAP, stated at the press conference that announced the new office, was to make treatment so available that no addicts could say they committed crimes because they could not get treatment. Although the Bureau of Narcotics and Dangerous Drugs (BNDD) had estimated that there were about a half million heroin users in the United States, in mid-1971 the true extent of the drug-abuse problem was unknown. The estimating techniques that were developed in the 1970s—the NATIONAL HOUSEHOLD SURVEY ON DRUG ABUSE, the DAWN system (or DRUG ABUSE WARNING NETWORK), and the HIGH SCHOOL SENIOR SURVEY—did not yet exist, but the rising rate of heroin-related deaths in several major cities and the thousands of addicts waiting for treatment because there was not enough treatment capacity gave stark evidence for the growing size of the heroin problem. There were drug OVERDOSE (OD) deaths among U.S. troops in Vietnam also. Surveys generally indicated widespread drug use among U.S. servicemen in Vietnam, with the extent of the problem estimated at 15 to 30 percent, but it was not known if these estimates were of drug users or of addicts.

In addition to the mandate to coordinate all the demand side drug-abuse activities of the federal bureaucracy so as to reduce overlap and redundancy and to expand treatment capacity, some of the additional tasks of the office included overseeing and coordinating the Vietnam drug-abuse intervention; creating a new federal agency with competence to develop national policy; creating the data systems by which the effectiveness of national policy could be evaluated; creating a science base so that research might lead to better ways to treat and prevent addiction; and developing a formal, written National Strategy for drug-abuse treatment and prevention.

Four major policy changes helped the agency achieve its objectives. The first was made by the president when the Vietnam testing and treatment program was initiated: Drug use was no longer a court martial offense. The second was having the federal government take responsibility for developing and funding treatment. The third made METHA-DONE-MAINTENANCE treatment, already being used for 20,000 people, an established and acceptable treatment method rather than an experiment. The fourth had to do with changes that were made in the thinking, language, and means by which treatment was supported.

A central effort for SAODAP was the expansion of treatment capacity, increasing not only the number of programs, but also their actual capacity and geographic distribution. In addition, recipients of funding for treatment programs became accountable for what they provided, such as the number of treatment slots and the type of treatment. While legitimizing methadone-maintenance treatment and developing regulations for its use were highly visible and highly controversial activities, they were only incidental to the overall mission of making effective treatment central to the nation's response to the drug problem. Within the first 18 months of SAODAP's efforts, the number of communities with federally supported drug-treatment programs increased from 54 to 214, and the number of programs grew to almost 400. More federally supported treatment capacity was developed within two years than over the previous fifty years.

Some of the other projects SAODAP initiated, funded, or grappled with were the Vietnam drug intervention and the Vietnam drug intervention follow-up study; the development of confidentiality regulations to protect the medical records of people seeking treatment; funding clinical research on new pharmacological treatments for drug dependence; initiating with other agencies projects such as TREATMENT ALTERNATIVES TO STREET CRIME (TASC), research centers for clinical and basic research on drug abuse and addiction, the Career Teachers program that incorporated drug abuse into medical school curricula, and a National Training Center. SAODAP introduced formula or block grants that gave money through the NATIONAL INSTITUTES ON MENTAL HEALTH (NIMH) to the states for treatment and prevention programs; it also introduced management concepts and language into treatment sys-

tems. SAODAP played a major role in improving drug-abuse treatment in the Veterans Administration; establishing laboratory standards for urine-testing facilities; and initiating several of the epidemiological tools that continue to shape policy, such as the National Household Survey of Drug Abuse and the Drug Abuse Warning Network (DAWN) system. Many of the programs and activities developed with interagency cooperation were implemented by the agencies involved in the collaboration. Many of the activities are ongoing in the mid-1990s. SAODAP also produced the first written national strategy, entitled "Federal Strategy for Drug Abuse and Drug Traffic Prevention."

Since the baseline funding for drug-abuse treatment, prevention, and research was so low in 1971, the new resources given to SAODAP for the task represented a manifold increase—and in some instances were the very first resources available for the purpose. The same legislation that authorized SAODAP provided for the establishment of the National Institutes on Drug Abuse (NIDA); in addition, the resources and policies for an invigorated research effort were put into place over the three budgetary cycles that preceded NIDA's creation. Dr. Robert Dupont, who succeeded Dr. Jaffe as director of SAODAP, became the first director of NIDA. Dr. Peter Bourne and Mr. Lee Dogoloff, both of whom worked at SAODAP during the first two years, later became key advisors on drug policy to President Jimmy Carter.

A noted researcher, Dr. Solomon Snyder, credits the SAODAP support he received with enabling him to discover the opiate RECEPTOR a year or two later. This discovery forms the basis for much of the neuroscience research into understanding the biology of drug dependence.

SAODAP was able to change the national response to illicit drug use by developing an infrastructure for treatment that is largely still in place, one that recognizes the heterogeneity of the drug-using population, their need for several different types of treatment, and the need for research on the efficacy of treatment. For a brief period after SAODAP's mandate expired in 1975, drug-abuse policy was coordinated by a smaller office within the Office of Management and Budget (OMB) under President Gerald R. Ford, and then by the Drug Abuse Policy Office within the White House under presidents Jimmy Carter and Ronald W. Reagan. However, until

President George H. Bush established the Office of National Drug Control Policy (ONDCP), there was no formal agency with substantial authority for coordinating federal drug policy.

(SEE ALSO: *Industry and Workplace, Drug Use in*)

FAITH K. JAFFE
JEROME H. JAFFE

Substance Abuse and Mental Health Services Administration (SAMHSA) This agency was established by the U.S. Congress (Public Law 102-321) on October 1, 1992, in a move to strengthen the nation's prevention- and treatment-services delivery system for persons with mental and addictive disorders. SAMHSA is the newest agency of the Public Health Service in the U.S. Department of Health and Human Services.

SAMHSA is the successor of the Alcohol, Drug Abuse, and Mental Health Administration (ADAMHA), which since 1974 had encompassed all research and services activities of the federal government aimed at reducing and controlling these public-health problems.

The three research institutes formerly in ADAMHA—the National Institute on Alcohol Abuse and Alcoholism (NIAAA), the National Institute on Drug Abuse (NIDA), and the National Institute of Mental Health (NIMH)—were transferred to the National Institutes of Health (NIH). SAMHSA's mission is to: (1) Reduce the incidence and prevalence of mental disorders and substance abuse in the United States and improve treatment outcomes for persons suffering from these disorders. (2) Improve access to high quality, effective prevention and treatment services for all those in need. (3) Ensure the widespread application of new knowledge from scientific studies and from new program models to prevent and treat addictive and mental disorders.

To help reduce, prevent, and treat pressing public-health problems, SAMHSA administers a variety of programs and activities whose four principal objectives are: (1) to increase access to prevention and treatment services for those who need them; (2) to improve the quality of those services; (3) to empower individuals, families, and communities to take actions that address one or more factors that cause or sustain substance abuse and mental disorders; and

(4) to monitor progress through objective surveys and conduct evaluation studies to improve prevention and treatment effectiveness.

SAMHSA carries out its activities primarily through three centers—the Center for Substance Abuse Prevention (CSAP); the Center for Substance Abuse Treatment (CSAT); and the Center for Mental Health Services. CMHS leads the national effort to prevent and treat mental illnesses and to promote mental health. The Center seeks to improve the treatment, care and rehabilitation of people with mental health problems and to assist their families by funding demonstration projects. The Center also keeps the fields up to date on the latest treatment findings and has established a new National Clearinghouse for Mental Health Information. A national data system, operated in collaboration with the States, tracks the incidence and prevalence of, and progress against, mental illness in the United States.

CSAT funds demonstration grants to expand treatment and rehabilitation services for people with substance abuse problems, particularly for high-risk populations. Other special CSAT grant programs support treatment system demonstration projects in "target cities" with especially severe drug abuse problems; for substances abusers in the criminal justice system, homeless shelters, and public housing; and for ethnic and racial minorities and women who are underserved by the current treatment system.

CSAP connects people and resources with innovative ideas, strategies, and programs designed to reduce and eliminate alcohol and other drug problems in our society. CSAP's High-Risk Youth Demonstration program funds community based organizations to develop and field-test innovative approaches for preventing AOD use. The Community Partnerships Demonstration program fosters public/private partnerships to address substance abuse prevention within communities. CSAP also operates an array of information and education projects that support prevention programs and disseminate the latest available knowledge about alcohol and other drug problems including: The National Clearinghouse for Alcohol and Drug Information (NCADI), the Regional Alcohol and Drug Awareness Resource (RADAR) Network; public information, education, and media campaigns; and training and technical assistance that increase State and community capacity to develop prevention programs tailored to specific needs and requirements. Each of these supports prevention or treatment demonstration projects (or both) at the state and local levels. This is accomplished principally through two annual block grants to each state, one for substance-abuse services and another for mental-health services, but also via targeted grants and contracts for special initiatives.

The centers also provide technical assistance in regard to provision of alcohol, drug-abuse, and mental-health services; make training available for many persons, ranging from treatment program managers to volunteer prevention workers; and carry out a variety of other treatment- and prevention-related activities. In addition, the centers collaborate to pursue solutions to complex health problems such as co-occurring addictive and mental disorders. They cooperate in ways for dealing with multilevel problems, such as how to deliver services to injecting drug users (IDUs) in populations that are highly susceptible to HIV infection but are hard to reach, such as homeless people.

Through an Office of Applied Studies, SAMHSA collects, analyzes, and disseminates national data on the extent and consequences of alcohol and other drugs of abuse as well as on mental illness, and on the nation's treatment systems for these disorders. This office also coordinates evaluation studies to determine what works in prevention and treatment.

SAMHSA also serves as a conduit for integrating new scientific-research discoveries into prevention and treatment programs for addictive and mental disorders. Its goal is to forge links between mental-health and substance-abuse programs and primary health-care providers, such as family physicians and nurse practitioners, to bring the most appropriate and most cost effective care possible to persons in need of substance-abuse and/or mental-health services.

From the beginning, SAMHSA identified three currently underserved populations for immediate attention: (1) women, (2) children and adolescents, and (3) persons with or at risk for contracting HIV or AIDS. An Office on AIDS has been created to bring the agency's full resources to bear on the problem of HIV and AIDS among substance abusers and the mentally ill. An Office for Women's Services has also been established; it coordinates and energizes efforts throughout the agency on behalf of women who suffer from mental or addictive disorders. Two special posts have been created to coordinate and promote the work of the centers on other important issues—one for Alcohol Prevention and Treatment Policy and one for Minority Concerns.

In 1994, the total budget of SAMSHA WAS $2.2 billion, and the staff numbered 755. For further information, write to SAMSHA Office of Communications, Room 13CO5, 56 Fishers Lane, Rockville, MD 20857.

ELAINE JOHNSON

U.S. Customs Service The U.S. Customs Service (USCS), in the Department of the Treasury, is the principal border-enforcement agency. Customs conducts a wide range of statutory and regulatory activities ranging from interdicting and seizing contraband entering the United States to intercepting illegal export of high-technology items. Customs officers also assist over forty other federal agencies with border-enforcement responsibilities, including public-health threats, terrorists, agricultural pests, and illegal aliens.

With a fiscal year 1993 budget of over $1.6 billion and 18,000 employees, Customs is a major revenue-producing agency; it collected $21.5 billion in duty, taxes, and fees in 1993.

CUSTOMS ROLE IN DRUG ENFORCEMENT

Customs is both a leader and a major player in stopping drug contraband from entering the United States. Approximately $570 million of the 1993 Customs budget was related to antidrug operations. Customs' inspection and control function is directed at stopping illegal entry of drugs and other contraband while accommodating the normal traffic of persons and cargo entering the United States and enforcing export laws.

As the federal lead agency at U.S. ports of entry, Customs inspects individuals, conveyances, mail, and cargo entering the United States at these ports (land, sea, and air). Customs has broad search and seizure authority at the U.S. borders and handles enormous workloads; for example, some 450 million international travelers arrive at U.S. borders each year. Customs operates a comprehensive computerized border information system and uses other domestic and international drug-intelligence networks. Priority efforts are targeted on illegal traffic in precursor chemicals, improving interdiction intelligence, and special high-intensity enforcement operations, particularly along the southwest border.

As a large, multipurpose border-control agency, Customs has considerable flexibility in determining the most effective means to meet its responsibilities. The traditional approach involves the physical presence of uniformed officers at the border to detect and seize violators and contraband. Customs emphasizes development of the best possible detection capabilities and information systems, including drug-sniffing DOGS, electronic chemical detectors, advanced computer systems, and sophisticated surveillance equipment. Reflecting the high priority for drug interdiction, over 650 National Guard personnel in twenty-seven states have been assigned to assist Customs with inspection of containerized cargo, vessels, and aircraft.

Customs has also developed major aviation and marine interdiction programs since the 1970s. Initially dependent on aircraft borrowed from the Department of Defense (DOD) and seized from smugglers, Customs now operates over 130 aircraft and 150 vessels. Customs supports a series of Command, Control, Communications, and Intelligence Centers (known as C3I) to provide coordinated tactical control for air interdiction. Using sophisticated aircraft, helicopters, and vessels, Customs works closely with the U.S. Coast Guard and U.S. military forces in providing surveillance, interception, and deterrence against drug smuggling by air and sea.

In addition to the tactical interdiction program, Customs conducts investigations of financial reporting and smuggling violations, developing both criminal and civil cases. USCS is represented in various interagency enforcement task forces.

Customs is an active participant in developing federal drug policy and has used its high public visibility to contribute to national drug-abuse prevention efforts, emphasizing "user responsibility" and drug education. Historically, Customs has provided staff assistance to executive and congressional drug-policy offices and committees. The Customs commissioner was included in the Executive Office of the President (EOP) drug-policy coordinating activities, including the Principals' Group, the Oversight Working Group, the National Narcotics Border Interdiction System, and others. The commissioner of Customs chairs the Office of National Drug Control Policy's (ONDCP) Border Interdiction Committee, with subcommittees that develop and guide the implementation of strategies for air, land, and sea interdiction. Customs also works with the international Customs Coordinating Council in developing new procedures and techniques.

(SEE ALSO: *Anslinger, Harry J., and U.S. Drug Policy; Drug Interdiction; International Drug Supply Systems; Operation Intercept; Zero Tolerance*)

BIBLIOGRAPHY

PRICE, C. E., & KELLER, M. (1989). *The U.S. Customs Service, a bicentennial history*. Washington, DC: Department of the Treasury, U.S. Customs Service. (An overview of 200 years of Customs history; a chapter on drug enforcement.)

U.S. EXECUTIVE OFFICE OF THE PRESIDENT, Office of Drug Abuse Policy. (1977) *Border management and interdiction—an interagency review*. Washington, DC. (Description of borders and border responsibilities.)

U.S. EXECUTIVE OFFICE OF THE PRESIDENT, Office of National Drug Control Policy. (1992). *National drug control strategy*. Washington, DC.

U.S. EXECUTIVE OFFICE OF THE PRESIDENT, Office of National Drug Control Policy. (1992). *National drug control strategy budget summary*. Washington, DC.

RICHARD L. WILLIAMS

U.S. Public Health Service Hospitals In 1929, President Herbert C. Hoover signed a law enacted by the U.S. Congress to establish two federal institutions for treatment of narcotic addiction. The principal purpose of the institutions was to confine and treat persons addicted to narcotic drugs who had been convicted of offenses against the United States. However, the law also provided for voluntary admission and treatment of addicts who were not convicted of any offense. The two institutions were named U.S. public health service hospitals. One was opened in 1935 at Lexington, Kentucky, and the other in 1938 at Fort Worth, Texas. The Lexington hospital had a capacity of 1,200 patients; the Fort Worth hospital could accommodate 1,000 patients. From opening to closure in 1974, the hospitals admitted over 60,000 narcotic addicts; because of readmissions, the total admissions exceeded 100,000. Most of the admissions were voluntary. The term *narcotic addiction* has been replaced in modern diagnostic terminology by the term *opioid dependence*, but in this discussion the older term is retained because it was regularly used during the era reviewed here. The history of the hospitals is divided into three periods.

FIRST PERIOD, 1935–1949

From the start, the hospitals were designed to treat not only the physical dependence but also the mental and emotional problems thought to be related to addiction. This was an advanced conception, for treatment of narcotic addiction until then had been focused almost exclusively on the PHYSICAL DEPENDENCE. The initial treatment programs at both hospitals emphasized residence in a drug-free environment for at least six months, during which time the patient could not only recover from the physical dependence but perhaps also overcome the mental difficulties or learn to adapt to them without using drugs. While all patients received psychological help in the form of encouragement and persuasion, only small numbers received formal psychotherapy. That was because few of the staff were trained in psychotherapy. All patients considered physically able had work assignments, and all had access to educational and vocational services, recreation, and religious activities. Treatment of voluntary patients was hindered because most left during or shortly after WITHDRAWAL treatment (often to return to lower doses of their drug—before readmission). In 1948, the research division of the Lexington hospital reported that a new synthesized narcotic drug called METHADONE was effective in the treatment of opiate withdrawal. Methadone substitution followed by a gradual decrease of its dose subsequently became the standard treatment for morphine and heroin withdrawal in the United States. Also in 1948 the research division of the Lexington hospital was administratively separated from the hospital, renamed the Addiction Research Center (ARC) and made a part of the National Institute of Mental Health (NIMH).

SECOND PERIOD, 1950–1966

After World War II, the prevalence of HEROIN addiction in the United States markedly increased. Heroin replaced morphine as the primary narcotic used. Annual admissions to the two hospitals doubled from the 1940s to the 1950s. The prewar addicts differed from their postwar counterparts. More of the postwar addicts came from large cities, and

more came from minority groups (mainly black and Hispanic).

While residence in a drug-free environment continued as a major feature, new psychosocial treatments were made a part of the program. Psychoanalytically oriented PSYCHOTHERAPY was offered, but few patients seemed willing or able to engage in this form of therapy. Group therapy, however, seemed more acceptable, and most patients participated in it to some extent. Influenced by new concepts of the therapeutic community, staff members tried to improve the quality of the patients' psychosocial experience in the hospital.

THIRD PERIOD, 1967–1974

In 1967, a research mission was assigned to the two hospitals, and each was renamed a National Institute of Mental Health Clinical Research Center. Before the research mission could be developed, however, a new clinical mission was assigned to the two institutions. The NARCOTIC ADDICT REHABILITATION ACT (NARA), enacted in 1966, provided for the CIVIL COMMITMENT of addicts instead of prosecution on a criminal charge, or sentence after conviction, or by petition with no criminal charge. The law authorized the Public Health Service to enter into contracts with any public or private agencies to provide examination or treatment of addicts committed under the NARA, but it was decided to use the two clinical research centers to implement the act quickly. Admission of prisoners and voluntary patients was phased out, and the centers concentrated on service to the NARA patients. From 1967 through 1973, over 10,000 NARA patients were admitted to the two centers. Nearly all were admitted under the provision of the law that permitted commitment with no federal criminal charge.

The NARA civil commitment seemed a promising way to eliminate the problem of voluntary patients who signed out prematurely. In practice, it only reduced the problem. Patients learned that commitment could be avoided or terminated if they refused to participate in treatment activities or engaged in disruptive or antagonistic behavior. Only about one-third of the NARA patients completed a six-month period of institutional treatment.

The NARA program led to the closure of the two centers. As more contracts were made with local facilities for examination and treatment of NARA patients, admissions to the two centers decreased. In addition, a new federal program, started in the late 1960s, of grants to states and communities for drug-abuse treatment programs made the centers less needed. The Fort Worth Center was closed in 1971 and the Lexington Center in 1974. The facilities were transferred to the Federal Bureau of Prisons and were converted into correctional institutions.

HISTORIC ROLES OF THE HOSPITALS

For approximately three decades, from the 1930s into the 1960s, the two Public Health Service hospitals were almost the only institutions in the United States engaged in the study and treatment of narcotic addiction. They became international centers of expertise. Staff members published many reports on the psychosocial characteristics of the addicts, the treatment programs, treatment outcomes, and related topics. Many clinicians and investigators who worked at Lexington and Fort Worth left these institutions to become leaders in treatment of or research on narcotic addiction at other locations. Despite great efforts, however, the hospitals failed to develop an enduring cure for narcotic addiction. Hospital treatment often produced a temporary remission in the addiction, but relapse within a year was the typical outcome.

(SEE ALSO: *Opioid Dependence; Treatment, History of; Wikler's Pharmacologic Theory of Drug Addiction*)

BIBLIOGRAPHY

LEUKEFELD, C. G., & TIMS, F. M. (EDS.) (1988). *Compulsory treatment of drug abuse: Research and clinical practice.* National Institute on Drug Abuse Research Monograph 86. DHHS Publication no. (ADM) 88-1578. Rockville, MD: U.S. Department of Health and Human Services.

MARTIN, W. R., & ISBELL, H. (EDS.) (1978). *Drug addiction and the U.S. Public Health Service.* DHEW Publication no. (ADM) 77-434. Rockville, MD: U.S. Department of Health, Education, and Welfare.

JAMES F. MADDUX

VALIUM   *See* Benzodiazepines.

VALUES AND BELIEFS: EXISTENTIAL MODELS OF ADDICTION   Existential models of addiction focus on beliefs, attitudes, and values of the drug users. For example, psychologists have found that problem drinkers and alcoholics anticipate greater benefits and more powerful effects from drinking than do other drinkers. These beliefs *precede* actual drinking experiences (Miller, Smith, & Goldman, 1990).

Beliefs about oneself and about the role of drugs or alcohol in one's life are sometimes called existential models (Greaves, 1980). Khantzian (1985) has proposed that addicts use drugs to offset or address specific problems they believe they have, such as a lack of confidence in social-sexual dealings, a view sometimes referred to as the adaptive model of addiction. According to Peele (1985), the individual becomes addicted to a substance because it fulfills essential intrapsychic, interpersonal, and environmental needs.

Views about oneself in regard to a substance-abuse problem are crucial for dealing with this problem. If the client and treatment personnel see the problem differently, in viewing it as a disease or not, for example, treatment will generally not succeed.

CULTURAL BELIEFS IN ADDICTION

Cultural differences are among the most powerful determinants of the patterns of substance use and the proclivity to addiction (Heath, 1982). For example, moderate drinking is inculcated as an early and firm cultural style among Mediterranean ethnic groups, the JEWS and the CHINESE. Such cultural socialization incorporates beliefs about the power of ALCOHOL and the nature of those who overindulge or misbehave when drinking. Groups such as the Irish, which invest alcohol with the power to control and corrupt their behavior, have high levels of ALCOHOLISM (Vaillant, 1983). In contrast, Jews, Italians, and Chinese believe that those who overdrink are displaying poor self-control and/or psychological dependence, rather than responding to the power of the alcohol itself (Glassner & Berg, 1984). Similar cultural variations occur in views toward drugs such as MARIJUANA, NARCOTICS, PSYCHEDELICS, and COCAINE.

Cultural recipes for moderate consumption of alcohol and other drugs have been developed, although systematic cross-cultural empirical support for these models is weak. One cross-cultural survey of addictive (loss-of-control) behavior is MacAndrew and Edgerton's (1969) *Drunken Comportment*, which describes cultural beliefs that encourage overconsumption and drunken excesses. Yet cultural attitudes about alcohol and other drugs in relation to their misuse are generally regarded as cultural oddities, rather than scientifically meaningful factors in models of addiction.

VALUES

If individual and cultural beliefs have been given short shrift in addiction theories, then values have been considered in such models primarily as illustrations of moralistic prejudice.

Whereas a layperson might condemn the values of a mother who uses drugs or drinks excessively during pregnancy or of a person who assaults others when drunk or using drugs, some pharmacologically based theorists instead emphasize the potency of the drug and the irrevocable need of the person to obtain the drug at the cost of any other consideration whatsoever.

Peele (1987) turned this model on its head—claiming that people become addicted due to a failure of other values that maintain ordinary life involvements. In Peele's view, personal values influence whether people use drugs, whether they use them regularly, whether they become addicted, and whether they remain addicted. These values included prosocial behavior (including achievement, concern for others, and community involvement), self-awareness and intellectual activity, moderation and healthfulness, and self-respect. Evidence for the role of values in addiction are the explicit values people cite as reasons for giving up addictions to cocaine, alcohol, and nicotine (Reinarman, Waldorf, & Murphy, 1991).

(SEE ALSO: *Addiction: Concepts and Definitions; Adjunctive Drug Taking; Asia, Drug Use in; Causes of Substance Abuse; Expectancies; Religion and Drug Use*)

BIBLIOGRAPHY

GLASSNER, B., & BERG, B. (1984). Social locations and interpretations: How Jews define alcoholism. *Journal of Studies on Alcohol, 45,* 16–25.

GREAVES, G. B. (1980). An existential theory of drug dependence. In D. J. Lettieri, M. Sayers, & H. W. Pearson (Eds.), *Theories on drug abuse* (pp. 24–28). Washington, DC: U.S. Government Printing Office (DHHS Pub. No. ADM 80–967).

HEATH, D. B. (1982). Sociocultural variants in alcoholism. In E. M. Pattison & E. Kaufman (Eds.), *Encyclopedic handbook of alcoholism* (pp. 426–440). New York: Gardner Press.

KHANTZIAN, E. J. (1985). The self-medication hypothesis of addictive disorders: Focus on heroin and cocaine dependence. *American Journal of Psychiatry, 142,* 1259–1264.

MACANDREW, C., & EDGERTON, B. (1969). *Drunken comportment: A social explanation.* Chicago: Aldine.

MILLER, P. M., SMITH, G. T., & GOLDMAN, M. S. (1990). Emergence of alcohol expectancies in childhood. *Journal of Studies on Alcohol, 51,* 343–349.

PEELE, S. (1987). A moral vision of addiction: How people's values determine whether they become and remain addicts. In S. Peele (Ed.), *Visions of addiction* (pp. 201–233). Lexington, MA: Lexington Books/Heath.

PEELE, S. (1985). *The meaning of addiction: Compulsive experience and its interpretation.* Lexington, MA: Lexington Books/Heath.

REINARMAN, C., WALDORF, D., & MURPHY, S. (1991). *Cocaine changes: The experience of using and quitting.* Philadelphia: Temple University Press.

VAILLANT, G. E. (1983). *The natural history of alcoholism.* Cambridge: Harvard University Press.

STANTON PEELE

VIETNAM: DRUG USE IN   In the spring of 1971, two members of Congress (John Murphy and Robert Steele) released an alarming report alleging that 15 percent of U.S. servicemen in Vietnam were addicted to HEROIN. The armed forces were attempting to cope with the drug problem by combining military discipline with "amnesty." Anyone found using or possessing illicit drugs was subject to court martial and dishonorable discharge from the service; but drug users who voluntarily sought help might be offered "amnesty" and brief treatment. This policy apparently was having little impact, since heroin use had increased dramatically over the preceding year and a half.

Because the United States was trying to negotiate settlement of the war, military forces in Vietnam were being rapidly reduced. About 1,000 men were being sent back to the United States each day, many of them to be discharged shortly thereafter to civilian life. If the reported rate of heroin addiction among servicemen were accurate, this rapid reduction in force meant that hundreds of active heroin addicts were being sent home each week. Concerned about the social problems that could ensue from such an in-flux of addicts, President Richard M. Nixon charged his staff with seeking an effective response. Domes-tic Council staff members Jeffrey Donfeld and Egil Krogh, Jr., sought advice from Dr. Jerome H. Jaffe, then on the faculty of the University of Chicago, who had previously prepared a report for the president on the development of a national strategy for the treat-ment of drug dependence. Dr. Jaffe recommended a radical change in the policy for responding to the problem of drug use in the military. The suggested plan included urine testing, to detect heroin use, and treatment rather than court martial when drug use was detected. President Nixon endorsed the plan and the military responded with such remarkable rapid-ity that, on June 17, 1971, less than six weeks from the time it was proposed, the plan was initiated in Vietnam.

In fact, there was no way to know whether the new approach would be better than the old one, no reliable information on the actual extent of drug use and addiction, and no solid information on which to base estimates of how many servicemen would re-quire additional treatment after discharge. To obtain information on the extent of drug use, the effective-ness of treatment, and the relapse rates it would be necessary to find and interview the servicemen at time of discharge and at various intervals after dis-charge.

In June 1971, President Nixon also announced the formation of the SPECIAL ACTION OFFICE FOR DRUG ABUSE PREVENTION (SAODAP) charged with coordinating the many facets of the growing drug problem and named Dr. Jaffe as its first director. One of the first tasks of the office was to evaluate the re-sults of the new drug policy for the military, espe-cially as it was implemented in Vietnam. SAODAP arranged for Dr. Lee Robins, of Washington Univer-sity in St. Louis, to obtain records from the Depart-ment of Defense and the Veterans Administration to conduct the study. The findings on drug use prior to and during service are summarized here. The drug-using behaviors of the servicemen after their return to civilian life are described in a separate article (see VIETNAM: FOLLOW-UP STUDY).

Around 1970, before going overseas, about half the army's enlisted men had had some experience with illicit drugs. However, only 30 percent had tried any drug other than MARIJUANA. At that time, the most common civilian drugs other than marijuana were BARBITURATES and AMPHETAMINES. Before

going to Vietnam, only 11 percent of soldiers had tried an OPIATE, and those who did so generally took cough syrups containing CODEINE, not heroin or OPIUM.

The men sent to Vietnam had either been drafted or had enlisted. Toward the end of the war, when drug use in the United States was highest, draftees were chosen by a lottery designed to make selection less susceptible to social-class biases. This produced draftees who were a reasonably representative sam-ple of young American men. Those who enlisted vol-untarily, however, who made up about 40 percent of the armed forces, were disproportionately school dropouts. Many of them enlisted before reaching draftable age because of their limited occupational opportunities. They also arrived in Vietnam with considerably more drug experience than the draf-tees.

Men who were sent to Vietnam before 1969 found marijuana plentiful but little else in the way of illicit drugs (Stanton, 1976). Some amphetamines were available—in part, because the military issued them to help men stay alert on reconnaissance missions. In 1969, heroin and opium began to arrive on the scene, and by 1970–1971 these opiates were very widely available. Marijuana was still the most com-monly used illicit drug, but opiates outstripped am-phetamines and barbiturates in availability. Heroin and opium were relatively cheap and very pure, so pure that the soldiers could get ample effect by smoking heroin in combination with TOBACCO or marijuana. This made opiates appealing to men who would have been reluctant to inject them.

At the height of the use of opiates, in 1971, almost half the army's enlisted men had tried them; of those who tried them, about half used enough to develop the hallmarks of addiction—TOLERANCE and WITH-DRAWAL symptoms (Robins et al., 1975). Marijuana use was even more common; about two-thirds of these soldiers used it. The estimates come from an independent survey of a random sample of army en-listed men eight to twelve months after their return from Vietnam, after the great majority had been dis-charged (Robins et al., 1975). Previous studies in Vietnam (Stanton, 1972; Roffman & Sapol, 1970; Char, 1972) or among men still in service after re-turn (Rohrbaugh et al., 1974) were less reliable, be-cause of difficulties in collecting a random sample, use of questionnaires rather than interviews (which can lead to careless responses or failure to answer completely), and because the surveys were being done by the army itself, while the men were still subject to possible disciplinary action.

The standard tour of duty for Vietnam soldiers was twelve months. Drug use typically began soon after arrival in Vietnam, showing that it was not at all difficult to find a supplier. Older men used less than younger soldiers, career soldiers less than those serving their first term. Drug experience before induction was a powerful predictor of use in Vietnam (Robins et al., 1980). Essentially all those with drug experience before enlistment used drugs in Vietnam. Of course, there were also some soldiers who used drugs there for the first time.

One interesting observation was that men who drank ALCOHOL in Vietnam tended not to use opiates, and opiate users tended not to drink (Wish et al., 1979). This is a very different pattern from the one seen in the same men both before and after Vietnam, when drinkers were much more likely to use illicit drugs than abstainers.

Soldiers who used drugs had more disciplinary problems, on average, than those who abstained. However, the great majority of drug users received little or no disciplinary action and were honorably discharged. Although there were instances in which drug use impaired a soldier's combat readiness, evidence is lacking that it had much impact on soldiers' ability to carry out orders or wage war.

(SEE ALSO: *Addiction: Concepts and Definitions; Drug Testing and Analysis; Military, Drug and Alcohol Abuse in the U.S.*)

BIBLIOGRAPHY

CHAR, J. (1972). Drug abuse in Vietnam. *American Journal of Psychiatry, 129*(4), 123–125.

ROBINS, L. N., HELZER, J. E., & DAVIS, D. H. (1975). Narcotic use in Southeast Asia and afterward: An interview study of 898 Vietnam returnees. *Archives of General Psychiatry, 32*(8), 955–961.

ROBINS, L. N., HELZER, J. E., HESSELBROCK, M., & WISH, E. (1980). Vietnam veterans three years after Vietnam: How our study changed our view of heroin. In L. Brill and C. Winick (Eds.), *Yearbook of substance use and abuse.* New York: Human Science Press.

ROFFMAN, R. A., & SAPOL, E. (1970). Marijuana in Vietnam: A survey of use among Army enlisted men in the two Southern corps. *International Journal of the Addictions, 5*(1), 1–42.

ROHRBAUGH, M., EADS, G., & PRESS, S. (1974). Effects of the Vietnam experience on subsequent drug use among servicemen. *International Journal of the Addictions, 9*(1), 25–40.

STANTON, M. D. (1976). Drugs, Vietnam, and the Vietnam veteran: An overview. *American Journal of Drug & Alcohol Abuse, 3*(4), 557–570.

STANTON, M. D. (1972). Drug use in Vietnam: A survey among Army personnel in the two Northern corps. *Archives of General Psychiatry, 26*(3), 279–286.

WISH, E. D., ROBINS, L. N., HESSELBROCK, M., & HELZER, J. E. (1979). The course of alcohol problems in Vietnam veterans. In M. Galanter (Ed.), *Currents in alcoholism.* New York: Grune & Stratton.

LEE N. ROBINS

VIETNAM: FOLLOW-UP STUDY In the summer of 1971, the U.S. military forces in Vietnam were being rapidly reduced. To deplete the forces there quickly, many men were being sent home before the usual tour of twelve months was complete. A urine-screening program was established in July to detect the recent use of illicit drugs by men scheduled to depart Vietnam for the United States. Those detected as positive were kept for DETOXIFICATION for about seven days, retested, and sent home only if they had a negative test. The urine screening was initiated in response to great concern that many members of the military had become addicted to HEROIN in Vietnam. The fear was that they might continue their addiction in the United States. Because the great majority of those returning were due for discharge on return, the MILITARY would have no further control over them. They might present overwhelming problems to the legal system and to veterans' hospitals.

To learn whether this fear was justified, the SPECIAL ACTION OFFICE FOR DRUG ABUSE PREVENTION (SAODAP) launched a follow-up study with the collaboration of the Department of Defense, the Veterans Administration, the National Institute of Mental Health, and the Department of Labor. The goal was to learn how many men had actually been addicted in Vietnam, whether those addicted would continue to use heroin after return and how many would be readdicted after return. The study was conducted by Washington University in St. Louis, with Lee N. Robins, Ph.D., as principal investigator (Robins, 1973, 1974; Robins et al., 1975).

The group believed to be most at risk of addiction was army enlisted men, who spent their whole tour of duty on Vietnam soil, rather than on ships or in the air like men in the navy or air force. Thus, two groups of 500 army enlisted men were selected for the follow-up, a random sample of men returning in September 1971, and a sample of men whose urines had been positive when tested just prior to departure for the United States that month. The overlap between the two groups selected made it possible to estimate what proportion of all army enlisted men had tested positive. Military records of all those selected were reviewed to verify the date of their departure from Vietnam and to obtain a civilian address and the names of close relatives who would know where to contact them. Records were also used to verify the men's reports of drug problems in the service. To protect from subpoena the confidentiality of the information given by the men, a certificate of confidentiality was obtained. Then each interview was identified only by a randomly selected number placed on its mailing envelope but not on the interview proper. The interview was then mailed to another country, where a second random identification number was selected to replace the original one. A list connecting the first number to identifiers was held in the United States, and a list linking the first number to the second one was kept abroad, so that no one in either country could link names to interviews.

Almost 900 men were personally interviewed eight to twelve months after their return from Vietnam. The response rate was extraordinary: 96 percent of the sample initially selected were personally interviewed. The men were extremely frank—97 percent of men whose military record showed drug use had reported it to the interviewer. Two findings were especially surprising. First, use of narcotics in Vietnam was much more common than the military had estimated. Almost half (43%) of the army enlisted men had used heroin or opium in Vietnam, and 20 percent had been addicted to narcotics there. Second, only a tiny proportion (12%) of those addicted in Vietnam became readdicted in the year after return (Robins et al., 1974). Follow-up again two years later showed that this low rate of readdiction continued (Robins et al., 1980). During their second and third years home, addiction rates among men drafted were not significantly greater than among men who qualified for the draft but did not serve. This surprisingly low rate of relapse could not be attributed to abstention from narcotics after return; half of those addicted in Vietnam did use again after return. Those who went back to narcotics were predominantly men who had used drugs before they entered the service.

Although the principal finding of this study was that heroin addiction in Vietnam had a much better outcome than expected, there were men whose addiction continued on return home. Treatment for them was no more effective than for men who developed addiction in the United States (Robins, 1975).

(SEE ALSO: *Addiction: Concepts and Definitions; Drug Testing and Analysis; Opioid Dependence; Treatment; Vietnam: Drug Use in*)

BIBLIOGRAPHY

ROBINS, L. N. (1975). Drug treatment after return in Vietnam veterans. *Highlights of the 20th annual conference, Veterans Administration Studies in Mental Health and Behavioral Sciences.* Perry Point, MD: Central NP Research Laboratory.

ROBINS, L. N. (1974). *The Vietnam drug user returns,* Special Action Office Monograph, Series A, No. 2. Washington, DC: U.S. Government Printing Office

ROBINS, L. N. (1973). *A follow-up of Vietnam drug users,* Special Action Office Monograph, Series A, No. 1. Washington, DC: Executive Office of the President.

ROBINS, L. N., DAVIS, D. H., & NURCO, D. N. (1974). How permanent was Vietnam drug addiction? *American Journal of Public Health, 64*(Suppl), 38–43.

ROBINS, L. N., HELZER, J. E., & DAVIS, D. H. (1975). Narcotic use in Southeast Asia and afterward: An interview study of 898 Vietnam returnees. *Archives of General Psychiatry, 32*(8), 955–961.

ROBINS, L. N., HELZER, J. E. HESSELBROCK, M., & WISH, E. (1980). Vietnam veterans three years after Vietnam: How our study changed our view of heroin. In L. Brill & C. Winick (Eds.), *Yearbook of substance use and abuse.* New York: Human Science Press.

LEE N. ROBINS

VIOLENCE AND SUBSTANCE ABUSE
See Crime and Drugs; Family Violence and Substance Abuse; Gangs and Drugs; International Drug Supply Systems.

VITAMINS Vitamins are organic substances that are required in small amounts for normal functioning of the body. Lack of adequate quantities of vitamins results in well-known deficiency diseases, such as scurvy from Vitamin C deficiency and rickets from Vitamin D deficiency in childhood. For the most part, vitamins are not synthesized by the body but are found in a variety of foods, hence the need for a well-balanced diet or supplementation by taking the vitamins separately.

In the United States, daily minimum requirements for vitamins are recommended, and periodically reassessed, by the Food and Nutrition Board of the National Academy of Science—National Research Council. Some professionals advocate taking larger amounts of certain vitamins is for better health or for disease prevention or therapy. The question of whether vitamins are drugs is, in one sense, a semantic issue. Sometimes, very high doses of a vitamin can actually be used as a medication. For example, in very high doses—twenty or more times higher than needed to prevent the vitamin deficiency disease pellagra—niacin, a member of the B vitamin complex, lowers blood levels of cholesterol and triglycerides and niacin is commonly prescribed for this purpose.

It is possible to OVERDOSE and have serious side effects from large quantities of certain vitamins, such as vitamins A and D. Therefore, taking larger than needed amounts of vitamins should be done only with the advice of a physician. Deficiencies in vitamin intake can occur under a variety of situations including poverty, dieting, or certain disease states where antibiotics or other factors reduce vitamin absorption. Individuals who drink large quantities of ALCOHOL, for example, without adequate attention to diet often become deficient in some vitamins, such as B_1 (thiamine), and may require their administration to avoid serious and permanent toxicity. Prolonged serious shortages of Vitamin B_1 can cause the death of certain NEURONS in the brain, a situation that leads to confusion and severe impairment of short-term memory (the Wernicke-Korsakoff syndrome).

(SEE ALSO: *Complications*)

BIBLIOGRAPHY

MARCUS, R., & COULSTON, A. M. (1990). The vitamins. In A. G. Gilman et al. (Eds.), *Goodman and Gilman's the pharmacological basis of therapeutics*, 8th ed. New York: Pergamon.

MICHAEL J. KUHAR

VULNERABILITY AS CAUSE OF SUBSTANCE ABUSE This section contains some articles that discuss one of several *Causes of Substance Abuse*—vulnerability. In addition to an *Overview* article, the following topics are discussed as vulnerability factors: *Gender; Genetics;* the *Psychoanalytic Perspective; Race; Sensation Seeking; Sexual and Physical Abuse;* and *Stress.* For more information, see *Comorbidity and Vulnerability, Families and Drug Use,* and *Poverty and Drug Use.*

An Overview There are marked individual differences in drug use and abuse. Some people never use drugs although drugs may be readily available to them. Others use drugs sporadically or regularly for years but never escalate their use to drug DEPENDENCE. Others become chronic, compulsive users and have difficulty functioning without drugs. These individual differences in drug-use patterns are the result of a combination of environmental and genetic factors. Environmental factors include the experiences of an individual, such as family and social conditions, as well as other conditions under which the person lives. Genetic factors refer to the genes that are passed down from parent to child and which are shared in part by other family members.

Environmental and genetic factors combine to produce risk factors, which are influences that increase the likelihood of drug use. They may also combine to produce protective factors, which are influences that decrease the likelihood of drug use. Vulnerability refers to the sum total of an individual's risk and protective factors. It defines the overall likelihood of drug use. Individuals with many risk factors and few protective factors are more likely than individuals with few risk factors and many protective factors to use drugs.

GOALS OF VULNERABILITY RESEARCH

In vulnerability research, attempts are made to identify risk and protective factors for both drug use and drug dependence, refine existing risk and protective factors by enhancing their specificity in predicting drug use, reduce the number of risk and protective factors to their most fundamental number, and understand the environmental and genetic influences (i.e., mechanisms) that underlie risk and protective factors.

Risk-Factor Identification. A large number of risk factors for substance abuse have been reported (Table 1). They include characteristics that fall within the demographic, environmental, sociocultural, family, personality, behavioral, psychiatric, and genetic domains. Among these are POVERTY, unemployment, poor quality of education, racial discrimination, ready availability of drugs, family discord, family alcohol and drug use, sexual abuse, lack of family rituals, neuropsychological deficits, childhood aggressiveness, low self-esteem, teenage pregnancy, rebelliousness, delinquency, drug use by peers, mental health problems, and cultural alienation.

A number of protective factors for substance abuse have also been reported (Table 2); however, these are considerably fewer than the reported number of risk factors, primarily because less attention has been focused on their identification. In general, the protective factors that have been reported are the opposite of known risk factors. As such, they include an adequate income, high-quality schools, positive self-esteem, and the like.

Given the fact that a large number of risk factors are commonly present in modern society, many people possess multiple risk factors for drug use. Becoming a drug user is not an inevitable outcome for these people, however, since many individuals with multiple risk factors do not become drug users. Similarly, some individuals who are drug users or drug dependent have few risk factors.

Risk-Factor Specificity. Unfortunately, many risk factors are so broadly defined that they are not useful as predictors. For example, we know that males are more likely than females to use illicit drugs and that underemployed people are more likely than employed people to become HEROIN addicts. Being male or being underemployed, however, is not a useful predictor of drug use. Most males do not use illicit drugs and most underemployed people are not heroin addicts. Combining GENDER and employment status into a single risk factor (i.e., the risk factor of being an underemployed male) increases specificity somewhat, and combining these factors with other risk factors (e.g., having an ANTISOCIAL PERSONALITY disorder) increases the predictive value even more.

The problem with lack of specificity is that it leads to overinclusion of people in risk groups. Many people are thus included in a risk group who are not actually at risk of becoming drug users. For example, although being male and being underemployed are factors statistically associated with heroin addiction, it is important to remember that this is only a statistical association. Most individuals with these characteristics never become heroin addicts. Thus, underemployed males represent a category that includes a large number of individuals who are not actually at risk for heroin addiction. Increasing specificity in risk factors is important because it allows the resources for PREVENTION to be directed toward the people in greatest need. Specificity also minimizes the problem of inappropriately stigmatizing people because they have a characteristic that is statistically associated with drug use.

Fundamental Risk Factors. Because of their current lack of etiological specificity, concern has been expressed about the usefulness of the large number of risk factors that have been reported for drug use. Over seventy risk factors for drug use have been reported to date, but it is not clear if they are all independent factors. Some reported risk factors may be the product of other risk factors. For example, neuropsychological deficits may precipitate learning problems, which in turn may lead to excessive CHILDHOOD aggressiveness. Similarly, family alcohol and drug use may result in family discord, and poor-quality schools may contribute both to underemployment and HOMELESSNESS.

Other risk factors may reflect different manifestations of more basic factors. For example, rebelliousness, DELINQUENCY, and aggressiveness may reflect a more basic personality characteristic or be the result of common genetic influences. Although the actual number of basic risk factors in drug use is not known, they are certain to be fewer than the large number of risk factors reported to date. The large number of reported risk factors probably reflects the highly interrelated nature of the influences involved in drug use.

Underlying-mechanism Identification. A risk factor may itself be a product of the interaction among environmental and genetic influences, or it may only be correlated with those influences. In either case, it is useful for predicting drug use. To most efficiently prevent drug use, however, it is necessary to understand the basic mechanisms that control drug use. As one increases the specificity of risk

TABLE 1
Risk Factors in Substance Abuse

Ecological Environment
  Poverty
  Living in economically depressed area with:
    High unemployment
    Inadequate housing
    Poor schools
    Inadequate health and social services
    High prevalence of crime
    High prevalence of illegal drug use
  Minority status involving:
    Racial discrimination
    Culture devalued in American society
    Differing generational levels of assimilation
    Cultural and language barriers to getting adequate
      health care and other social services
    Low educational levels
    Low achievement expectations from society

Family Environment
  Alcohol and other drug dependency of parent(s)
  Parental abuse and neglect of children
  Antisocial, sexually deviant, or mentally ill parents
  High levels of family stress, including:
    Financial strain
    Large, overcrowded family
    Unemployed or underemployed parents
    Parents with little education
    Socially isolated parents
    Single female parent without family/other support
    Family instability
    High level of marital and family conflict and/or fam-
      ily violence
  Parental absenteeism due to separation, divorce, or
    death
  Lack of family rituals
  Inadequate parenting and little parent/child contact
  Frequent family moves

Constitutional Vulnerability of the Child
  Child of an abuser of alcohol or other drugs
  Less than 2 years between the child and its older/
    younger siblings
  Birth defects, including possible neurological and neu-
    rochemical dysfunctions
  Neuropsychological vulnerabilities
  Physical handicap
  Physical or mental health problems
  Learning disability

Early Behavior Problems
  Aggressiveness combined with shyness
  Aggressiveness
  Decreased social inhibition
  Emotional problems
  Inability to express feelings appropriately
  Hypersensitivity
  Hyperactivity
  Inability to cope with stress
  Problems with relationships
  Cognitive problems
  Low self-esteem
  Difficult temperament
  Personality characteristics of ego undercontrol:
    Rapid tempo, inability to delay gratification, overre-
      acting, etc.

Adolescent Problems
  School failure and dropping out
  At risk of dropping out
  Delinquency
  Violent acts
  Gateway drug use
  Other drug use and abuse
  Early unprotected sexual activity
  Teenage pregnancy/teen parenthood
  Unemployment or underemployment
  At risk of unemployment
  Mental health problems
  Suicidal

Negative Adolescent Behavior and Experiences
  Lack of bonding to society (family, school, and com-
    munity)
  Rebelliousness and nonconformity
  Resistance to authority
  Strong need for independence
  Cultural alienation
  Fragile ego
  Feelings of failure
  Present versus future orientation
  Hopelessness
  Lack of self-confidence
  Low self-esteem
  Inability to form positive close relationships
  Vulnerability to negative peer pressure

SOURCE: Adapted from Goplerud, E. N. (Ed.). (1990), *Breaking new ground for youth at risk: Program summaries*. (DHHS Publication No. [ADM] 89-1658). Washington, DC: Office for Substance Abuse Prevention.

TABLE 2
Protective Factors in Substance Abuse

Ecological Environment
Middle or upper class
Low unemployment
Adequate housing
Pleasant neighborhood
Low prevalence of neighborhood crime
Good schools
School climate that promotes learning, participation,
   and responsibility
High-quality health care
Easy access to adequate social services
Flexible social service providers who put clients' needs
   first

Family Environment
Adequate family income
Structured and nurturing family
Promotion of learning by parents
Fewer than four children in family
2 or more years between siblings
Few chronic stressful life events
Multigenerational kinship network
Nonkin support network—e.g., supportive role models,
   dependable substitute child care
Warm, close personal relationship with parent(s)
   and/or other adult(s)

Family Environment
Little marital conflict
Family stability and cohesiveness
Plenty of attention during first year of life
Sibling as caretaker/confidant

Constitutional Strengths
Adequate early sensorimotor and language develop-
   ment
High intelligence
Physical robustness
No emotional or temperamental impairments

Traits of the Child
Affectionate/endearing personality
Easy temperament
Autonomy
Adaptability and flexibility
Positive outlook
Healthy expectations
Self-esteem
Self-discipline
Internal locus of control
Problem-solving skills
Social adeptness
Tolerance

SOURCE: Adapted from Goplerud, E. N. (Ed.). (1990), *Breaking new ground for youth at risk: Program summaries.* (DHHS Publication No. [ADM] 89-1658). Washington, DC: Office for Substance Abuse Prevention.

factors and reduces them to their most fundamental number, one comes ever closer to identifying the specific environmental and genetic mechanisms involved.

At present, most risk factors are hypothetical constructs and only conceptually defined. Consequently, the risk factor does not identify the mechanisms responsible for drug use. To understand how the risk factor increases the likelihood of drug use, one must identify the mechanisms involved. For example, having drug-using peers is recognized as a risk factor for drug use (because drug use by ADOLESCENTS is frequently associated with having drug-using peers). Although the specific mechanisms mediating this influence are not definitely known, it is likely that the influence is mediated in part through drug-using peers increasing drug availability and providing social reinforcement for drug use. Similarly, coming from an impoverished environment is thought to be a risk factor for drug use because it fails to provide reinforcers as an alternative to drug use.

GENETIC influences may also underlie many risk factors for both drug use and dependence. These influences may contribute to drug use through personality characteristics (e.g., SENSATION SEEKING, risk taking) that increases the likelihood of drug use and that may be genetically determined. Genetic influences may also contribute to the development of drug dependence by altering the effects of a drug (e.g., causing greater euphoria in some people than in others). In addition, they may contribute to both drug use and dependence by being responsible for the absence of normal protective factors (e.g., failure to experience a hangover after excessive alcohol use). The specific genetic mechanisms involved will be the genes (as yet unidentified) that contribute to personality development, drug response, and other important components.

The specific mechanisms that control drug use are undoubtedly the same environmental and genetic mechanisms that control human behavior in general. The mechanisms responsible for the initial drug use and for the progression to regular use and possibly drug dependence may not be the same. Once these mechanisms are understood, however, it will be possible to more directly address risk factors for drug use by means of intervention measures. The ultimate goal of those engaged in vulnerability research is to develop efficient, cost-effective prevention programs that specifically target individuals at risk for both drug use and drug dependence.

VULNERABILITY RESEARCH STRATEGIES

A variety of strategies are available for achieving the goals of vulnerability research. They include both epidemiological and experimental studies, genetic studies, and ANIMAL RESEARCH.

Cross-sectional Epidemiological Studies. Risk factors are initially identified through their statistical association with drug use. Most of the risk and protective factors reported to date have been identified by comparing drug abusers and controls on the basis of currently existing characteristics or reports of conditions existing prior to onset of drug use. For example, individuals are divided into drug users and non-drug users on the basis of a survey, and compared as to demographic characteristics and other traits. The factors that distinguish the drug users from the non-drug users are then identified as risk factors for drug use.

This strategy permits the inexpensive identification of a large number of possible risk factors for drug use. The ability of the strategy to detect possible risk factors is limited only by the selection of characteristics to be compared. With this strategy, however, it is sometimes not clear if a characteristic existed prior to onset of drug use or developed as a consequence of drug use. Since, moreover, the reports of the preexisting conditions are often based on retrospective recall, people's memory problems as well as their attempts to justify their drug use may confound the accuracy of the self-reports. Finally, inappropriate control groups are sometimes employed whose subjects differ from drug users in important aspects (e.g., demographic and clinical features), and this confounds the research design.

Longitudinal Epidemiological Studies. A better research method for identifying risk and protective factors in drug use is the longitudinal study design. With this design, individuals are assessed for various characteristics prior to the age of risk for drug abuse and then followed over time to determine those who do and those who do not become drug users. After drug users have been identified, earlier characteristics that distinguished them from nonusers can be determined.

The advantages of this method are that the drug users and nonusers are drawn from the same population and therefore constitute appropriate comparison groups. Furthermore, because the study design is prospective, it does not rely on the retrospective recall of events or conditions that might have existed prior to the onset of drug use and therefore might be confounded by incorrect memory or other problems. Finally, because this design provides for initial assessment of the subjects prior to the onset of drug use, preexisting conditions can be separated from the consequences of drug use. This design has not been widely employed, however, owing to the expense and time required to conduct the studies. There is also the problem of sample bias that might occur as a result of the attrition of subjects. For example, drug users with severe dependence or psychiatric disorders might be lost in the longitudinal follow-up process, thus leaving only the less severe drug users in the subject sample.

In general, both cross-sectional and longitudinal epidemiological strategies are useful in identifying risk factors for drug use and dependence. They are also both useful in increasing the predictive specificity of risk factors and in allowing fundamental features of various risk factors to be identified by use of sophisticated statistical modeling.

One problem that may affect both types of epidemiological studies is the failure to define risk factors operationally or objectively. This occurs less often when the risk factor involves direct measurement of the individual or use of standardized tests than when individuals are asked about a trait and no definition or operational criteria for the trait is given. For example, if subjects are asked to report on their current level of self-esteem (i.e., whether it is low, medium, or high), failure to define the concept operationally may cause confusion over its presence or absence in a given individual, and this confusion will also increase its variability across individuals.

Experimental Laboratory Studies. This strategy (termed the high-risk design) is aimed at determining the mechanism by which risk factors exert their effects. It compares two groups of individuals who are distinguished by the presence or absence of a particular risk factor. For example, the two groups might consist of children of substance abusers and children of non-substance abusers, or individuals who are depressed and individuals who are not depressed. The two groups are then compared on the basis of various dependent measures, which may include baseline characteristics (e.g., personality) or response to experimental manipulations (e.g., reaction to stress). If the two groups respond differently on a dependent measure, this suggests that the measure is a possible mechanism by which the trait is related to drug use.

This strategy has several advantages. Because it entails selecting subjects on the basis of a specific characteristic, it affords a high degree of control over extraneous factors that might confound the interpretation of epidemiological studies. It also allows researchers to measure subjects' responses directly under standard environmental conditions, rather than relying on self-reports of past events. In addition, it permits the experimental manipulation of test conditions, which in turn allows the generality of an observed effect to be determined. It also enhances the probability that the observed effect is due to the experimental manipulation. Finally, it permits mechanisms underlying the risk factors to be identified and explored, a process that can only be assessed correlationally through statistical modeling in epidemiological studies.

In contrast to epidemiological strategies, however, the high-risk strategy can only address one risk factor per study. It is further restricted by the appropriateness of criteria used for subject selection and the experimental measures employed. For example, inappropriate subject inclusion criteria may exclude the subjects at risk, or inappropriate response measures may fail to detect group differences that are present. Laboratory studies also typically employ only a relatively small number of subjects. This small number increases the likelihood that a biased sample will result, thus making for reduced generalizability of the findings.

Genetic Studies. A number of strategies are available to determine if genetic influences are involved in drug use and dependence. Family studies determine if drug use or dependence "run in families." If higher rates of drug use are found in the relatives of drug users than in the relatives of non-drug users, then genetic influences may be involved.

To separate the effects of genes and environment, however, requires doing adoption or twin studies. In adoption studies, evidence of genetic influences is provided by adoptees having higher rates of drug use if their biological parents were drug users than if their biological parents were not drug users. In twin studies, since identical (monozygotic) twins have more of their genes in common than do fraternal (dizygotic) twins, evidence of genetic influence is suggested by higher concordance rates for drug use or dependence in identical than in fraternal twins.

Other types of genetic strategies are also available. The purpose of linkage and association studies is to identify specific genes involved in drug use and dependence. In linkage studies, different generations of FAMILIES are examined to determine if a genetic marker is inherited along with a disorder (e.g., substance abuse). In association studies, individuals with and without a disorder are compared to determine the association of the disorder with a genetic marker. The previously described high-risk study designs are frequently employed in genetic research. In these studies, subjects who are not yet substance abusers are typically divided into two groups on the basis of their known risk for substance abuse (e.g., having or not having a family history of substance abuse). The two groups are then compared to identify factors that may contribute to their differences in risk for substance abuse.

Most of these genetic strategies have the same strengths and limitations previously described in regard to epidemiological and experimental laboratory studies. In addition, twin and adoption studies are based on certain assumptions about the nature of the genetic influence and parental mating characteristics that may affect interpretation of the results.

Animal Studies. Certain factors contributing to drug use and dependence can be studied experimentally only in animals. For example, it would be unethical to make a human being dependent on drugs in order to study the process of becoming drug dependent. In animals, this process can be brought under experimental control and studied directly. In human beings, drug use or dependence typically becomes evident to researchers only after it has occurred, and then the process can be studied only retrospectively.

A number of strategies are available for studying drug taking by animals. The most common of these are the animal drug self-administration methods.

With these methods, animals are equipped with small tubes (catheters) that run directly from the animal's bloodstream to an injection pump located outside the cage. By pressing a lever, the animal automatically activates the injection pump and receives a predetermined amount of drug solution injected directly into the bloodstream. Similar methods are available to study self-administration of drugs by other routes. By means of these methods, it has been found that animals self-administer essentially the same drugs that humans abuse, and this has resulted in the methods being used to predict the abuse potential of new drugs before they are marketed. Keeping drugs with high dependence potential off the market is also an effective strategy for reducing people's vulnerability to drug use and dependence.

Animal drug self-administration methods can also be used to study factors that contribute to a person's acquiring the problem of drug use and dependence. With these methods, factors thought to influence vulnerability can be experimentally manipulated and studied under controlled laboratory conditions. As a result of the research, a large number of factors have been identified with animal drug self-administration methods that are relevant to the development of human drug dependence. Among these are the reinforcing property of the drug itself, the speed with which a drug is injected, the schedule of drug delivery, the availability of other reinforcers, and the aversiveness of the environment. The knowledge gained from the research can be applied directly to human drug abuse prevention efforts.

Animal methods make possible the experimental study of factors that influence the acquiring of the habit of drug use and dependence, a process that cannot be ethically studied with human beings. Animals, however, differ from human beings in many ways that may be important in the etiology of drug abuse, and therefore care must be taken in generalizing the results of animal studies to human beings. In addition, although animal models provide an excellent way of studying behavioral and environmental factors in drug use, the approach cannot readily be used to study other risk factors (i.e., psychosocial and cultural influences) that are believed to be important in the development of drug abuse by human beings.

(SEE ALSO: *Abuse Liability of Drugs: Testing in Animals; Addiction: Concepts and Definitions; Adjunc-*

tive Drug Taking; Comorbidity and Vulnerability; Complications: Mental Disorders; Conduct Disorder and Drug Use; Disease Concept of Alcoholism and Drug Abuse; Epidemiology of Drug Abuse; Ethnicity and Drugs; Research, Animal Model; Wikler's Pharmacologic Theory of Drug Addiction)

BIBLIOGRAPHY

GLANTZ, M., & PICKENS, R. (1992). *Vulnerability to drug abuse.* Washington, DC: American Psychological Association.

HAWKINS, J. D., CATALANO, R. F., & MILLER, J. Y. (1992). Risk and protective factors for alcohol and other drug problems in adolesence and early adulthood: Implications for substance abuse prevention. *Psychological Bulletin, 112,* 64–105.

KAHN, H. A., & SEMPOS, C. T. (1989). *Statistical methods in epidemiology.* New York: Oxford University Press.

OFFICE OF SUBSTANCE ABUSE PREVENTION. (1991). Breaking new ground for youth at risk: Program summaries. OSAP Technical Report 1, DHHS Publication No. (ADM) 91-1658. Washington, DC: U.S. Government Printing Office.

ROY W. PICKENS
DACE S. SVIKIS

Gender   Apart from the use of TOBACCO (cigarettes) and PSYCHOACTIVE DRUGS, men show a consistently higher rate of drug use than do WOMEN, especially with reference to ALCOHOL and to MARIJUANA and other illicit drugs (Substance Abuse and Mental Health Services Administration, 1992; Anthony, 1991; Robins et al., 1984; Kandel & Yamaguchi, 1985; Windle, 1990; Robbins, 1989). Women are more likely than men to use the drugs prescribed by a physician, especially psychotrophic drugs (Cafferata et al., 1983), and although men still have a higher rate of CIGARETTE use, this difference is decreasing (Kandel & Yamaguchi, 1985; National Institute on Drug Abuse, 1989 & 1991; SAMSA, 1992).

Gender differentiation in society occurs at many levels and in the major institutions such as government, family, the economy, education, and religion, as well as in face-to-face interpersonal interaction (Giele, 1988). It is therefore not surprising that drug use behavior differs for men and women. Because of the pervasive way in which gender roles affect most aspects of people's lives, it remains a complex task to understand gender differences in patterns of drug use. It is expected that gender will influence patterns of substance use and consequences of substance abuse, in part because men and women are socialized according to different behavior patterns and values. Normative expectations for men include self-reliance and physical effectiveness. By contrast, women are taught to value close relationships and to define themselves in terms of those relationships. With regard to substance use, the literature shows that gender (a) is associated with use of alcohol and drugs; (b) is associated with a variety of psychosocial characteristics that are themselves associated with alcohol and drug use; (c) and may be associated with different etiologies of alcohol and drug use—and with different consequences of substance use and treatment outcomes. The role of gender in drug use has been demonstrated in a number of studies conducted in the United States; several of these have provided comprehensive comparisons of the psychological, social, and biological characteristics of male and female drug users (Kaplan & Johnson, 1992; Lex, 1991; Gomberg, 1986; Ray and Braude, 1986).

According to the convergence hypothesis, the increasing similarity of roles and activities of men and women, as illustrated by the increasing participation of women in the paid labor force, will result in the drug and alcohol behaviors of women increasingly approximating those of men (see Adler, 1975; Bell, 1980). Although there is some evidence that male and female ADOLESCENTS have similar drug-use behaviors, recent epidemiological data indicate that alcohol and drug problems are still more common among men than among women (Anthony, 1991). Lennon (1987) found no support for the hypothesis that women in "male" jobs resembled men in terms of their levels of drinking. In the case of cigarettes, the increasing similarity of men's and women's behavior has been the result of both women increasing and men decreasing their use of cigarettes. There is little evidence to support the theory of increasing convergence of substance use, although it should be noted that many of the early studies of alcohol or drug use included only men, so that little is known about trends in women's use (Robins & Smith, 1980; see Vannicelli & Nash [1984] for an analysis of sex bias in alcohol studies).

The various perspectives that can be used to explain gender differences in drug and alcohol use include: (1) gender role explanations; (2) the social control theory; and (3) biological explanations. Explanations that draw on gender role theories to explain male-female differences refer to normative expectations and rules regarding the behavior of males and females. According to one hypothesis, there are distinctive gender styles in expressing pathology (Dohrenwend & Dohrenwend, 1976). The male style features acting-out behaviors (including drug and alcohol use), whereas the female style involves the internalization of distress. A finding consistent with this hypothesis was that of several researchers, who observed that for females, conformity to the female identity was related to higher psychological distress and lower substance use than was observed in males (Horowitz & White, 1987; Huselid & Cooper, 1992; Snell, Belk, & Hawkins, 1987; Koch-Hattem & Denman, 1987). The evidence for males has been inconsistent, however. Although there was more alcohol and drug use among males than among females, ascribing to the conventional masculine role did not necessarily lead to more alcohol or drug problems for males.

A second explanation for gender differences in alcohol and drug use is that societal expectations differ for men and women, with the result that using illicit substances for pleasure is more acceptable in men than it is in women (Landrine, Bardwell, & Dean, 1988; Lemle & Mishkind, 1989; Gomberg, 1986). Women are more likely to use substances for therapeutic reasons, specifically for the relief of mental and physical distress, whereas men are more likely to use drugs for recreation. Surveys in which it was found that men use more illicit drugs, primarily for recreation, and women use more psychotherapeutic drugs have borne out this theory.

A closely related hypothesis that is particularly relevant to the higher use of psychotropic drugs by women is that society permits women to perceive more illness (morbidity) and to use more medical care than it does men, who are expected to be stoic in the face of illness. Survey results seem to confirm the behavioral differences suggested by this hypothesis. In a review of morbidity and mortality studies, Verbrugge (1985) found that women consulted physicians more often than men, assumed the patient's role more readily, and appeared to take better care of themselves in general. These behaviors would make women more inclined than men to use prescription drugs and less inclined to use other drugs. The increasing use of cigarettes by younger women, however, is one behavior that runs counter to this hypothesis.

According to the social control theory, those who have strong ties to societal institutions such as family, school, or work are less likely to have a problem with use of substances. This perspective stems from Emile Durkheim's classic study of SUICIDE (1898). Umberson (1987) applied Durkheim's perspective to health behaviors and showed that social ties affect the health behaviors of individuals (e.g., physical activity, alcohol consumption, compliance with doctor's recommendations, etc.) and that consequently they affect health status and mortality rates. Social ties, according to this argument, affect drug use behaviors in two ways. First, there is an increased likelihood that the behavior of those with strong social ties will be monitored by family members and friends, and this would tend to decrease use of illicit or unhealthy substances. Second, the responsibility and obligation entailed in an individual sharing strong ties and frequent activities with family and friends make for more self-regulation of behavior. Marriage and being a parent represent important social ties that may affect people's use of substances, especially in the case of women, because of their traditional roles in nurturing and maintaining family relationships.

Several studies have shown the increased vulnerability to drug use of women in relation to social ties. Kaplan and Johnson (1992) showed that the attenuation of interpersonal ties resulting from initial drug use caused women, but not men, to increase their drug use. Similarly, Kandel (1984) reported that interpersonal factors were more significant for women than for men in explaining marijuana use. Ensminger, Brown, and Kellam (1982) showed that strong family bonds inhibited drug use in female adolescents but not in male adolescents.

Physiological differences may also be important in accounting for gender differences in patterns of substance use. Mello has (1986) suggested that a woman's use of drugs and alcohol may be influenced by menstrual cycle phases (Mello, 1986), although little evidence exists for this hypothesis. Halbreich et al. (1982) examined the scores on the Premenstrual Assessment Form and found that women who increased their marijuana use at the premenstruum reported significantly greater DEPRESSION, ANXIETY, mood changes, anger, and impaired social functioning than did women whose marijuana use decreased or stayed the same.

The relatively low rate of consumption of drugs by women may be related to biological differences in the ways drugs are cleared from the body in women versus men. The lower ratio of water to total body weight in women causes them to metabolize alcohol and drugs differently (Mello, 1986; Straus, 1984). This and other biological factors may cause women to have higher BLOOD-ALCOHOL CONCENTRATIONS (BACs) than men at equal dosages (Corrigan, 1985; McCrady, 1988). Drugs that are deposited in body fat, such as marijuana, may be slower to clear in women than in men because of the higher ratio of fat in women (Braude & Ludford, 1984).

Gender roles are the major roles in human society, and they influence almost every aspect of an individual's life. Despite the evidence for gender differences in patterns of drug use, little attention has been given either to the potential strategic advantages that this observation presents for furthering our understanding of drug and alcohol use patterns in males and females, or for determining how prevention and treatment programs might be redesigned.

(SEE ALSO: *Comorbidity and Vulnerability; Conduct Disorder and Drug Use; Epidemiology; Gender and Complications of Substance Abuse*)

BIBLIOGRAPHY

ADLER, F. (1975). *Sisters in crime.* Prospect Heights, IL: Woreland.

ANTHONY, J. C. (1991). The epidemiology of drug addiction. In N. S. Miller (Ed.), *Comprehensive handbook of drug and alcohol addiction.* New York: Marcel Dekker.

BELL, D. S. (1980). Dependence on psychotropic drugs and analgesics in men and women. In O. J. Kalant (Ed.), *Alcohol and drug problems in women.* New York: Plenum.

BRAUDE, M. C., & LUDFORD, J. P. (1984). *Marijuana effects on the endocrine and reproductive systems: A RAUS review report* (NIDA Research Monograph 44). Rockville, MD: National Institute on Drug Abuse.

CAFFERATA, G. L., KASPER, J., & BERNSTEIN, A. (1983, June). Family roles, structure, and stressors in relation to sex differences in obtaining psychotropic drugs. *Journal of Health and Social Behavior, 24,* 132–143.

CORRIGAN, E. M. (1985). Gender differences in alcohol and other drug use. *Addictive Behaviors, 10,* 313–317.

DOHRENWEND, B. P., & DOHRENWEND, B. S. (1976). Sex differences in psychiatric disorders. *American Journal of Sociology, 81,* 1447–1454.

DURKHEIM, E. (1898). *Suicide: A study in sociology.* (J. A. Spaulding & G. Simpson, trans.). New York: Free Press.

ENSMINGER, M. E., BROWN, C. H., & KELLAM, S. G. (1982). Sex differences in antecedents of substance use among adolescents. *Journal of Social Issues, 38*(2), 25–42.

GIELE, J. Z. (1988). Gender and sex roles. In N. J. Smelser (Ed.), *Handbook of sociology.* Newbury Park, CA: Sage Publications.

GOMBERG, E. S. L. (1986). Women: Alcohol and other drugs. In *Drugs and society.* Binghamton, NY: Haworth Press.

HALBREICH, U., ENDICOTT, J., SCHACHT, S., & NEE, J. (1982). The diversity of premenstrual changes as reflected in the Premenstrual Assessment Form. *Acta Psychiatrica, 65,* 46–65.

HOROWITZ, A. V., & WHITE, H. R. (1987, June). Gender role orientations and styles of pathology among adolescents. *Journal of Health and Social Behavior, 28,* 158–170.

HUSELID, R. F., & COOPER, M. L. (1992). Gender roles as mediators of sex differences in adolescent alcohol use and abuse. *Journal of Health and Social Behavior, 33,* 348–362.

KANDEL, D. B. (1984). Marijuana users in young adulthood. *Archives of General Psychiatry, 41,* 200–209.

KANDEL, D. B., & YAMAGUCHI, K. (1985). Developmental patterns of the use of legal, illegal, and medically prescribed psychotropic drugs from adolescence to young adulthood. In *Etiology of drug abuse: Implications for prevention* (NIDA Research Monograph Series No. 56, DHHS Publication No. ADH 85–1335). Washington, DC: U.S. Government Printing Office.

KAPLAN, H. B., & JOHNSON, R. J. (1992). Relationships between circumstances surrounding initial illicit drug use and escalation of drug use: Moderating effects of gender and early adolescent experiences. In M. Glantz & R. Pickens (Eds.), *Vulnerability to drug abuse.* Washington, DC: American Psychological Association.

KOCH-HATTEM, A., & DENMAN, D. (1987). Factors associated with young adult alcohol abuse. *Alcohol and Alcoholism, 22,* 181–192.

LANDRINE, H., BARDWELL, S., & DEAN, T. (1988). Gender expectations for alcohol use: A study of the significance of the masculine role. *Sex Roles, 19,* 703–712.

LEMLE, R., & MISHKIND, M. E. (1989). Alcohol and masculinity. *Journal of Substance Abuse Treatment, 6,* 213–222.

LENNON, M. C. (1987). Sex differences in distress: The impact of gender and work roles. *Journal of Health and Social Behavior, 28,* 290–305.

LEX, B. W. (1991). Gender differences and substance abuse. In N. K. Mello (Ed.), *Advances in substance abuse* (Vol. 4). London: Jessica Kingsley.

McCRADY, B. S. (1988). Alcoholism. In E. A. Blechman & K. O. Brownell (Eds.), *Handbook of behavioral medicine for women.* New York: Pergamon.

NATIONAL INSTITUTE ON DRUG ABUSE. (1991). *National Household Survey on Drug Abuse: Main Findings 1990.* Washington, DC: U.S. Department of Health and Human Services, Public Service, Alcohol, Drug Abuse, and Mental Health Administration.

NATIONAL INSTITUTE ON DRUG ABUSE. (1989). *National Household Survey on Drug Abuse: Highlights 1988.* Washington, DC: U.S. Department of Health and Human Services, Public Service, Alcohol, Drug Abuse, and Mental Health Administration.

RAY, B. A., & BRAUDE, M. C. (EDS.). (1986). *Women and drugs: A new era for research* (NIDA Research Monograph No. 65, DHHS Publication No. ADM 86–1447). Washington, DC: U.S. Government Printing Office.

ROBBINS, C. (1989, March). Sex differences in psychosocial consequences of alcohol and drug abuse. *Journal of Health and Social Behavior, 30,* 117–130.

ROBINS, L. N., & SMITH, E. M. (1980). Longitudinal studies of alcohol and drug problems: Sex differences. In O. J. Kalant (Ed.), *Alcohol and drug problems in women: Research advances in alcohol and drug problems* (Vol. 5). New York: Plenum.

ROBINS, L. N., ET AL. (1984). Lifetime prevalence of specific psychiatric disorder in three sites. *Archives of General Psychiatry, 41,* 929–958.

SNELL, W. E., JR., BELK, S. S., & HAWKINS, R. C., II. (1987). Alcohol and drug use in stressful times: The influence of the masculine role and sex-related personality attributes. *Sex Roles, 16,* 359–373.

STRAUS, R. (1984). The need to drink too much. *Journal of Drug Issues, 14,* 125–136.

SUBSTANCE ABUSE AND MENTAL HEALTH SERVICES ADMINISTRATION. (1993). *National Household Survey on Drug Abuse: Population estimates 1992.* Washington, DC: U.S. Department of Health and Human Services, Public Health Service.

UMBERSON, D. (1987). Family status and health behaviors: Social control as a dimension of social integration. *Journal of Health and Social Behavior, 28,* 306–319.

VANNICELLI, M., & NASH, L. (1984). Effect of sex bias on women's studies on alcoholism. *Alcohol Clinical and Experimental Research, 8,* 334–336.

VERBRUGGE, L. M. (1985, September). Gender and health: An update on hypotheses and evidence. *Journal of Health and Social Behavior, 26,* 156–182.

WINDLE, M. (1990). A longitudinal study of antisocial behaviors in early adolescence as predictors of late adolescence substance use: Gender and ethnic group differences. *Journal of Abnormal Psychology, 99*(1), 86–91.

MARGARET E. ENSMINGER
JENNEAN EVERETT

Genetics Genes are passed from parent to child in the process of sexual reproduction. These genes determine some of the features of the individual and contribute directly and indirectly to many more. The possibility of genetic influences in substance abuse has received considerable attention. Evidence that genetic influences may be involved comes from family studies, where substance abuse has been found to run in families. For example, alcoholics have been found to have more relatives who are alcoholic than would be expected from the base rate for ALCOHOLISM in the general population. Similarly, higher rates of HEROIN and COCAINE abuse are also seen in the relatives of heroin and cocaine abusers than occur in the general population.

Both twin and family studies have been conducted to separate genetic from environmental influences in the familial transmission of substance abuse. Most of the research has involved ALCOHOL. There is general agreement that genetic influences are involved in both alcohol use and alcoholism, at least for males. Twin studies of males from the general population have found that if one pair member drinks alcohol, the other pair member is more likely to drink (i.e., they are concordant for this behavior) if the two members shared all the same genes (if they are monozygotic or identical twins) than if they share only about half of their genes (if they are dizygotic or fraternal twins). Similar studies on clinical patients have found higher concordance for alcoholism among men who are monozygotic rather than dizygotic twins. Adoption studies have found that sons of alcoholic biological parents were more likely to be alcoholic as adults than sons of nonalcoholic biological parents, when both groups were adopted out early in life and raised by nonalcoholic adoptive parents. Among men, estimates of the proportion of variance in alcohol-dependence liability due to genetic influences (i.e., heritability) range from 0.50 to

0.60, depending on the subject population and subtype of alcoholism.

For women, the role of genetic factors in alcohol use and alcoholism is less convincing. This is primarily because women have been studied less often than men and in smaller numbers. One reason for this discrepancy is that women are less likely to have alcohol problems, and this fact itself may reflect the greater role of nongenetic influences for women. In twin and adoption studies involving women, evidence of genetic influence has been found less consistently than has been found for men, with heritabilities for women ranging from 0.00 to 0.56, depending on the study. Nevertheless, women have similar percentages of same- and opposite-sex alcoholic relatives as do men, and this suggests that there is no differential heritability related to gender.

Although less frequently studied, genetic influences for other forms of drug use and dependence have also been shown, but only males have typically been studied in this context. Heritabilities reported for tobacco smoking range from 0.28 to 0.84 and are not affected by other factors that may contribute to differences in concordance rates in twins. Heritabilities reported for other types of illicit drug use (but not necessarily drug dependence) range from 0.4 to 0.6. Heritability for any substance abuse or dependence (excluding alcohol and tobacco) in alcoholic probands is 0.31.

Linkage and association studies permit the identification of specific genes involved in substance abuse. In linkage studies, different generations of families are examined to determine if a genetic marker is inherited along with a disorder (e.g., substance abuse). In association studies, individuals with and without a disorder are compared to determine the association of the disorder with a genetic marker. To date, no specific gene for alcoholism or for other types of drug dependence has been identified.

Animal models have also been employed to study genetic influences in substance abuse. Evidence of significant genetic influence has been found in the characteristics of many drug responses relevant to drug abuse (e.g., drug preference), and chromosomal loci have been identified that mediate at least some of these effects. To the extent that the genetic structure of mice is similar to that of human beings, the findings derived from animal models suggest testable hypotheses to be explored in human-association studies. In strains of rats that were bred in laboratories to study their preference for alcohol, the strain that developed a strong preference for alcohol had lower brain levels of the NEUROTRANSMITTER serotonin compared to the strain that did not prefer alcohol. This is of interest because alterations in SEROTONIN neurotransmission have also been noted in studies of impulsive aggressive human males (who have a higher likelihood of developing alcohol or drug problems) compared to human males without those behaviorial traits.

(SEE ALSO: *Attention Deficit Disorder; Causes of Substance Abuse; Conduct Disorder and Drug Use; Disease Concept of Alcoholism and Drug Addiction; Epidemiology of Drug Abuse*)

BIBLIOGRAPHY

GLANTZ, M., & PICKENS, R. (1992). *Vulnerability to drug abuse.* Washington, DC: American Psychological Association.

HAWKINS, J. D., CATALANO, R. F., & MILLER, J. Y. (1992). Risk and protective factors for alcohol and other drug problems in adolesence and early adulthood: Implications for substance abuse prevention. *Psychological Bulletin, 112,* 64–105.

KAHN, H. A., & SEMPOS, C. T. (1989). *Statistical methods in epidemiology.* New York: Oxford University Press.

OFFICE OF SUBSTANCE ABUSE PREVENTION. (1991). Breaking new ground for youth at risk: Program summaries. DHHS Publication No. (ADM) 91-1658. Washington, DC: U.S. Government Printing Office.

DACE S. SVIKIS
ROY W. PICKENS

Psychoanalytic Perspective   Increased vulnerability to ALCOHOL and drugs is related to the coming together of a number of influences, each of which is itself of varying strength. Our biologies, our individual social and cultural settings and backgrounds, our personal idiosyncratic life experiences, and the persons we become as a result of all these may contribute to the likelihood of our using drugs—and then of our continuing to use them. We are neither vulnerable nor invulnerable to using drugs or alcohol, nor to using them to excess; vulnerability is a continuum, ranging from least to most vulnerable. Under the right, or the wrong, circumstances, many of us will use drugs.

ALCOHOLISM runs in families; if an individual's parent, grandparent, or sibling is alcoholic, that individual's own risk is significantly increased. It seems certain that an important contributor to this in many families is GENETIC. While we find a similar increase in the frequency of substance abuse in the children of parents who use all sorts of drugs, we do not yet have evidence that this too is genetic. Certainly, another contributor to this familial pattern is the exposure that a developing child has to the sight and experience of a parent or other important figure in the environment using alcohol and/or other drugs. It tells the child that this is acceptable behavior, particularly if the surrounding social culture echoes that opinion. Cultures and subcultures that traditionally control drinking generally produce people who drink in a controlled way; cultures and subcultures that condone excess also reproduce themselves.

It is important to remember, however, that even those with a strong genetic loading for alcoholism can only become a "practicing" alcoholic if they have alcohol available. Despite its many problems, Prohibition (1920–1933) reduced the number of alcoholics; successful interdiction of drugs would reduce the number of substance abusers. However, growing up in an area where drugs are freely available increases the likelihood of trying them and—assuming community complacence or peer approval and encouragement—of continuing to take them. For example, during the war in VIETNAM, many U.S. soldiers who had not been OPIATE addicts found themselves in the war zone, exposed to STRESS and personal danger, and surrounded by cheap available HEROIN in a context that condoned its use. Many became addicted. On their return home, however, almost all gave up their drug use with relative ease.

We also know that the person one is—the kind of *personality* one has—also plays a role in one's susceptibility to using and misusing drugs. A number of studies suggest that maladjustment precedes the use of illicit drugs; the closer one is in style to an Eagle Boy Scout, the less likely one is to use drugs. Rebelliousness, stress on independence, apathy, pessimism, DEPRESSION, low self-esteem, and low academic aspirations and motivation make the use of illicit drugs more likely. Delinquent and deviant behavior come before the drug use; they are not the result of it.

(SEE ALSO: *Causes of Substance Abuse: Psychological (Psychoanalytic) Perspective; Comorbidity and Vulnerability; Conduct Disorder and Drug Use; Families and Drug Use; Religion and Drug Use*)

BIBLIOGRAPHY

BOHMAN, M., SIGVARDSSON, S., & CLONINGER, C. R. (1981). Maternal inheritance of alcohol abuse. *Archives of General Psychiatry, 38,* 965–969.

CHEIN, I., ET AL. (1964). *The road to H: Narcotics, delinquency, and social policy.* New York: Basic Books.

CLONINGER, C. R., BOHMAN, M., & SIGVARDSSON, S. (1981). Inheritance of alcohol abuse. *Archives of General Psychiatry, 38,* 861–868.

WILLIAM A. FROSCH

Race Long-standing conceptual difficulties in defining *race* and *ethnicity* pose a challenge to the interpretation of racial and ethnic differences in drug use and drug-abuse treatment. The validity of the concept of race has been questioned, and biological notions of race (subspeciation) may be confused with cultural and social notions of ETHNICITY (Hahn 1992; Jones, LaVeist, & Lillie-Blanton 1991; Lillie-Blanton, Anthony, & Schuster 1993; Fullilove 1993).

The categories used to describe race/ethnicity may include people of diverse origins and practices; for example, in the United States the term *Hispanic* refers to people of Mexican, Puerto Rican, and Cuban origin, as well as the other Latin American countries, and people from these countries may differ substantially in their drug-use behaviors (Oetting & Beauvais 1990; Bachman et al., 1991; National Institute on Drug Abuse, 1988); yet they are often lumped together for analytic purposes. Moreover, within these HISPANIC subgroups, generational status and level of acculturation will result in further differentiation of drug-use rates (Oetting & Beauvais, 1990; Velez & Ungemack, 1989). Similarly, the category *Native American* refers to members of diverse tribes and nations located throughout the United States, and they may have very different drug-use patterns (May, 1986).

As Hahn (1992) documents, the terms *race* and *ethnicity* have not been consistently defined by federal data-collection agencies, are often understood differently by respondents in surveys, and are not necessarily perceived consistently from one survey to the next. The following brief review of the data pertaining to ethnicity, race, and drug use should be read with these caveats in mind.

Patterns of drug use by race and ethnicity are complex, and they seem to vary by age. Survey data suggest that prevalence rates for African-American and Hispanic youth are lower than they are for white youth, whereas data on morbidity and treatment in adulthood seem to indicate that adult problems with drug use are more evident in African Americans than in whites. The two surveys that have been used most often to characterize adolescent substance use during the 1970s and 1980s (Oetting & Beauvais, 1990) are the national HIGH SCHOOL SENIOR SURVEY, administered as part of the Monitoring the Future Project (Bachman et al., 1991; Johnston, O'Malley, & Bachman, 1989), and the NATIONAL HOUSEHOLD SURVEY ON DRUG ABUSE, sponsored by the NATIONAL INSTITUTE ON DRUG ABUSE (NIDA, 1990). Both surveys monitored ADOLESCENT substance use from the mid-1970s until the early 1990s. Methodologically, the surveys differ in the mode of data collection and the populations surveyed, but both have shown consistent differences across ethnic and racial subgroups of adolescents.

The two surveys and other school-based and general population studies show the lowest rates of MARIJUANA, COCAINE, and ALCOHOL use to be by black and Asian youth; Hispanic and non-Hispanic whites are in the middle in terms of use, and Native American youth have the highest rates (Bachman et al., 1991; Barnes & Welte, 1986; NIDA, 1990; Oetting & Beauvais, 1990; Windle, 1990). Methodological criticism of these results has largely focused on the reliability and validity of self-report data (Mensch & Kandel, 1988), the populations excluded from the school-based or community data (Bachman et al., 1991; Wallace & Bachman, 1991), or the possible underreporting of deviant activities by members of minority groups (Mensch & Kandel, 1988). Persons in drug treatment, homeless people, and incarcerated people are not represented in population surveys. Dropouts or those who are frequently absent from school are not included in school-based surveys. Since minority-group members make up a disproportionate number of these populations and it is among these populations that drug use is believed to be more pervasive, surveys may be an inadequate means for generalizing about the full range of drug problems among members of minority groups.

As indicated by measures that show alcohol and drug problems, African-American adults do not appear to be characterized by lower drug use. Although the National Household Survey on Drug Abuse shows there is lower drug use among black youth than among white youth, by early adulthood the differences have become smaller and by middle adulthood the drug-use rates are slightly higher for most drugs among black people (NIDA, 1990). According to the 1988 NIDA survey, CRACK-cocaine smoking is more common among African Americans and Hispanic Americans than it is among white Americans for those older than twenty-six years, and the highest rates of HEROIN use are seen in African Americans aged twenty-six to thirty-four (NIDA, 1990). In addition, Robbins (1989) presented data from the National Household Survey that showed that although at younger ages there were fewer problems with substance abuse among black youth than there were among white and Hispanic youth, both male and female black adults over the age of twenty-six had more substance-abuse problems than did either white or Hispanic adults of the same age group. Data drawn from treatment settings and mortality records also show higher rates of problems among black adults than among white adults. For example, Herd (1986) reported that the figures for mortality due to cirrhosis among nonwhite people were twice as high as those for whites. Black people appear to be about three times more likely than white people to be in treatment for drug-related problems (NIDA, 1986). These data, however, are not adequate to describe all the patterns of drug use by race and ethnicity because both access to treatment and mortality are affected by socioeconomic conditions and ethnicity as well as by drug use (Gittelsohn, Halpern, & Sanchez 1991; Jaynes & Williams, 1989).

Thus there appear to be two pictures of drug use patterns by race and ethnicity. Except for Native American youths, there are fewer reports of drug use among adolescents of minority groups than among white adolescents, whereas there is some evidence that African-American young and middle-age adults have more drug problems than do white adults. Several possible explanations might account for these reversals in rates. First, because of methodological issues relating to the data, the surveys of youth focusing on in-school or household populations might be missing high-risk minority youth who are more

likely to have high rates of drug use; similarly, the cited data drawn from adults' treatment and health records may come disproportionately from public programs, thereby making clients with fewer economic resources more visible and those with more resources less visible. Because members of minority groups tend to have lower incomes, their problems may be more visible in the data. Second, some of the decline in drug use observed during the period from adolescent to young adulthood could be attributed to people's increasing involvement in marriage and employment during this stage of life (Sampson & Laub, 1990). In view of the high rates of unemployment and low rates of marriage among young African-American adults (Wilson, 1987), these sources of social bonds and informal social control may be more evident for white people than they are for African-Americans.

Because of the small differences in the level of substance use among ethnic groups and the considerable heterogeneity within these subgroups, few obvious implications can be drawn for prevention or intervention measures. Understanding language and cultural differences as well as the ways people classify themselves is important in designing prevention and treatment programs. Given the lower economic resources, the discrimination, and the residential segregation of many ethnic groups, the design of services must take into account not only cultural differences but also economic and social ones.

(SEE ALSO: *Asia, Drug Use in; Chinese Americans, Alcohol and Drug Use among; Ethnic Issues and Cultural Relevance in Treatment; Ethnicity and Drugs; Hispanics and Drug Use; Poverty and Drug Use*)

BIBLIOGRAPHY

BACHMAN, J. G., ET AL. (1991). Racial/ethnic differences in smoking, drinking, and illicit drug use among American high school seniors, 1976–89. *American Journal of Public Health, 81*(3), 372–377.

BARNES, G. M., & WELTE, J. W. (1986). Patterns and predictors of alcohol use among 7–12th grade students in New York state. *Journal of Studies on Alcohol, 47*, 53–62.

FULLILOVE, M. T. (1993). Perceptions and misperceptions of race and drug use. *Journal of the American Medical Association, 269*(8), 1034.

GITTELSOHN, A. M., HALPERN, J., & SANCHEZ, R. L. (1991). Income, race, and surgery in Maryland. *American Journal of Public Health, 81*(11), 1435–1441.

HAHN, R. A. (1992). The state of federal health statistics on racial and ethnic groups. *Journal of the American Medical Association, 267*(2), 268–271.

HERD, D. (1986). A review of drinking patterns and alcohol problems among U.S. blacks. In *Report of the Secretary's Task Force on Black and Minority Health: Vol. 7. Chemical dependency and diabetes.* Washington, DC: U.S. Department of Health and Human Services.

JAYNES, G. D., & WILLIAMS, R. M. JR. (1989). Black Americans' health. In *A common destiny: Blacks and American society.* Washington, DC: National Academy Press.

JOHNSTON, L. D., O'MALLEY, P. M., & BACHMAN, J. G. (1989). *Drug use, drinking and smoking: National survey results from high school, college, and young adult populations, 1975–1988* (National Institute on Drug Abuse DHHS Pub. No. [ADM] 89–1638). Washington, DC: Government Printing Office.

JONES, C. P., LaVEIST, T. A., & LILLIE-BLANTON, M. (1991). "Race" in the epidemiologic literature: An examination of the *American Journal of Epidemiology, 1921–1990. American Journal of Epidemiology, 134*(10), 1079–1084.

LILLIE-BLANTON, M., ANTHONY, J. C., & SCHUSTER, C. R. (1993). Probing the meaning of racial/ethnic group comparisons in crack cocaine smoking. *Journal of the American Medical Association, 269*(8), 993–997.

MAY, P. A. (1986). Alcohol and drug abuse prevention programs for American Indians: Needs and opportunities. In *Report of the Secretary's Task Force on Black and Minority Health: Vol. 7. Chemical dependency and diabetes.* Washington, DC: U.S. Department of Health and Human Services.

MENSCH, B. S., & KANDEL, D. B. (1988). Underreporting of substance use in a national longitudinal youth cohort. *Public Opinion Quarterly, 52,* 100–124.

NATIONAL INSTITUTE ON DRUG ABUSE. (1990). *National Household Survey on Drug Abuse: Main findings* (U.S. Department of Health and Human Services Publication No. [ADM] 90-1682). Washington, DC: U.S. Government Printing Office.

NATIONAL INSTITUTE ON DRUG ABUSE. (1988). First national data on drug use among Hispanics released. *NIDA Notes,* 23–24.

NATIONAL INSTITUTE ON DRUG ABUSE. (1986). *Demographic characteristics and patterns of drug abuse treatment programs in selected states: Annual data 1983.* Rockville, MD: NIDA, Division of Epidemiology and Statistical Analysis.

OETTING, E. R., & BEAUVAIS, F. (1990). Adolescent drug use: Findings of national and local surveys. *Journal of Consulting and Clinical Psychology, 58,* 385–394.

ROBBINS, C. (1989, March). Sex differences in psychosocial consequences of alcohol and drug abuse. *Journal of Health and Social Behavior, 30,* 117–130.

SAMPSON, R. J., & LAUB, J. H. (1990, October). Crime and deviance over the life course: The salience of adult social bonds. *American Sociological Review, 55,* 609–627.

VELEZ, C. N., & UNGEMACK, J. A. (1989). Drug use among Puerto Rican youth: An exploration of generational status differences. *Social Science and Medicine, 29,* 779–789.

WALLACE, J. M., JR., & BACHMAN, J. G. (1991). Explaining racial/ethnic differences in adolescent drug use: The impact of background and lifestyle. *Social Problems, 38*(3), 333–357.

WILSON, W. J. (1987). *The truly disadvantaged.* Chicago: University of Chicago Press.

WINDLE, M. (1990). A longitudinal study of antisocial behaviors in early adolescence as predictors of late adolescent substance use: Gender and ethnic group differences. *Journal of Abnormal Psychology, 99*(1), 86–91.

MARGARET E. ENSMINGER
SION KIM
JENNEAN EVERETT

Sensation Seeking The term *sensation seeking* was defined (Zuckerman, 1979) as "the need for varied, novel, and complex sensations and experiences and the willingness to take physical and social risks for the sake of such experience." The trait is usually assessed using the Sensation Seeking Scale (SSS), which contains four subscales:

1. Thrill and Adventure Seeking—the seeking of excitement through certain kinds of risky physical activities;
2. Experience Seeking—the seeking of sensation through music, art, travel, and an unconventional lifestyle;
3. Disinhibition—the seeking of sensations through partying, drinking, sex, etc.;
4. Boredom Susceptibility—an aversion to lack of stimulation; restlessness.

The SSS and most of its subscales have been related to smoking, drinking, and the use of illegal drugs in adolescent and adult populations in the United States and many other countries. The SSS scores of preadolescents predict subsequent drug and ALCOHOL use, as well as other kinds of socially deviant behavior. In studies using the SSS with other personality scales, sensation seeking is much more predictive of early alcohol and drug use than any other personality trait.

Why is this trait such an important predictor of drug use? Most young people say (Zuckerman, 1983) that they first try drugs like MARIJUANA, LYSERGIC ACID DIETHYLAMIDE (LSD), and others out of curiosity, "for a new experience." Experience seeking is their major motive. Since high-sensation seekers are also thrill and adventure seekers, the physical risks do not deter them and they are not afraid of DRIVING, for example, under the influence of alcohol or other drugs. The drug "scene" evolved in the 1960s and 1970s to include rock music—at concerts and parties—where even today there is often disinhibited behavior, including easy sex. The seeking of sexual variety is also a form of sensation seeking, and both high-sensation seeking and drug use are associated with sexual experience. This association of drugs and sex reinforces social drug use among high-sensation seekers. Since sensation seeking itself has relatively strong genetic determination, it is not surprising to find (Zuckerman, Buchsbaum, & Murphy, 1980) that it has a number of biological-trait correlates. Some of these have also been found to be characteristic of alcohol and drug users, suggesting (Zuckerman 1987) a link in common biological mechanisms.

While sensation seeking explains why certain people are vulnerable to drug abuse, it does not explain eventual drug DEPENDENCE. Drug dependence involves, to some degree, the development of TOLERANCE for the drugs and both psychological and physical discomfort produced by WITHDRAWAL. Thus, according to some theories of dependence, drug users start by using the drugs to produce pleasure or to "get high"; they end by using them to avoid pain or just to "feel normal." Even after the tolerance and physical dependence has been treated, the sensation-seeking trait may still contribute to relapse.

Recognition of the role of sensation seeking in drug involvement may be a clue to effective treatment: The drug user has to be convinced that one can lead an exciting life without the use of drugs. Exciting kinds of sports, travel, parties, music, and

sex do not require drugs to be enjoyed. The sensation seeker does not have to give up sensation seeking along with drugs. In fact, sensations and experiences are usually sharper and more intense without mind alteration or by dampening with alcohol and other depressant drugs. The choice of an interesting job is especially important, since routine, unchallenging kinds of work create a special stress in boredom-susceptible sensation seekers.

(SEE ALSO: *Adolescents and Drug Use; Conduct Disorder and Drug Use; Comorbidity and Vulnerability; Prevention*)

BIBLIOGRAPHY

ZUCKERMAN, M. (1987). Biological connection between sensation seeking and drug abuse. In J. Engel et al. (Eds.), *Brain reward systems and abuse: Proceedings of the seventh international Berzelius symposium.* New York: Raven Press.

ZUCKERMAN, M. (1983). Sensation seeking: The initial motive for drug abuse. In E. Gotheil et al. (Eds.), *Etiological aspects of alcohol and drug abuse.* Springfield, IL: Charles C. Thomas Publishers.

ZUCKERMAN, M. (1979). *Sensation seeking: Beyond the optimal level of arousal.* Hillsdale, NJ: Erlbaum.

ZUCKERMAN, M., BUCHSBAUM, M. S., & MURPHY, D. L. (1980). Sensation seeking and its biological correlates. *Psychological Bulletin, 88,* 187–214.

MARVIN ZUCKERMAN

Sexual and Physical Abuse   An increased recognition of the experience of physical and sexual abuse in the lives of many children and ADOLESCENTS has led to the increased interest in the impact of such abuse on drug use (Cavaiola & Schiff, 1989; Straus & Gelles, 1990; Dembo et al., 1988). In their 1985 survey of over 6,000 families in the United States, Straus and Gelles (1990) report that 23 per 1,000 children (2.3%) are seriously assaulted every year. Data from a 1991 telephone national survey of women indicate that about 20 per 100 (20%) of the sample reported one or more childhood sexual-abuse experiences (Wilsnack et al., 1994). Few research studies have focused specifically on the question of whether children who are physically and sexually abused are at increased risk of substance abuse. Dembo et al. (1988) suggest three reasons why child abuse has not been included in the conceptual schemes examining the process by which youths become involved in drug use. First, CHILD ABUSE has only recently (in the 1980s) surfaced as an issue receiving research and policy attention. Second, both child-abuse experiences and illicit drug use are often hidden phenomena, so that any covariation in their occurrence is difficult to observe. Third, the focus on social-psychological and sociocultural factors left little opportunity for child-abuse variations to be considered. Throughout the 1980s and into the 1990s, there has been increasing recognition of the potential importance of abuse to the child's and adolescent's emotional development and the potential connection to substance use and other problem behaviors (Widom, 1991; Zingraff et al., 1993). The central hypothesis guiding research is that physically and sexually abused children and adolescents may use illicit drugs to help cope with the emotional difficulties caused by their negative self-perceptions or other internal difficulties that result from the abuse (Cavaiola & Schiff, 1989; Singer, Petchers, & Hussey, 1989; Dembo et al., 1988).

Much existing research has concentrated on cohorts of adolescents. The rationale for the vulnerability of childhood victims of abuse to drug dependence in adolescence includes first, the ramifications of abuse for lowering self-image and self-esteem, while increasing self-hatred. Based on Kaplan, Martin, and Robbins' (1984) proposition that self-derogation leads to drug use, this model suggests that the abuse of children is related to illicit drug use, both directly and as mediated by self-derogation (Dembo et al., 1988). Second, drugs may provide emotional or psychological escape and self-medication for young abuse victims; they may turn to drugs to chemically induce forgetting or to cope with feelings of ANXIETY (Miller, 1990). Third, drug use may provide abused children or adolescents with a peer group, in the form of a drug culture, hence reducing feelings of isolation and loneliness (Singer, Petchers, & Hussey, 1989; Widom, 1991).

Methodological limitations have prevented the existing research from giving a definitive answer. According to Widom (1991), most studies of the association between illicit drug use and childhood victimization have focused on sexually or physically abused children in clinical or institutional settings, making it difficult to generalize to other populations; the studies are often cross-sectional in design, in-

clude only retrospective information about childhood-abuse experiences, and do not utilize control groups. Therefore, the validity and reliability of these data have been criticized. Since abuse-related consequences can vary across the life span, cross-sectional studies may miss important ramifications of abuse and it may be impossible to determine the developmental-causal sequence (Briere, 1992; Dembo et al., 1988). Furthermore, most of the studies do not control for other childhood characteristics that may mediate the effects of abuse. Studies focusing on the abuse victims as adults run further methodological risks. When asked about abuse from their childhood, these adults may forget, redefine events in terms of the present, or repress certain thoughts and events.

In one of the earliest reviews of the impact of sexual abuse in childhood, Browne and Finkelhor (1986) reported that adult WOMEN victimized as children were more likely to manifest DEPRESSION, self-destructive behavior, anxiety, feelings of isolation, poor self-esteem, and substance abuse than their nonvictimized counterparts. They distinguished initial effects—identified as the manifestations within two years of termination of abuse—from long-term effects.

In a carefully designed study, Widom (1992) followed two groups in arrest records for fifteen to twenty years. One group of 908 individuals with court-substantiated cases of childhood abuse or neglect was matched according to sex, age, race, and socioeconomic status with a comparison group of 667 children not officially recorded as abused or neglected. As indicated by arrest records, the behavior of those who had been abused or neglected was worse than those with no reported abuse—abused or neglected children were more likely to be arrested as juveniles, as adults, and for a violent CRIME. With regard to drug use, as adults, the abused and neglected females were more likely to be arrested for drug offenses compared to the nonabused females. In a large sample (N=3018) of Alabama 8th and 10th graders, Nagy et al. (1994) found that about 10 percent (13% of females and 7% of males) of the students reported being sexually abused. Sexual abuse was defined to include one or more episodes of forced intercourse. Both sexually abused males and sexually abused females reported a higher use of illegal drugs in the past month than those students who did not report sexual abuse. While the associations were strong, the analyses did not attempt to control for confounding variables and were cross-sectional rather than longitudinal, so that causality cannot be inferred.

Wilsnack et al. (1994), using a national sample of adult women, examined the abuse of alcohol and drugs by women who reported retrospectively on whether they had been sexually abused as children. They found strong positive associations between being abused sexually as a child and six different measures of drinking behaviors and two summary drug-use measures. While these analyses are considered preliminary by the authors, because they do not attempt to control for confounding variables, the findings do suggest that early sexual trauma may be an important risk factor for substance abuse later in life.

In a retrospective study, Miller (1990) compared forty-five alcoholic women with forty women chosen randomly from the same community. The relationships between child abuse by the father and the development of alcoholism was examined by controlling on the parents' alcohol problems, family structure during childhood, income source, and age. Higher levels of negative verbal interaction and higher levels of moderate and serious violence were both predictive of those who were found in the alcoholic group.

In their review and synthesis of empirical studies regarding the impact of sexual abuse on children, Kendall-Tackett, Williams, and Finkelhor (1993) found that poor self-esteem was a frequently occurring consequence of sexual abuse. They also conclude that substance abuse, while being a common behavior for sexually abused adolescents, is not an inevitable outcome. In a residential treatment center, Cavaiola and Schiff compared with two control groups the self-esteem of 150 physically or sexually abused, chemically dependent adolescents. The results showed that abused chemically dependent adolescents had lower self-esteem than the two comparison groups; they found negligible difference between those who had been sexually abused and those who had been physically abused.

In two populations of youths studied in a juvenile detention center, Dembo et al. (1988, 1989) compared the lifetime drug use between detainees and a comparable age group in an adjacent county. The studies showed that the detainees' sexual victimization and their physical-abuse experiences related significantly to their lifetime use of illicit drugs. Sexual victimization had a direct effect on the frequency of lifetime drug use, whereas physical abuse had both a direct and an indirect effect on drug use, mediated by the adolescents' feelings of self-derogation. These findings were based on multiple-regression analyses that included family background, other risks for drug use, race, and sex.

CONCLUSION

Despite methodological issues, the body of available evidence suggests that involvement in substance use as an adolescent or adult is linked to an increased likelihood of having experienced physical or sexual abuse as a child. Owing to limitations in the retrospective, cross-sectional, and correlational designs of the research, causal linkages cannot be definitively attributed, and as Briere (1992) notes, while much of the existing research is flawed in its design, it has set the stage for the development of more tightly controlled and methodologically sophisticated studies that will be able to better disentangle the antecedents, correlates, and impacts of sexual and physical abuse.

Further research is needed to examine questions in which our knowledge is meager. First, are there different effects from physical abuse, sexual abuse, or neglect on substance use or dependence? Do other psychosocial factors lead to substance abuse? Second, does the perpetrator of the abuse matter for the impact? Third, does continuity or duration of the abuse matter? Fourth, and perhaps most important, what are the links between suffering maltreatment as a child and later alcohol or drug problems?

(SEE ALSO: *Families and Drug Use; Family Violence and Substance Abuse*)

BIBLIOGRAPHY

BRIERE, J. (1992). Methodological issues in the study of sexual abuse effects. *Journal of Consulting and Clinical Psychology, 60,* 196–203.

BROWNE, A., & FINKELHOR, D. (1986). Impact of child sexual abuse: A review of the research. *Psychological Bulletin, 99*(1), 66–77.

CAVAIOLA, A. A., & SCHIFF, M. (1989). Self-esteem in abused chemically dependent adolescents. *Child Abuse and Neglect, 13,* 327–334.

DEMBO, R., ET AL. (1989). Physical abuse, sexual victimization, and illicit drug use: Replication of a structural analysis among a new sample of high-risk youths. *Violence and Victims, 4*(2), 121–138.

DEMBO, R., ET AL. (1988). The relationship between physical and sexual abuse and tobacco, alcohol, and illicit drug use among youths in a juvenile detention center. *The International Journal of the Addictions, 23*(4), 351–378.

KAPLAN, H. B., MARTIN, S. S., & ROBBINS, C. (1984). Pathways to adolescent drug use: Self-derogation, peer influence, weakening of social controls, and early substance use. *Journal of Health and Social Behavior, 25*, 270–289.

KENDALL-TACKETT, K., WILLIAMS, L. M., & FINKELHOR, D. (1993). Impact of sexual abuse on children: A review and synthesis of recent empirical studies. *Psychological Bulletin, 113*(1), 164–180.

KINGERY, P. M., PRUITT, B. E., & HURLEY, R. S. (1992). Violence and illegal drug use among adolescents: Evidence from the U.S. National Adolescent Student Health Survey. *The International Journal of the Addictions, 27*(12), 1445–1463.

MILLER, B. A. (1990). The interrelationship between alcohol and drugs and family violence. *NIDA Research Monograph*, No. 103. Rockville, MD: National Institute on Drug Abuse.

NAGY, S., ADCOCK, A. G., & NAGY, M. C. (1994). A comparison of risky health behaviors of sexually active, sexually abused, and abstaining adolescents. *Pediatrics, 93*(4), 570–575.

STRAUS, M. A., & GELLES, R. J. (EDS.). (1990). *Physical violence in American families*. New Brunswick, NJ: Transaction.

SINGER, M. I., PETCHERS, M. K., & HUSSEY, D. (1989). The relationship between sexual abuse and substance abuse among psychiatrically hospitalized adolescents. *Child Abuse and Neglect, 13*, 319–325.

WIDOM, C. S. (1991). "Childhood victimization and adolescent problem behaviors." Paper for the National Institute of Child Health and Human Development conference on "Adolescent Problems and Risk-taking Behaviors." April 13–16, Berkeley Springs, West Virginia.

WIDOM, C. S. (1992). *The cycle of violence*. Washington, DC: National Institute of Justice Research in Brief: 1–6, October.

WILSNACK, S. C., ET AL. (1994). "Childhood sexual abuse and women's substance abuse: National survey findings." Paper for the American Psychological Association conference on "Psychosocial and Behavioral Factors in Women's Health: Creating an Agenda for the 21st Century." May 12–14, Washington, DC.

ZINGRAFF, M. T., ET AL. (1993). Child maltreatment and youthful problem behavior. *Criminology, 31*(2), 173–202.

MARGARET E. ENSMINGER
COLLEEN J. YOO

Stress The term *stress* is frequently defined as a process involving perception, interpretation, and response to harmful, threatening, or challenging events (Lazarus & Folkman, 1984). This kind of conceptualization allows the separate consideration of (1) the events that cause stress (stressors), (2) the cognitive processes that may be applied to evaluating these stressors, and (3) the selection of ways of coping with them (appraisal)—and with responses to the precipitating stressors (including actual coping). Other depictions of stress put more or less emphasis on appraisal mechanisms and attempt to characterize stressors more precisely, or operationally; however, the model of an organism responding to substantial threat or danger is basic to most theories of stress (e.g., Cohen et al., 1986; Mason, 1975; Selye, 1976).

Stress may best be thought of as the negative emotional state that is the product of appraisal, extremely demanding events, situational and psychological factors, and the impetus for coping. Sympathetic arousal, activation of the pituitary–adrenocortical axis, and endogenous opioid-peptide release are among the mechanisms that support coping responses aimed at reducing emotional distress (Baum, 1990).

Stress is important in understanding drug use at several levels. Stress is believed to be a motivating state, in that people are usually eager to reduce it. Thus, stress should theoretically initiate complex coping responses directed at reducing the stressor's effects or altering sources of stress. According to some, drug use is one way of coping with stress and may evolve from stress-related drug use to chronic problems unrelated to stress. Emotional disorders, for example (DEPRESSION, ANXIETY, chronic anger, or SCHIZOPHRENIA), may themselves be stressors—ongoing or recurrent stressors—rather than the results of stressors (or genetics). Complex interactions exist among stressors and mental disorders.

Several forms of coping with stress are possible. Lazarus (1966) identified two primary classes of coping: (1) direct action, which is usually behavioral and involves activity aimed at altering the source of stress or one's relationship to it, and (2) palliation, focused on managing one's emotional responses rather than causes of stress. Palliative coping may be behavioral or cognitive; it may include denial, withdrawal, taking drugs, and/or other forms of making oneself feel better (or less bad). Direct action is a manipulative response aimed at changing a stressor, while palliation is generally accommodative.

One reason that people use drugs is to enhance mood and, in the face of stress, motivation to do so may be high. The Tension Reduction Hypothesis (Conger, 1956), and more recent variants (e.g., Sher & Levenson, 1982; Powers, 1987) suggest that drugs become reinforcers by decreasing internally aversive drives or states. A drug may be used initially to modulate tension or distress; then with repeated success in doing so, it may become a more ubiquitous response to stress in a variety of settings. Thus, by decreasing tension or stress or because of the positive expectancies from drug effects, people may come to use drugs in the face of stress—that is, in anticipation of stress.

The majority of studies of stress as a factor in drug use have been retrospective, asking subjects to recall stressful events over a period of time. These studies use some type of stressful life events inventory, such as the Social Readjustment Rating Scale (SRRS) (Holmes & Rahe, 1967). Extensive research using various measures of stress has produced consistent but relatively small correlations with a variety of health indices.

Retrospective studies using these instruments have generally reported that substance abusers experienced more stressful life events prior to drug initiation than did nonabusers, and current abusers had higher stress levels compared to nonabusers; however, there are several limitations to retrospective correlational studies—causation cannot be determined, recall bias is not measurable, and a variety of other potential causes of either inaccurate drug use or life event reporting are uncontrolled.

A better approach to assessing the relationship between stress and drug use is the initiation of prospective studies. Prospective studies measure stressful events as they occur and use them to predict drug use in the future. These studies avoid many of the problems associated with retrospective studies (e.g., less recall bias) but have not yielded clear evidence of stress-induced drug use. These studies indicate that the relationship between stress and drug use is complicated by variables such as loss of control (perception of not being able to control events in one's life) and meaninglessness (life having no meaning, difficulty in planning future and approaching problems) (Newcomb & Harlow, 1986). Several of such studies have found that stress is associated with drug use (e.g., Wills, 1986; Gorman, 1988). Wills used a prospective design in two samples of adolescents, one with 675 subjects and the other with 901. Subjective stress and negative events predicted smoking and drinking behavior. Some studies indicate that stress increases susceptibility to substance abuse; it potentially increases and sustains ongoing drug use. Further research is needed using prospective methods to better understand the role of stress in the initiation and maintenance of substance abuse.

Several hypotheses explaining the relationship of drug use and stress have been offered beyond the obvious palliation-based notions. One mechanism that contributes to the development of drug tolerance and could result in increased drug use is an increase in the rate of drug excretion from the body (Grunberg & Baum, 1985; Jaffe, 1990). As a psychophysiological process initiated by central nervous system (CNS) activity and resembling a sympathetic neuronal readying response, stress involves changes in metabolic rate, liver function, respiratory and cardiovascular activity, and leads to the acidification of urine and to a variety of endocrine and immune changes. These changes are likely to affect the intensity and duration of drug effects, particularly if stress is sustained for long periods of time. For instance, Schachter (1977) found that as urinary pH decreased (became more acid) due to stress, NICOTINE excretion rates increased and SMOKING rates increased. If urinary pH was kept stable, stress had no effect on smoking (Schachter, 1977). Although this model has been supported by evidence in nicotine studies with animals (Grunberg, Morse, & Barrett, 1983), its application to other drugs has not been substantiated.

Stress may reduce the time needed to metabolize drugs, and if it increases the speed at which one processes and disposes of a drug, it could increase the frequency of need and overall drug use. Stress could also hasten symptoms of WITHDRAWAL, resulting in a composite of the two motivational states. Since symptoms of withdrawal are similar to those associated with stress, this may suggest other mechanisms invoking identification of causal factors. OPIOID addicts, as well as other drug addicts, may misattribute increased arousal associated with stress as withdrawal symptoms, thereby increasing their drug use to counter this withdrawal (Grunberg & Baum, 1985). Coupled with the possibility that stress increases both the speed of metabolism and the clearance of drugs, stress could also increase the level of withdrawal symptoms during treatment as well as frequency of relapse (Whitehead 1974). However, Hall, Havassay, and Wasserman (1990, 1991), found that while commitment to abstinence from nicotine, alcohol, cocaine, and opioid use was related to abstinence goals, stress levels did not predict relapse. Determinants of relapse are complex and include factors related to age, experience, and so on—and the relative role of stress has not been established.

Drug use, then, may be a primary form of accommodative coping, particularly when the source of stress is seen as being beyond one's ability to control it. This may depend on the particular drug one considers; while opiates may be closely associated with coping intended to regulate emotional distress, the use of TOBACCO, CAFFEINE, AMPHETAMINES, or COCAINE may also reflect attempts to bolster one's ability to deal with stressful demand. The alerting, energizing, and confidence-enhancing properties of these last four drugs may make them "useful" in a variety of settings. Consideration of the CAUSES of drug use as well as the complex reinforcement patterns associated with drug use and stress may help in treating or preventing drug use. For example, a drug such as amphetamine or cocaine has quantifiable effects of its own that can establish strong habits—when used as a means of coping with stress, either may acquire the powerful reinforcing properties of a stress-reducer as well. This is similar to the notion that drugs may be self-administered because they relieve hunger, affect eating in other ways, or assist in weight control (Grunberg, 1990). In both examples, the strength of the drug-use habit derives from more than just the primary psychoactive effects of the drug.

Despite the intuitive appeal to the idea that stress can lead to drug use, more research is needed to better understand its potential role in the initiation, maintenance, and withdrawal of substance abuse. Despite the limitations of current research, some conclusions can be drawn. Retrospective studies support the notion that stress makes people more susceptible to substance abuse and increases substance use in individuals already using drugs. Prospective studies generally show the same pattern. If high-risk groups can be identified (as, for example, groups exposed to trauma), intervention procedures may serve to decrease stress and the initiation of drug use. Similarly, a better understanding of the psychological and physiological effects of stress in relation to drug use may lead to more comprehensive treatment strategies and greater resistance to relapse.

(SEE ALSO: *Addiction: Concepts and Definitions; Comorbidity and Vulnerability; Complications; Endorphins; Epidemiology of Drug Abuse; Families and Drug Use; Family Violence and Substance Abuse; Poverty and Drug Use*)

BIBLIOGRAPHY

BAUM, A. (1990). Stress, intrusive imagery, and chronic distress. *Health Psychology, 9,* 653–675.
COHEN, S., ET AL. (1986). *Behavior, health, and environmental stress.* New York: Plenum.
CONGER, J. J. (1956). Reinforcement theory and the dynamics of alcoholism. *Quarterly Journal of Studies in Alcohol, 17,* 296–305.
GORMAN, D. M. (1988). Employment, stressful life events and the development of alcohol dependence. *Drug and Alcohol Dependence, 22,* 151–159.
GRUNBERG, N. E. (1990). The inverse relationship between tobacco use and body weight. In L. T. Kozlowski (Ed.), *Research advances in alcohol and drug problems,* Vol. 10. New York: Plenum.
GRUNBERG, N. E., MORSE, D. E., & BARRETT, J. E. (1983). Effects of urinary pH on the behavioral responses of squirrel monkeys to nicotine. *Pharmacology Biochemistry and Behavior, 19,* 553–557.
GRUNBERG, N. E., & BAUM, A. (1985). Biological commonalities of stress and substance abuse. In S. Shiffman & T. A. Wills (Eds.), *Coping and substance abuse.* New York: Academic.
HALL, S. M., HAVASSAY, B. E., & WASSERMAN, D. A. (1991). Effects of commitment to abstinence, positive moods, stress, and coping on relapse to cocaine use. *Journal of Consulting and Clinical Psychology, 54*(4), 526–532.
HALL, S. M., HAVASSAY, B. E., & WASSERMAN, D. A. (1990). Commitment to abstinence and acute stress in relapse to alcohol, opiates and nicotine. *Journal of Consulting and Clinical Psychology, 58,* 175–181.

HOLMES, T. H., & RAHE, R. H. (1967). The social readjustment rating scale. *Journal of Psychosomatic Research, 11*, 213–218.

JAFFE, J. H. (1990). Drug addiction and drug abuse. In A. G. Gilman et al. (Eds.), *Goodman and Gilman's the pharmacological basis of therapeutics*, 8th ed. New York: Pergamon.

LAZARUS, R. S. (1966). *Psychological stress and the coping process.* New York: McGraw-Hill.

LAZARUS, R. S., & FOLKMAN, S. (1984). *Stress, appraisal, and coping.* New York: Springer.

MASON, J. W. (1975). A historical view of the stress field. *Journal of Human Stress, 1*, 22–36.

NEWCOMB, M. D., & HARLOW, L. L. (1986). Life events and substance use among adolescents: Mediating effects of perceived loss of control and meaninglessness in life. *Journal of Personality and Social Psychology, 51*(3), 564–577.

POWERS, R. J. (1987). Stress as a factor in alcohol use and abuse. In E. Gottheil et al. (Eds.), *Stress and addiction.* New York: Brunner/Mazel.

RAHE, R. H. (1975). Life changes and near future illness reports. In L. Levii (Ed.), *Emotions: Their parameters and measurement.* New York: Raven.

SCHACHTER, S. (1977). Nicotine regulation in heavy and light smokers. *Journal of Experimental Psychology: General, 106*, 3–48.

SELYE, H. (1976). *The stress of life.* New York: McGraw-Hill.

SHER, K. J., & LEVENSON, R. W. (1982). Risk for alcoholism and individual differences in the stress-response-dampening effect of alcohol. *Journal of Abnormal Psychology, 91*, 350–368.

WHITEHEAD, C. C. (1974). Methadone pseudowithdrawal syndrome: Paradigm for a psychopharmacological model of opiate addiction. *Psychosomatic Medicine, 36*, 189–198.

WILLS, T. A. (1986). Stress and coping in early adolescence: Relationships to substance use in urban school samples. *Health Psychology, 5*(6), 503–529.

LORENZO COHEN
ANDREW BAUM

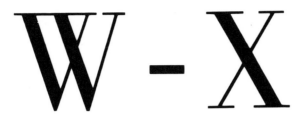

W - X

WALDEN HOUSE Walden House is a comprehensive THERAPEUTIC COMMUNITY (TC) in San Francisco. It consists of residential facilities for adults and adolescents, a day treatment program, outpatient services, and a nonpublic school and training institute. Walden House is a highly structured and well-supervised program designed to treat the behavioral, emotional, and family issues of substance abusers.

The heart of the TC is a long-term residential treatment program—three to eighteen months in duration—consisting of a series of phases from orientation to aftercare. Within the TC, all the household tasks, groups, individual counseling, meals, and seminars promote responsibility and emotional growth. The activities are part of an integrated array of therapeutic experiences, in which residents continuously see themselves in a context of mutual support. The philosophy of Walden House emphasizes self-help and peer support. For many residents, a stay at Walden House is their first positive experience of belonging in a family. In the daily process of working and growing together, residents learn to trust others, take responsibility, develop self-esteem, and construct a substance-free lifestyle.

HISTORY

Founded in 1969 by Walter Littrell as a response to the drug epidemic of the 1960s, Walden House has grown into one of the largest substance-abuse programs in California. Its first facility was an old Victorian mansion at 101 Buena Vista East in the Haight-Ashbury district of San Francisco. In 1971, Alfonso Acampora, current chief executive officer, joined Walden House, bringing concepts and clinical tools learned at SYNANON and Delancey Street. Acampora refined these tools, discarded those that degraded clients, and added important innovations to the TC, such as reentry into the community, family services, and a methadone-to-abstinence program. Walden House also pioneered the use of alternative treatments with substance abusers, for example, herbs, diet, and physical exercise.

A major turning point occurred in 1978, when Walden House purchased a convent at 815 Buena Vista West in San Francisco and opened a 100-bed residential facility. With this expansion, Walden House added degreed professionals to its staff of recovering counselors and began to treat clients with more severe psychosocial and medical problems. A new component, Supportive Clinical Services, was formed to enhance the TC by providing specialized psychological, medical, and social services for individuals with severe mental disorders as well as drug and alcohol problems (dual-diagnostic clients) and clients with HIV (see AIDS). In 1985, Walden House achieved international recognition by hosting the Ninth World Conference of Therapeutic Communities in San Francisco, the theme of which was "Bridging Services." In recognition of the alarming increase of substance abuse among youth, Walden House opened its 40-bed adolescent facility at 214 Haight Street in 1986 and offered a program combining psychiatric clinical treatment with the TC model.

CLINICAL PROGRAM

Walden House has several components—the two 100-bed residential facilities for adults, one short term, the other long term; the 40-bed adolescent program with an on-campus school that provides special education for severely emotionally disturbed youths; and the multiservice center that houses central intake and an outpatient day treatment program. Table 1 shows the various treatment modalities of the TC at Walden House. In addition, Walden House has designed many special programs to treat particular populations, including dual-diagnostic clients, clients with AIDS, homeless people, minorities, pregnant women, women with children, and clients referred from the criminal-justice system as an alternative to incarceration.

Walden House has adapted the TC model to changing times. In 1991, it developed a six-month short-term treatment program for homeless substance abusers and for parolees released from prison. It has also created a new component, Women's Services, that includes a therapeutic day-care center for the children of women in treatment. Finally, the Walden Institute of Training (WIT) initiated a multilevel training and certification program for counselors, and it collaborated with the University of California in a National Institute on Drug Abuse research project to compare day treatment with residential treatment.

(SEE ALSO: *Treatment*)

BIBLIOGRAPHY

ACAMPORA, A., & NEBELKOPF, E. (1986). *Bridging services: Proceedings of the Ninth World Conference of Therapeutic Communities.* San Francisco: World Federation of Therapeutic Communities.

ACAMPORA, A., & STERN, C. (1992). Evolution of the therapeutic community. In *Drugs & society: Proceedings of the Fourteenth World Conference of Therapeutic Communities.* Montreal: World Federation of Therapeutic Communities.

NEBELKOPF, E. (1989). Innovations in drug treatment and the therapeutic community. *International Journal of Therapeutic Communities,* 10(1), 37–49.

NEBELKOPF, E. (1987). Herbal therapy in the treatment of drug use. *International Journal of the Addictions,* 22(8), 695–717.

NEBELKOPF, E. (1986). The therapeutic community and human services in the 1980s. *Journal of Psychoactive Drugs,* 18(3), 283–286.

SORENSEN, J., ACAMPORA, A., & ISCOFF, D. (1984). From maintenance to abstinence in the therapeutic community: Clinical treatment methods. *Journal of Psychoactive Drugs,* 16(3), 229–239.

TABLE 1
Walden House Therapeutic Community

Treatment Modality	Target Population	Duration
Long-term residential	Adult substance abusers	6–18 months
	Dual-diagnostic substance abusers	
	HIV+ substance abusers	
Short-term residential	Homeless substance abusers	6 months
	Parolees with substance abuse history	
	Recovering addicts in relapse	
Day treatment	Adult substance abusers	12 months
	Adults on residential waiting list	
Outpatient aftercare	Recovering addicts, postresidential	6 months
Group home	Adolescent substance abusers, 13–18	18 months
	Emotionally disturbed adolescents, 13–18	
Nonpublic school	Adolescents in group home, 13–18	6–18 months
Methadone to abstinence	Substance abusers tapering from methadone in residential treatment	6–18 months
Women's services	Pregnant women in treatment	1 year
	Women with children in treatment	

SORENSEN, J., DEITCH D., & ACAMPORA, A. (1984). Treatment collaboration of methadone maintenance programs and therapeutic communities. *American Journal of Drug and Alcohol Abuse*, 10(3), 347–359.

ALFONSO ACAMPORA
ETHAN NEBELKOPF

WAR ON DRUGS *See* Epidemics of Drug Abuse; Treatment, History of; U.S. Government, The Organization of U.S. Drug Policy; Zero Tolerance.

WASHINGTONIAN TEMPERANCE SOCIETY/WASHINGTONIANS *See* Temperance Movement; Treatment: History of; Women's Christian Temperance Union.

WERNICKE'S SYNDROME *See* Alcoholism; Complications: Neurological.

WIKLER'S PHARMACOLOGIC THEORY OF DRUG ADDICTION Abraham Wikler (died 1981) was one of the first researchers who, in the late 1940s, strongly advocated the idea that drug abuse and relapse following treatment are influenced by basic learning processes. Early in his career, Wikler became interested in reports from relapsed heroin addicts that despite being free of withdrawal symptoms during treatment and upon discharge, they experienced withdrawal symptoms and craving when they returned to their drug-use environments—and that these feelings were responsible for their return to drug use.

Based on these and other anecdotes, Wikler—who was familiar with the recent work of Russian physiologist Ivan Petrovich Pavlov (1849–1936) on conditioning—proposed that events which reliably signal drug self-administration or drug withdrawal elicit conditioned responses (CRs) that take the form of withdrawal and drug craving. According to Wikler, these CRs motivate further drug use, which, by terminating negative withdrawal feelings, perpetuates the cycle of drug dependency.

At the heart of Wikler's model lies the notion that classical conditioning mechanisms are activated when events surrounding drug use *reliably* begin to signal upcoming drug administration. These events may be external cues (e.g., the sight of a syringe) or internal states (e.g., depression) that *consistently* precede drug use. In nondependent users (who take drugs infrequently), Wikler proposed that the unconditioned response (UR) elicited by the drug consists of direct effects of that drug on the nervous sytem. In such individuals, stimuli that signal drug use would then come to evoke druglike responses; however, a different set of CRs are thought to occur in long-term drug users who have become physically dependent on the drug. These individuals experience withdrawal symptoms as the drug effect wanes and consequently, stimuli associated with drug withdrawal in these individuals evoke withdrawal reactions.

The aversive symptoms produced by withdrawal in dependent users provide motivation to self-administer the drug. Through a process of operant conditioning, drug taking is rewarded by the termination of the negative withdrawal symptoms. These reward experiences further strengthen the tendency of the drug user to turn to drug use when experiencing withdrawal symptoms. Likewise, stimuli paired temporally with withdrawal may also acquire the ability to elicit drug taking. Because Wikler invoked both classical and operant conditioning mechanisms as contributors to drug use, his model has often been characterized as a two-process model of drug use.

Wikler's model also provides for a powerful account of relapse following treatment for drug use. Because some treatment programs separate the abuser from the drug-use environment, the patient never learns to deal with drug-related events. Upon returning home following treatment, even though no longer physically dependent, the patient encounters drug signals, experiences conditioned withdrawal reactions, and eventually turns to drug use to reduce the negative feelings. Since conditioned responses show little spontaneous decay over time, the drug-use patient is at risk even following an extended treatment program. According to Wikler, treatment programs need to address conditioned responses directly. One suggested approach involves having subjects go through their usual drug-preparation ritual in a protected setting, where drugs are not available. Such exposures should serve to extinguish drug-use responses by failing to reinforce them with relief from withdrawal. Extinction training as well as other techniques for reducing the role of conditioned responses in relapse are currently being explored.

(SEE ALSO: *Behavioral Tolerance; Causes of Substance Absue: Learning; Naltrexone; Research, Animal Model: Learning, Conditioning and Drug Effects*)

BIBLIOGRAPHY

WIKLER, A. (1977). The search for the psyche in drug dependence. *Journal of Nervous and Mental Disease, 165,* 29–40.

WIKLER, A. (1973). Dynamics of drug dependence: Implications of a conditioning theory for research and treatment. *Archives of General Psychiatry, 28,* 611–616.

WIKLER, A. (1965). Conditioning factors in opiate addiction and relapse. In D. I. Wilner & G. G. Kassenbaum (Eds). *Narcotics.* New York: McGraw-Hill.

WIKLER, A. (1948). Recent progress in research on the neurophysiologic basis of morphine addiction. *American Journal of Psychiatry, 105,* 329–338.

STEVEN J. ROBBINS

WINE *See* Alcohol; Fermentation.

WITHDRAWAL This section contains the articles on withdrawal syndromes, each of which describes and discusses withdrawal signs, symptoms, and treatment. The following substances are covered: *Alcohol; Alcohol, Beta Blockers; Benzodiazepines; Cocaine; Nicotine (Tobacco);* and *Nonabused Drugs.* For descriptions and discussions of withdrawal from Amphetamines, see *Amphetamine;* Anabolic Steroids, see *Anabolic Steroids;* Barbiturates, see *Barbiturates;* Caffeine, see *Caffeine;* Cannabis, see *Cannabis,* see also *Marijuana;* for Heroin, Opiates/Opioids, see *Opioid Complications and Withdrawal.* For additional information, see also *Treatment.*

Alcohol The nervous system undergoes adaptation in response to the chronic consumption of alcohol (ethanol). If consumption is heavy enough (adequate dose) and occurs for a long enough time period (duration), a withdrawal syndrome will ensue following a rapid decrease or sudden cessation of drinking. This occurs in association with readapta-

tion of the nervous system to a drug-free state. The dose and duration of alcohol consumption required to produce a withdrawal syndrome in a given population or even a given individual are difficult to predict, since no well-controlled studies have been conducted (or are likely to be, for ethical reasons). Such studies have been done in animals. The goals of treatment are to relieve discomfort and to prevent complications.

In the nondrinker or social drinker who consumes alcohol to the point of legal intoxication, an acute withdrawal syndrome may ensue ("hangover"). Symptoms occur in inverse relation to the fall in BLOOD ALCOHOL CONCENTRATION (BAC). These consist of insomnia, headache, and nausea. Usually no treatment is required and there are no serious consequences of this acute withdrawal. The withdrawal syndrome following chronic long-term alcohol consumption (usually months to years), however, is a more serious disorder.

The natural history of alcohol dependence to the point of requesting or clearly requiring detoxification services is usually fifteen to twenty years. The average age of persons admitted to detoxification units is around 42 years. (That is not to say that persons as young as 20 or as old as 80 do not require detoxification services.) The withdrawal syndrome seen in persons requiring detoxification ranges from a mild degree of discomfort to a potentially life-threatening disorder.

The severity of the withdrawal syndrome is dependent on both the dose and duration of alcohol exposure. This is clearly demonstrated in animal studies (rats) where a severe withdrawal syndrome can be demonstrated following high-level exposure to alcohol in a vapor chamber in as short a time period as a week. Administration of alcohol into the stomach is associated with a longer time period for acquisition of physical dependence. In humans also, the severity of withdrawal depends on the amount of alcohol consumed and the time period during which it has been consumed. For practical purposes this means the amount taken on a daily basis for the weeks and months preceding detoxification. One study of inpatients (who were federal prisoners and narcotic users) demonstrated that the consumption of 442 grams of alcohol or 32 standard drinks (a standard drink being 13.6 gm of alcohol—12 oz. of beer, 5 oz. of wine, or 1.5 oz. of liquor) per day for about two months results in a major withdrawal syndrome in all subjects, whereas the consumption of

280 to 377 grams (21 to 28 standard drinks) per day results in a mild syndrome of anxiety and tremor (Isbell et al., 1955). Other studies that involve patients (as opposed to research subjects) have not been able to demonstrate a consistent relationship between recent alcohol consumption and severity of the withdrawal syndrome (Shaw et al., 1981). This in part relates to the lack of accurate recall of exact quantities consumed within a given time period. Furthermore, in the real world there are different patterns of consumption (e.g., some drinkers consume alcohol in a binge pattern, whereas others drink in a more regular pattern), and different drinkers have varying durations of lifetime exposure to alcohol. One drinker may take two or three years to become dependent, another fifteen years, and yet another forty years. In addition, a person who has previously experienced significant alcohol withdrawal may be at higher risk for developing repeat withdrawal, both in terms of the severity of the syndrome and the rate of reacquisition of physical dependence (since it takes a shorter time to become re-addicted). This more rapid reacquisition has been attributed to sensitization (or "kindling") of the central nervous system (Linnoila et al., 1987). Other factors that may be implicated in the severity of the withdrawal syndrome include age, nutritional status, and presence of concurrent physical disorders or illness (e.g., pancreatitis or pneumonia) (Sullivan & Sellers, 1986). Alcoholics are at increased risk for these and other medical disorders.

The symptoms and signs of alcohol withdrawal appear in inverse relation to the elimination of alcohol from the body. Many alcoholics note this phenomenon on a daily basis—they require a drink in the morning to "steady the nerves," to suppress tremor and anxiety. The following are some of the more common symptoms of alcohol withdrawal: anxiety, agitation, restlessness, insomnia, feeling shaky inside, anorexia (loss of appetite), nausea, changes in sensory perception (tactile: skin itchy; auditory: sounds louder; visual: light brighter), headache, and palpitations. Common signs include vomiting, sweating, increase in heart rate, increase in blood pressure, tremor (shakiness of hands and sometimes face, eyelids, and tongue), and seizures. More severe withdrawal is associated with intensification of the above symptoms and signs together with progression to hallucinations (tactile: feeling things that are not there; auditory: hearing things that are not there; visual: seeing things that are not there), disorientation, and

confusion (DELIRIUM TREMENS, DTs). After stopping alcohol, the more common and milder symptoms usually peak at 12 to 24 hours and have mostly subsided by 48 hours (Sellers & Kalant, 1976). More severe or late withdrawal usually peaks later, 72 to 96 hours, and is potentially life threatening. Less than 5 percent of persons withdrawing from alcohol (depending on how they are selected) are estimated to develop a severe reaction. With appropriate drug treatment, an even lower percentage are estimated to develop a major withdrawal reaction. Under ideal circumstances there should be almost no mortality from this disorder on its own, so overall mortality ought to be similar to that of any concurrent medical disorder.

Assessment of the severity of withdrawal can be accomplished on the basis of clinical experience or with the assistance of various rating instruments. One of the simplest and easiest to administer is the Clinical Institute Withdrawal Assessment for Alcohol-revised (CIWA-Ar). This consists of ten items that can be scored at frequent intervals (Sullivan et al., 1989). The health-care provider can administer this instrument in less than a minute (see Figure 1).

TREATMENT

Treatment for the alcohol withdrawal syndrome consists of supportive care, general drug treatment, and specific drug treatments. *Supportive care* consists of reassurance, reality orientation, reduced sensory stimuli (dark, quiet room), attention to fluids, nutrition, physical comforts, body temperature, sleep, rest and positive encouragement toward long-term rehabilitation. The majority of patients can be treated with supportive care alone; however, it is impossible to be able to predict which patients will or will not require more intensive care. *General drug treatment* includes the B vitamin thiamine, which should be given to all patients. This is given to prevent the brain damage that occurs commonly in alcoholics who are thiamine deficient. Occasionally magnesium may be given if there is a severe deficiency and there are potential cardiac problems. Intravenous fluids may be required in uncommon circumstances.

Specific drug treatments may also be given to suppress the signs and symptoms of withdrawal. While over a hundred drug treatments have been suggested as useful in the treatment of alcohol withdrawal, very few adequate scientific studies have been con-

Addiction Research Foundation Clinical Institute Withdrawal Assessment for Alcohol (CIWA-Ar)

Patient _____   Date | ___ | ___ | ___ | Time _____ : _____
                                   y   m   d   (24-hour clock, midnight = 00:00)

Pulse or heart rate, taken for one minute: _____   **Blood pressure:** _____/_____

NAUSEA AND VOMITING—Ask "Do you feel sick to your stomach? Have you vomited?" Observation.

0 no nausea and no vomiting
1 mild nausea with no vomiting
2
3
4 intermittent nausea with dry heaves
5
6
7 constant nausea, frequent dry heaves and vomiting

TREMOR—Arms extended and fingers spread apart. Observation.

0 no tremor
1 not visible, but can be felt fingertip to fingertip
2
3
4 moderate, with patient's arms extended
5
6
7 severe, even with arms not extended

PAROXYSMAL SWEATS—Observation.

0 no sweat visible
1 barely perceptible sweating, palms moist
2
3
4 beads of sweat obvious on forehead
5
6
7 drenching sweats

TACTILE DISTURBANCES—Ask "Have you any itching, pins and needles sensations, any burning, any numbness or do you feel bugs crawling on or under your skin?" Observation.

0 none
1 very mild itching, pins and needles, burning or numbness
2 mild itching, pins and needles, burning or numbness
3 moderate itching, pins and needles, burning or numbness
4 moderately severe hallucinations
5 severe hallucinations
6 extremely severe hallucinations
7 continuous hallucinations

AUDITORY DISTURBANCES—Ask "Are you more aware of sounds around you? Are they harsh? Do they frighten you? Are you hearing anything that is disturbing to you? Are you hearing things you know are not there?" Observation.

0 not present
1 very mild harshness or ability to frighten
2 mild harshness or ability to frighten
3 moderate harshness or ability to frighten
4 moderately severe hallucinations
5 severe hallucinations
6 extremely severe hallucinations
7 continuous hallucinations

ducted—the main reasons being that appropriate studies are difficult to conduct and that many patients do very well with placebo and/or supportive care alone. Nevertheless, appropriate and effective specific treatments are available and consist of drugs belonging to the same general class as alcohol (central nervous system depressants). The drugs of choice are the longer-acting benzodiazepines (usually diazepam [Valium], but others include chlordiazepoxide [Librium], lorazepam [Ativan], and oxazepam [Serax]), or occasionally a long-acting barbiturate like phenobarbital. The specific drug treatment is usually given either before most withdrawal has occurred (substitution or prophylactic treatment) or after significant symptoms and signs manifest themselves (suppressive treatment). The

advantages of substitution treatment include the prevention of potential discomfort and the possible prevention of more severe withdrawal. The disadvantages include an unnecessary treatment for some patients. The advantages of suppression treatment include more appropriate titration of dose of medication, according to a given patient's needs. The disadvantages include unnecessary patient discomfort, at least initially, possibly the development of more severe withdrawal, and sometimes drug-seeking behavior from patients and unnecessary drug withholding from staff.

BENZODIAZEPINES have been well demonstrated to prevent complications (Sellers et al., 1983) of serious withdrawal, such as seizures, HALLUCINATIONS, and cardiac arrhythmias. In general, high doses of

ANXIETY—Ask "Do you feel nervous?" Observation.
0 no anxiety, at ease
1 mildly anxious
2
3
4 moderately anxious, or guarded, so anxiety is inferred
5
6
7 equivalent to acute panic states as seen in severe delirium or acute schizophrenic reactions

AGITATION—Observation.
0 normal activity
1 sommewhat more than normal activity
3
4 moderately fidgety and restless
5
6
7 paces back and forth during most of the interview, or constantly thrashes about

VISUAL DISTURBANCES—Ask "Does the light appear to be too bright? Is its color different? Does it hurt your eyes? Are you seeing anything that is disturbing to you? Are you seeing things you know are not there? Observation.
0 not present
1 very mild sensitivity
2 mild sensitivity
3 moderate sensitivity
4 moderately severe hallucinations
5 severe hallucinations
6 extremely severe hallucinations
7 continuous hallucinations

HEADACHE, FULLNESS IN HEAD—Ask "Does your head feel different? Does it feel like there is a band around your head?" Do not rate for dizziness or lightheadedness. Otherwise, rate severity.
0 not present
1 very mild
2 mild
3 moderate
4 moderately severe
5 severe
6 very severe
7 extremely severe

ORIENTATION AND CLOUDING OF SENSORIUM—Ask "What day is this? Where are you? Who am I?"
0 oriented and can do serial additions
1 cannot do serial additions or is uncertain about date
2 disoriented for date by no more than 2 calendar days
3 disoriented for date by more than 2 calendar days
4 disoriented for place and/or person

Total CIWA-A Score _____
Rater's Initials _____
Maximum Possible Score 67
This scale is not copyrighted and may be used freely.

these benzodiazepines (with medium to long half-lives) are provided early in treatment, to cover the patient for the time period of acute withdrawal (usually 24 to 48 hours). Some patients require very large doses of drug (e.g., several hundred milligrams of diazepam) to suppress symptoms and signs. Patients with histories of withdrawal seizures (convulsions) or those that have epilepsy are always treated prophylactically, usually with benzodiazepines and any other anticonvulsant drug (medication) that they are prescribed on a regular basis. Patients who develop hallucinations are given (in addition to benzodiazepines) a phenothiazine (neuroleptic or antipsychotic drug). Typical drugs from this class include halo-peridol (Haldol), and chlorpromazine (Thorazine). These drugs are effective in the treatment of hallucinations.

SUMMARY

In summary, alcohol withdrawal syndrome is a constellation of symptoms and signs that accompany the detoxification and readaptation of the nervous system to a drug-free state in chronic users. In most cases, these signs and symptoms are a source of mild discomfort and run a self-limited course. Occasionally, more severe withdrawal occurs or patients have concurrent complications (e.g., seizures). Under these circumstances appropriate drug treatment is mandatory to relieve symptoms and prevent complications.

(SEE ALSO: *Withdrawal: Alcohol, Beta Blockers*)

BIBLIOGRAPHY

ISBELL, H., ET AL. (1955). An experimental study of "rum fits" and delirium tremens. *Quarterly Journal of Studies on Alcohol, 16,* 1–33.

LINNOILA, M., ET AL. (1987). Alcohol withdrawal and noradrenergic function. *Annals of Internal Medicine, 107,* 875–889.

SELLERS, E. M., & KALANT, H. (1976). Alcohol intoxication and withdrawal. *New England Journal of Medicine, 294,* 757–762.

SELLERS, E. M., ET AL. (1983). Oral diazepam loading: Simplified treatment of alcohol withdrawal. *Clinical Pharmacology and Therapeutics, 34,* 822–826.

SHAW, J. M., ET AL. (1981). Development of optimal tactics for alcohol withdrawal. 1. Assessment and effectiveness of supportive care. *Journal of Clinical Psychopharmacology, 1,* 382–389.

SULLIVAN, J. T., & SELLERS, E. M. (1986). Treating alcohol, barbiturate, and benzodiazepine withdrawal. *Rational Drug Therapy, 20,* 1–8.

SULLIVAN, J. T., ET AL. (1989). Assessment of alcohol withdrawal: The revised Clinical Institute Withdrawal Assessment Scale for Alcohol (CIWA-Ar). *British Journal of Addiction, 84,* 1353–1357.

JOHN T. SULLIVAN

Alcohol, Beta Blockers A person who stops ingesting alcohol after prolonged and excessive drinking may experience a syndrome of alcohol withdrawal. This syndrome is characterized by many features that are similar to the flight-or-fright response, for example, high blood pressure, rapid pulse rate, and agitation. It has long been known that the flight-or-fright response occurs because of a surge in the release of the hormones called catecholamines.

Scientists studying the alcohol-withdrawal state have demonstrated that there is also an increase in the blood levels of catecholamines during this period. This fact has led some investigators to explore the possibility that certain drugs that are known to block the effects of catecholamines (beta-blocking

drugs) might be beneficial in the treatment of the alcohol-withdrawal state.

Propranolol. Many clinical research scientists have evaluated beta blockers in the treatment of alcohol withdrawal syndrome (AWS). Carlsson and Johansson (1971) demonstrated the effectiveness of propranolol, a beta blocker, in reducing tension-related symptoms during alcohol withdrawal. Sellers and his research associates (1977) published the results of a study that was designed to determine whether using a beta blocker, propranolol, together with a BENZODIAZEPINE in the treatment of AWS was more effective than the use of either drug alone. They concluded the following: (1) Propranolol could be given to patients in AWS who had no history of adverse reactions to, or hypersensitivity to, beta blockers; (2) Propranolol may be helpful as adjunctive treatment for selected patients with severe tremors or rapid abnormal heart rhythms (tachyarrhythmias); and (3) In severe alcohol withdrawal, a benzodiazepine, chlordiazepoxide, should be the first drug used with the beta blocker propranolol as additional therapy to ameliorate the signs and symptoms of AWS.

Zilm and associates (1980) studied the effects of propranolol alone and in combination with a benzodiazepine, chlordiazepoxide, on cardiac arrhythmias. They determined that propranolol was more effective in controlling cardiac arrhythmias and was less effective in treating associated hallucinations. They concluded that the combination of the two medicines performed best in diminishing cardiac arrhythmias and hallucinations.

Timolol. Potter and his colleagues (1984) evaluated the effectiveness of a beta blocker, timolol, when added to standardized therapy, which included a sedative hypnotic, chlomethiazole, in patients undergoing alcohol withdrawal. Those who were treated additionally with timolol, compared to those treated additionally with placebo, required less sedation with chlomethiazole and had a greater decrease in their systolic blood pressure and pulse rate during alcohol withdrawal.

Atenolol. Kraus and associates (1985) studied the potential usefulness of another beta blocker, atenolol, in a large number of patients who were undergoing alcohol withdrawal as inpatients in a community hospital. In that study, a randomized controlled trial, they noted that compared with patients taking placebo, the patients taking atenolol had a significant reduction in their overall length of hospital stay and that their vital signs (blood pressure, pulse rate, and temperature) normalized more rapidly.

The atenolol-treated patients' abnormal behavior (anxiety, agitation, and irritability) and clinical characteristics also resolved more rapidly. Additionally, the atenolol-treated patients required significantly less sedative medication.

This led Horwitz, Gottlieb, and Kraus (1989) to conduct another randomized controlled trial to further study the potential usefulness of beta blockers in patients who were undergoing alcohol withdrawal in a community hospital outpatient (ambulatory) alcohol detoxification program. Again, it was found that the atenolol-treated patients had more rapid normalization of their vital signs and abnormal behavioral characteristics and that treatment failure rates were reduced in those patients receiving atenolol. Furthermore, levels of alcohol craving were strongly associated with treatment failure, and reduction in the level of craving was especially associated with the group of patients who received atenolol.

SUMMARY

Thus, beta blockers play a useful role in the treatment of AWS when used as adjunct therapy (that is, in combination with a supportive environment, counseling, hydration, vitamin supplementation, adequate nutrition, and sedative-hypnotic medication, used as necessary) in carefully selected patients.

BIBLIOGRAPHY

CARLSSON, C., & HAGGENDAL, J. (1967). Arterial noradrenaline levels after ethanol withdrawal. *Lancet, 2,* 889.

CARLSSON, C., & JOHANSSON, T. (1971). The psychological effects of propranolol in the abstinence phase of chronic alcoholics. *British Journal of Psychiatry, 119,* 605–606.

HORWITZ, R. I., GOTTLIEB, L. D., & KRAUS, M. L. (1989). The efficiency of atenolol in the outpatient management of the alcohol withdrawal syndrome. *Archives of Interim Medicine, 149,* 1089–1093.

KOCH-WESNER, J., SELLERS, E. M., & KALAUT, H. (1976). Alcohol intoxication and withdrawal. *New England Journal of Medicine, 294,* 757–762.

KRAUS, M. L., ET AL. (1985). Randomized clinical trial of atenolol in patients with alcohol withdrawal. *New England Journal of Medicine, 313,* 905–909.

POTTER, J. F., BANNAN, L. T., & BEEVERS, D. G. (1984). The effect of nonselective lepophilic beta blockers on the blood pressure and noradrenalin, vasopressin, cortisol, and renin release during alcohol withdrawal. *Clinical Experience Hyperteus* [A]6, 1147–1160.

SELLERS, E. M., ZILM, D. H., & DEAGANI, N. C. (1977). Comparative efficacy of propranolol and chlordiazepoxide in alcohol withdrawal. *Journal of the Study of Alcohol, 38,* 2096–2108.

ZILM, D. H., ET AL. (1980). Propranolol and chlordiazepoxide effects on cardiac arrhythmias during alcohol withdrawal. *Alcohol Clinical Experience Resource, 4,* 440–450.

MARK L. KRAUS
LOUIS D. GOTTLIEB
RALPH I. HORWITZ

Benzodiazepines Like many other drugs that alter central nervous system (CNS) NEUROTRANSMISSION, benzodiazepines may produce a withdrawal syndrome when the drugs are abruptly discontinued. These withdrawal symptoms, including increased ANXIETY and insomnia, are often the mirror image of the therapeutic effects of the drug. Since the term *withdrawal* is usually applied to drugs of abuse, these symptoms are sometimes called abstinence syndrome or discontinuance syndrome when associated with benzodiazepines, thereby distinguishing these substances from drugs such as ALCOHOL, OPIOIDS, COCAINE, and BARBITURATES.

ETIOLOGY

Not all patients who take benzodiazepines will experience a discontinuance syndrome when the drug is stopped. Several conditions must be present before the discontinuance syndrome is likely:

1. *Duration of treatment.* The benzodiazepine must be taken long enough to produce alterations in the CNS that will predispose to a discontinuance syndrome. When benzodiazepines are taken at therapeutic doses, the range of time that usually produces a discontinuance syndrome is from several weeks to several months. Taking benzodiazepines once or twice during a crisis, or even for

several weeks during a prolonged period of stress, ordinarily does not set the stage for discontinuance symptoms.

2. *Dose.* The amount of drug taken on a daily or nightly basis is also a critical factor. When higher-than-therapeutic doses are taken—for example, for treatment of panic disorder—then the period required before a discontinuance syndrome may develop is shortened.

3. *Abrupt discontinuance of the benzodiazepine.* Discontinuance symptoms arise because the level of drug at the CNS receptor sites is suddenly diminished. Since drug level in the CNS is proportional to the amount circulating throughout the body, an abrupt decline in CNS drug levels occurs when the blood level abruptly drops. Gradual tapering of benzodiazepines usually prevents the appearance or reduces the intensity of discontinuance symptoms.

4. *Type of benzodiazepine.* Benzodiazepines are classified into short and long half-life compounds. These terms refer to the time it takes for liver metabolism to remove (clear) benzodiazepines from the body. Short half-life benzodiazepines are cleared very rapidly, usually from 4 to about 16 hours, depending on the drug. In contrast, long half-life benzodiazepines may take anywhere from 24 to 100 or more hours to be cleared. Since the appearance of discontinuance symptoms depends, in part, on the rapidly diminishing blood level of the drug, abrupt cessation of the short half-life benzodiazepines is more likely to produce discontinuance symptoms. Controversy exists about whether other factors that distinguish one benzodiazepine from another are associated with the appearance of a discontinuance syndrome.

MANIFESTATIONS

Virtually all who experience discontinuance symptoms from benzodiazepines describe increased anxiety, restlessness, and difficulty falling asleep. These symptoms may be mild, little more than an annoyance for a few days, or they may be quite severe and even more intense than the symptoms of anxiety or insomnia for which the drugs were initially prescribed. The reappearance of the initial symptom, such as anxiety or insomnia, only in greater severity, is known as the *rebound symptom*. Rebound symptoms usually occur within hours to days of benzodi-

azepine discontinuance and then gradually fade. In some cases, however, they may be so intense that the patient resumes taking the benzodiazepine to avoid the discontinuance symptoms themselves. Thus a cycle of benzodiazepine dependence may begin— the patient is taking the drug primarily to treat or prevent rebound discontinuance symptoms from appearing, rather than treating an underlying anxiety or sleep disorder.

Benzodiazepines that are given to induce sleep may also be associated with the development of discontinuance symptoms. Rebound insomnia, the most common discontinuance symptom, typically occurs on the first night and sometimes the second night after discontinuance of short half-life benzodiazepines. Rebound insomnia may be so intense during these nights that the patient may be unwilling to risk another sleepless night and so returns to taking the benzodiazepine hypnotic. Rebound insomnia is less common with long half-life benzodiazepines.

If untreated, rebound symptoms may sometimes persist for many months. When this occurs it is difficult to determine whether the symptoms are still manifestations of discontinuance or are the result of the return of the problems (anxiety, insomnia) for which the drug was originally prescribed. Sometimes new symptoms that did not exist before the patient took benzodiazepine appear after discontinuance; these are termed true withdrawal symptoms, indicating a change in CNS functioning. Usual withdrawal symptoms include headache, anxiety, insomnia, restlessness, depression, irritability, nausea, loss of appetite, gastrointestinal upset, and unsteadiness. Patients may also experience increased sensitivity for sound and smell, difficulty concentrating, and a sense that events are unreal (depersonalization). Unusual withdrawal symptoms include psychosis and seizures.

OCCURRENCE OF SEIZURES

From a medical perspective, the most serious of all discontinuation symptoms is the development of withdrawal seizures. Seizures are generally grand mal in type (tonic-clonic; epileptic) and may threaten the life of the patient. They tend to occur only when higher-than-therapeutic doses are abruptly discontinued.

Withdrawal seizures almost always occur when the patient has been taking other drugs, such as AN-

TIDEPRESSANTS or ANTIPSYCHOTIC agents, together with a benzodiazepine.

COEXISTING PSYCHOPATHOLOGY

Apparently some people are more predisposed to develop the discontinuation syndrome than others. Those who have been previously dependent on benzodiazepines, alcohol, or other SEDATIVE-HYPNOTIC drugs, such as barbiturates, are more likely to experience discontinuance symptoms after the termination of benzodiazepine therapy. It is especially important, therefore, that such patients never stop taking their benzodiazepines abruptly.

TREATMENT

Although a variety of treatments have been proposed for the discontinuance syndrome, the best approach is to prevent its occurrence. Logically, prevention consists of a very gradual tapering of the benzodiazepine dose, with a firm rule never to discontinue these medications abruptly if they have been taken for more than a few weeks on a regular basis.

Even with gradual tapering, however, some patients may continue to experience rebound or withdrawal symptoms that are sufficiently disturbing to require treatment. Drugs that tend to reduce CNS hyperarousal states, such as anticonvulsants, have sometimes been employed to treat benzodiazepine discontinuance. Alternatively, benzodiazepine treatment is restarted using a long half-life compound that is then very gradually tapered.

CONCLUSION

For the great majority of patients, benzodiazepine discontinuance is a relatively benign and short-lived syndrome; many, if not most, patients have no difficulty. It is generally agreed that the therapeutic benefits of taking benzodiazepines far outweigh any problems with discontinuance when drug treatment is no longer necessary.

BIBLIOGRAPHY

RICKELS, K., ET AL. (1993). Maintenance drug treatment for panic disorders. *Archives of General Psychiatry*, 50, 61.

SALZMAN, C. (1991). The APA Benzodiazepine Task Force Report on dependency, toxicity, and abuse. *American Journal of Psychiatry*, 148, 151–152.

SALZMAN, C., ET AL. (1990). *American Psychiatric Association Task Force on benzodiazepine dependency, toxicity, and abuse.* Washington, DC: American Psychiatric Press.

CARL SALZMAN

Cocaine Withdrawal from cocaine was mentioned by H. W. Maier in his 1928 classic *Der Kokainismus* (Cocaine Addiction), but systematic efforts to describe and understand cocaine withdrawal did not begin until the 1980s, during the most recent epidemic.

The features of withdrawal from depressant drugs such as ALCOHOL and OPIOIDS are more robust and recognizable than from a stimulant drug such as cocaine—since the grossly observable pattern of physiologic disturbances seen in depressant withdrawal syndromes are not observed when a person stops using cocaine. This difference highlights and contrasts depressant withdrawal and stimulant withdrawal, such as is seen with cocaine.

In alcohol withdrawal, for example, the drinker may manifest all or several of the following set of symptoms and signs: tremulousness, elevated pulse and blood pressure, sweatiness, nervousness, and (rarely) seizure. Craving, or desire, for alcohol is typically high during this period, since the drinker knows it will quickly relive the withdrawal symptoms. These symptoms and signs will generally resolve within three to ten days of ceasing the intake of alcohol. Finally, the withdrawal syndrome is reproduceable—individuals tend to experience the same symptoms every time they withdraw from alcohol. Withdrawal from OPIATES such as HEROIN and MORPHINE similarly involves physiologic symptoms and signs—diarrhea, gooseflesh, changes in pulse and blood pressure, muscle cramps, stomach cramps, and anxiety.

In the cocaine abuser, the absence of early apparent physiologic symptoms and signs of cocaine withdrawal led to a widely held misperception (among the public and medical professions alike)—that cocaine was not an addicting drug. This misperception was based in part on cocaine's lack of a withdrawal syndrome that was as easy to characterize as those associated with alcohol or opioids.

If cocaine withdrawal does not evidence physiologic symptoms and signs, then how can it be recognized? The concept has been advanced that cocaine withdrawal is mediated through the central nervous system, that observable symptoms are limited to subjective states such as depression, lack of energy, agitation, and craving for cocaine. Evidence that neurophysiologic dysfunction may underlie reported symptoms consists of electroencephalogram (EEG) changes, neurohormonal dysregulation, and dopamine-receptor alteration (Satel et al., 1993).

In 1986, Gawin and Kleber were among the first to describe the clinical course of the symptoms following cocaine cessation, and they proposed a three-phase model of cocaine abstinence. Although this triphasic model has gained wide acceptance, other recent data suggest the model may not be applicable in all clinical situations, as will be discussed below.

The triphasic model postulated by F. Gawin and H. Kleber on the basis of interviews with outpatients comprises three phases that occur after cocaine cessation: (1) crash, (2) withdrawal, and (3) extinction. The *crash* is described as an extreme state of exhaustion that follows a sustained period of cocaine use (binge); it can last between nine hours and four days. The beginning of the crash is marked by craving, irritability, dysphoria, and agitation; the middle is characterized by yearning for sleep; and the late crash by hypersommnolence (excessive sleep). Certain individuals may experience especially severe depressed mood in the early stages of cocaine abstinence and are at risk for suicidal ideation and action at this time. This may be particularly true for those who are struggling with ongoing problems with depression. When alcohol is used with cocaine, depressed mood can intensify. Also alcohol-induced reduction of impulse control, combined with cocaine crash-related despair, creates a high-risk situation for suicide.

As depression and desire for sleep increase, craving subsides. Upon awakening from a lengthy sleep, the individual enters a brief euthymic (normal) period with mild craving. This is followed by a protracted period of milder *withdrawal*, lasting 1 to 10 weeks, during which time craving reemerges and anhedonia (loss of pleasure) prevails. This is succeeded by an indefinite period of *extinction*, marked by euthymic mood and episodic craving.

According to the triphasic model, protracted withdrawal is represented by phase 3, thus beginning after two weeks or more. These clinical phenomena are believed to reflect disturbances in central catecholamine (neurotransmitter) function produced by long-term cocaine use. The crash phase, however, can occur even in first-time stimulant users—if their initial episode is of sufficient duration and dose.

Recently, two groups of investigators have observed a mild constellation of subjective features of the post-crash cocaine abstinence syndrome as described by Gawin and Kleber, but without the phases those investigators described. Weddington et al. (1990) documented the absence of cyclic or phasic changes in mood states, cocaine craving, or interrupted sleep in twelve cocaine-dependent inpatients examined during a four-week period. All had abstained from continuous cocaine use within the preceding forty-eight hours. No euthymic window was evident, although subjects reported significantly greater depressed mood than nondrug-using controls at admission. Subjective symptoms of mood, craving, and anxiety displayed a steady and gradual improvement during the course of the study. By the end of week 4, the cocaine users and the nondrug-using controls had comparable scores. Thus, withdrawal had been completed over the course of one month.

Similar subjective findings emerged from a study by Satel and coworkers (1991), in which 22 newly abstinent COCAINE-dependent males were observed during a 21-day hospitalization. Over the 21 days, both subjective and objective ratings of mood and arousal showed gradual improvement. Although all subjects had consumed cocaine within twenty-four hours of admission, some claimed that they had slept prior to admission and thus the crash phase may have been missed in both studies.

The major differences between the triphasic model and the reports made by the two groups of investigators who actually observed cocaine users during withdrawal reside in the *euthymic interval*, the severity of symptoms, and the time-to-recovery of mood and craving. Nevertheless, all three studies are consistent with at least a mild postcessation syndrome. It may be important that the original conceptualization of the triphasic cocaine withdrawal was derived from observations of outpatients. The subsequent studies involved inpatients, who were largely protected from environmental cues.

Divergent findings with respect to a delineation between acute and protracted withdrawal is related to the difficulty in distinguishing acute cocaine withdrawal symptoms from those that characterize protracted withdrawal. (This distinction is less blurred in alcohol and opiate withdrawal, where the intense physiologic symptoms take place within the first week of ceasing usage—and the protracted syndromes, though uncomfortable, are considerably milder.) Conditioned withdrawal symptoms have been documented in opiate users and in alcoholics. These represent actual physiologic correlates of pharmacologic withdrawal (e.g., changes in skin temperature, gooseflesh, diarrhea, and cramps, accompanied by intense craving for the drug) elicited in *drug-free* individuals after they complete acute withdrawal and are exposed to reminders of drug use (e.g., visual or olfactory cues).

Conceivably, Gawin and Kleber's subjects may have experienced a delineated withdrawal, with a clear transition to a protracted state—because as outpatients they were constantly exposed to environmental cues and reminders of drug use. In inpatients, symptoms of acute cocaine withdrawal may be less clearly delineated. Constant exposure to cues may intensify a clinically observable acute syndrome, making the acute-protracted distinction easier to recognize. Environmental influences on clinical withdrawal may determine, in part, the severity of the observable manifestations of changes in neuroreceptors and neurotransmitters that accompany chronic cocaine use. Clearly, the behavioral and subjective manifestations are variable.

In addition, it is possible that nonorganic factors play a role in the prolonged psychic distress following termination of the chronic use of cocaine. Indeed, the period of abstinence following heavy drug use is a time when addicts must squarely face the shambles of their lives—the destruction of their families, loss of jobs, financial ruin, insults to health and self-esteem. Cocaine craving during this period is likely triggered by negative mood states as well as a conscious desire to obliterate the psychological pain with more drug—a return to drug use.

Pharmacologic treatment for the crash phase of withdrawal has received attention, although most treatment centers do not use medicines to help detoxify crashing cocaine addicts. The two major drugs that have been reported useful during the crash phase are bromocriptine and AMANTADINE. The action of these two drugs is to enhance transmission of the NEUROTRANSMITTER dopamine. Indeed, drugs that have this action were specifically chosen by investigators for use in treatment trials, because they assumed such drugs would reverse the reduction in dopamine levels in the brain that normally follows

cocaine binging. This reduction is presumed to account for the depression, irritability, agitation, and drug craving during the crash phase.

Pharmacotherapy for detoxifying cocaine addicts becomes especially important when a person is also dependent on alcohol or opioids. Such codependent states are very common. The usual choice for alcohol detoxification is a BENZODIAZEPINE drug (e.g., Librium); for opiate withdrawal, a choice exists for METHADONE, CLONIDINE, NALTREXONE, or combinations of these. Important interactions occur between cocaine and other drugs of abuse. For example, cocaine plus alcohol in the body produces a compound called COCAETHYLENE. This compound produces more intense and longer euphoria—but it also heightens the risk of death, due to cardiac arrythmia. Also, in methadone clinics, cocaine use has been noted to be of epidemic proportion; the opiate methadone mediates the jitteriness and paranoia that often accompanies cocaine use. Some evidence shows that cocaine addicts, who are also dependent on opiates, may have less severe opiate withdrawal than those who do not use cocaine.

Cocaine CRAVING is the major cause of relapse in individuals trying to attain and sustain abstinence. Such craving is typically most severe in the early stages of withdrawal from cocaine, although, as Gawin and Kleber noted in their model, cocaine addicts are extremely cue-responsive; reminders of drug use in the community (old copping areas, people with whom they used to get high, etc.) can stimulate craving at any stage of abstinence. Thus, people with severe addiction trying to relinquish cocaine must often enter a rehabilitation program with an outpatient phase that lasts from one to two years, at minimum.

Ideally, a heavy cocaine user with good social support and resources could enter an inpatient program to undergo detoxification (when sustained craving is usually at its peak) for a minimum of one week, before beginning outpatient work. Individuals without social support or a stable living situation can often benefit from weeks to months in a residential-treatment setting. Since it appears that the immediate postcessation phase may be milder for inpatients, this might be a way for addicts to experience less distress and to better concentrate on therapy and education. It might also be a period of time when they feel a somewhat greater sense of control over themselves—control being especially difficult to achieve when craving for cocaine is high. It is critical to realize, however, that many patients can develop a false sense of control over the addiction because as inpatients they are protected from environmental cues that trigger craving. Thus gradual reintroduction to the ambulatory environment, psychological preparation of the patient for the likely return of craving, and therapy using relapse-prevention techniques (a form of cognitive therapy) are all necessary.

(SEE ALSO: *Amphetamine; Cocaine, Treatment Strategies*)

BIBLIOGRAPHY

GAWIN, F. H. (1991). Cocaine addiction: psychology and neuropsychology. *Science* (March 29), 1580–1585.

GAWIN, F. H., & KLEBER, H. D. (1986). Abstinence symptomatology and psychiatric diagnosis in chronic cocaine abusers. *Archives of General Psychiatry, 43*, 107–113.

MAIER, H. W. (1928/1987) *Der kokainismus* (Cocaine Addiction), O. J. Kalant (Trans.). Toronto: Addiction Research Foundation.

SATEL, S. L., ET AL. (1993). Should protracted withdrawal from drugs be included in DSM-IV? *American Journal of Psychiatry, 150*, 695–701.

SATEL, S. L., ET AL. (1991). Clinical phenomenology and neurobiology of cocaine abstinence: A prospective inpatient study. *American Journal of Psychiatry, 148*, 1712–1716.

WEDDINGTON, W. W., ET AL. (1990). Changes in mood, craving and sleep during short-term abstinence reported by male cocaine addicts: A controlled residential study. *Archives of General Psychiatry, 47*, 861–868.

SALLY L. SATEL
THOMAS R. KOSTEN

Nicotine (Tobacco) Many smokers find they have a difficult time getting through long airline flights with current smoking restrictions. They may experience NICOTINE withdrawal including difficulty thinking and concentrating. For the pilots, such cognitive impairment might have devastating consequences. Therefore, the cockpit is the one area where smoking restrictions are not presently mandated by the federal government. The impairments produced by nicotine withdrawal are of special concern for nicotine dependent people in occupations requiring quick and accurate decision making (e.g., pilots, air traffic controllers, and some military personnel), but they are also a major obstacle to any of the 15 to 20 million smokers who try to quit each year. This article will discuss the manifestations of nicotine withdrawal and current treatment approaches to alleviate withdrawal.

No longer viewed as a behavior resulting from a weak will or defective personality, drug dependence is now understood to involve a collection of cognitive, behavioral, and physiological consequences resulting from exposure to certain types of chemicals. The user has impaired control over substance use and continues to use despite adverse effects. Although people can develop some degree of dependence on a wide range of chemical types, those which most commonly produce addiction share several features. First, their administration alters central nervous system function, generally at specific receptors, and often changes structure as well, as increases (up regulation) or decreases (down regulation) in receptor numbers may occur, depending on the chemical. Second, repeated exposure to the drug results in TOLERANCE, and the individual must progressively self-administer higher doses of the drug to obtain the same effects that initially occurred after lower doses. Third, as cellular and neurological functioning adapt to the continuous presence of the drug during tolerance development, a state of physical or physiological dependence is produced, so that removal of the drug is accompanied by feelings of dysphoria and an inability to function normally. The individual comes to need continued drug intake to function normally. As consistently acknowledged by the WORLD HEALTH ORGANIZATION EXPERT COMMITTEES since 1964, tolerance and physiological dependence are neither necessary nor sufficient to produce a pattern of compulsive drug seeking; however, these processes can contribute to the severity of dependence and complicate treatment. Dependence-producing drugs alter mood, cognition, affect, and behavior and thus are considered psychoactive. Finally, a hallmark of dependence producing drugs is that they serve as biological reinforcers for animals, including humans. Some researchers believe that demonstrating that a drug can be discriminated, is psychoactive, and is effective as a reinforcer, is tantamount to demonstrating that exposing a person or animal to the drug under certain circumstances will lead to the development of compulsive drug-seeking behavior.

NICOTINE TOLERANCE
AND DEPENDENCE

Nicotine, the addictive chemical in tobacco, shares the features listed above with other drugs of abuse. Nicotine's ability to cause physical dependence and withdrawal is less overt than some other drugs of abuse (e.g., OPIATES and COCAINE); nicotine's physical dependence had been described starting in the 1930s, and evidence mounted into the 1960s, but not until the 1970s and 1980s did more controlled studies convince the majority of scientists. Although more subtle, nicotine withdrawal and dependence are no less clinically important in causing relapse than are withdrawal and dependence from other abused drugs. The development of tolerance as a result of repeated nicotine exposure is a crucial factor in the development of lung and other cancers from cigarette smoking. Essentially, this process allows people to self-administer much greater amounts of tobacco-delivered toxins than had they not become tolerant, and this has a profound effect on their health. As a whole, cancer death rates for smokers are twice as high as for nonsmokers, and heavier smokers have rates four times as high as nonsmokers. In the United States, 30 percent of all cancer deaths are directly attributable to tobacco use: These are roughly 85 percent of all lung cancer deaths; 80 percent of larynx, pharynx, oral, and lip cancer deaths; 75 percent of esophageal cancer deaths; 45 percent of urinary cancer deaths; and 20 percent of stomach cancer deaths.

Several forms of tolerance occur, and the time course of tolerance development and decay can be complex. For example, after rapidly smoking a cigarette, even a regular smoker deprived of cigarettes for several days may experience nausea and dizziness. This indicates some loss of tolerance. It would be extremely unusual, however, for such a person to experience the degree of symptoms observed in first-time smokers. One's level of nicotine tolerance will change throughout the course of a day. Some degree of tolerance is lost overnight, and smokers report that the first cigarettes of the day are the strongest and/or the best. As the day progresses, acute tolerance develops and later cigarettes produce less effect. Physiological measures, such as heart rate, also follow this pattern. The first few cigarettes of the day produce a large increase in heart rate that lasts for only several minutes after the cigarette. Later cigarettes do not have this effect, and the smoker's heart rate will follow the same diurnal pattern as the nonsmoker, with about a seven beat per minute elevation. These observations indicate that, as with other addictive drugs, the development of nicotine tolerance is only partial.

With development of PHYSICAL DEPENDENCE, the person comes to feel normal, comfortable, and most effective when taking nicotine and to feel dysphoric and ineffective when deprived of nicotine. In actuality, cigarette smokers in this seemingly normal state have elevated metabolism, heart rate, and sympathetic nervous system–activating hormones but lower bodyweights than nonsmokers. Postmortem studies of smokers have revealed increased numbers of nicotinic receptors and other brain changes caused by cigarette smoking. The process of dependence development weakens the ability of the person to achieve and sustain even short-term abstinence. Thus, in the nicotine-dependent person, "normal" function depends on nicotine, and removal of nicotine leads to impaired function.

Exposure to nicotine produces distinct effects on the physiological functioning of the heart, brain, hormonal system, and other organs. These effects are mediated by the nicotinic subtype of the ACETYLCHOLINE receptor, which also affects mood, behavior, and cognitive functioning. After a period of daily exposure to nicotine (generally assumed to be at least a few weeks), the body adapts to these effects of nicotine (i.e., tolerance develops), and the effects of nicotine exposure become less pronounced. Many people continue to increase their smoking rates for more than eight years. First exposures to cigarettes or smokeless tobacco may produce sickness, intoxication, and severe disruptions in physiological functioning and behavior (i.e., "turning green"). As exposure continues and tolerance develops, there is a conditioning of self-administration behavior and physical dependence occurs. Although there are differences between people in the degree of dependence produced by similar rates of smoking, dependence is related both to the dose and the pattern of nicotine delivery (nicotine alone, without tobacco, can produce tolerance and dependence). Perhaps more important from a clinical perspective, abrupt elimination or reduction in nicotine intake will produce physiological and behavioral disruption, a withdrawal syndrome. In many respects, the withdrawal responses are opposite to those of nicotine administration; hence they are sometimes referred to as rebound responses.

NICOTINE WITHDRAWAL

Reductions in nicotine intake of 50 to 60 percent will precipitate withdrawal symptoms, which include CRAVING for nicotine, ANXIETY, irritability, difficulty controlling ones' emotions and dealing with stress, increased appetite, impaired cognition, difficulty concentrating, and decreased heart rate. The severity of symptoms depends on the time since the last nicotine dose, the level of nicotine intake before quitting, and the severity of nicotine dependence. Mild symptoms may in fact occur after switching to a low tar and nicotine brand of cigarettes. There is, however, considerable variability across individuals. The onset of the withdrawal syndrome is within a few hours after the last tobacco use. Symptoms peak within the first few days and begin to subside within a few weeks. For some people, symptoms may persist for months or more. The cognitive deficits experienced by many smokers who quit are correlated with disruptions in brain electrophysiologic function. Figure 1 shows that deficits on an arithmetic task follow a similar time course as changes in the brain's electrical activity. These effects begin within a few hours after the last smoke (dose of nicotine), peak during the first few days of abstinence (when smokers trying to quit are most likely to relapse), and mostly subside in a few weeks.

Although the withdrawal syndrome varies among individuals, some symptoms are universal (e.g., cognitive impairment). Other symptoms may occur in a few individuals and may involve the emergence of psychological traits suppressed, controlled, or altered by tobacco use. For example, those with histories of major DEPRESSION may experience a bout of severe depression. Most symptoms of nicotine withdrawal do not return to pre-abstinence levels for two to four weeks. A clinically important but poorly understood component of the nicotine withdrawal syndrome is craving, or urges, for nicotine. This has been described as a major obstacle to successful abstinence. It may be positively ("pleasure seeking") or negatively ("avoiding withdrawal") motivated. Increased craving for tobacco may be very persistent, lasting for six months or longer. Rate of relapse is related to the degree of dependence, with most people resuming smoking within about one week of quitting.

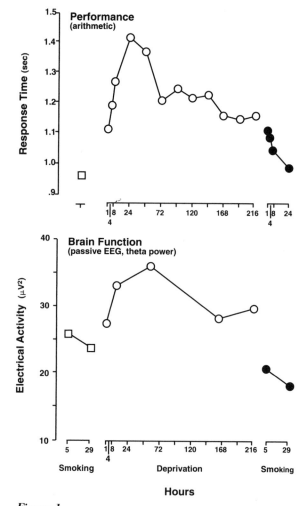

Figure 1

Cognitive Performance and an Electrophysiological Measure of Brain Function during Smoking and Abstinence.

The tobacco withdrawal syndrome is pharmacologically based on nicotine; however, behavioral-conditioning factors do have an effect on the degree of dependence and the severity of withdrawal symptoms. For example, cigarette smoking becomes a deeply ingrained habit as the smoker engages in a great deal of repetitive behavior to obtain daily nicotine intake. Because environmental stimuli, tobacco smoking behavior, and nicotine reinforcement occur together dozens of times per day and hundreds of thousands of times per year in the person smoking only a pack of cigarettes each day, these elements become powerfully associated. The feel, smell, and sight of a cigarette, cigarette advertisements, stimuli usually present during smoking (such as friends who smoke, the ringing of the telephone, an alcoholic beverage, or a cup of coffee), and the ritual of obtaining, handling, lighting, and smoking the cigarette become cues which signal smoking and may be highly pleasurable to the smoker in their own right. Being in the presence of such cues may increase chances the abstinent smoker will relapse. In some people, environmental stimuli may not only signal smoking but also elicit powerful cravings for smoking that occur throughout the day, even for the person who has been smoking regularly.

TREATMENT OF NICOTINE WITHDRAWAL SYMPTOMS

Two major tacks may be taken to relieve the nicotine withdrawal syndrome—nicotine replacement or symptomatic treatment. *Nicotine replacement* is designed to substitute a safer, more controllable, and less addictive form of nicotine to relieve withdrawal and aid in quitting. Currently, NICOTINE GUM and transdermal patches are the only approved forms of nicotine replacement in the U.S. Other forms under investigation include a nasal vapor inhaler, nasal nicotine spray (gel droplets), and smoke-free nicotine cigarettes. The forms differ by providing various doses and speeds of dosing. These parameters may be important to provide the smoker with levels of nicotine necessary to alleviate withdrawal and cravings.

Symptomatic treatments are drug therapies designed to relieve the discomforts and mood changes associated with withdrawal. Various types of symptomatic treatment—including sedatives, tranquilizers, anticholinergics, sympathomimetics, and anticonvulsants—have all been tried at one time, generally with limited success. Two agents that may hold promise in relieving withdrawal discomfort are the antihypertension drug clonidine and the benzodiazepine tranquilizer alprazolam. Both CLONIDINE and alprazolam have been shown to reduce anxiety, irritability, restlessness, and tension, but only clonidine relieves craving for cigarettes (and only in women). Before recommending clonidine to smokers, potential side effects must be considered. Abruptly stopping clonidine treatment has sometimes led to severe hypertension and, in very rare circumstances, to death. More commonly, it may cause drowsiness, which is potentially dangerous to someone operating machinery or driving. More study is necessary on whether these drugs are effective in helping people to quit smoking.

Cigarettes are among the most addicting substances known. In fact, some people who abuse multiple drugs report that to quit smoking is more difficult than giving up alcohol or cocaine. Nicotine withdrawal symptoms, some lasting several months, contribute to the difficulty. The pharmacological tools to treat withdrawal are continually improving, which may make smoking cessation a possible, if not easy, task.

(SEE ALSO: *Nicotine Delivery Systems for Smoking Cessation; Tobacco: Dependence; Tobacco: Smoking Cessation and Weight Gain*)

BIBLIOGRAPHY

ACKWORTH, W. B., HERNING, R. I., & HENNINGFIELD, J. E. (1989). Spontaneous EEG changes during tobacco abstinence and nicotine substitution in human volunteers. *Journal of Pharmacology and Experimental Therapeutics, 251,* 976–982.

AMERICAN PSYCHIATRIC ASSOCIATION. (1987). *Diagnostic and statistical manual of mental disorders-3rd edition-revised.* Washington, DC: Author.

BALFOUR, D. J. (1991). The influence of stress on psychopharmacological responses to nicotine. *British Journal of Addiction, 86,* 489–493.

GLASSMAN, A. H., ET AL. (1984). Cigarette craving, smoking withdrawal, and clonidine. *Science, 226,* 864–866.

GRITZ, E. R., FIORE, M. C., & HENNINGFIELD, J. E. (1993). Smoking and cancer. In *Textbook of clinical oncology.* New York: American Cancer Society.

HENNINGFIELD, J. E., COHEN C., & SLADE, J. D. (1991). Is nicotine more addictive than cocaine? *British Journal of Addiction, 86,* 565–569.

SNYDER, F. R., DAVIS, F. C., & HENNINGFIELD, J. E. (1989). The tobacco withdrawal syndrome: Performance decrements assessed on a computerized test battery. *Drug and Alcohol Dependence, 23,* 259–266.

U.S. DEPARTMENT OF HEALTH AND HUMAN SERVICES. (1990). *The health benefits of smoking cessation. A report of the surgeon general.* DHHS Publication No. (CDC) 90-8416. Washington, DC: U.S. Government Printing Office.

U.S. DEPARTMENT OF HEALTH AND HUMAN SERVICES. (1989). *Reducing the health consequences of smoking: 25 years of progress. A report of the surgeon general.* DHHS Publication No. (CDC) 89-8411 (1989). Washington, DC: U.S. Government Printing Office.

U.S. DEPARTMENT OF HEALTH AND HUMAN SERVICES. (1988). *The health consequences of smoking: Nicotine addiction: A report of the surgeon general.* DHHS Publication No. (CDC) 88-8406. Washington, DC: U.S. Government Printing Office.

JACK E. HENNINGFIELD
LESLIE M. SCHUH

Nonabused Drugs Although drug withdrawal is often considered synonymous with matters relating to drug abuse, a number of drugs which have no abuse potential and are prescribed for medical illness are associated with clear symptoms of withdrawal when their use is abruptly discontinued. It is useful to compare these withdrawal states to those which occur with a prototype drug of abuse—ALCOHOL—which is also one of the most common drugs of abuse.

CARDIOVASCULAR DRUGS

Beta-Adrenergic Blockers (Prototype Propranolol). These drugs are taken by many people to treat hypertension (high blood pressure), angina pectoris (chest pain from heart muscle deprived of oxygen), heart arrhythmias following heart attack, and for migraine headache. The mechanism for each of these effects is related to the drug occupying the beta-adrenergic receptors in the blood vessels and the heart. When a patient abruptly stops taking propranolol or another beta blocker, particularly when angina pectoris is the symptom being treated, a marked increase in the frequency and/or severity of angina pectoris may occur. This occurs within the first few days of discontinuing the beta blocker; it may be prevented by slowing decreasing the drug dose over several days before completely stopping the drug. This is probably related to an increased sensitivity of the beta receptor for the body's own hormones NOREPINEPHRINE and epinephrine, when its antagonist, the beta blocker, is suddenly removed. The withdrawal syndrome disappears in a few days, consistent with the time required for beta-adrenergic receptor reregulation.

Clonidine. This drug is used for hypertension and to treat withdrawal from opiate narcotics. Its mechanism of effect is stimulation of alpha(type 2)-adrenergic receptors in the central nervous system, which results in decreased stimulation of nerves that release norepinephrine and epinephrine in blood vessels. When CLONIDINE is abruptly stopped, blood pressure increases to well above baseline levels and may become dangerously high. This occurs within one to two days after stopping the drug and is prevented by slowly (over several days) decreasing the drug dose before stopping it completely. This may be due to a "rebound" overstimulation of norepinephrine and epinephrine releasing nerves in blood vessels. This rebound hypertension disappears within a few days, again consistent with the time required for alpha-adrenergic receptor reregulation.

Nitroglycerin and Other Nitrates. These drugs are taken to treat angina pectoris. They cause the relaxation of blood vessels by the activation of an intracellular enzyme, guanylyl cyclase, which catalyzes formation of cyclic GMP (guanosine monophosphate). The coronary arteries (blood vessels which supply heart muscle) relax when exposed to nitrates. If the coronary arteries are blocked by atherosclerosis, causing insufficient blood supply to the heart, angina pectoris can occur. Relaxation of these arteries improves blood supply to the heart and the chest pain rapidly disappears. When nitrates are taken continuously for relief of chest pain, then abruptly discontinued, rebound angina pectoris which is more frequent or more severe than the angina experienced pretreatment may occur. This begins within a few hours of the last nitrate dose and in a time course consistent with the metabolism and removal of the nitrate drug from the body. If the nitrate dose is slowly decreased before discontinuation, the rebound angina may be prevented. The mechanism for this withdrawal syndrome is not certain, however, it is probably related to loss of the chronic activation of guanyl cyclase during nitrate therapy and abnormal regulation of the contractile apparatus in the blood vessel muscle, leading it to have rebound contraction.

NEUROPSYCHOPHARMACOLOGICAL DRUGS

Heterocyclic Antidepressants. These drugs are used to treat major depressive illness. Therefore they are frequently administered daily for periods of weeks or months. Abrupt discontinuation of ANTIDEPRESSANTS may result in symptoms referable to the digestive tract; they include nausea, vomiting, abdominal pain, and diarrhea. In addition some patients complain of a flulike illness consisting of weakness, chills, fatigue, and muscle aches. Central nervous system dysfunction characterized by difficulty falling asleep, anxiety, or jitteriness can also occur. The mechanism of withdrawal may result from up-regulation and increased sensitivity of the muscarinic receptor, which is blocked by these drugs. During chronic heterocyclic-antidepressant treatment, muscarinic-receptor sensitivity increases. When receptor blockade is suddenly stopped, over-activity of these receptors in the digestive tract and brain causes the withdrawal symptoms. The newer antidepressant agents which have little, if any, blocking effect on muscarinic receptors do not appear to be associated with a withdrawal syndrome.

Major Tranquilizers. NEUROLEPTIC agents are commonly used in psychiatric practice for the treatment of psychotic disorders such as schizophrenia. These agents all block brain dopaminergic receptors—the basis for their effectiveness in treating psychotic illness. These agents also inhibit emesis (vomiting), which is caused by dopaminergic blockade in the brain as it affects the perception and initiation of vomiting. Chronic blockade results in increased numbers of these receptors. The abrupt discontinuation of this class of drugs, then, results in the overactivity of the dopaminergic system—causing nausea, vomiting, and headaches.

Monoamine Oxidase Inhibitor (MAOI) Antidepressants. These drugs interfere with the enzymatic breakdown of NEUROTRANSMITTERS (such as norepinephrine) in the brain. Sudden discontinuation after high chronic dosing has been associated with psychosis and delirium—consisting of visual hallucinations as well as mental confusion. Milder symptoms consisting of anxiety, vivid dreaming, or nightmares may also occur. The exact mechanism of withdrawal has not been well studied, but it may relate to the way nerve cells regulate the release of neurotransmitters in the brain. Presynaptic receptors serve to provide a message to nerve cells about how much neurotransmitter is present in the synapse—the space between two nerve cells where messages, in the form of neurotransmitters, flow between cells. When activated, these types of receptors—present on the surface of the nerve cell releasing the message—inhibit any further release of neurotransmitter. As a result of treatment with MAOI, decreases in the number of presynaptic receptors occur, resulting in larger amounts of neurotransmitter being released before the cell shuts down release. The increase in the amount of neurotransmitter may result in with-

drawal symptoms that abate over a period of days after discontinuation.

OTHER DRUGS

Baclofen. As a muscle relaxant, this drug is used to treat muscle spasticity associated with certain paralytic states. It acts as an agonist (mimic) of the inhibitory neurotransmitter in the spinal cord, called GAMMA-AMINOBUTYRIC ACID (GABA). Therefore baclofen inhibits excitatory neural pathways, which are modulated by GABA and which ultimately stimulate skeletal muscles to contract. This is a rather selective effect as there are two types of GABA receptors and pathways, GABA-A and GABA-B, of which baclofen only acts on GABA-B receptors. When baclofen is used to treat muscle spasm, the excitatory pathways of the spine are chronically modulated and inhibited. When baclofen is abruptly discontinued, this inhibition is released and, within a few hours as is consistent with the rate of disappearance of baclofen, the excitatory pathways rebound—probably due to a transient unregulated state. The symptoms experienced by a person suddenly discontinuing baclofen may include auditory and visual hallucinations, severe anxiety, increased heart rate and blood pressure, and generalized seizures. Such clinical symptoms are consistent with the impaired modulation of neural-excitatory pathways. When baclofen dosage is gradually reduced before discontinuation, these symptoms either do not occur or are attenuated, indicating that the inhibitory/excitatory-neural-pathway balance, which has been disturbed by the excessive inhibitory modulus of baclofen, has the capacity to reregulate over a few days.

Corticosteroids. The prototype drug prednisone will be discussed; however, the biological changes that result in withdrawal phenomena after discontinuation of long-term prednisone treatment hold for all members of the glucocorticoid group. When, for example, a significant dose (5–10 mg daily) of prednisone is taken for a period of several weeks, a series of feedback regulatory events occurs resulting in the patient becoming functionally adrenally insufficient. Specifically, in mimicking the endogenous corticosterone cortisol, prednisone signals the pituitary gland to stop the synthesis and release of the adrenocorticotrophic hormone (ACTH) and, perhaps, the hypothalamus to stop the release of the corticotropin-releasing hormone (CRH). ACTH release from the pituitary, which normally stimulates the adrenal glands to produce corticosterones and which is modulated by the hypothalamic CRH, is blocked by the drug prednisone when ingested in the above dose or greater. Not only does adrenal production of cortisol decrease but also the adrenal glands atrophy. When prednisone therapy is abruptly discontinued, the atrophic adrenal glands no longer respond to ACTH stimulation, so the patient has symptoms of adrenal insufficiency. Clinically, this is manifested by fatigue, weakness, electrolyte imbalance, and the lack of many bodily responses to stress. If an individual remains in this state for more than a few hours, severe illness and death can be expected. When the adrenal glands become atrophic during long-term prednisone treatment, if the prednisone is to be discontinued, it must be done with slowly decreasing doses over many weeks to permit the adrenal glands sufficient time to regrow to their normal size under the influence of ACTH stimulation and to have sufficient stores of the body's own cortisol to respond to stress in a physiologically appropriate manner.

COMPARISONS WITH DRUGS OF ABUSE

If alcohol withdrawal is used as a basis for comparison, marked similarity in effect is noted when considering the cardiovascular drugs (beta-blockers, clonidine, nitrates) and baclofen. Alcohol, a nonspecific central nervous system depressant, leads to an ill-defined reregulated state, allowing habituated individuals some level of function during their chronic alcohol-induced depressive state. Abrupt cessation of alcohol consumption results in loss of the depressive state, with a rebound state of psychic and physical excitation. This is not unlike the cardiovascular drugs and baclofen; there, the withdrawal syndrome is the clinical manifestation of a neural- or cellular-regulatory system that has reached a new homeostatic state under the influence of the drug and the sudden drug removal leaves insufficient time for physiological reregulation. In the case of corticosteroids, the reverse of this mechanism occurs. Here the physiological regulation which has occurred during prednisone therapy leads to loss of the capacity to have a physiological response, instead of an overresponse.

Human physiology is characterized by the coordinated and finely tuned operation of multiple messaging systems, exhibiting both positive and negative feedback regulation, with multiple levels of control. All the drugs mentioned exert both their desired and undesired effects by interfering with these systems. In the drug-treated individual, homeostasis is maintained by counteracting some of the drug effects at the cellular level. Such adaptation is not without cost. The sudden discontinuation of a drug to which the system has adapted results in a period of disequilibrium between the affected messaging systems. The disturbed physiology is expressed by specific withdrawal symptoms.

The striking similarities for withdrawal syndromes of both abused and nonabused drugs may raise a question of why the individual selects one drug in preference to another. For each of the nonabused drugs, with the possible exception of the corticosteroids, initial drug use and chronic drug ingestion produces no positive psychic or euphoric effect. In several cases (beta-blockers, clonidine, nitrates, baclofen, antidepressants, major tranquilizers), many individuals experience negative psychic effects instead—such as lethargy, nightmares, impotence, headache, and dry mouth. The result is therefore negative rather than positive reinforcement for continued drug use.

(SEE ALSO: *Anabolic Steroids*; *Withdrawal: Alcohol, Beta Blockers*)

BIBLIOGRAPHY

BYYNY, R. L. (1976). Withdrawal from glucocorticoid therapy. *New England Journal of Medicine, 295,* 30–32.

CEDERBAUM, J. M., & SCHLEIFER, L. S. (1990). Drugs for Parkinson's disease, spasticity, and acute muscle spasms. In A. G. Goodman et al. (Eds.), *Goodman & Gilman's the pharmacological basis of therapeutics,* 8th ed. New York: Pergamon.

HOUSTON, M. C., & HODGE, R. (1988). Beta-adrenergic blocker withdrawal syndromes in hypertension and other cardiovascular diseases. *American Heart Journal, 116,* 515–523.

PARKER, M., & ATKINSON, J. (1982). Withdrawal syndromes following cessation of treatment with antihypertensive drugs. *General Pharmacology, 13,* 79–85.

SHATAN, C. (1966). Withdrawal symptoms after abrupt termination of imipramine. *Canadian Psychiatric Association Journal, 2,* 150–157.

DARRELL R. ABERNETHY
PAOLO DEPETRILLO

WOMEN AND SUBSTANCE ABUSE

There are gender differences in the prevalence of substance abuse.

ALCOHOL AND TOBACCO USE

General population studies indicate that fewer women drink than men, and women who do drink consume less alcohol than men. Of the estimated 15 million alcohol-abusing or alcohol-dependent individuals in the United States, fewer than one-third are women. In the 1993 NATIONAL HOUSEHOLD SURVEY on Drug Abuse (NHSDA), 57 percent of men reported they drank alcoholic beverages in the previous month, compared with 43 percent of women. The NHSDA defines heavy alcohol use as 5 or more drinks per day on each of 5 or more days in the past 30 days. By this definition, in 1993 men were much more likely than women to be heavy drinkers (10 and 2 percent, respectively).

It has been suggested that male and female sex roles, and therefore drinking norms, have become more similar in recent years. Some sex-role changes that could increase opportunities for, and acceptability of, female drinking include greater female labor force participation, delayed marriage and childbearing, and more equitable sex-role attitudes. According to this convergence thesis, greater sex-role equality may cause PROBLEM DRINKING and ALCOHOLISM to increase among women. However, recent epidemiological data reveal little evidence of increased female alcoholism or problem drinking. Changing female drinking patterns have resulted more in a reduction in female abstainers than an increase in problem drinkers. Nevertheless, there is some evidence for convergence in the youngest cohorts, with the smallest sex differences in heavy drinking being for youths aged twelve to seventeen (2 percent of boys and 1 percent of girls in 1993). Among adults aged thirty-five and older, men are eight times as likely as women to be heavy drinkers (8 percent compared with 1 percent).

There is greater evidence of sex-role convergence in TOBACCO use. In 1955, 52 percent of adult men smoked, compared with 25 percent of adult women. Since then, the proportion of men who smoke has decreased markedly while rates among women have held fairly steady. Among adults aged 35 and older in 1993, 27 percent of men and 21 percent of women were current smokers. Among youths aged twelve to seventeen, girls have surpassed boys in their rates of current cigarette use (10 percent of girls compared with 9 percent of boys in 1993). Because boys are more likely than girls to use smokeless tobacco products, however, their overall rates of nicotine addiction still exceed girls' rates.

Biener (1987) reviews factors that have contributed to the convergence in male and female smoking. Product developments such as filtered and low-tar cigarettes have made smoking easier for women to tolerate physically. Tobacco companies have targeted ADVERTISING to make smoking attractive to young women. Once tobacco use is initiated, women are less likely than men to quit smoking and, compared with men who have quit smoking, women quitters are more likely to relapse.

The convergence in male and female smoking rates has been accompanied by a convergence in smoking-related health problems. For example, lung cancer deaths among women have increased markedly since the 1970s, and lung cancer now surpasses breast cancer as the leading cause of CANCER deaths among women.

ILLICIT DRUG USE

Males are far more likely than females to be arrested for possessing or selling illicit drugs. In 1992, for example, the Federal Bureau of Investigation reported that only 16 percent of those arrested for drug-abuse violations were female. At all ages, males are more likely than females to use illicit drugs. Gender differences are smallest among adolescents aged twelve to seventeen and among adults aged thirty-five and older, and largest among young adults aged eighteen to thirty-four, the age range in which illicit-drug use is most prevalent. In the 1993 NHSDA, 11 percent of men, compared with 6 percent of women, aged twenty-six to thirty-four reported they had used some illicit drug in the previous month. Nineteen percent of men and 8 percent of women reported current (i.e., past month) illicit-drug use in 1993. Among both men and women, marijuana is the most frequently used illicit substance, with 16 percent of men and 6 percent of women aged eighteen to twenty-five reporting current use.

COCAINE use has decreased since the mid-1980s, and is rare compared with marijuana use. Sex differences in regular cocaine use are small. In the young adult age group, where use is most common, 1.7 percent of men and 1.4 percent of women reported cocaine use in the past month. In 1993, among youths aged twelve to seventeen, boys and girls were equally likely to report cocaine use in the past month (0.4 percent).

Prior to the HARRISON NARCOTICS ACT of 1914, the typical OPIATE addict in the United States was a white, middle-aged, middle-class housewife who had become addicted to medically prescribed drugs or nonprescription PATENT MEDICINES. Following criminalization of most opiate use through the Harrison Act and subsequent legislation and court interpretations, overall levels of opiate use declined dramatically. When HEROIN addiction reemerged as a social problem in the 1950s and 1960s, the typical opiate addict was a nonwhite urban male from a lower socioeconomic class. Although the VIETNAM war exposed a broader spectrum of young American men to heroin use, and although many servicemen tried opiates and even became addicted in Vietnam, most were able to discontinue use when they returned to the United States.

In the 1970s and 1980s, heroin use decreased and became quite rare in the United States. In 1993, only about one in 1,000 Americans aged twelve and older reported use of heroin in the past year, and the majority of users were men. An increase in drug seizures, arrests, and heroin-related emergency room episodes in the early 1990s led to assertions that heroin was making a comeback and that women would be especially vulnerable to addiction. Although these trends merited watching, such speculation was premature, given current evidence.

MEDICAL DRUG USE

In the 1970s feminist scholars drew attention to possible overmedication of women with PSYCHOACTIVE DRUGS. These early critiques derived from content analyses of sex-stereotyped advertisements in medical publications. Most of the ads depicted woman patients, and survey research on representative populations confirmed that women were using more prescription psychoactive drugs than were men.

Critics of these patterns are concerned that drugs are being used beyond traditional medical psychiatric concepts of disease. For example, medical ads suggested prescribing TRANQUILIZERS and ANTIDEPRESSANTS to alleviate normal life transitions, such as menopause, starting college, or a woman's adult children moving out. It has been suggested that prescribing psychoactive drugs is a subtle form of social control that diffuses or channels women's discontent with limiting and inequitable sex roles.

Some of the prescription psychoactives have dangerous side effects and a high potential for producing dependency. Further, since women also use more OVER-THE-COUNTER medications and women's alcohol problems are often undetected by physicians, use of prescription psychoactive drugs may make women especially vulnerable to adverse drug interactions. Alcohol in combination with other substances is the most frequent cause of emergency-room episodes in the DRUG ABUSE WARNING NETWORK (DAWN) system. Although women drink less and are less likely to use illicit drugs, they have equaled or exceeded men in drug-related emergency room episodes since the mid-1980s. This is because more women needed emergency treatment related to tranquilizer, sedative, and nonnarcotic analgesic use.

GENDER DIFFERENCES IN THE ETIOLOGY OF SUBSTANCE ABUSE

Studies of ADOLESCENTS generally find similar correlates of substance abuse among both boys and girls. The strongest predictor of adolescent alcohol, tobacco, and illicit-drug use is having friends who use alcohol, tobacco, and drugs. Other factors that predict substance abuse by boys and girls include parental substance abuse, poor academic performance, and low commitment to educational pursuits.

Researchers, however, have identified some gender differences in the development of alcohol and drug problems. Relationship issues are particularly salient in the etiology of female substance abuse. For example, alcoholism in women is more strongly correlated with a family history of drinking problems than is alcoholism in men. Girls and women are likely to be introduced to alcohol or illicit drugs by a boyfriend or spouse, and female alcohol or drug dependence frequently develops in a relationship with an alcohol- or drug-dependent male partner.

Alcohol and drug abuse are more often associated with DEPRESSION in girls and women compared with males, but it is not clear whether depression is more likely to cause female substance abuse or is a more typical consequence of substance abuse among girls and women. Women in treatment for substance abuse are more likely than men to say their problem drinking or drug abuse developed after a life crisis or tragedy, such as the death of a family member.

Also, a sizable proportion of women in treatment report histories of sexual abuse. Men are more likely to say their problem drinking or drug abuse developed out of social or recreational use.

Some believe these different attributions and recollections reflect genuine sex differences in the etiology of substance abuse. Others caution, however, that the greater stigma attached to female substance abuse may motivate women to develop an explanation for their problem drinking or drug use, and that personal crises and emotional difficulties serve as socially acceptable reasons.

The course of problem drinking and drug addiction varies by gender. Women entering treatment for alcoholism or drug abuse tend to have begun heavy drinking or drug use at a later age, on average, compared with men entering treatment. The term "telescoping" has been used to describe a more rapid progression from controlled alcohol or drug use to alcohol and drug dependency in women, compared with men.

GENDER DIFFERENCES IN THE CONSEQUENCES OF SUBSTANCE ABUSE

It is generally presumed that alcohol and drug abuse will produce more deleterious consequences among women than among men. This expectation is grounded both in biological differences and in social-role expectations.

From a biological standpoint, it is frequently noted that the lower ratio of water to total body weight in women causes them to metabolize alcohol and drugs differently than men. Even when body weight is controlled, given equivalent alcohol consumed, women pass more alcohol into the bloodstream and reach higher peak BLOOD ALCOHOL CONCENTRATIONS than men, in part because of differences in enzyme activity in the intestinal wall. Drugs such as marijuana that are deposited in body fat may be slower to clear in women than in men. Slow clearance rates create a potential for cumulative toxicity and adverse drug and alcohol interactions.

The behavioral telescoping of women's uncontrolled drinking and drug use is paralleled by a telescoping of some physical health consequences of alcohol and drug use. Alcoholic liver disease progresses more rapidly in women compared with men.

Women also seem to be more prone to alcohol-related brain damage. They show physical brain abnormalities after a shorter drinking history and at lower peak alcohol consumption. Women also exhibit cognitive deficits on psychological tests of memory, speech, and perceptual accuracy with a shorter drinking history than that of men.

Women diagnosed as alcoholic have very high mortality rates relative to both the general population of women and to alcoholic men. A follow-up study of alcoholic women in St. Louis, found that, 11 years after treatment, they had lost an average of 15 years from their expected life span. Another study of 1,000 female and 4,000 male alcoholics in Sweden found the excess mortality was higher for the women (5.2 times the expected rate) than for the men (3 times the expected rate).

Deaths due to drugs other than alcohol and tobacco are relatively uncommon among women. Men are far more likely than women to die from drug use. The higher male death rates are largely explained by males' greater drug use rather than by sex differences in vulnerability among drug users. In 1990, medical examiners in twenty-seven U.S. metropolitan areas reported 5,830 deaths involving illicit and/or legally obtained drugs. Of those who died from drug-related causes (e.g., OVERDOSE, accidental injury), 71 percent were male.

The HIV virus that causes AIDS is transmitted primarily via infected blood and semen. Sharing needles and having sexual relations with intravenous (IV) drug users places both men and women at risk for contracting that incurable disease. Although most AIDS cases have resulted from transmission of HIV during intimate sexual contact between men, about 12,000 of the 43,000 people reported to have AIDS in 1990 were IV drug users. Most of these AIDS cases involving IV drug use were male. When women contract AIDS, the most common route of transmission is through their own IV drug use or sexual contact with a partner who is an IV drug user.

Women's reproductive function increases alcohol- and drug-related health risks to themselves and to their unborn children. Alcohol and drug abuse are associated with numerous disorders of the female reproductive system, including breast cancer, amenorrhea, failure to ovulate, atrophy of the ovaries, miscarriage, and early menopause. Men also experience reproductive and sexual difficulties as a result of alcohol and drug abuse, including impotence, low testosterone levels, testicular atrophy, breast enlargement, and diminished sexual interest.

Infants born to women who used alcohol, tobacco, or other drugs during PREGNANCY can experience numerous health problems, including low birth weight, major congenital malformations, neurological problems, mental retardation, and withdrawal symptoms. Although substance abuse at any time during pregnancy can cause birth defects, the very rapid cell division in the first weeks of embryonic development means the teratogenic effects of alcohol and drugs are generally greatest early in pregnancy, before a woman even realizes she is pregnant.

As the medical and social costs of prenatal alcohol and drug exposure become more apparent, so does public pressure for action. Many advocate termination of parental rights in cases where a newborn tests positive for drug or alcohol exposure. In some jurisdictions, mothers who used alcohol or drugs during pregnancy have been charged with child abuse or delivering a controlled substance to a minor. Critics of these policies charge that alcohol and drug screening will discourage substance-abusing women from obtaining necessary prenatal care. Legally, it may be difficult to establish criminal intent if substance abuse occurred early in an unintended and unrecognized pregnancy. Further, it is often difficult to causally disentangle alcohol or drug effects from other adverse conditions the mother may have experienced, such as poor nutrition, acute or chronic illness, and inadequate prenatal care. As currently practiced, prenatal drug-use detection procedures raise important questions of fairness. Hospitals and clinics serving largely poor and minority patient populations are more likely to detect prenatal substance abuse despite evidence that substance abuse occurs in all socioeconomic categories.

The tendency of female problem drinking and drug abuse to develop in a relationship with a substance-abusing male partner may shield women from some consequences of their substance abuse. For example, women alcoholics and addicts are less vulnerable to arrest if their partner procures drugs for the couple or drives when they are intoxicated. On the other hand, substance-abusing partners increase some other risks for alcohol- and drug-dependent women compared with men. Women with substance-abusing partners are vulnerable to domestic VIOLENCE. Also, a substance-abusing partner can be an impediment to women's seeking or complying with alcohol and drug treatment.

Despite women's biophysical vulnerability and the stigma associated with female alcohol and drug abuse, men are more likely than women to experience some problems related to heavy drinking and illicit drug use. Substance abuse is more strongly related to intrapsychic problems among women, and to problems in social functioning (employment difficulties, financial problems, unsafe driving, arrest) among men.

These gender differences may be related to sex-role differences in drinking and drug use. Male substance use is less socially controlled—occurring more often in recreational contexts, public places, and all-male settings—whereas female substance use is more likely to occur in the home, with a male partner, and under medical auspices. Sex roles may also allow males to exercise less personal control while drinking or using drugs. For example, male episodes of intoxication are more often associated with rapid ingestion, blackouts, and AGGRESSION.

GENDER AND SUBSTANCE ABUSE TREATMENT

Men outnumber women in drug and alcoholism treatment units. The 1991 National Drug and Alcoholism Treatment Unit Survey (NDATUS) found 213,681 women in some type of treatment, compared with 562,388 men (U.S. Department of Health and Human Services, 1992). Self-reports of treatment experience indicate a somewhat smaller sex difference. In the 1991 NHSDA, 1.8 percent of males aged twelve and older reported they were treated for substance abuse in the previous year, compared with 0.9 percent of females. The discrepancy may occur because women are less likely to report informal help, such as pastoral counseling or SELF-HELP groups, as TREATMENT.

Among alcoholics and addicts, a greater percentage of women are parents, and among substance-abusing parents, more women have child custody. Parenting considerations are a major barrier to women seeking substance-abuse treatment. Few residential treatment programs make provisions for pregnant women or mothers. Many women are unable to find caregivers for their children if they enter residential treatment, and fear permanent loss of custody if their children enter the foster care system.

Substance-abuse treatment programs have been geared more to the problems and needs of male clients. Some contend that only sex-segregated treatment can meet the unique needs of female clients. Even those advocating integrated programs acknowledge the need for greater attention to women's issues. In addition to parenting responsibilities, it is urged that treatment programs address women's histories of physical and sexual abuse, domestic violence, and relationships with substance-abusing partners. Burman (1994) also suggests that treatment programs for women should emphasize skills such as problem solving, assertiveness, self-advocacy, and LIFE SKILLS (including parenting and job seeking).

(SEE ALSO: *Addicted Babies; Complications: Endocrine and Reproductive Systems; Family Violence and Substance Abuse; Gender and Complications of Substance Abuse; Injecting Drug Users and HIV; Stress; Treatment; Vulnerability As Cause of Substance Abuse*)

BIBLIOGRAPHY

BIENER, L. (1987). Gender differences in the use of substances for coping. In R. C. Barnett, L. Biener, & G. K. Baruch (Eds.), *Gender and Stress*. New York: Free Press.

BLUME, S. B. (1990). Chemical dependency in women: Important issues. *American Journal of Drug and Alcohol Abuse, 16,* 297–307.

BRAUDE, M. C., & LUDFORD, JACQUELINE P. (1984). *Marijuana effects on the endocrine and reproductive systems.* National Institute on Drug Abuse Research Monograph 44. Rockville, MD: National Institute on Drug Abuse.

BURMAN, S. (1994). The disease concept of alcoholism: Its impact on women's treatment. *Journal of Substance Abuse Treatment, 11,* 121–126.

CICERO, T. J. (1980). Sex differences in the effects of alcohol and other psychoactive drugs on endocrine function: Clinical and experimental evidence. In O. J. Kalant (Ed.), *Alcohol and drug problems in women: Research advances in alcohol and drug problems,* Vol. 5. New York: Plenum.

COOPERSTOCK, R. (1978). Sex differences in psychotropic drug use. *Social Science and Medicine, 12,* 179–186.

DUNNE, F. (1988). Are women more easily damaged by alcohol than men? *British Journal of Addiction, 83,* 1135–1136.

FREZZA, M. ET AL. High blood alcohol levels in women: The role of decreased gastric alcohol dehydrogenase activity and first-pass metabolism. *New England Journal of Medicine, 322,* 95–99.

INCIARDI, J. A. (1986). *The war on drugs.* Mountain View, CA: Mayfield.

LEX, B. (1991). Some gender differences in alcohol and polysubstance abusers. *Health Psychology, 10,* 121–132.

LEX, B. (1985). Alcohol problems in special populations. In J. H. Mendelson & N. K. Mello (Eds.), *The diagnosis and treatment of alcoholism* (2nd ed.). New York: McGraw-Hill.

MERLO, A. V. (1993). Pregnant substance abusers: The new female offender. R. Muraskin and T. Alleman (Eds.), *It's a crime: Women and justice.* Englewood Cliffs, NJ: Regents/Prentice-Hall.

ROBBINS, C. (1989). Sex differences in psychosocial consequences of alcohol and drug abuse. *Journal of Health and Social Behavior, 30,* 117–130.

ROBBINS, C., & CLAYTON, R. R. (1989). Gender-related differences in psychoactive drug use among older adults. *Journal of Drug Issues, 19,* 207–219.

ROBBINS, C., & MARTIN, S. S. (1993). Gender, styles of deviance, and drinking problems. *Journal of Health and Social Behavior, 34,* 302–321.

ROBINS, L. N. (1973). *The Vietnam drug user returns.* Washington, DC: U.S. Government Printing Office.

ROSENBAUM, M. (1981). Sex roles among deviants: The woman addict. *International Journal of the Addictions, 16,* 89–97.

SANDMAIER, M. (1980). *The invisible alcoholics.* New York: McGraw-Hill.

SUBSTANCE ABUSE AND MENTAL HEALTH SERVICES ADMINISTRATION. (1994). *Preliminary estimates from the 1993 National Household Survey on Drug Abuse.* Rockville, MD: U.S. Department of Health and Human Services.

SUBSTANCE ABUSE AND MENTAL HEALTH SERVICES ADMINISTRATION. (1993). *National Household Survey on Drug Abuse: Population estimates 1992.* Rockville, MD: U.S. Department of Health and Human Services.

SUBSTANCE ABUSE AND MENTAL HEALTH SERVICES ADMINISTRATION. (1992). *Annual emergency room data, 1992.* Rockville, MD: U.S. Department of Health and Human Services.

U.S. DEPARTMENT OF HEALTH AND HUMAN SERVICES. (1992). *Highlights from the 1991 National Drug and Alcoholism Treatment Unit Survey (NDATUS).* Washington, DC: U.S. Department of Health and Human Services.

U.S. DEPARTMENT OF JUSTICE. (1993). *Crime in the United States, 1992*. Washington, DC: U.S. Government Printing Office.

U.S. DEPARTMENT OF JUSTICE. (1992). *Drugs, crime, and the justice system*. Washington, DC: U.S. Government Printing Office.

WINDLE, M., & BARNES, G. M. (1988). Similarities and differences in correlates of alcohol consumption and problem behaviors among male and female adolescents. *International Journal of the Addictions, 23*, 707–728.

ZELLMAN, G. L., JACOBSEN, P. D., DuPLESSIS, H., & DiMATTEO, M. R. (1993). Detecting prenatal substance exposure: An exploratory analysis and policy discussion. *Journal of Drug Issues, 23*, 375–387.

CYNTHIA ROBBINS

WOMEN FOR SOBRIETY (WFS)

This organization (P.O. Box 618, Quakertown, PA 18951-0618; 215-536-8026 or 1-800-333-1606) began providing services to women alcoholics in 1975. The organization now has approximately 300 chapters or local units and approximately 5,000 members. Founded by sociologist Jean Kirkpatrick, WFS, like SECULAR ORGANIZATIONS FOR SOBRIETY (SOS), stresses individual responsibility and positive thinking. Its single-sex approach to recovery is justified by the view that women alcoholics lack self-esteem, are too dependent on alcohol, husbands, and other influences, and must develop self-confidence, self-esteem, and a sense of self-worth to achieve and maintain SOBRIETY. These goals, it is believed, are most comfortably achieved with the support of other women in the absence of men. In Kirkpatrick's view, AA precepts create special problems for women because they encourage dependencies. The WFS New Life program is intended to meet the special emotional needs of women and is frequently used along with traditional AA programs.

Based upon a Thirteen Statement Program of positive thinking that encourages emotional and spiritual growth, the New Life program shows women how to change their way of thinking to overcome depressions, guilt, and feelings of low self-esteem, and to use the support of the group to overcome drinking problems. Groups are small (usually consisting of fewer than ten women), and most are run by comoderators. Confidentiality is assured, and the small size of the group provides an opportunity for all who wish to speak to do so. Commitment to the New Life program is considered the key to each woman's sobriety; this entails a member's commitment to the other women in the group, to herself emotionally and spiritually, and to the WFS New Life program financially. Each member of the group is expected to make a financial contribution at every meeting she attends. WFS derives its operational money from these contributions, the sale of literature, speaking engagements, workshops, and charitable donations.

(SEE ALSO: *Alcoholics Anonymous; Rational Recovery; Treatment Types: Self-Help and Anonymous Groups; Women and Substance Abuse*)

FAITH K. JAFFE

WOMEN'S CHRISTIAN TEMPERANCE UNION

The nineteenth century was a time of drastic changes in the way many Americans viewed ALCOHOL. Early in the century, on average, U.S. citizens each consumed approximately 7 gallons of alcohol annually, the equivalent of about 2.5 ounces of pure alcohol daily. Concern that the United States would turn into a "nation of drunkards" led to the TEMPERANCE MOVEMENT of the early nineteenth century. This movement was loosely organized, consisting of the following diverse factions: (1) the neorepublicans, who were concerned with a host of problems that threatened the nation's security; (2) temperance societies, such as the Washingtonians, which served as the forerunners of modern-day self-help groups; and (3) physicians, who came to view habitual drunkenness as a disease. The goals of these groups varied; they ranged from helping habitual drunkards, to discouraging the use of alcoholic beverages, to advocating the prohibition of alcoholic beverages.

This first wave of temperance activists met with some success—thirteen states passed prohibition laws by 1855, and average alcohol consumption rates dropped to less than 3 gallons per person annually—but this was stopped by the growing national concern surrounding the approaching Civil War. Although the role of women was nearly nonexistent during this first temperance movement, the early movement set the stage for the post–Civil War temperance movement, in which women played a crucial part.

The years following the Civil War were a somewhat chaotic time. With the onset of the urban-industrial revolution and the concomitant changes witnessed in postbellum America, many people sought what Lender and Martin (1982, p. 92) term "a search for order." This search found a home in various social-reform movements. Broad-based reform movements attacked a number of issues thought to threaten American society, including education reform, women's rights, and intemperance.

Aaron and Musto (1981) refer to this period as the second great prohibition wave. Many local temperance societies survived the Civil War, as did the American Temperance Union. In 1869, the National Prohibition party was formed. This group supported the abolition of alcohol and recruited women into the anti-liquor fight. The National Prohibition party advocated complete and unrestricted suffrage for women, and their enlistment of women into the temperance movement marked the first public involvement of women in the temperance effort.

The post–Civil War Progressive movement also influenced the issue of temperance. The Progressives believed that alcohol was "the enemy of industrial efficiency, a threat to the working of democratic government, the abettor of poverty and disease" (Bordin, 1981, p. xvi). To the Progressives, temperance reform was a means for confronting genuine social problems. Business leaders increasingly came to view the use of alcohol as incongruous with the new technological society that America was becoming. Alcohol symbolized wastefulness, rampant pluralism, individualism, and potential social disorder.

At the same time, a growing number of physicians and temperance workers were coming to regard habitual drunkenness as a disease. At the core of the conception of this disease was its inherently progressive nature. Moderate drinking inevitably led to addiction, according to temperance workers, who proposed that as long as liquor was available to entice people to drink, and as long as moderate drinkers were around to act as models, then there would be drunkards. Increasingly, the blame for such addiction to alcohol was placed less on the individual and more on the society that permitted the sale of liquor and condoned drinking.

Some of the other factors that contributed to the milieu in which the women's temperance movement developed included better education for women, fewer children to care for, and the growing urbanization of America. As more household appliances became available and fewer women had to work around the clock at home or on the farm, they gained more leisure time. In addition, women came to be viewed as the protectors of the home—while, increasingly, alcohol was seen as a threat to the security of the home. These factors, in combination with an increased middle class and better communications, set the stage for the first mass movement of women into U.S. politics.

DIO LEWIS AND THE WOMEN'S CRUSADE

Ironically, the direct origins of the movement in which women gained entry into the political arena can be traced back to a man—Dio Lewis. By the 1870s, Lewis, a trained homeopathic physician, had given up his practice of medicine to embark on a career as an educator and lecturer. In December 1873, Lewis's lecture circuit included the cities and small towns of Ohio and New York. In each of them, he agreed to deliver an additional lecture as well as his scheduled talk related to women's issues—the topic of his extra speech was the duty of Christian women in temperance work. As an immediate result of his temperance lectures, women in each of these cities organized and marched on saloons and liquor distributors. Praying and singing hymns, the women were able to convince many proprietors of alcohol establishments to pledge themselves to stop selling liquor.

This grass-roots movement, which came to be known as the Women's Crusade, quickly moved through Ohio and into neighboring states. Typically, the women of a community would call a meeting eliciting support from other women. After praying over their cause, they would organize their efforts, which included asking local ministers to preach on the topic of temperance. They also sought pledges of support from local political leaders. Finally, they would take to the streets, marching on distributors of liquor as they attempted to persuade them to cease their sales of alcohol.

HISTORY

By November 1874, the Women's Crusade had grown to the point where a national convention was called. Sixteen states were represented at this convention, out of which the Woman's Christian Temperance Union (WCTU) emerged. Annie Wittenmeyer was named the first president of the WCTU, and a platform of action was agreed upon including the principle of total abstinence for WCTU members. Other plans involved committing the organization to (1) strongly promote the introduction of temperance education in both Sunday schools and public schools; (2) continue to use the evangelical methods, mass meetings, and prayer services that had been successful during their crusades; (3) urge the newspapers to report on their activities; and (4) distribute literature informing people of their cause. Although these first program commitments were later expanded, the convention's first set of resolutions provided the direction the WCTU would initially follow.

1874–1879. Under the leadership of Annie Wittenmeyer, the primary commitment of the WCTU was to gospel temperance. Wittenmeyer contended that the WCTU program should stress personal reform of the drunkard and of the whole liquor industry by moral suasion. She supported conversion to Christianity, religious commitment, acknowledgment of sin, and willingness to abandon evil ways as methods to reform those who drank. She shied away from seeking out legislative mandates as the solution to intemperance, however, and intentionally distanced herself from the women's suffrage movement; she feared possible repercussions for women *in the home*, should they campaign for the right to vote.

Although Wittenmeyer was instrumental in the early success of the WCTU, Frances Willard is recognized as the most influential leader of the women's temperance movement. Willard was chosen to be secretary at the first convention. Her views were often more radical than those of Wittenmeyer, particularly regarding women's rights. In 1879, she was elected president of the WCTU and served in that role until her death in 1898. Twentieth-century observers of the women's temperance movement may be more familiar with the name of Carrie Nation, who was known for raiding saloons armed with axes and hatchets; however, militant individuals such as she constitute a small fringe element of the WCTU. During the latter part of the nineteenth century, the true spirit of the WCTU was embodied in the person of Frances Willard.

1879–1898. While Wittenmeyer's primary commitment was to moral suasion, from the beginning of Willard's involvement in the WCTU, women's rights commanded her deeper loyalty. This commitment would be seen in the direction the WCTU

would take after 1879 (and was even evident while Willard served as secretary, as she subtly pushed for commitment to broader political programs). In 1876, Willard had introduced the concept of "home protection" to the WCTU. Building on earlier arguments that made use of women's traditional roles within the home and the need to defend and protect those roles, Willard proposed extending women the right to vote on prohibition issues as a means of further protecting women. At the time of this proposal, the idea of granting women the right to vote based on their natural or political right to do so was *not* palatable to many people, women and men alike. By introducing the suffrage issue under the guise of home protection, Willard was able to introduce the right-to-vote issue within the WCTU with less opposition than if she had sought solely to address women's suffrage.

As president, Willard ran the WCTU as a "well-oiled reform machine." Emphasizing organization at the local level, Willard was able to establish the mass base necessary for effective action. By 1880 the WCTU easily outstripped other women's organizations in both size and importance. Bordin (1981) estimates that there were 1,200 local unions with 27,000 WCTU members by the time Willard became president.

Under the leadership of Willard, the WCTU continued many of the programs that were adopted while Wittenmeyer was president. A number of states passed compulsory temperance-education laws, in large part due to the influence of the WCTU. In addition, the omnipresent push for abstinence from alcoholic beverages continued to typify the movement's goals—as is evidenced by the brief alliance forged between the WCTU and the Prohibition party. The WCTU of the 1880s, however, also departed from its roots on a variety of issues. It evolved from a temperance praying society to an activist organization. Whereas Wittenmeyer sought for change through moral suasion, Willard saw the advantages of political solutions to both the problems caused by intemperance as well as the problems facing women. Willard supported federal constitutional prohibition as the most effective way to deal with alcohol abuse, and she endorsed the temperance ballot for women as the surest way to achieve prohibition.

By the mid-1880s, the WCTU had expanded to every U.S. state and territory, and its platform had undergone similar expansion. Willard adopted the slogan "Do Everything" to describe the focus of the

WCTU under her guidance; initially, she had coined this phrase to depict the lengths to which she was willing to go to support the prohibition cause. By the late 1880s, however, she was committed to broader societal changes. Willard's strongest commitment remained to women's rights, and she argued as well for equal rights.

The membership of the WCTU in the early 1890s grew to an estimated 150,000 dues-paying members, with an additional 150,000 in affiliated groups. The WCTU had reached out to women of all social classes and minority groups. The growing influence of the WCTU was evident in the passage of several state prohibition laws in the 1880s, as well as in the growing support for a federal constitutional prohibition of liquor.

Although the number of women involved in the WCTU would continue to grow to approximately 1.5 million in the early twentieth century, as the nineteenth century drew to a close, the WCTU began losing its power and importance. Most notably, Willard became less visible in the years preceding her death. In her absence, conflicts arose among other leaders of the movement as to the organization's proper direction. In addition, as older leaders died or withdrew from active participation, fewer young women joined the WCTU to replace them.

1898–Present. As other organizations endorsing women's rights and/or prohibition were developed, membership in the WCTU slowly dwindled. Following Willard's death in 1898, the WCTU returned to a single-issue approach, focusing solely on prohibition. Although the ultimate goal of prohibition would eventually be achieved, it was not until the growth of the Anti-Saloon League (established 1896) that national prohibition would be realized. The Eighteenth Amendment to the U.S. Constitution was proposed and sent to the states December 18, 1917, and was ratified by three quarters of the states by January 16, 1919; it became effective January 16, 1920, establishing that the manufacture, sale, or transportation of intoxicating liquors, for beverage purposes, was prohibited. During the 1920s, it was clear that enforcement of the alcohol-beverage industry was almost impossible and that Americans would not give up drinking easily. The Repeal of Prohibition began as a movement that culminated in the Twenty-first Amendment to the U.S. Constitution; it was proposed and sent to the states February 20, 1933, and was ratified December 5, 1933.

Small groups of WCTU members can still be found in, for the most part, rural areas of the United States. The organization is based in Evanston, Illinois, and listed about 100,000 members in 1990.

(SEE ALSO: *Alcohol; Disease Concept of Alcoholism and Drug Abuse; Treatment, History of*)

BIBLIOGRAPHY

AARON, P., & MUSTO, D. (1981). Temperance and prohibition in America: A historical overview. In M. H. Moore & D. R. Gerstein (Eds.), *Alcohol and public policy.* Washington, DC: National Academy Press.

BLOCKER, J. S. (1985). *"Give to the winds thy fears": The women's temperance crusade, 1873–1874.* Westport, CT: Greenwood Press.

BORDIN, R. (1986). *Frances Willard: A biography.* Chapel Hill, NC: University of North Carolina Press.

BORDIN, R. (1981). *Woman and temperance: The quest for power and liberty, 1873–1900.* Philadelphia: Temple University Press.

ESTEP, B. (1992). Losing its bite. *Lexington Herald-Leader,* January 19, pp. 1, 11.

LENDER, M. E., & MARTIN, J. K. (1982). *Drinking in America: A history.* New York: Free Press.

LEVINE, H. G. (1984). The alcohol problem in America: From temperance to alcoholism. *British Journal of Addiction, 79,* 109–119.

MENDELSON, J. H., & MELLO, N. K. (1985). *Alcohol: Use and abuse in America.* Boston: Little, Brown.

GARY BENNETT

WOOD ALCOHOL (METHANOL) Methanol (methyl alcohol, CH_3OH) is the simplest of the alcohols. It is the natural by-product of wood distillation—an older method of producing drinking ALCOHOL (ethanol). Chemically synthesized methanol is a common industrial solvent found in paint remover, cleansing agents, and antifreeze. It is used to denature the ethanol found in some of these solutions and thereby render them unfit for drinking.

Methanol ingestion is usually accidental, but some alcoholics resort to the desperate measure of consuming methanol when they cannot obtain the beverage ethanol. Persons working in poorly ventilated areas can suffer ill effects from inhaling methanol-containing products, and ingestion of methanol is considered a medical emergency. Methanol is metabolized to formaldehyde and formic acid by the same liver enzymes that break down ethanol (these are alcohol dehydrogenase and aldehyde dehydrogenase). The formaldehyde and formic acid are toxic metabolites responsible for the symptoms of methanol poisoning; these appear several hours or days after methanol ingestion. Blurred vision, leading to permanent bilateral blindness, is characteristic of methanol poisoning. The accumulation of formic acid results in severe metabolic acidosis, which can rapidly precipitate coma and death. Other symptoms of methanol toxicity include dizziness, headaches, cold clammy extremities, abdominal pain, vomiting, and severe back pain.

The treatment for methanol poisoning is sodium bicarbonate, given to reverse the acidosis. In more serious cases, dialysis may be required; in addition, ethanol is given intravenously because it competitively binds to alcohol dehydrogenase, thereby slowing the production of toxic metabolites and allowing unchanged methanol to be excreted in the urine.

BIBLIOGRAPHY

KLAASSEN, C. D. (1990). Non-metallic environmental toxicants: Air pollutants, solvents and vapours, and pesticides. In A. G. Gilman et al. (Eds.), *Goodman and Gilman's the pharmacological basis of therapeutics*, 8th ed. New York: Pergamon.

LIPPMAN, M., & RUMLEY, W. (1989). Medical emergencies. In W. Dunagon & M. Ridner (Eds.), *Manual of medical therapeutics*, 26th ed. Boston: Little, Brown.

MYROSLAVA ROMACH
KAREN PARKER

WORKPLACE, DRUGS IN THE *See* Employee Assistance Programs; Industry and Workplace, Drug Use in.

WORLD HEALTH ORGANIZATION EXPERT COMMITTEE ON DRUG DEPENDENCE The World Health Organization (WHO) originated from a proposal at the first United Nations (U.N.) conference held in San Francisco in 1945 that "a specialized agency be created to deal with all matters related to health." This proposal resulted in a draft WHO constitution signed by sixty-one governments at an international health conference held in New York City in 1946. The constitution was subsequently ratified by the twenty-six member states of the U.N. and came into force on April 7, 1948. The enormous proposed scope of WHO led to the early concept of "Expert Committees," and they have become an essential part of the machinery of the organization. Their function is to give technical advice to WHO. Members of these committees are "appointed by the Director-General, in accordance with regulations established by the Executive Board." The members are chosen for their "abilities and technical experience" with "due regard being paid to adequate geographical distribution." Reports of expert committees can only be published with the authorization of the World Health Assembly or the WHO executive board.

One of the first tasks of the U.N. and WHO was to pick up the regulatory work on addiction-producing drugs that had been initiated and carried out by the League of Nations. Thus, the Expert Committee on Habit-Forming Drugs was established in 1948 to provide expert technical advice to the U.N. Permanent Central Opium Board and Drug Supervisory Body and the Division of Narcotic Drugs. The first meeting of the expert committee was held January 24–29, 1949, at the Palais des Nations in Geneva, Switzerland, where it continued to meet until the WHO building was opened in 1961. The expert committee, in its report on the second session, felt that the expression "habit forming" was no longer appropriate and recommended that the designation of the committee be changed to Expert Committee on Drugs Liable to Produce Addiction. This change was adopted by the WHO executive board at its fifth session and remained until 1964, when it was altered to Expert Committee on Dependence Producing Drugs and finally in 1968 to its present designation, Expert Committee on Drug Dependence.

In its early years, the expert committee reported directly to the director-general of WHO through its own secretary. In 1965, it became part of the Division of Pharmacology and Toxicology. During much of the period from its inception to 1972, the Seretariat was in the hands of Dr. Hans Halbach. In 1977, the expert committee became part of the Division of Mental Health, under the direction of Dr. Inayat Khan, where it remained until 1990 when a new Programme on Substance Abuse was created.

The early meetings of the expert committee were mainly devoted to the opioids—including the natural products, semisynthetics, and synthetics. Notifications on specific compounds by individual nations were responded to and recommendations as to international control were communicated to the secretary-general of the U.N. The beginnings of often recurring discussions were initiated concerning definitions, methods for evaluating dependence liability in animals and humans, the need for accurate epidemiological data concerning the extent of abuse and public health problems associated with drugs in general and of specific compounds in particular. During this period, the expert committee had an important consultative role in the development of a new international drug-control treaty, which resulted in an international conference held in New York City in January 1961. From this Conference emerged the SINGLE CONVENTION ON NARCOTIC DRUGS, 1961. This convention was amended in 1972, again with strong input from the expert committee, and remains the current instrument for the international control of the opioids, cocaine, and cannabis (marijuana).

The committee's concern for the potential abuse of the newly emerging ataractics (tranquilizing drugs) began in the mid-1950s and was soon joined in the 1960s by discussions of the problems created by amphetamines, amphetamine-like drugs, and hallucinogens. The difficulties associated with controlling these new heterogeneous groups of drugs under the Single Convention of 1961 became apparent and, at its seventeenth meeting in 1969, the committee began discussions of a draft Protocol on Psychotropic Substances, developed by the U.N. Commission on Narcotic Drugs, which formalized a classification of psychotropic drugs developed by the expert committee at its sixteenth meeting in 1968. The increasingly serious international public-health problems created by these drugs led the United Nations to hold a conference for the Adoption of a Protocol on Psychotropic Substances held in Vienna in February 1971; this resulted in the Convention on Psychotropic Substances, 1971, which the United Nations finally ratified in 1976. One important feature of this convention is that it mandates a WHO assessment of a substance prior to control and states that WHO's "assessments shall be determinative as to medical and scientific matters." This mandate added great responsibility to the functional role of the expert committee.

Only two meetings of the expert committee were held between the adoption of the Convention on Psychotropic Substances in 1971 and its ratification in 1976. The nineteenth meeting in 1972 was mainly devoted to a review of the current status of the epidemiological study of drug dependence. This meeting was also the last attended by Dr. Nathan B. Eddy, before his death in 1973. Dr. Eddy, a giant in the study of drug abuse and dependency, was at all the first nineteen meetings and served as chairman or rapporteur for most of them. The twentieth meeting of the committee was essentially devoted to the topic of prevention and resulted in a thorough review of the literature and a series of conclusions and recommendations, which were of considerable influence in the future development of the field.

The twenty-first meeting of the committee was held in 1977. It was entirely concerned with consideration of the Convention on Psychotropic Substances, and how WHO would handle its obligations under the treaty. This included consideration of appropriate pharmacological studies in animals and humans, assessment of public-health and social problems, assessment of therapeutic usefulness, the problem of chemically generic extensions to the list of scheduled substances, and the decision-making process. The meeting resulted in a number of recommendations that were mainly concerned with international cooperation in the development and collection of the relevant data needed to make rational decisions on controlling substances under the convention.

The expert committee did not meet formally again until 1985. In the interim, however, a number of WHO ad-hoc committees met to consider various aspects of the implementation of the treaty. In 1980, an extensive review of the Assessment of Public Health and Social Problems Associated with the Use of Psychotropic Drugs was carried out. To assist WHO, the U.S. National Institute on Drug Abuse, in collaboration with the Committee on Problems of Drug Dependence, published a monograph on "Testing Drugs for Physical Dependence Potential and Abuse Liability," which updated a similar WHO report published a decade earlier. A particularly difficult section of the psychotropic convention concerns exempt preparations. This involves thousands of pharmaceutical products and how to handle them, and it has still not been completely resolved despite three meetings of WHO advisory groups in 1977, 1982, and 1984.

Initially, to handle WHO's necessary functions under the conventions, it was decided to use ad-hoc advisory groups rather than to call formal meetings of the expert committee. The first of these was held in 1978. In 1979, specific compounds were considered under both conventions and the recommendation was made that, in the future, compounds proposed for control under the psychotropic convention be considered by class. In 1980, nine anorectic substances (things that cause loss of appetite) were reviewed and recommendations as to control were forwarded. Discussions concerning KHAT and its active principals, cathine and cathinone, were begun and research was initiated by a widespread group of laboratories. In 1981, the mixed opioid AGONIST-ANTAGONIST drugs were reviewed, and in 1981 and 1982 the BENZODIAZEPINES as a class were reviewed and recommendations for control were sent to the U.N. Also during this period a more formal method for review emerged from discussions with the U.N. Commission on Narcotic Drugs and the WHO Executive Board. Detailed critical reviews of substances to be considered for control were developed and the Programme Planning Working Group was formed to review these and suggest future classes of compounds for review by the expert committee. Two additional ad hoc advisory committee meetings were held in 1983 and 1984 to consider a variety of individual compounds and exempt preparations.

The twenty-second meeting of the expert committee was held in Geneva in April 1985. The committee adopted the new procedures for review of substances recently approved by the WHO Executive Board. These guidelines mandated a procedural sequence and schedule for the review. WHO was to obtain detailed information on each substance from a wide variety of sources including individual experts, research groups (e.g., WHO Collaborating Centers), the pharmaceutical industry, and relevant publications. It should be noted that this was the first time that the pharmaceutical industry was included in deliberations concerning regulatory control of their products. The twenty-second meeting was held, primarily, to consider twenty-eight phenethylamines for control under the Psychotropic Convention. A large number of groups and individuals was involved in preparing the critical review of these substances. Many of the substances considered were recommended for control under various schedules of the Psychotropic Convention. Some were not considered to need control, and no recommendation was made on these. Among the recommendations emerging from this meeting were requests for more and better data, particularly epidemiological, and more consideration of structure-activity relationships, isomeric state, and drug metabolism.

The twenty-third meeting in 1986 was nearly entirely devoted to the review of thirty-one BARBITURATES. A number of new factors were considered in the deliberations on this group of drugs. These included therapeutic indication (e.g., ultrashort-acting intravenous anesthetics, intermediate-acting sedative-hypnotics, and anticonvulsants), therapeutic usefulness, and demonstrable international public-health and social problems. Particular concern was expressed concerning PHENOBARBITAL, an inexpensive, effective antiepileptic widely used in developing countries, since it was felt by some that international control might lead to the use of more expensive and less safe medications. The committee also noted a lack of data on many compounds concerning dependence potential from either animals or controlled clinical studies and recommended that this be systematically collected by WHO prior to consideration for control.

The twenty-fourth meeting in 1987 discussed the control of seven nonbarbiturate sedative hypnotics. None of these were recommended for control. The committee also considered the marked increase in the illicit traffic in SECOBARBITAL and recommended that it be moved from Schedule III to Schedule II of the Psychotropic Convention. Finally, the committee recommended control of a number of fentanyl and MEPERIDINE analogs under the Single Convention.

The twenty-fifth meeting in 1988 considered the control of an additional four nonbarbiturate sedative-hypnotics including METHAQUALONE, which had been suggested for control in Schedule I of the Psychotropic Convention at the twenty-fourth expert committee meeting. Of these compounds, only methaqualone was recommended for control. The committee did not recommend rescheduling to Schedule I but urged the secretary-general of WHO that "every effort should be made to urge all countries whether or not they are signatories to the Convention on Psychotropic Substances, 1971, to stop producing methaqualone and to ban its import or export." The expert committee also revisited the opioid agonist-antagonist analgesics and recommended that BUPRENORPHINE and pentazocine be controlled under Schedule III of the Psychotropic Convention. This was a significant departure and was the first time that compounds with some opioid-like properties were considered for control under this convention rather than the Single Convention, 1961. A number of other compounds were considered for control, the most interesting being propylhexadrene. This substance was the first to be considered for decontrol under the Psychotropic Convention. The committee recommended that additional epidemiological data be collected and the substance reviewed again in two years. This was done in 1990, and a recommendation to remove propylhexadrene from control was forwarded to the U.N. secretary-general.

The twenty-sixth meeting of the committee in 1989 considered four additional uncontrolled benzodiazepines and recommended control for only one. The remainder were held over for the twenty-seventh meeting, in which the 33 benzodiazepines already under control were to be reviewed. This meeting also recommended the control of a number of "DESIGNER DRUGS," including analogs of fentanyl, tenamfetamine (MDA), and aminorex. Also considered was the notification from the government of the United States to transfer delta-9-tetrahydrocannabinol, the active principle of MARIJUANA, from Schedule I to Schedule II of the Convention on Psychotropic Substances. The committee so recommended, with the exception of two members who felt the decision should be deferred for additional data concerning therapeutic usefulness.

The twenty-seventh and last meeting to date of the expert committee was held in 1990 and was essentially devoted to the scheduling of the benzodiazepines as a class. Of particular interest was the conclusion that differential scheduling of the benzodiazepines was possible. Thus, the committee recommended that of the thirty-three substances currently under control, nineteen were appropriately controlled under Schedule IV. Thirteen of the substances had moderate to high therapeutic usefulness and few or no reports of abuse or illicit activity, and the committee declared that WHO should "monitor these compounds to amass enough data to determine whether or not they should be placed under critical review to consider descheduling." Two compounds, diazepam and flunitrazepam, "showed a continuing higher incidence of abuse and association with illicit activity." It was recommended that WHO keep these compounds under surveillance "to determine whether or not they merit being placed under critical review to consider appropriate scheduling."

As a result of structural changes within WHO and the creation of the new Programme on Substance Abuse, it is clear that in the future the expert committee will change its focus from reviewing substances for control under the international conventions to a broader consideration of the issues of prevention and reduction of demand.

(SEE ALSO: *Abuse Liability of Drugs: Testing in Humans*)

BIBLIOGRAPHY

Encyclopaedia Britannica, vol 13, 232–233.
Handbook of resolutions and decisions of the World Health Organization assembly and the executive board, vols. I and II. Geneva: World Health Organization, 1985.
W.H.O. What it is, what it does. Geneva: World Health Organization, 1988.

LOUIS HARRIS

XTC *See* Slang and Jargon.

Y

YIPPIES When large numbers of individuals with shared values engage in certain patterns of drug use, the political consequences can be serious. The Yippies of the late 1960s and early 1970s provide such an example.

Rather than quietly retreating from society as part of the baby-boom's countercultural (hippie) revolution, the Yippies shocked those with conventional values in the United States through spectacular media events. Thousands of young Americans shared the antimaterialistic values of Yippie leaders Abbie Hoffman and Jerry Rubin. In 1967, Hoffman dumped dollar bills from the visitors' gallery onto the floor of the New York Stock Exchange. In 1968, another protest event was staged—the Chicago Yippie Convention—timed to coincide with the Chicago Democratic Presidential Convention and considered an opportunity to protest the VIETNAM War.

Yippies challenged the establishment with a Festival of Life and invited drug-using hippies to attend; it included LSD seminars, rock shows, light shows, films, marches, love-ins, put-ons, guerrilla theater, and bizarre stunts—such as nominating a pig named Pigasus for president. The protest escalated into a confrontation with Chicago authorities; the mayor called out the police; and, in a rioting atmosphere, Yippies were beaten and imprisoned; the presidential convention was disrupted; Yippie leaders were tried in a case that became known as the Chicago Seven; and the Democrats lost the 1968 election.

During that time, a team of scientists surveyed the drug-use activity of 432 Yippies (Hughes et al. 1969). These showed a strong preference for hallucinogenic substances. Weekly MARIJUANA use was reported by 79 percent, HASHISH by 40 percent, LYSERGIC ACID DIETHYLAMIDE (LSD) by 29 percent, MESCALINE by 10 percent, PSILOCYBIN by 5 percent, and PEYOTE by 3 percent. Weekly use of nonhallucinogens was low—ALCOHOL 34 percent, COCAINE 4 percent, and HEROIN 3 percent.

It may be too simplistic to attribute the 1968 political events to marijuana and LSD. Yet we do know that certain chemicals help free users from conventional values and ways of perceiving reality. Researchers need to further examine this issue in future outbreaks of antiestablishment protest.

(SEE ALSO: *Epidemics of Drug Abuse; Hallucinogens*)

BIBLIOGRAPHY

FIEGELSON, N. (1970). *The underground revolution: Hippies, Yippies, and others.* New York: Funk & Wagnalls.

HOFFMAN, A. (1989). On to Chicago. In Daniel Simon (Ed.), *The best of Abbie Hoffman.* New York: Four Walls Eight Windows.

HUGHES, P., ZAKS, M., JAFFE, J., & BALLOU-DOLKART, M. (1969). The Chicago Yippie convention of 1968—drug use patterns. *Scientific Proceedings of the One Hun-*

dred *Twenty-Second Annual Meeting.* Washington, DC: American Psychiatric Association.

ZAKS, M., HUGHES, P., JAFFE, J., & BALLOU-DOLKART, M. (1969). Chicago Yippie convention, 1968: Socio-cultural drug use and psychological patterns. *American Journal of Orthopsychiatry, 39*(2):188–190.

PATRICK H. HUGHES

YOUTH AND SUBSTANCE ABUSE *See* Adolescents and Drugs; Gangs and Drugs; Prevention Programs; Treatment.

YUPPIES *See* Slang and Jargon.

Z

ZERO TOLERANCE This was a federal drug policy initiated during the War on Drugs campaign of the Reagan and Bush administrations (1981–1993). Under this policy, which was designed to prohibit the transfer of illicit drugs across U.S. borders, no possession, import, or exportation of illicit drugs was tolerable, and possession of any measurable amount of illicit drugs was subject to all available civil and criminal sanctions. Zero Tolerance, which represented a criminalistic perspective on drug use, was an example of a criminal-justice approach to drug control. Under such an approach, the control of drugs rests within the domain of the criminal-justice system, and the use of drugs is regarded as a criminal act, with legal sanction as the consequence.

Zero tolerance is a "user-focused" strategy of drug control, according to which law-enforcement agents target users of illicit drugs as opposed to dealers or transporters. The rationale for this approach is that the users of illicit substances create the demand for drugs and constitute the root cause of the drug problem. If, therefore, demand for drugs can be curbed by exacting harsh penalties on users, the supply of drugs into the country will slow.

The zero-tolerance policy was initiated by the U.S. CUSTOMS SERVICE, in conjunction with the U.S. Attorney's office in San Diego, California, as part of an effort to stop drug trafficking across the U.S.–Mexican border. Individuals in possession of illicit drugs were arrested and charged with both a misdemeanor and a felony offense. Customs Service

officials believed the policy to be successful at reducing the flow of drugs across the border and recommended that it be implemented nationwide. Subsequently, the National Drug Policy Board, in conjunction with the White House Conference on a Drug-Free America had all federal drug-enforcement agencies implement zero tolerance on March 21, 1988, at all U.S. points of entry (United States Congress, 1988).

The policy did not involve enacting new laws or regulations; it only entailed instituting strict interpretation and enforcement of existing laws. In practice, it meant that any type of vehicle—including bicycles, transfer trucks, and yachts—would be confiscated and the passengers arrested upon the discovery of any measurable amount of illicit drugs. The U.S. Coast Guard and the U.S. Customs Service began to crack down on all cases of drug possession on the water and at all borders. If, during the course of their regular patrols and inspections, Coast Guard personnel boarded a vessel and found one marijuana cigarette, or even the remnants of a marijuana cigarette, they arrested the individual and seized the boat. Before this policy was instituted, the Coast Guard had either looked the other way or issued fines when "personal-use" quantities of illicit substances were discovered (United States Congress, 1988).

Zero tolerance was criticized because federal agencies expended substantial resources to identify individual drug users instead of concentrating their resources on halting the influx of major quantities of drugs into the country for street sale. The policy of seizing boats upon the discovery of trace amounts of drugs was also controversial. Some believed the policy to be an unfair and unusually harsh punishment; seizing a commercial boat that was the sole source of income for an individual or family was denounced as being too severe a penalty for possession of "one marijuana cigarette." There were some highly publicized cases of commercial fishing boats being seized on scant evidence that the boat owner was responsible for the illicit drugs found.

The term *zero tolerance* has a broader application than that indicated by the foregoing. It describes a perspective on drug use according to which it is maintained that the use of any amount of illicit drugs is harmful to the individual and society and that the goal of drug policy should be to prohibit any and all illicit drug use. According to the contrasting viewpoint, the simple use of drugs is distinguishable from problem drug use and although absence of all drug use is desirable, the resources of government would be used more efficiently if they targeted individuals who demonstrated problem use or if they addressed problems related to or caused by illicit drug use.

(SEE ALSO: *Drug Interdiction; Operation Intercept; U.S. Government: The Organization of U.S. Drug Policy*)

BIBLIOGRAPHY

UNITED STATES CONGRESS. HOUSE COMMITTEE ON MERCHANT MARINE AND FISHERIES. SUBCOMMITTEE ON COAST GUARD AND NAVIGATION. (May 26, 1988). "Zero Tolerance" drug policy and confiscation of property: Hearing before the Subcommittee on Coast Guard and Navigation of the Committee on Merchant Marine and Fisheries (House of Representatives, 100th Congress, 2nd session). Washington, DC: U.S. Government Printing Office.

AMY WINDHAM